Local loads in plates and shells

Monographs and textbooks on mechanics of solids and fluids

editor-in-chief: G. Æ. Oravas

Mechanics of surface structures

editor: W. A. Nash

1. P. Seide
   Small elastic deformations of thin shells

2. V. Panc
   Theories of elastic plates

3. J. L. Nowinski
   Theory of thermoelasticity with applications

4. S. Łukasiewicz
   Local loads in plates and shells

Local loads in plates and shells

Stanisław Łukasiewicz

Warsaw Technical University
Institute of Aeronautical Technology and
Applied Mechanics

SIJTHOFF & NOORDHOFF

Alphen aan den Rijn, The Netherlands

PWN–POLISH SCIENTIFIC PUBLISHERS

Warszawa

Sijthoff & Noordhoff International Publishers B. V.,
Alphen aan den Rijn, The Netherlands
and
PWN–Polish Scientific Publishers,
Warszawa, Poland

First English edition based on
Obciążenia skupione w płytach, tarczach i powłokach
published in 1976 by Państwowe Wydawnictwo Naukowe, Warszawa

Translation: *Stanisław Łukasiewicz*

ISBN 90 286 00 477

Printed in Poland by D.R.P.

Contents

Introduction

Thin walled structures so extensively used nowadays in industry and civil engineering are usually loaded by very complex systems of forces acting on their edges or over their surfaces. In calculating the strength of a structure we replace real loads by certain idealized loads distinguishing between typical surface loads distributed over a great area of the structure and loads acting over a small area. The latter are called concentrated loads. When the area under the load is very small in comparison with the dimensions of the surface of the structure, for example, when the diameter of the loaded area is smaller than the wall thickness, the load can be considered as a single force or a moment acting on the structure at one point only. The real loads which are met in practice can always be replaced by a combination of components such as forces normal and tangential to the wall as well as bending and twisting moments. Knowing the distribution of the stresses in the structure produced by each component, we can find it under any arbitrary load using the principle of superposition.

There are two main reasons for the appearance of the concentration of stresses in the structure. It can be produced by notches, rapid changes of the cross-section, holes, cutouts, etc. on one hand and by concentrated loads resulting from the interaction of the elements of the structure on the other. Only the latter are the subject of the present book. The main part of it is devoted to the analysis of the cases where the load is applied to the structure by means of a single concentrated force or bending moment at one point of the surface. It should be noticed here that such loads do not exist in practice. However, if we calculate the stresses far from the point of application of the load, we can replace the real load by the load applied at one point of the surface of the structure. This load is only a convenient idealization of the real load.

The problem is different if we are interested in the stresses in the vicinity of the small loaded region of the structure. Then the correct answer can be obtained only by taking into account the fact that the load acts on the finite surface and by considering the manner of distribution over this surface. Usually, the surface loads do not involve important difficulties in determining the stresses and displacements in thin-walled structures. Analysing the problem by means of infinite Fourier series we get the result in the form of rapidly convergent series. In order to obtain sufficient accuracy of the results it is adequate to take into account only a few terms of the series. On the contrary, the calculation for thin-walled structures loaded by a concentrated force come into serious difficulties in computation of the stresses. These difficulties are brought about by the fact that the equations of the theory of elasticity have singular solutions at the points of application of the loads. In order to obtain the stresses in the vicinity of the loading point a great number of terms of the series must be calculated which makes the amount of work involved prohibitive and decreases the accuracy of the results.

Therefore in the case of concentrated loads the solutions by means of series should be avoided. Here the solutions in closed form are desirable.

The aim of this book is to acquaint the reader with the methods of calculation of stresses and displacements in thin-walled structures subject to concentrated loads. The book considers also the effects produced by these loads in the structure, namely, the state of stress and displacement accompanying typical, concentrated loads.

Singularities of the functions determining the stresses have also been considered. Diagrams enabling quick and direct calculation of many values helpful for designing thin-walled structures have been given. Optimum design of a structure under concentrated loads is also the subject of the book.

The author tried to avoid complex mathematical methods mostly by reducing the solutions to the evaluation of the integrals or solving the partial differential equations with constant coefficients. Unfortunately, it is not possible to get far with elementary mathematics only, for example some solutions have been obtained in special Thomson's functions and in some cases the calculus of residues has been indispensable for evaluation of integrals.

The book has been written for engineers designing thin-walled structures, students and research workers interested in the problems of concentrated loads.

Notation

a, b, c, d	numerical coefficients, segments
c	radius of loaded circular area
$a_{\alpha\beta}$	metric tensor of the middle surface
a^{ijkl}	tensor of elastic constants of the material
$b_{\alpha\beta}$	second fundamental metric tensor of the surface
$c_{\alpha\beta}$	third fundamental metric tensor of the surface
$d_{\alpha\beta}$	permutation symbol
$g_{\alpha\beta}$	metric tensor of the surface
$\mathbf{g}_i$	base vectors
h	thickness of the plate or shell
k	coefficient
$k_R = l^2/R^2$	coefficient characterizing thinness of the shell
l	characteristic length
n, s	directions: normal and tangential
n	surface tension
r	radius, radius vector
$\mathbf{n}$	normal vector
$\mathbf{u}$	displacement vector
u, v, w	components of the displacement vector
$\bar{x}, \bar{y}, \bar{z}$	rectangular dimensional coordinates
$x = \bar{x}/l,\ y = \bar{y}/l,\ r = \bar{r}/l$	dimensionless coordinates
r, ϑ, φ	polar coordinates
q	load distributed of the surface
A	surface of the cross-section
B_{ij}, C_{ij}	tensors of the elastic constants
$D_{\alpha\beta}$	bending rigidity of the plate or shell
$D_{\alpha\beta}, D_{xx}, D_{yy}, D_{xy}, D_{12}$	bending rigidity of the orthotropic shell
G	shear modulus
E	Young modulus of elasticity
F	auxiliary function
$\bar{D}$	dissipation of the strain energy on the unit surface
C_i	integration constants

J_i	Bessel functions
$I_{m,n}, H_{2n}, K_{m,n}$	integrals
N	normal and tangential membrane forces
Γ_{ij}^k	Christoffel's symbol
K	Gauss curvature, Green function
L	work of the external load, wave length
$M_{\alpha\beta}$	bending and twisting moments
$\tilde{M}$	twisting moment
Q_α	shear forces
P_n, Q_n	Legendre functions
P, T, N	concentrated forces; transverse, tangential and normal
M	concentrated bending moment
R	Riemann–Christoffel curvature tensor
R_n	normal radius of curvature
R_i	principal radii of curvature
$S, \tilde{S}$	tangential forces
X_i	body forces per unit volume or surface
X, Y, Z	components of body forces parallel to the directions of coordinates x, y, z
U	strain energy
$L(\ )$	differential operator
$\Delta(\ )$	Laplace operator
$D_k(\ ), I(\ )$	differential operators
Φ	stress function
α, β, γ	variables of integration
γ_{ij}	shear angles
γ_0	0.5772 Euler's constant
α_i	curvilinear coordinates on the middle surface of the shell
$\beta_{\dot{\alpha}}$	angles of rotations of lateral sides of an element of plate or shell
θ_i	curvilinear coordinates
$\varepsilon, \bar{\varepsilon}$	coefficients expressing effects of stress normal to middle surface
η	coefficient expressing effect of transverse shear deformation
$\delta(x, y)$	Dirac's function

$\delta_{\alpha\beta}$	Kronecker's delta
$e_{\alpha\beta}$	strain tensor
$\varepsilon_{\alpha\beta}$	strain tensor of the middle surface
ν	Poisson's ratio
$\varkappa_{\alpha\beta}$	tensor of the change of the curvature of the middle surface
$\varkappa_{\alpha}$	principal changes of curvature
$\varkappa$	coefficient characterizing geometry of the shell
$\varkappa_{c}$	coefficient characterizing properties of the core of the sandwich plate or shell
μ, λ	coefficients
$\lambda_R = \dfrac{R_2}{R_1}$	coefficient, ratio of the principal curvatures
φ_R, χ_R, μ_R	coefficients characterizing effect of the non-shallowness of the shell
σ_{ij}	components of the stress tensor
σ_{pl}, σ_0	yield stress

1

Basic equations of the theory of plates and shells

A body bounded by the two curved surfaces is called a *shell* when the average distance between these surfaces is small in comparison with its other dimensions. The same body is called a *plate* or *sheet* if the surfaces become planes. The surface created by the points lying at an equal distance from both surfaces is called a *middle surface* and in the case of the plate: a *middle plane*. Usually shells and plates are loaded by the forces both perpendicular and parallel to their middle surface. If a plate is loaded only in its middle plane, it is called a *sheet*.

This chapter is devoted to the derivation of the fundamental equations of the theory of plates and shells permitting the determination of the stresses and displacements in the loaded structure. In order to obtain better accuracy of the stresses produced by concentrated loads there are also taken into account the effects of transverse normal and shear deformations. Both these effects are neglected in the classical theory, but in the case of concentrated loads they may be of greater importance.

At first we shall consider the laws concerning the behaviour of the three-dimensional elastic body, then we shall derive fundamental equations of the theory of plates and shells. Since the equations of the theory of plates can easily be obtained from the equations of the theory of shells by a simple assumption that the radii of curvature of the shell increase to infinity, we shall not derive them separately.

1 Basic equations of the theory of plates and shells

1.1 A short introduction to the tensor calculus

In contemporary papers concerning the theory of plates and shells the tensor calculus is applied very often because it enables a short and easy description of complex physical relations. Therefore some elements of the tensor calculus are introduced here.

At first sight, the tensor calculus may seem involved, but a little study will soon reveal its simplicity. Let us begin with the matter of notations. The notations employed here correspond to those usually applied. In tensor analysis one makes extensive use of indices.

For example, a set of variables $x_1, x_2, \ldots, x_n$ is denoted by x_i, $i = 1, 2, \ldots, n$. A set of variables x^1, x^2, ..., x^n by x^i. We emphasize that x^i are not the powers of the variable x. The sum of the series can be written as

$$S = a_1 x^1 + a_2 x^2 + a_3 x^3 = \sum_{i=1}^{3} a_i x^i.$$

Using the summation convention, we can write this equation in the simple form:

$$S = a_i x^i, \quad i = 1, 2, 3. \tag{1.1}$$

The convention is as follows: The repetition of an index (whether superscript or subscript) in a term will denote a summation with respect to that index over its range. An index that is summed over is called a *dummy index*. The others are called *free indices*.

We introduce a symbol δ_{ij} which is called the *Kronecker delta*. This symbol has the following properties:

$$\begin{aligned} &\delta_{11} = \delta_{22} = \delta_{33} = 1, \\ &\delta_{12} = \delta_{21} = \delta_{13} = \delta_{31} = \delta_{23} = \delta_{32} = 0. \end{aligned} \tag{1.2}$$

Using this symbol, we may write, for example, the square of the length of the line element:

$$\mathrm{d}s^2 = \mathrm{d}x^2 + \mathrm{d}y^2 + \mathrm{d}z^2.$$

If we define

$$\mathrm{d}x^1 = \mathrm{d}x, \quad \mathrm{d}x^2 = \mathrm{d}y, \quad \mathrm{d}x^3 = \mathrm{d}z,$$

we obtain

$$(\mathrm{d}s)^2 = \delta_{ij} \mathrm{d}x^i \mathrm{d}x^j, \quad i, j = 1, 2, 3. \tag{1.3}$$

The other symbol often used in the tensor calculus is the permutation symbol d_{rst} which is defined by the equations

$$d_{111} = d_{222} = d_{333} = d_{112} = d_{113} = d_{221} = d_{223} = d_{331} = d_{332} = 0,$$
$$d_{123} = d_{231} = d_{312} = 1, \quad d_{213} = d_{321} = d_{132} = -1.$$

The following determinant illustrates its application:

$$\begin{vmatrix} a_{11} & a_{12} & a_{13} \\ a_{21} & a_{22} & a_{23} \\ a_{31} & a_{32} & a_{33} \end{vmatrix} = |a_{ij}| = d_{rst}\, a_{r1}\, a_{s2}\, a_{t3}.$$

The Kronecker delta and the permutation symbol are connected by the identity

$$d_{ijk}\, d_{ist} = \delta_{js}\, \delta_{kt} - \delta_{jt}\, \delta_{ks}. \tag{1.4}$$

The summation convention is extended to symbols of differentiation. Let $f(x^i)$ be a function of i variables x^i; then its differential shall be written

$$\mathrm{d}f = \frac{\partial f}{\partial x^i}\, \mathrm{d}x^i.$$

The partial derivatives can also be denoted in the following way:

$$\frac{\partial f}{\partial x^i} = f_{,i}, \qquad \frac{\partial^2 f}{\partial x^i \partial x^j} = f_{,ij}. \tag{1.5}$$

The most important advantage of the tensor calculus is the simplicity of description of any physical relations independent of the assumed system of coordinates.

Let us calculate, for example, the length of the line element in two coordinate systems x^i and θ^i in a three-dimensional Euclidean space. Let

$$x^i = x^i(\theta^1, \theta^2, \theta^3), \qquad i = 1, 2, 3, \tag{1.6}$$

be an admissible transformation of coordinates from the rectangular Cartesian coordinates x^i to some general coordinates θ^i. The inverse transformation

$$\theta^i = \theta^i(x^1, x^2, x^3)$$

is assumed to exist. Then we have

$$\mathrm{d}s^2 = \mathrm{d}x^i \mathrm{d}x^i = \delta_{ij} \mathrm{d}x^i \mathrm{d}x^j. \tag{1.7}$$

On differention of (1.6) we have

$$\mathrm{d}x^i = \frac{\partial x^i}{\partial \theta^k}\, \mathrm{d}\theta^k. \tag{1.8}$$

Substituting (1.8) into (1.7), we obtain

$$ds^2 = \sum_{i=1}^{3} \frac{\partial x^i}{\partial \theta^k} \frac{\partial x^i}{\partial \theta^m} d\theta^k d\theta^m.$$

Letting

$$\sum_{i=1}^{3} \frac{\partial x^i}{\partial \theta^k} \frac{\partial x^i}{\partial \theta^m} = g_{km}(\theta^1, \theta^2, \theta^3),$$

the square of the length of the line element takes the form

$$ds^2 = g_{km} d\theta^k d\theta^m. \tag{1.9}$$

The summation here should be with respect to all combinations of the indices k and m.

The functions g_{km} are called the *components of the metric tensor in the coordinate system* $\theta^1, \theta^2, \theta^3$.

The *law of the transformation* of the components (1.8) of some physical value, by introducing the new system of coordinates, is a characteristic property of the tensor. The physical quantity is the tensor only if it is transformed according to this specific law of transformations.

There are two kinds of vectors and tensors whose names are connected with the transformation law. Let us assume two systems of coordinates x^i and θ^i, the transformation $x^i = x^i(\theta^k)$, and the inverse transformation $\theta^i = \theta^i(x^k)$.

A vector **u** is called a *contravariant vector field* or *contravariant tensor field of rank one* if it has three components $u^i(x^i)$ and three components $u^i(\theta^i)$ related by the *characteristic law*

$$u^i(x^1, x^2, x^3) = u^k(\theta^1, \theta^2, \theta^3) \frac{\partial x^i}{\partial \theta^k}. \tag{1.10}$$

This vector is indicated by a superscript called a *contravariant index*.

A vector **u** is called a *covariant vector field* or *covariant tensor field of rank one* if it has three components $u_i(\theta^i)$ and three components $u_i(x^i)$ related by the characteristic law

$$u_i(x^1, x^2, x^3) = u_k(\theta^1, \theta^2, \theta^3) \frac{\partial \theta^k}{\partial x^i}. \tag{1.11}$$

Acovariant vector is indicated by a subscript called a *covariant index*.

In a similar way we distinguish the tensors of rank two.

If the system of coordinates is orthogonal, then the components of the metric tensor ($g_{km} = 0$ for $m \neq k$) are called *Lamé's coefficients.* We define

$$g_{kk}(\theta^1, \theta^2, \theta^3) = H_k^2 = \left(\frac{\partial x}{\partial \theta^k}\right)^2 + \left(\frac{\partial y}{\partial \theta^k}\right)^2 + \left(\frac{\partial z}{\partial \theta^k}\right)^2.$$

For example, in the case of cylindrical coordinates (Fig. 1) we have

$$\theta^1 = R, \quad \theta^2 = \varphi, \quad \theta^3 = z,$$

$$x = R\cos\varphi, \quad y = R\sin\varphi, \quad z = z,$$

$$H_1 = 1, \quad H_2 = R, \quad H_3 = 1.$$

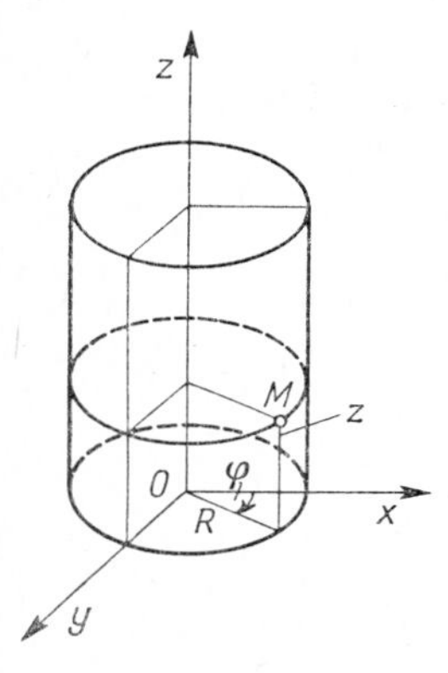

Fig. 1. Cylindrical coordinates.

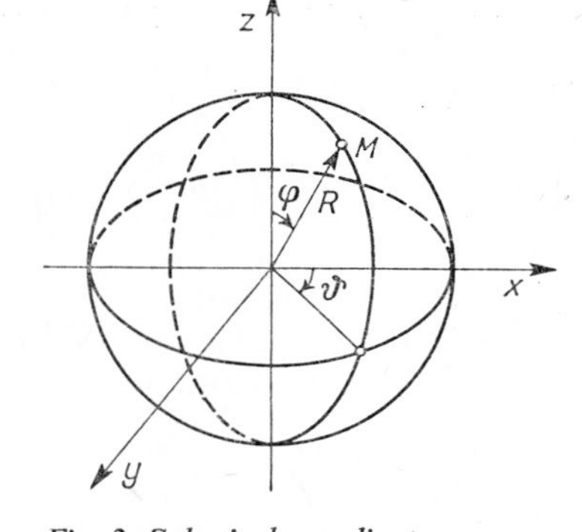

Fig. 2. Spherical coordinates.

For the spherical coordinates Lamé's coefficients take the form (Fig. 2)

$$\theta^1 = R, \quad \theta^2 = \varphi, \quad \theta^3 = \vartheta;$$

$$x = R\sin\varphi\cos\vartheta, \quad y = R\sin\varphi\sin\vartheta, \quad z = R\cos\varphi.$$

Then

$$H_1 = 1, \quad H_2 = R, \quad H_3 = R\sin\varphi.$$

In orthogonal Cartesian coordinates we have

$$\mathrm{d}s^2 = \mathrm{d}x^2 + \mathrm{d}y^2 + \mathrm{d}z^2;$$

then $H_1 = H_2 = H_3 = 1$.

A partial derivative of the tensor in the system of curvilinear coordinates is not a tensor. In order to obtain another tensor field after differentiation of the vector or tensor field, we introduce certain corrections which are linear functions of the coordinates. If u^i is a vector, then only the following combination can be a tensor:

$$\frac{\partial u^i}{\partial x^j}+\Gamma(i,j,\alpha)u^\alpha. \tag{1.12}$$

The quantity $\Gamma(i,j,\alpha)$ is a certain function with three indices, called the *Euclidean Christoffel symbol*, which is a certain combination of the derivatives of the metric tensor g_{ij}. The whole expression (1.12) is called the *covariant derivative of the contravariant vector* u^i. We shall use the following notation:

$$u^i|_\alpha = \frac{\partial u^i}{\partial x^\alpha}+\Gamma^i_{j\alpha}u^j. \tag{1.13}$$

The *covariant derivative of the covariant vector* is denoted by

$$u_i|_\alpha = \frac{\partial u^i}{\partial x^\alpha}-\Gamma^j_{i\alpha}u_j.$$

The *Christoffel symbols* are defined as follows:

$$\Gamma^j_{\alpha\beta} = \tfrac{1}{2}g^{ij}\left(\frac{\partial g_{i\beta}}{\partial x^\alpha}+\frac{\partial g_{\alpha i}}{\partial x^\beta}-\frac{\partial g_{\alpha\beta}}{\partial x^i}\right)$$

where

$$g^{ij} = \frac{\partial \theta^i}{\partial x^m}\cdot\frac{\partial \theta^j}{\partial x^m}.$$

In the Cartesian system of coordinates the components of the metric tensor g_{ij} are constant. Then the Christoffel symbols are equal to zero. As previously mentioned the covariant derivative of a tensor is a tensor. It means that if its components in one coordinate system are equal to zero, then in any admissible systems of coordinates they are also equal to zero. The Christoffel symbols in the Cartesian coordinates vanish and the covariant derivatives of a tensor field reduce to partial derivatives of the tensor field.

Then for an arbitrary vector u^k and the arbitrary system of coordinates θ_i of the three-dimensional Euclidean space we have

$$u^k|_{st}-u^k|_{ts} = 0. \tag{1.14}$$

The above relation is denoted by the symbol

$$R^k_{p\ t}u^p = u^k|_{st}-u^k|_{ts}$$

and is called the *Riemann–Christoffel tensor*.

It can be proved that

$$R^k_{pst} = \frac{\partial \Gamma^k_{ps}}{\partial x^t} - \frac{\partial \Gamma^k_{pt}}{\partial x^s} + \Gamma^r_{ps}\Gamma^k_{rt} - \Gamma^r_{pt}\Gamma^k_{rs}. \tag{1.15}$$

Conditions (1.14) are necessary and sufficient for the system of coordinates θ^k to be considered as a system of three-dimensional Euclidean space. Conditions (1.14) contain only six independent equations. In the orthogonal system of coordinates we obtain three equations of the form

$$\frac{\partial}{\partial \theta_1}\left(\frac{1}{H_1}\frac{\partial H_2}{\partial \theta_1}\right) + \frac{\partial}{\partial \theta_2}\left(\frac{1}{H_2}\frac{\partial H_1}{\partial \theta_2}\right) + \frac{1}{H_3^2}\frac{\partial H_1}{\partial \theta_3}\frac{\partial H_2}{\partial \theta_3} = 0 \tag{1.16}$$

and three equations of the form

$$\frac{\partial^2 H}{\partial \theta_2 \partial \theta_3} - \frac{1}{H_2}\frac{\partial H_2}{\partial \theta_3}\frac{\partial H_1}{\partial \theta_2} - \frac{1}{H_3}\frac{\partial H_2}{\partial \theta_2}\frac{\partial H_1}{\partial \theta_3} = 0.$$

The remaining equations can easily be obtained by the change of indices according to the scheme

$$\begin{array}{ccc} 1 & \longleftarrow & 3 \\ & \searrow \quad \nearrow & \\ & 2 & \end{array}$$

1.2 Strains and displacements of a three-dimensional body

Let us consider an arbitrary point M of the body in the orthogonal Cartesian coordinates x_i $(i = 1, 2, 3)$. Subsequently, the body is deformed to a new configuration and the point M is moved to N with coordinates θ_i $(i = 1, 2, 3)$ with respect to a new coordinate system $\theta_1, \theta_2, \theta_3$. The coordinate system x_i may be curvilinear. Let us assume that there exist the equations of the transformation

$$\theta_i = \theta_i(x_1, x_2, x_3) \tag{1.17}$$

and the unique inverse transformation

$$x_i = x_i(\theta_1, \theta_2, \theta_3)$$

for every point in the body. The functions $\theta_i(x_i)$ and $x_i(\theta_i)$ are assumed to be continuous and differentiable. We shall be concerned with the description of the strain of the body. Let us consider an infinitely small line element connecting the point M to a neighbouring point $M^1(x_1+dx_1, x_2+dx_2, x_3+dx_3)$. The square of the length ds_0 of MM^1 in the original configuration is given by

$$ds_0^2 = a_{ij}\,dx^i dx^j,$$

where a_{ij}, evaluated at the point M, is the Euclidean metric tensor for the coordinate system x_i. If x_i are Cartesian coordinates $a_{ii} = 1$, then $a_{ij} = 0$. When the points M, M^1 are deformed to the points N, N^1, the square of the length ds of the new element NN^1 is

$$ds^2 = g_{ij}\,d\theta^i d\theta^j,$$

where g_{ij} is the Euclidean metric tensor for the coordinates θ^i.

If both the systems of coordinates are Cartesian coordinates, then $a_{ij} = g_{ij} = \delta_{ij}$. Taking into consideration relations (1.17), we may write

$$ds_0^2 = a_{ij}\frac{\partial x_i}{\partial \theta_l}\frac{\partial x_j}{\partial \theta_m}\,d\theta^l d\theta^m, \tag{1.18}$$

$$ds^2 = g_{ij}\frac{\partial \theta_i}{\partial x_l}\frac{\partial \theta_j}{\partial x_m}\,dx^l dx^m.$$

Let us calculate now the difference between the squares of the lengths of the elements. This difference appears as the effect of the deformation of the body, and may be written

$$ds^2 - ds_0^2 = \left(g_{\alpha\beta}\frac{\partial \theta_\alpha}{\partial x_i}\frac{\partial \theta_\beta}{\partial x_j} - a_{ij}\right)dx^i dx^j \tag{1.19}$$

or

$$ds^2 - ds_0^2 = \left(g_{ij} - a_{\alpha\beta}\frac{\partial x_\alpha}{\partial \theta_i}\frac{\partial x_\beta}{\partial \theta_j}\right)d\theta^i d\theta^j.$$

We define the functions

$$e_{ij} = \frac{1}{2}\left(g_{\alpha\beta}\frac{\partial \theta_\alpha}{\partial x_i}\frac{\partial \theta_\beta}{\partial x_j} - a_{ij}\right), \tag{1.20}$$

$$\bar{e}_{ij} = \frac{1}{2}\left(g_{ij} - a_{\alpha\beta}\frac{\partial x_\alpha}{\partial \theta_i}\frac{\partial x_\beta}{\partial \theta_j}\right) \tag{1.21}$$

as the strain tensors. The strain tensor e_{ij} is called *Green's strain tensor*. The tensor $\bar{e}_{ij}$ is known as *Almansi's strain tensor*. The first describes the state of strain in Lagrangian coordinates and the second, $\bar{e}_{ij}$, in Eulerian coordinates, i.e. in coordinates connected with the deformed body.

Let us introduce now the displacement vector

$$u_i = \theta_i - x_i. \tag{1.22}$$

Let us substitute the above relation in (1.20) and (1.21). If we use rectangular Cartesian coordinates, then the strain tensors have the following simple forms:

$$e_{ij} = \frac{1}{2}\left(\frac{\partial u_i}{\partial x_j} + \frac{\partial u_j}{\partial x_i} + \frac{\partial u_k}{\partial x_i}\frac{\partial u_k}{\partial x_j}\right) \quad \text{(Green)}, \tag{1.23}$$

$$\bar{e}_{ij} = \frac{1}{2}\left(\frac{\partial u_i}{\partial \theta_j} + \frac{\partial u_j}{\partial \theta_i} - \frac{\partial u_k}{\partial \theta_i}\frac{\partial u_k}{\partial \theta_j}\right) \quad \text{(Almansi)}. \tag{1.24}$$

Non-indicial notations have the typical terms

$$\begin{aligned} e_{xx} &= \frac{\partial u}{\partial x} + \frac{1}{2}\left[\left(\frac{\partial u}{\partial x}\right)^2 + \left(\frac{\partial v}{\partial x}\right)^2 + \left(\frac{\partial w}{\partial x}\right)^2\right], \\ e_{xy} &= \frac{1}{2}\left(\frac{\partial u}{\partial x} + \frac{\partial v}{\partial y} + \frac{\partial u}{\partial x}\frac{\partial u}{\partial y} + \frac{\partial v}{\partial x}\frac{\partial v}{\partial y} + \frac{\partial w}{\partial x}\frac{\partial w}{\partial y}\right). \end{aligned} \tag{1.25}$$

We have similar relations for the Almansi strain tensor. If the components of the displacement vector $\mathbf{u}$ and their first derivatives are so small that the squares and products of the partial derivatives of $\mathbf{u}$ are negligible, the differences between Green's and Almansi's tensors vanish, because it is immaterial whether the derivatives are calculated at the position of a point before or after deformation.

We obtain the following six relations in the curvilinear orthogonal coordinate system: three equations of the form

$$e_{11} = \frac{1}{H_1}\frac{\partial u_1}{\partial \theta_1} + \frac{1}{H_1 H_2}\frac{\partial H_2}{\partial \theta_2} u_2 + \frac{1}{H_1 H_3}\frac{\partial H_1}{\partial \theta_3} u_3$$

and three equations of the form

$$e_{12} = \frac{1}{2}\left(\frac{H_1}{H_2}\frac{\partial}{\partial \theta_2}\left(\frac{u_1}{H_1}\right) + \frac{H_2}{H_1}\frac{\partial}{\partial \theta_1}\left(\frac{u_2}{H_2}\right)\right). \tag{1.26}$$

The remaining equations can be obtained by a change of indices according to the scheme

$$\begin{array}{ccc} 1 & \longleftarrow & 3 \\ & \searrow \; 2 \; \nearrow & \end{array}.$$

The geometrical meaning of the strain tensor can be illustrated by considering the line element of length $dx^1 = ds_0$, $dx^2 = dx^3 = 0$. We have

$$e_1 = \frac{ds - ds_0}{ds_0}.$$

From Eq. (1.20) we have $ds^2 - ds_0^2 = 2e_{11}(dx^1)^2$; then $(1+e_1)^2 - 1 = 2e_{11}$ or $e_1 = \sqrt{1+2e_{11}} - 1$; when e_{11} is small, $e_1 = e_{11}$.

In a similar way we can prove that for small e_{12} the angle $\gamma_{12} = 2e_{12}$.

The strain tensor determines the six relations between the components of the strain tensor and three components of the displacement vector. Eliminating the u_i from this equation, we obtain the so-called *compatibility conditions*. In the case of linear equations and Cartesian coordinates we obtain by suitable differentiation the following condition:

$$e_{ij,kl} - e_{kl,ij} - e_{ik,jl} - e_{jl,ik} = 0 \tag{1.27}$$

where the comma , denotes partial differentiation. Of the 81 equations represented by (1.27) only six are essential.

In the case of nonlinear relations a more successful method of derivation uses the concept that the compatibility conditions say that the body before and after deformation remains in the Euclidean space. Then it is adequate to introduce in Eq. (1.14) the components of the metric tensor $\bar{g}_{ij}$ of the deformed body in place of the components of the metric tensor g_{km}.

1.3 Stresses and equilibrium equations of a three-dimensional body

The state of stress at an arbitrary point M of the body is defined by the symmetrical stress tensor σ_{ij} with six components. Considering the equilibrium of an infinitesimal parallelepiped element of the body in Cartesian coordinates, we obtain an equation which can be represented in the following form:

$$\frac{\partial \sigma_{ij}}{\partial x_j} + X_i = 0, \tag{1.28}$$

where X_i are components of body forces. In order to obtain equilibrium equations in the arbitrary system of curvilinear coordinates it is only necessary to notice that these equations must be expressed in tensor form. Thus, the equations of equilibrium must be

$$\sigma^{ij}|_j + X^i = 0, \tag{1.29}$$

where $|_j$ means covariant derivative. These equations for the particular case of Cartesian coordinates (x_i) take the form (1.28). In the coordinates x_i there is no difference between the covariant and contravariant vectors and tensors, therefore the stress tensor can be written as σ_{ij} or σ^{ij}. In the curvilinear coordinates the components σ_{ij} and σ^{ij} are different in spite of the fact that they represent the same physical quantity.

The conditions at the surface of the body can be written as

$$\sigma^{ij}\nu_j = T^i, \tag{1.30}$$

where ν_j are the components of the unit vector normal to the external surface of the body and T^i represents the force acting on the surface of the body.

If the material of the body is elastic, we have the following linear tensor equation relating the components of stress to the components of strain, called the *generalized Hooke law*:

$$\sigma^{ij} = a^{ijkl}e_{kl} \tag{1.31}$$

where a^{ijkl} is the tensor of elastic constants of the material. For the isotropic material, Hooke's law takes the form

$$\sigma_{ij} = \bar{\lambda}e^{\alpha}_{\alpha}\delta_{ij}+2Ge_{ij}$$

or

$$e_{ij} = \frac{1+\nu}{E}\sigma_{ij}-\frac{\nu}{E}\sigma^{\alpha}_{\alpha}\delta_{ij}, \qquad i,j = 1,2,3,$$

where $\bar{\lambda} = 2G\nu/(1-2\nu)$, $G = E/2(1+\nu)$ is the shear modulus, E is Young's modulus of elasticity, and ν is called *Poisson's ratio.*

1.4 Fundamental assumptions in the theory of plates and shells

The analysis of a three-dimensional body, like a plate and shell, we reduce to the analysis of its middle surface, i.e. to that of a two-dimensional body. It requires the determination of the method of conversion from the three-dimensional state of strain and stress to the two-dimensional state. The first papers concerning this problem date from the XIX century. A. Cauchy [1.1] and S. Poisson [1.2] developed the components of the state of stresses in the power series with respect to the variable z determining the distance of an arbitrary point from the middle surface of the shell. Another method was given by A. Love [1.3]. It consisted in making use of two simplified Kirchhoff's assumptions of the theory of plates. Love assumed that the straight lines perpendicular to the middle surface of the shell remain straight lines perpendicular to the deformed middle surface. In accordance with the second assumption the normal stresses perpendicular to the middle surface can be neglected as compared with other stresses. These assumptions correspond to the two following:

1. The thickness of the plate or shell is small in comparison with the smallest radius of curvature and the dimensions of the middle surface.
2. The strains and displacements are small in comparison with the thickness of the plate or shell.

This theory is often called the *theory of the first approximation* or the *theory of Kirchhoff–Love*. Recently, new papers have appeared in which the assumptions of Kirchhoff–Love have not been adopted. Having in view the achievement of a better accuracy of the description of the state of stress and strain in the shell, the authors took into account the effects of transverse shear and normal stresses. The following works should be mentioned: F. B. Hildebrand, E. Reissner and G. B. Thomas [1.4], A. E. Green and W. Zerna [1.5], A. I. Lure [1.6] and E. Reissner [1.7]. The latter deals with sandwich shells of revolution. In [1.8] P. M. Naghdi formulated the relations between displacements and stresses and the corresponding boundary conditions. Some other works continued the direction pointed out by Cauchy and Poisson consisting in the developing of the functions in the power series with respect to the variable z, [1.9] and [1.10].

In some other papers the development into series of Legendre polynomials or trigonometrical series have been performed. The above-mentioned papers lead to new variants of the theory of shells called sometimes the *theory of second approximation*.

In our further considerations the theory based on the Kirchhoff–Love assumptions, where the effect of transverse shear and normal stresses is not taken into account, we shall call the classical theory.

The principal aim of this book is to acquaint the reader with the effects produced in thin-walled structure by concentrated loads. Because of this, most attention will be given to the engineering theory of plates and shells important from the point of view of applications.

The engineering theory of shells in the original version has been worked out by L. H. Donnell [1.11], Kh. M. Mushtari [1.13], and V. Z. Vlasov [1.12]. Their equations are very convenient for solving many technical problems owing to their simplicity. Their advantage is that they become the equations of the theory of plates when the curvatures of the shell decrease to zero. However, the results obtained by means of these equations are accurate only for shallow shells. For non-shallow shells of slowly varying curvatures a better accuracy can be obtained by making use of the equations derived by S. Łukasiewicz [1.14], [1.15].

Introducing relatively small corrections, the accuracy of the results for non-shallow shells of slowly varying curvatures sufficient for engineering purposes have been obtained. This can be proved by the following example. The deflection of an infinitely long circular cylindrical shell under the action of two opposed radial concentrated forces, acting along a diameter, calculated on the basis of the Donnell–Vlasov equations differs by about 30 per cent from the result obtained by the accurate equations due to W. Flügge [1.16] or A. L. Goldenveizer [1.17]. The deflection of the same shell calculated by means of the equations of the improved theory differs from them only by about 1 per cent.

These equations are equivalent to L. S. D. Morley [1.18] equations in the case of the cylindrical shell. Morley obtained his equations from the accurate equations of W. Flügge by disregarding some small terms. In the case of the spherical shell these equations are identical with W. T. Koiter [1.19] and Vlasov [1.12] equations. The engineering theory has also been improved by taking into consideration the effects of transverse shear and normal stresses [1.39]. The following hypothesis has been adopted here, enabling us to replace the real state of displacements along the thickness of the shell by certain substitute simplified displacements. Namely, it has been assumed that straight lines normal to the middle surface of the shell before deformation remain straight after deformation but can be inclined to the middle surface [1.30]. This hypothesis reduces to the Kirchhoff–Love [1.21] hypothesis in the case of neglecting the transverse shear deformations. In the derived equations the geometrically non-linear terms resulting from large displacements have been taken into consideration.

Besides the above-mentioned assumptions the following basic assumptions of the engineering theory of shells have been adopted:

1. The normal displacement of the shell may be of the same order as the thickness of the shell, but must be much smaller than the dimensions of its middle surface.

2. Displacements tangential to the middle surface are much smaller than the displacements in the normal direction.

3. The components of the state of strain are small.

4. The material of the shell is elastic and satisfies Hooke's law.

1.5 Some fundamental results from the theory of surfaces

1.5.1 *First metric tensor*

From the geometrical point of view the shell is characterized, first of all, by its middle surface. It is therefore justified to give at the beginning some information concerning the theory of surfaces.

Let us consider the system of curvilinear coordinates θ^1, θ^2 on a surface (Fig. 3). The lines θ^1 = const and θ^2 = const create two families of curves

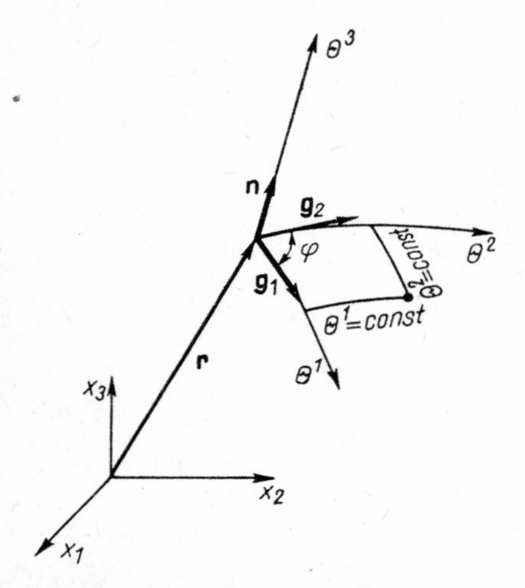

Fig. 3. System of curvilinear coordinates.

on the surface. Every point M on the surface can be considered as the intersection point of two coordinate lines θ^1 and θ^2. Let $\mathbf{r}(\theta^\alpha)$ ($\alpha = 1, 2$) denote the radius vector from a fixed origin of the Cartesian coordinates x_i to a generic point on the middle surface of the undeformed shell as a vector function of the surface coordinates.

$$\mathbf{r} = \mathbf{r}(\theta^1, \theta^2). \tag{1.32}$$

Between the cordinates θ^α and the Cartesian coordinates x_i we have the following relations

$$\theta^\alpha = \theta^\alpha(x_1, x_2, x_3).$$

Let us consider now the vector $\mathbf{r}+\Delta\mathbf{r}$ corresponding to the point M' of the line $\theta^2 = \text{const}$. If we consider the ratio $\Delta\mathbf{r}/\Delta\theta^1$ if $\Delta\theta^1 \to 0$, we obtain the partial derivative of the vector with respect to the coordinate θ^1

$$\mathbf{g}_1 = \lim_{\Delta\theta^1 \to 0} \left(\frac{\Delta\mathbf{r}}{\Delta\theta^1} \right) = \frac{\partial\mathbf{r}}{\partial\theta^1} = \mathbf{r}_{,1}\,. \tag{1.33}$$

The direction of the vector $\partial\mathbf{r}/\partial\theta^1$ follows the direction of the line $\theta^1 = \text{const}$ at the point M. The second vector $\mathbf{g}_2 = \partial\mathbf{r}/\partial\theta^2$ is directed tangentially to the line $\theta^2 = \text{const}$.

The vectors $\mathbf{g}_\alpha$ are called the *base vectors*. The plane S given by two vectors $\mathbf{g}_1$, $\mathbf{g}_2$ is a plane tangential to the surface at the point M. The straight line, perpendicular to the surface, is given by the vector product of the base vectors. The unit vector $\mathbf{n}$ will be obtained by dividing it by its own length.

Thus:

$$\mathbf{n} = \frac{\mathbf{g}_1 \times \mathbf{g}_2}{|\mathbf{g}_1 \times \mathbf{g}_2|}\,. \tag{1.34}$$

If the lines of the system of coordinates cross each other with the angle φ, then from the scalar product of the vectors $\mathbf{g}_i$ we have $\mathbf{g}_1 \cdot \mathbf{g}_2 = |\mathbf{g}_1| \cdot |\mathbf{g}_2| \cos\varphi$; then

$$\cos\varphi = \sqrt{\frac{a_{12}^2}{a_{11} \cdot a_{22}}}\,, \tag{1.35}$$

where we denoted $a_{\alpha\beta} = \mathbf{g}_\alpha \cdot \mathbf{g}_\beta$.

If the scalar product of the vectors $\mathbf{g}_\alpha$ is equal to zero, then the coordinates are called *orthogonal coordinates*, and $a_{12} = 0$. In this case we shall denote them by α_1, α_2. Since $\sin^2\varphi = 1 - \cos^2\varphi$, from (1.35) there results

$$\sin\varphi = \sqrt{\frac{a}{a_{11} \cdot a_{22}}} \qquad \text{where} \qquad a = a_{11}a_{22} - a_{12}^2\,.$$

Then we have

$$\mathbf{n} = \frac{\mathbf{g}_1 \times \mathbf{g}_2}{\sqrt{a}}\,.$$

Let us calculate now the square of the length of the line element on the surface, which is given by the two points $M(\theta^1, \theta^2)$, $M(\theta^1 + \mathrm{d}\theta^1, \theta^2 + \mathrm{d}\theta^2)$. If we consider $\mathbf{dr}$ as the increase of the radius vector $\mathbf{r}$ by moving from the point M to M', the square of the length of the line element is defined by the scalar product

$$\mathrm{d}s^2 = \mathrm{d}\mathbf{r} \cdot \mathrm{d}\mathbf{r} = \mathbf{g}_\alpha \cdot \mathbf{g}_\beta\, \mathrm{d}\theta^\alpha \mathrm{d}\theta^\beta = a_{\alpha\beta}\, \mathrm{d}\theta^\alpha\, \mathrm{d}\theta^\beta \tag{1.36}$$

or

$$ds^2 = a_{11}(d\theta^1)^2 + 2a_{12}(d\theta^1)(d\theta^2) + a_{22}(d\theta^2)^2.$$

The relation (1.36) is called the *first quadratic form* of the surface. The components $a_{\alpha\beta}$ are unique functions of the coordinates θ^α. The first quadratic form of the surface has important meaning in the determination of the strains in the middle surface of the shell.

The quantities $a_{\alpha\beta}$ in (1.35) and (1.36) are called the *components* of the first metric tensor of the surface. Then we have the definition:

The *symmetric covariant metric tensor* of the undeformed middle surface is defined by the scalar product of the base vectors:

$$a_{\alpha\beta} = \mathbf{r}_{,\alpha} \cdot \mathbf{r}_{,\beta}. \tag{1.37}$$

Its determinant is denoted by a.

The symmetric contravariant metric tensor of the undeformed middle surface is defined by the equations

$$a^{\alpha\delta} a_{\delta\beta} = \delta^\alpha_\beta.$$

If the radius vector $\mathbf{r}$ is defined in Cartesian coordinates $x_1 = x$, $x_2 = y$ and $x_3 = z$, then we have

$$a_{\alpha\beta} = \frac{\partial x_k}{\partial \theta^\alpha} \cdot \frac{\partial x_k}{\partial \theta^\beta}. \tag{1.39}$$

In the case of orthogonal coordinates we have

$$ds^2 = a_{11}(d\theta^1)^2 + a_{22}(d\theta^2)^2. \tag{1.40}$$

Also other relations are very often introduced in the theory of shells. Namely, the components of the first quadratic form are denoted by

$$A_1 = \sqrt{a_{11}}, \quad A_2 = \sqrt{a_{22}}.$$

The so-called *raising* or *lowering of indices* will be performed by means of the metric tensors $a^{\alpha\beta}$ and $a_{\alpha\beta}$. For example, for an arbitrary tensor T we have the relations

$$T_{\alpha\beta} = T^\lambda_\alpha a_{\beta\lambda} = T^{\lambda\mu} a_{\alpha\mu} a_{\beta\lambda},$$

$$T^{\alpha\beta} = T^\alpha_\lambda a^{\beta\lambda} = T_{\lambda\mu} a^{\alpha\mu} a^{\beta\lambda}.$$

For the metric tensor we have the relations

$$a^{11} = \frac{a_{22}}{a}, \quad a^{22} = \frac{a_{11}}{a}, \quad a^{12} = -\frac{a_{12}}{a}.$$

Let us introduce the *permutation tensor* of the surface $d_{\alpha\beta}$ and $d^{\alpha\beta}$ which is specified by a single independent component

$$d_{12} = \sqrt{a}, \quad d^{12} = \frac{1}{\sqrt{a}},$$

$$d^{11} = d^{22} = d_{11} = d_{22} = 0,$$

$$d_{21} = -\sqrt{a}, \quad d^{21} = -\frac{1}{\sqrt{a}}.$$

We shall make use of the indentities

$$d_{\alpha\beta}\, d^{\mu\lambda} = \delta^{\mu}_{\alpha}\,\delta^{\lambda}_{\beta} - \delta^{\mu}_{\beta}\,\delta^{\lambda}_{\alpha}, \quad d_{\alpha\mu}\, d^{\alpha\lambda} = \delta^{\lambda}_{\mu};$$

$$a_{\alpha\beta}\, d^{\alpha\mu}\, d^{\beta\lambda} = a^{\mu\lambda}, \quad a^{\alpha\beta} d_{\alpha\mu}\, d_{\beta\lambda} = a_{\mu\lambda}.$$

Let us consider now an arbitrary vector $\mathbf{u}$ fixed at the point M and tangential to the surface. This vector can be represented by the base vectors $\mathbf{g}_i$

$$\mathbf{u} = \mathbf{g}_i u^i \quad (i = 1, 2, 3). \tag{1.41}$$

Here, the components u^i are the contravariant components of the vector $\mathbf{u}$. The same vector can be determined by its projections u_i on the directions tangential to the curves θ^i at the point M.

We have

$$u_i = \mathbf{u} \cdot \mathbf{g}_i. \tag{1.42}$$

Expressing the vector $\mathbf{u}$ by (1.41), we obtain

$$u_i = \mathbf{g}_i \cdot \mathbf{g}_k u^k = a_{ik} u^k, \tag{1.43}$$

where u_k are the covariant components and a_{ik} are the components of the metric tensor of the surface. If we consider now relation (1.43) as the set of linear algebraic equations with respect to the components u^k, we obtain

$$u^k = a^{ik} u_i \quad \text{where} \quad a^{ik} = \frac{1}{a}\frac{\partial a}{\partial a_{ik}}. \tag{1.44}$$

In the case of orthogonal coordinates α_1, α_2 we have

$$a^{\alpha\alpha} = \frac{1}{a_{\alpha\alpha}}, \quad a^{\alpha\beta} = 0, \quad \alpha \neq \beta \text{ (not summed).} \tag{1.45}$$

The arbitrary vector can be presented by its covariant and contravarian components.

We have the relations

$$\mathbf{u} = u^k \mathbf{g}_k = u_k \mathbf{g}^k, \quad \text{where} \quad u^k = a^{kj} u_j,\; u_k = a_{kj} u^j.$$

The quantities $\mathbf{g}^k$ are called the *contravariant base vectors* and are connected with the base vectors $\mathbf{g}_k$ by the relations

$$\mathbf{g}_i \mathbf{g}^j = \delta_i^j, \quad \mathbf{g}^i = a^{ij}\mathbf{g}_j, \quad \mathbf{g}_i = a_{ij}\mathbf{g}^i.$$

The geometrical interpretation of the vectors $\mathbf{g}_i$ and $\mathbf{g}^i$ is presented in Fig. 4.

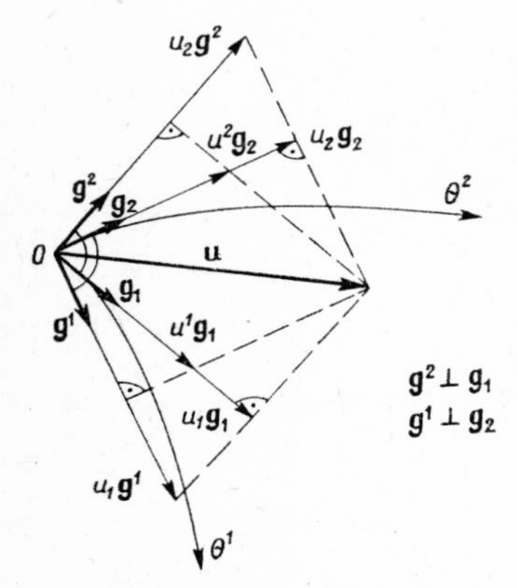

Fig. 4. Covariant and contravariant components of the vector $\mathbf{u}$ *on the plane* θ^1, θ^2.

The covariant derivative of the vector $\mathbf{u} = \mathbf{g}_i u^i$ is defined in the following way. Let us notice that while moving from the point M to the point M' we change not only the components of the vector $\mathbf{u}^i$ but also the base vectors $\mathbf{g}_i$. We have

$$\mathbf{u}|_j = \mathbf{g}_i \frac{\partial u^i}{\partial \theta^j} + u^i \frac{\partial \mathbf{g}_i}{\partial \theta^j}. \tag{1.46}$$

Writing

$$\frac{\partial \mathbf{g}_i}{\partial \theta^j} = \frac{\partial^2 \mathbf{r}}{\partial \theta^i \partial \theta^k} = \frac{\partial \mathbf{g}_j}{\partial \theta^i} = \Gamma_{ij}^k \mathbf{g}_k,$$

we obtain

$$\mathbf{u}|_j = \mathbf{g}_i \frac{\partial u^i}{\partial \theta^j} + u^i \Gamma_{ij}^k \mathbf{g}_k = \left(\frac{\partial u^k}{\partial \theta^j} + u^i \Gamma_{ij}^k\right) \mathbf{g}_k = u^k|_j \mathbf{g}_k,$$

where Γ_{ij}^k are the *Christoffel symbols* which are defined in the following way:

$$\Gamma_{ik}^j = \frac{1}{2} a^{lj} \left(\frac{\partial a_{kl}}{\partial \theta^i} + \frac{\partial a_{li}}{\partial \theta^k} - \frac{\partial a_{ik}}{\partial \theta^l}\right). \tag{1.47}$$

In a system of orthogonal coordinates α_1, α_2 with $a_{11} = A_1^2$, $a_{22} = A_2^2$, $a_{12} = 0$ we obtain

$$\Gamma^1_{11} = \frac{1}{A_1}\frac{\partial A_1}{\partial \alpha_1}, \qquad \Gamma^1_{12} = \Gamma^1_{21} = \frac{1}{A_1}\frac{\partial A_1}{\partial \alpha_2},$$

$$\Gamma^1_{22} = -\frac{A_2}{A_1^2}\frac{\partial A_2}{\partial \alpha_1}, \qquad \Gamma^2_{22} = \frac{1}{A_2}\frac{\partial A_2}{\partial \alpha_2}, \tag{1.48}$$

$$\Gamma^2_{12} = \Gamma^2_{21} = \frac{1}{A_2}\frac{\partial A_2}{\partial \alpha_1}, \qquad \Gamma^2_{11} = -\frac{A_1}{A_2^2}\frac{\partial A_1}{\partial \alpha_2}.$$

We have the following relations for the covariant derivatives of the tensor:

$$T_{ik}|_j = \partial_j T_{ik} - T_{sk}\Gamma^s_{ij} - T_{is}\Gamma^s_{jk},$$

$$T^{ik}|_j = \partial_j T^{ik} + T^{sk}\Gamma^i_{sj} + T^{is}\Gamma^k_{sj}.$$

The covariant derivatives of the metric tensor of the surface are equal to zero:

$$a_{\alpha\beta}|_k = 0, \qquad d_{\alpha\beta}|_k = 0.$$

1.5.2 *The physical components*

It is worthwhile to mention here that the covariant and contravariant tensor components of a vector or tensor do not have the same kind of physical significance in a curvilinear coordinate system as they have in a rectangular Cartesian system. In fact, they often have different dimensions. For example, the differential du in cylindrical coordinates (r, ϑ, z) has the contravariant components $(\mathrm{d}r, \mathrm{d}\vartheta, \mathrm{d}z)$ but $\mathrm{d}\vartheta$ does not have the same dimensions as the others. In this case the physical components are $(\mathrm{d}r, r\mathrm{d}\vartheta, \mathrm{d}z)$, the components of the infinitesimal vector differential du in a local rectangular Cartesian system with the axes tangent to the cylindrical surface at the point of consideration. When the system of coordinates is orthogonal the physical components of a vector or tensor will be just the Cartesian components in a local rectangular Cartesian system with axes tangent to the coordinate curves. In a general curvilinear coordinate system the physical components of a vector are parallel to the covariant base vectors.

Let us introduce non-dimensional unit vectors $\hat{\mathbf{e}}_i$ parallel to $\mathbf{g}_i$ $(i = 1, 2, 3)$

$$\hat{\mathbf{e}}_1 = \frac{\mathbf{g}_1}{|\mathbf{g}_1|}, \qquad \hat{\mathbf{e}}_2 = \frac{\mathbf{g}_2}{|\mathbf{g}_2|}, \qquad |\mathbf{g}_1| = \sqrt{g_{11}}, \qquad |\mathbf{g}_2| = \sqrt{g_{22}},$$

and the vector $\hat{\mathbf{e}}_3 = \dfrac{\mathbf{g}_3}{\sqrt{g_{33}}}$; then

$$u^i \mathbf{g}_i = u^1 \sqrt{g_{11}}\,\hat{\mathbf{e}}_1 + u^2 \sqrt{g_{22}}\,\hat{\mathbf{e}}_2 + u^3 \sqrt{g_{33}}\,\hat{\mathbf{e}}_3,$$

and the physical components $u^{\langle i\rangle}$ parallel to $\mathbf{g}_i$ are

$$u^{\langle i\rangle} = u^i \sqrt{g_{ii}} \qquad \text{(not summed).}$$

For orthogonal coordinates we have

$$u_i = \sqrt{g_{ii}}\, u^{\langle i\rangle} \qquad \text{and} \qquad u^{\langle i\rangle} = \frac{u_i}{\sqrt{g_{ii}}}. \tag{1.49}$$

The index in brackets, $\langle i\rangle$, denotes the physical component of the vector $\mathbf{u}$. The physical components of the symmetrical tensor of rank two in the curvilinear orthogonal coordinates can be presented in the form

$$T^{\langle ij\rangle} = \sqrt{g_{ii} g_{jj}}\, T^{ij} = \frac{T_{ij}}{\sqrt{g_{ii} g_{jj}}}.$$

In our further considerations the relations which do not have the tensor character will be expressed by the physical components of vectors and tensors. Then we shall omit the brackets $\langle\ \rangle$ comprising the indexes.

1.5.3 *The metric tensor of the curvature*

Let us determine now the radius of curvature of the middle surface at the arbitrary point M. This radius may be obtained as the radius of the curvature of the curve which is created by a normal plane crossing the surface. From Frenet's condition we have

$$-\frac{1}{R_n} = \frac{\mathrm{d}^2\mathbf{r}}{\mathrm{d}s^2}\cdot \mathbf{n}, \tag{1.50}$$

where $\mathrm{d}\mathbf{r}^2/\mathrm{d}s^2$ is the second derivative of the vector $\mathbf{r}$ with respect to the variable s along the crossing curve $\theta_i = \theta_i(s)$:

$$\frac{\mathrm{d}\mathbf{r}}{\mathrm{d}s} = \frac{\partial \mathbf{r}}{\partial \theta_1}\frac{\partial \theta_1}{\partial s} + \frac{\partial \mathbf{r}}{\partial \theta_2}\frac{\partial \theta_2}{\partial s} = \mathbf{g}_1 \frac{\partial \theta_1}{\partial s} + \mathbf{g}_2 \frac{\partial \theta_2}{\partial s},$$

$$\begin{aligned} \frac{\mathrm{d}\mathbf{r}^2}{\mathrm{d}s^2} &= \mathbf{g}_1 \frac{\partial^2 \theta_1}{\partial s^2} + \mathbf{g}_2 \frac{\partial^2 \theta_2}{\partial s^2} + \mathbf{g}_{1,1}\left(\frac{\partial \theta_1}{\partial s}\right)^2 + \\ &\quad + 2\mathbf{g}_{1,2}\frac{\partial \theta_1}{\partial s}\frac{\partial \theta_2}{\partial s} + \mathbf{g}_{2,2}\left(\frac{\partial \theta_2}{\partial s}\right)^2. \end{aligned} \tag{1.51}$$

Introducing the above relations into (1.50) and remembering that each of the vectors $\mathbf{g}_\alpha$ is orthogonal to the normal vector $\mathbf{n}$, we obtain

$$-\frac{1}{R_n} = b_{11}\left(\frac{\partial\theta_1}{\partial s}\right)^2 + 2b_{12}\frac{\partial\theta_1}{\partial s}\frac{\partial\theta_2}{\partial s} + b_{22}\left(\frac{\partial\theta_2}{\partial s}\right)^2; \tag{1.52}$$

$$b_{\alpha\beta} = \mathbf{n}\cdot\mathbf{r}_{\alpha,\beta},$$

where $b_{\alpha\beta} = b_{\beta\alpha}$ is called the *second metric tensor of the surface.*

The formula for the normal curvature may be presented also in the form

$$-\frac{1}{R_n} = \frac{b_{11}\,d\theta_1^2 + 2b_{12}\,d\theta_1\,d\theta_2 + b_{22}\,d\theta_2^2}{a_{11}\,d\theta_1^2 + 2a_{12}\,d\theta_1\,d\theta_2 + a_{22}\,d\theta_2^2}. \tag{1.53}$$

The expression in the numerator is called the *second quadratic form* of the surface. The components $b_{\alpha\beta}$ can be obtained by introducing relation (1.34) into $(1.52)_2$. We find

$$\begin{aligned} b_{11} &= \frac{1}{\sqrt{a}}\left(\frac{\partial^2\mathbf{r}}{\partial\theta_1^2}, \frac{\partial\mathbf{r}}{\partial\theta_1}, \frac{\partial\mathbf{r}}{\partial\theta_2}\right), \\ b_{12} &= \frac{1}{\sqrt{a}}\left(\frac{\partial^2\mathbf{r}}{\partial\theta_1\,\partial\theta_2}, \frac{\partial\mathbf{r}}{\partial\theta_1}, \frac{\partial\mathbf{r}}{\partial\theta_2}\right), \\ b_{22} &= \frac{1}{\sqrt{a}}\left(\frac{\partial^2\mathbf{r}}{\partial\theta_2^2}, \frac{\partial\mathbf{r}}{\partial\theta_1}, \frac{\partial\mathbf{r}}{\partial\theta_2}\right). \end{aligned} \tag{1.54}$$

Introducing the relation $\mathbf{r} = f(x, y, z) = x\mathbf{i} + y\mathbf{j} + z\mathbf{k}$, we obtain

$$b_{11} = \frac{1}{\sqrt{a}}\begin{vmatrix} \dfrac{\partial^2 x}{\partial\theta_1^2} & \dfrac{\partial^2 y}{\partial\theta_1^2} & \dfrac{\partial^2 z}{\partial\theta_1^2} \\ \dfrac{\partial x}{\partial\theta_1} & \dfrac{\partial y}{\partial\theta_1} & \dfrac{\partial z}{\partial\theta_1} \\ \dfrac{\partial x}{\partial\theta_2} & \dfrac{\partial y}{\partial\theta_2} & \dfrac{\partial z}{\partial\theta_2} \end{vmatrix},$$

$$b_{12} = \frac{1}{\sqrt{a}}\begin{vmatrix} \dfrac{\partial^2 x}{\partial\theta_1\,\partial\theta_2} & \dfrac{\partial^2 y}{\partial\theta_1\,\partial\theta_2} & \dfrac{\partial^2 z}{\partial\theta_1\,\partial\theta_2} \\ \dfrac{\partial x}{\partial\theta_1} & \dfrac{\partial y}{\partial\theta_1} & \dfrac{\partial z}{\partial\theta_1} \\ \dfrac{\partial x}{\partial\theta_2} & \dfrac{\partial y}{\partial\theta_2} & \dfrac{\partial z}{\partial\theta_2} \end{vmatrix}.$$

We have for b_{22} the analogous expression. If $d\theta_2 = 0$, then $1/R_{11} = -b_{11}/a_{11}$; we see that the components b_{11} and b_{22} characterize the normal curvatures of the coordinate lines θ_1 = constant and θ_2 = constant. The component b_{12} describes the twisting of the surface.

The magnitude of the radius of the curvature R_n given by Eq. (1.50) depends on the direction of the normal plane crossing the surface. From all values of the radii of the curvature at the point considered we can choose two, the first having minimum value and the second—maximum. These radii of curvature are called the *principal radii of curvature* and their directions the *directions of the principal curvatures.* The curves lying on the surface and given by the principal curvatures are called the *lines of curvature*. If the assumed system of coordinates θ_1 and θ_2 follows the lines of curvatures, then the components $b_{12} = 0$ and $a_{12} = 0$. We find the principal curvatures by looking for the extremum of the expression $1/R_n$.

Let us consider the dependence of the curvature $k_n = 1/R_n$ on the direction of the normal plane, i.e. on $d\theta_1/d\theta_2$. Let us present the relation (1.53) in the following form:

$$b_{11}\,d\theta_1^2+2b_{12}\,d\theta_1\,d\theta_2+b_{22}\,d\theta_2^2+ \\ +k_n(a_{11}\,d\theta_1^2+2a_{12}\,d\theta_1\,d\theta_2+a_{22}\,d\theta_2^2) = 0. \tag{1.55}$$

Let us find the extremum of k_n. We obtain the conditions for it by equating to zero the partial derivatives of the left-hand side of Eqs. (1.55) with respect to $d\theta_1$ and $d\theta_2$.

$$\begin{aligned} b_{11}\,d\theta_1+b_{12}\,d\theta_2+k_n(a_{11}\,d\theta_1+a_{12}\,d\theta_2) &= 0, \\ b_{12}\,d\theta_1+b_{22}\,d\theta_2+k_n(a_{12}\,d\theta_1+a_{22}\,d\theta_2) &= 0. \end{aligned} \tag{1.56}$$

On eliminating $d\theta_1$ and $d\theta_2$ from the above equations, we obtain the following equation for k_n:

$$(a_{11}a_{22}-a_{12}^2)k_n^2+(a_{11}b_{22}+a_{22}b_{11}-2a_{12}b_{12})k_n+b_{11}b_{22}-b_{12}^2 = 0.$$

The roots of this equation are the extremum values of the curvature k_n. These values are called the *principal curvatures* of the surface. The corresponding principal directions can be found by determining the value of $d\theta_1/d\theta_2$. Eliminating k_n from Eqs. (1.56), we find

$$(a_{12}b_{22}-a_{22}b_{12})\left(\frac{d\theta_2}{d\theta_1}\right)^2+(a_{11}b_{22}-a_{22}b_{11})\frac{d\theta_2}{d\theta_1}+ \\ +a_{11}b_{12}-a_{12}b_{11} = 0,$$

Other notations are often used in the theory of shells, viz.:

$$b_{\langle 11\rangle} = \frac{b_{11}}{a_{11}} = -\frac{1}{R_{11}},$$

$$b_{\langle 22\rangle} = \frac{b_{22}}{a_{22}} = -\frac{1}{R_{22}},$$

$$b_{\langle 12\rangle} = \frac{b_{12}}{\sqrt{a_{11}a_{22}}} = \frac{1}{R_{12}},$$

where R_{11} and R_{22} are the radii of curvature of the surfaces along the lines of coordinates. The geometrical meaning of R_{12} is more complex. It can be proved that this quantity is connected with the principal curvature by the relation

$$\frac{1}{R_{12}} = \frac{\sin 2\chi}{2}\left(\frac{1}{R_1} - \frac{1}{R_2}\right),$$

where χ is the angle between the direction of the line α_1 and the direction of the principal curvature $1/R_1$; R_1 and R_2 are the principal radii of curvature.

The *mean* and *Gaussian curvatures* H and K of the surface are defined by the invariants

$$\begin{aligned} K &= \frac{1}{R_1 R_2} = \frac{b_{11}b_{22}-b_{12}^2}{a_{11}a_{22}-a_{12}^2} = \tfrac{1}{2}d^{\alpha\lambda}d^{\beta\mu}b_{\alpha\beta}b_{\lambda u} \\ &= b_1^1 b_2^2 - b_1^2 b_1^1 = \frac{1}{R_{11}}\frac{1}{R_{22}} - \frac{1}{R_{12}^2}, \end{aligned} \tag{1.57}$$

$$H = -\frac{1}{2}\left(\frac{1}{R_1} + \frac{1}{R_2}\right) = \frac{-2a_{12}b_{12}+a_{11}b_{22}+a_{22}b_{11}}{2(a_{11}a_{22}-a_{12}^2)} = \tfrac{1}{2}b.$$

The mean curvature H has a constant value at any point of the shell

$$2H = -\left(\frac{1}{R_1} + \frac{1}{R_2}\right) = -\left(\frac{1}{R_{11}} + \frac{1}{R_{22}}\right),$$

where R_{11} and R_{22} denote the radii of curvature in the two arbitrary mutually perpendicular directions.

Let us calculate the components $a_{\alpha\beta}$ and $b_{\alpha\beta}$ for the surface given by $z = f(x, y)$. Assuming that $\theta_1 = x$, $\theta_2 = y$, we obtain from the previous formulae

$$a = 1+\left(\frac{\partial z}{\partial x}\right)^2 + \left(\frac{\partial z}{\partial y}\right)^2,$$

$$a_{11} = 1+\left(\frac{\partial z}{\partial x}\right)^2, \quad a_{12} = \frac{\partial z}{\partial x}\cdot\frac{\partial z}{\partial y}, \quad a_{22} = 1+\left(\frac{\partial z}{\partial y}\right)^2, \tag{1.58}$$

$$b_{11} = -\frac{1}{\sqrt{a}}\frac{\partial^2 z}{\partial x^2}, \quad b_{12} = \frac{1}{\sqrt{a}}\frac{\partial^2 z}{\partial x \partial y}, \quad b_{22} = -\frac{1}{\sqrt{a}}\frac{\partial^2 z}{\partial y^2}.$$

If we consider a shallow shell, we may assume that

$$\left(\frac{\partial z}{\partial x}\right)^2 \ll 1, \quad \left(\frac{\partial z}{\partial y}\right)^2 \ll 1, \quad \text{and} \quad a = 1.$$

The Gaussian curvature takes the form

$$K = \frac{\partial^2 z}{\partial x^2} \cdot \frac{\partial^2 z}{\partial y^2} - \left(\frac{\partial^2 z}{\partial x \partial y}\right)^2.$$

For the axially-symmetrical shells given by the equation $z = z(r)$, we have $x = r\cos\varphi$, $y = r\sin\varphi$, $z = z$. Assuming $\theta^1 = r$, $\theta^2 = \varphi$, we find from (1.39) and (1.54)

$$a_{22} = r^2, \quad a_{12} = 0, \quad a_{11} = 1+z'^2, \quad \sqrt{a} = r\sqrt{1+z'^2},$$

$$b_{11} = -\frac{z''}{\sqrt{1+z'^2}}, \quad b_{12} = 0, \quad b_{22} = -\frac{rz'}{\sqrt{1+z'^2}},$$

where z' and z'' are the derivatives of z with respect to the variable r. The Gaussian curvature is

$$K = \frac{z''z'}{r(1+z'^2)^2} \tag{1.59}$$

and the principal curvatures are

$$\frac{1}{R_1} = \frac{z''}{(1+z'^2)^{3/2}}, \quad \frac{1}{R_2} = \frac{z'}{r\sqrt{1+z'^2}}.$$

The Gaussian curvature has an important meaning while considering the local form of the shell. This curvature is positive when both radii of curvature have the same sign. The surface is then convex and both curvature centres lay on the one side of the surface. When the Gaussian curvature is equal to zero, the surface considered is conical or cylindrical. For the surface of negative Gaussian curvature it is characteristic that the curvature centres lay on different sides of the surface and the radii of curvature have opposite signs.

The first and second metric tensors define uniquely the geometry of the surface. However, these tensors are not arbitrary functions of the coordinates θ^1, θ^2, they have to satisfy three equations called the *Codazzi–Gauss equations*, because they describe the surface in the three-dimensional Euclidean space. We can obtain them from the Riemann–Christoffel tensor (1.16) which can be presented in the form

$$R^{\lambda}_{\alpha\beta\mu} = \bar{R}^{\lambda}_{\alpha\beta\mu} + \Gamma^3_{\alpha\mu}\Gamma^{\lambda}_{3\beta} - \Gamma^3_{\alpha\beta}\Gamma^{\lambda}_{3\mu},$$

where $\overline{R}^{\lambda}_{\alpha\beta\mu}$ is the Riemann–Christoffel tensor of the surface and

$$\Gamma^{3}_{\alpha\beta} = b_{\alpha\beta}, \quad \Gamma^{\alpha}_{\beta 3} = -b^{\alpha}_{\beta}, \quad \Gamma^{3}_{\alpha 3} = 0,$$

$$\Gamma^{\alpha}_{\beta\mu} = \mathbf{g}^{\alpha} \cdot \mathbf{g}_{\mu,\beta} = \mathbf{g}^{\alpha} \cdot \mathbf{g}_{\beta,\mu} = -\mathbf{g}_{\mu} \cdot \mathbf{g}^{\alpha}_{,\beta}$$

are Christoffel symbols for the surface.

Since in Euclidean space $R^{\lambda}_{\alpha\beta\mu} = 0$, we have $\overline{R}^{\lambda}_{\alpha\beta\mu} = \Gamma^{3}_{\alpha\beta}\Gamma^{\lambda}_{3\mu} - \Gamma^{3}_{\beta\mu}\Gamma^{\lambda}_{3\alpha}$ which gives on lowering the index

$$\overline{R}_{\alpha\beta\lambda\mu} = b_{\alpha\beta}b_{\lambda\mu} - b_{\alpha\mu}b_{\beta\lambda} = Kd_{\alpha\beta}d_{\lambda\mu}$$

and

$$b^{\alpha}_{\lambda|\mu} - b^{\alpha}_{\mu|\lambda} = 0.$$

1.6 The geometry of the shell

Let us consider an isotropic shell of constant thickness. Let us apply the system of curvilinear coordinates θ^1, θ^2 lying on the middle surface of the shell. The third direction ξ is given by the straight line perpendicular to the middle surface.

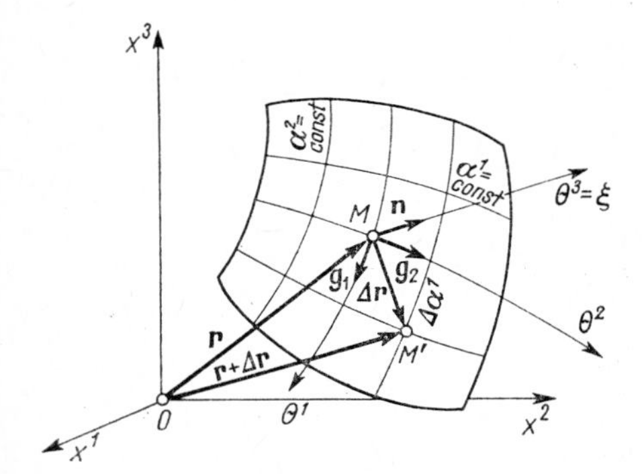

Fig. 5. Geometry of surface.

Let us assume that between the Cartesian coordinates x^i and the curvilinear coordinates θ^α there exists the relations $x^i = x^i(\theta)$. The position of an arbitrary point M is defined by the radius vector $\mathbf{r}(x^i)$ (Fig. 5)

$$\mathbf{r} = \mathbf{r}(\theta^1, \theta^2) + \xi\mathbf{n}(\theta^1, \theta^2) \tag{1.60}$$

where the coordinate ξ changes between the limits $-\frac{1}{2}h < \xi < \frac{1}{2}h$. Let us define now the components of the metric tensor in space corresponding to this triple of coordinates $(\theta^1, \theta^2, \xi)$.

On a similar treatment as in the previous section, we obtain the base vectors

$$\mathbf{g}_\alpha = \mathbf{r}_{,\alpha} + \xi \mathbf{n}_{,\alpha}.$$

The components of the metric tensor are defined by the scalar product of the base vectors $\mathbf{g}_\alpha$

$$g_{\alpha\beta} = \mathbf{g}_\alpha \cdot \mathbf{g}_\beta = a_{\alpha\beta} - 2\xi b_{\alpha\beta} + \xi^2 c_{\alpha\beta}, \qquad g_{\alpha 3} = 0, \quad g_{33} = 1, \tag{1.61}$$

where $a_{\alpha\beta} = \mathbf{r}_{,\alpha}\mathbf{r}_{,\beta}$ are the components of the first metric tensor of the middle surface $a_{ij} = (g_{ij})\xi = 0$, g_{ij} are the components of the metric tensor of the space of the shell, and

$$b_{\alpha\beta} = -\mathbf{r}_{,\beta} \cdot \mathbf{n}_{,\alpha} = -\mathbf{r}_{,\alpha} \cdot \mathbf{n}_{,\beta};$$

$$c_{\alpha\beta} = \mathbf{n}_{,\alpha} \cdot \mathbf{n}_{,\beta} = b^\delta_\alpha b_{\delta\beta} = 2Hb_{\alpha\beta} - Ka_{\alpha\beta},$$

where b_{ij}, c_{ij} are the second and third fundamental metric tensors of the underformed middle surface.

The length of the line element $\mathrm{d}s$ in the space of the shell is given by

$$(\mathrm{d}s)^2 = (a_{\alpha\beta} - 2\xi b_{\alpha\beta} + \xi^2 c_{\alpha\beta})\mathrm{d}\theta^\alpha \mathrm{d}\theta^\beta + (\mathrm{d}\xi)^2. \tag{1.62}$$

The components $b_{\alpha\beta}$ may be presented in a different way. Since the vectors $\mathbf{r}_{,\alpha}$ and $\mathbf{n}$ are orthogonal, we have

$$\mathbf{r}_{,\alpha} \cdot \mathbf{n}_{,\alpha} = \frac{\partial(\mathbf{r}_{,\alpha} \cdot \mathbf{n})}{\partial\theta^\beta} - \mathbf{n} \cdot \mathbf{r}_{,\alpha\beta} = -\mathbf{n} \cdot \mathbf{r}_{,\alpha\beta};$$

then $b_{\alpha\beta} = \mathbf{n} \cdot \mathbf{r}_{,\alpha\beta}$, $b_{\alpha\beta} = b_{\beta\alpha}$. Comparing the above formulae with the relations in (1.52), we find that the coefficients $b_{\alpha\beta}$ are the components of the curvature tensor of the middle surface of the shell.

The Christoffel symbols with the index 3, corresponding to the normal directions, depend on the components of the curvature tensor:

$$\Gamma^3_{\alpha\beta} = b_{\alpha\beta}, \quad \Gamma^k_{3\alpha} = \Gamma^k_{\alpha 3} = -b^k_\alpha, \quad \Gamma^k_{33} = \Gamma^3_{13} = \Gamma^3_{23} = 0. \tag{1.63}$$

If the system of coordinates θ^i is orthogonal and their directions follow the directions of the principal curvatures of the middle surface, we have the relations

$$b_{11} = -\frac{a_{11}}{R_1}, \quad b_{12} = 0, \quad b_{22} = -\frac{a_{22}}{R_2}.$$

The length of the line element in the orthogonal coordinates α_1, α_2, α_3 takes the form

$$(\mathrm{d}s)^2 = (H_1)^2(\mathrm{d}\alpha_1)^2 + (H_2)^2(\mathrm{d}\alpha_2)^2 + (H_3)^2(\mathrm{d}\alpha_3)^3 \tag{1.64}$$

where $H_1 = A_1(1+\xi/R_1)$, $H_2 = A_2(1+\xi/R_2)$, $H_3 = A_3 = 1$. Those are called the *Lamé coefficients* of the curvilinear coordinates of an arbitrary point of the shell. A_i are the coefficients of the first quadratic form of the middle surface ($\xi = 0$):

$$(A_1)^2 = a_{11} = \frac{1}{a^{11}}, \qquad (A_2)^2 = a_{22} = \frac{1}{a^{22}}.$$

Introducing the Lamé coefficients H_i with $\xi = 0$ into Eq. (1.16), we obtain the *Codazzi–Gauss conditions*:

$$\left[\frac{A_{2,1}}{A_1}\right]_{,1} + \left[\frac{A_{1,2}}{A_2}\right]_{,2} = -\frac{A_1 A_2}{R_1 R_2},$$
$$\left(\frac{A_1}{R_1}\right)_{,2} = \frac{A_{1,2}}{R_2}, \qquad \left(\frac{A_2}{R_2}\right)_{,1} = \frac{A_{2,1}}{R_1} \tag{1.65}$$

which yields also

$$\frac{H_{2,1}}{H_1} = \frac{A_{2,1}}{A_1}, \qquad \frac{H_{1,2}}{H_2} = \frac{A_{1,2}}{A_2}.$$

The above conditions are expressed here in physical components.

1.7 Deformation of the shell

Let us consider now the deformed shell. An arbitrary point of the shell displaces a certain distance which can be called the *displacement vector* $\mathbf{u}^*$. The radius vector of the generic point in the new position is given by (Fig. 6)

$$\bar{\mathbf{r}}^*(\theta^1, \theta^2, \xi) = \mathbf{r}^*(\theta^1, \theta^2, \xi) + \mathbf{u}^*(\theta^1, \theta^2, \xi). \tag{1.66}$$

The displacement vector $\mathbf{u}^*$ can be represented by means of the undeformed base vectors $\mathbf{g}_i$ and the unit normal vector to the undeformed middle surface:

$$\mathbf{u}^* = u^\alpha \mathbf{g}_\alpha + u^3 \mathbf{n}$$

where u^α and u^3 constitute a contravariant space vector.

On the other hand we may consider u^α as a contravariant surface vector and u^3 as a surface invariant. Also the covariant space tensor of valence 2, at the middle surface may be decomposed into one covariant surface tensor $T_{\alpha\beta}$, two covariant surface vectors $T_{\alpha 3}$, $T_{3\alpha}$, and one surface invariant T_{33}.

The rules connecting the covariant spatial derivatives of space vectors and tensors and the covariant surface derivatives of the surface representation of these vectors and tensor are the following:

$$
\begin{aligned}
u^{\alpha}||_{\beta} &= u^{\alpha}_{,\beta}+\Gamma^{\alpha}_{\delta\beta}u^{\delta}+\Gamma^{\alpha}_{3\beta}u^{3} = u^{\alpha}|_{\beta}-b^{\alpha}_{\beta}u^{3},\\
u^{3}||_{\beta} &= u^{3}_{,\beta}+\Gamma^{3}_{\delta\beta}u^{\delta} = u^{3}_{,}|_{\beta}+b_{\delta\beta}u^{\delta} = u^{3}|_{\beta}+b_{\delta\beta}u^{\delta},\\
u_{\alpha}||_{\beta} &= u_{\alpha,\beta}-\Gamma^{\delta}_{\alpha\beta}u_{\delta}-\Gamma^{3}_{\alpha\beta}u_{3} = u_{\alpha}|_{\beta}-b_{\alpha\beta}u_{3},\\
u_{3}||_{\beta} &= u_{3,\beta}-\Gamma^{\delta}_{3\beta}u_{\delta} = u_{3,\beta}+b^{\delta}_{\beta}u_{\delta} = u_{3}|_{\beta}+b^{\delta}_{\beta}u_{\delta},\\
T_{\alpha\beta}||_{k} &= T_{\alpha\beta,k}-\Gamma^{\delta}_{\alpha k}T_{\delta\beta}-\Gamma^{\delta}_{\beta k}T_{\alpha\delta}-\Gamma^{3}_{\alpha k}T_{3\beta}-\Gamma^{3}_{\beta k}T_{\alpha 3}\\
&= T_{\alpha\beta}|_{k}-b_{\alpha k}T_{3\beta}-b_{\beta k}T_{\alpha 3}, \quad \alpha,\beta = 1,2.
\end{aligned}
\tag{1.67}
$$

Here we introduce the notation $||_{\beta}$ for the *covariant spatial derivative* and $|_{\beta}$ for the *covariant surface derivative.* For the normal displacement $u^3 = w$ we have $w|_{\beta} = w_{,\beta}$.

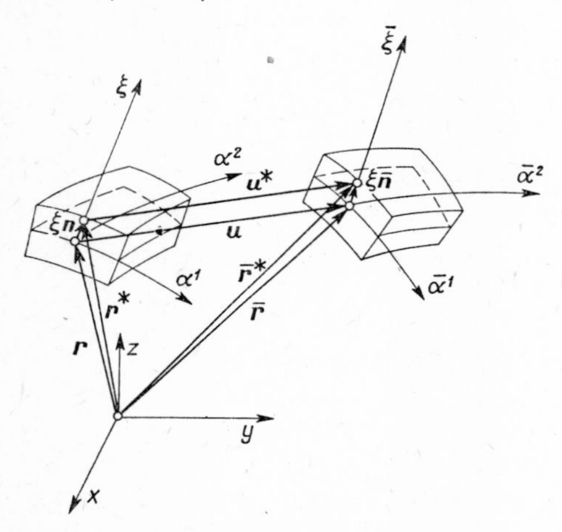

Fig 6. Displacements of a shell element.

Let us calculate now the Green strain tensor (1.23) defining the difference of the length of the line element in the initial and deformed configuration.

From (1.20), (1.23) we have

$$2e_{ij} = u^{*}_{j}||_{i}+u^{*}_{i}||_{j}+u^{*}_{\delta}||_{j}\cdot u^{*\delta}||_{i}, \quad i,j = 1,2,3. \tag{1.68}$$

According to the kinematic hypothesis we assume that the components of the displacement vector can be written in the following form:

$$u^{*}_{\alpha} = u_{\alpha}+\beta_{\alpha}\xi, \quad u^{*}_{3} = u_{3} = w, \quad \alpha = 1,2. \tag{1.69}$$

Here u_{α} ($u_1 = u, u_2 = v$) are the components of the displacement vector of the middle surface. The quantities $\beta_{\alpha}(\theta^1,\theta^2)$ are the angles of rotation of the lateral sides of the shell element during deformation. The positive directions of the displacements follow the positive directions of the vectors $\mathbf{g}_\alpha$ and $\mathbf{n}$.

The angles β_α together with angle Ω_n of the rotation around the normal compose a vector $\mathbf{\Omega}$:

$$\mathbf{\Omega} = \beta^1 \mathbf{g}_2 - \beta^2 \mathbf{g}_1 + \Omega_n \mathbf{n};$$

$\mathbf{\Omega}$ is the vector of the rotation angles of the point of the middle surface. The above formula can be presented in the form

$$\mathbf{\Omega} = d^{\beta\alpha}\beta_\beta \mathbf{g}_\alpha + \mathbf{\Omega}_n \mathbf{n}.$$

Let us present the components of the strain tensor as a power series of the variable ξ. If we take into account only two terms, we have

$$e_{\alpha\beta} = \varepsilon_{\alpha\beta} + \xi \varkappa_{\alpha\beta}, \qquad \alpha, \beta = 1, 2. \tag{1.70}$$

Introducing relations (1.69), (1.70) into Eq. (1.68), we find

$$\varepsilon_{\alpha\beta} = \tfrac{1}{2}(u_\alpha||_\beta + u_\beta||_\alpha + u_\delta||_\alpha u^\delta||_\beta + w||_\alpha w||_\beta),$$
$$\varkappa_{\alpha\beta} = \tfrac{1}{2}(\beta_\alpha||_\beta + \beta_\beta||_\alpha + u_\delta||_\beta \beta^\delta||_\alpha + \beta_\delta||_\beta u^\delta|_\alpha|).$$

Since from (1.67)

$$w||_\alpha = w|_\alpha + b^\delta_\alpha u_\delta = w_{,\alpha} + b^\delta_\alpha u_\delta,$$

we have

$$\begin{aligned}\varepsilon_{\alpha\beta} = {} & \tfrac{1}{2}(u_\alpha|_\beta + u_\beta|_\alpha) - b_{\alpha\beta} w + \tfrac{1}{2}(w_{,\alpha} w_{,\beta} + b^\delta_\alpha b_{\delta\beta} w^2) + \\ & + \tfrac{1}{2} u^\delta|_\alpha u_\delta|_\beta - \tfrac{1}{2}(b_{\delta\beta} u^\delta|_\alpha + b^\delta_\alpha u_\delta|_\beta) w + \tfrac{1}{2}(b^\delta_\alpha w|_\beta + b^\delta_\beta w|_\alpha) u_\delta + \\ & + \tfrac{1}{2} b^\delta_\alpha u_\delta b^\gamma_\beta u_\gamma.\end{aligned} \tag{1.71}$$

From the definition of the vector $\mathbf{\Omega}$ it follows that

$$d^{\delta\alpha}\beta_\delta||_\beta = d^{\delta\alpha}\beta_\delta|_\beta - b^\alpha_\beta \Omega_n,$$
$$\beta_\alpha||_\beta = \beta_\alpha|_\beta - b^\delta_\beta d_{\alpha\delta} \Omega_n.$$

Then we find

$$\begin{aligned}\varkappa_{\alpha\beta} = {} & \tfrac{1}{2}(\beta_\alpha|_\beta + \beta_\beta|_\alpha) - \tfrac{1}{2}(b^\delta_\alpha d_{\beta\delta} + b^\delta_\beta d_{\alpha\delta})\Omega_n + \\ & + u_\delta|_\beta \beta^\delta|_\alpha - d^\delta_\gamma (b^\gamma_\alpha u_\delta|_\beta + b^\gamma_\beta u_\delta|_\alpha) - (b^\delta_\alpha \beta_\delta|_\beta + b^\delta_\beta \beta_\delta|_\alpha) w + \\ & + d^\delta_\gamma (b_{\delta\beta} b^\gamma_\alpha + b^\gamma_\beta b_{\alpha\delta}) w \Omega_n.\end{aligned} \tag{1.72}$$

Here, $\varepsilon_{\alpha\beta}$ is symmetric and is called the *middle surface strain tensor*. The quantity $\varkappa_{\alpha\beta}$ is called the *tensor of the changes of curvature*.

The angle of rotation around the normal Ω_n is defined by

$$\Omega_n = \tfrac{1}{2} d^{\beta\alpha} u_\alpha|_\beta = \frac{1}{2\sqrt{a}}(u_2|_1 - u_1|_2).$$

The nonlinear terms in the above relations will be neglected in further considerations, except those terms containing the products and squares of the derivatives of the component w and products of the components $b_{\alpha\beta}$ and the component w. That is justified by the assumption that the displacements of the shells considered tangential to the middle surface are much smaller than the displacement in the normal direction.

In conventional notations in physical components, introducing the orthogonal coordinates (α_1, α_2), we have

$$\varepsilon_{11} = \frac{u_{,1}}{A_1} + \frac{A_{1,2}v}{A_1 A_2} + \frac{w}{R_{11}} + \frac{1}{2}\left[\left(\frac{w_{,1}}{A_1}\right)^2 + w^2\left(\frac{1}{R_{11}^2} + \frac{1}{R_{12}^2}\right)\right],$$

$$\varepsilon_{22} = \frac{v_{,2}}{A_2} + \frac{A_{2,1}u}{A_1 A_2} + \frac{w}{R_{22}} + \frac{1}{2}\left[\left(\frac{w_{,2}}{A_2}\right)^2 + w^2\left(\frac{1}{R_{22}^2} + \frac{1}{R_{12}^2}\right)\right], \quad (1.71a)$$

$$\varepsilon_{12} = \frac{1}{2}\left[\frac{A_1}{A_2}\left(\frac{u}{A_1}\right)_{,2} + \frac{A_2}{A_1}\left(\frac{v}{A_2}\right)_{,1} - \right.$$
$$\left. - \frac{2w}{R_{12}} + \frac{1}{A_1 A_2} w_{,1} w_{,2} + \left(\frac{1}{R_{11}} + \frac{1}{R_{22}}\right)\frac{w^2}{R_{12}}\right].$$

The curvature tensor takes the form

$$\varkappa_{11} = \frac{1}{A_1}\beta_{1,1} + \frac{A_{1,2}}{A_1 A_2}\beta_2 - \frac{\Omega_n}{R_{12}},$$

$$\varkappa_{22} = \frac{1}{A_2}\beta_{2,2} + \frac{A_{2,1}}{A_1 A_2}\beta_1 + \frac{\Omega_n}{R_{12}}, \quad (1.72a)$$

$$\varkappa_{12} = \frac{1}{2}\left\{\frac{A_1}{A_2}\left(\frac{\beta_1}{A_1}\right)_{,2} + \frac{A_2}{A_1}\left(\frac{\beta_2}{A_2}\right)_{,1} - \left(\frac{1}{R_{11}} - \frac{1}{R_{22}}\right)\Omega_n\right\},$$

where

$$\Omega_n = \frac{1}{2A_1 A_2}[(A_2 v)_{,1} - (A_1 u)_{,2}].$$

The above relations result from (1.71) if we replace the covariant derivatives by relations (1.13) and (1.48) and if we introduce the physical components. They are almost in agreement with the forms given by Goldenveizer (by $\beta_\alpha = -w\|_\alpha$) and differ slightly in $\varkappa_{12}$. Some different relations for the changes of curvature are given in papers concerning the theory of shells, such as [1.20]. However, the differences are within the limits of the initial assumptions of the theory of thin shells. The coefficient 1/2 in the expressions for $\varkappa_{12}$ results from the different definition of these quantities.

W. T. Koiter [1.22] considered the spatial metric tensor in the deformed and undeformed shell and obtained a different form for the curvature tensor, viz.:

$$\begin{aligned}\bar{\varkappa}_{\alpha\beta} &= \tfrac{1}{2}(\beta_\alpha|_\beta+\beta_\beta|_\alpha)+c_{\alpha\beta}w-\tfrac{1}{2}(b^\delta_\alpha u_\delta|_\beta+b^\delta_\beta u_\delta|_\alpha)\\ &= \varkappa_{\alpha\beta}-\tfrac{1}{2}(b^\delta_\alpha \varepsilon_{\delta\beta}+b^\delta_\beta \varepsilon_{\delta\alpha}),\end{aligned} \tag{1.73}$$

where $\beta_\alpha = -(w|_\alpha+b^\delta_\alpha u_\delta)$ because the effect of transverse shear deformations was not taken into account.

The difference consists only in the last term in brackets and is negligible within the framework of shell theory based on the assumption of approximately plane stress. The two measures for the changes of curvature $\varkappa_{\alpha\beta}$ and $\bar{\varkappa}_{\alpha\beta}$ are equivalent and we may, as we please, employ either $\varkappa_{\alpha\beta}$ or $\bar{\varkappa}_{\alpha\beta}$ as the tensor of changes of curvature. Each of the definitions has certain advantages. The tensor $\varkappa_{\alpha\beta}$ seems to be more attractive for the discussion of the general equations of shell theory and we shall formulate our equations in terms of $\varkappa_{\alpha\beta}$.

The translation of the tensor notation into the conventional notations in physical components may introduce certain difficulties. Therefore the example of this operation is presented here. In the case of the strain ε_{11} (1.71) we have

$$\begin{aligned}\varepsilon_{11} &= u_1|_1-b_{11}w+\frac{1}{2}\left(\frac{\partial w}{\partial\alpha_1}\right)^2+\frac{1}{2}(2Hb_{11}-Ka_{11})w^2\\ &= \frac{\partial u_1}{\partial\alpha_1}-\Gamma^1_{11}u_1-\Gamma^2_{11}u_2-b_{11}w+\frac{1}{2}\left[\left(\frac{\partial w}{\partial\alpha_1}\right)^2+(2Hb_{11}-Ka_{11})w^2\right].\end{aligned}$$

Introducing the notations $u_1 = u$, $u_2 = v$ and the Christoffel symbols from (1.49), we obtain

$$\begin{aligned}\varepsilon_{11} &= \frac{\partial u}{\partial\alpha_1}-\frac{1}{A_1}\frac{\partial A_1}{\partial\alpha_1}u+\frac{A_1}{A_2^2}\frac{\partial A_1}{\partial\alpha_2}v-b_{11}w+\\ &\quad+\frac{1}{2}\left[\left(\frac{\partial w}{\partial\alpha_1}\right)^2+(2Hb_{11}-Ka_{11})w\right].\end{aligned}$$

Let us introduce now the physical components. We have

$$u = u^{\langle\rangle}A_1,\quad v = v^{\langle\rangle}A_2,\quad w = w^{\langle\rangle},$$

$$2Hb_{11}-Ka_{11} = \left(\frac{1}{R_{11}^2}+\frac{1}{R_{12}^2}\right)A_1^2,\qquad \varepsilon_{11} = \varepsilon^{\langle 11\rangle}A_1^2.$$

On substitution and differentiation we find

$$A_1^2\varepsilon^{\langle 11\rangle} = A_1\frac{\partial u^{\langle\rangle}}{\partial\alpha_1}+\frac{A_1}{A_2}\frac{\partial A_1}{\partial\alpha_2}v^{\langle\rangle}+\frac{A_1^2}{R_1^2}w^{\langle\rangle}+$$

$$+\frac{1}{2}\left[\left(\frac{\partial w}{\partial \alpha_1}\right)^2 + A_1^2\left(\frac{1}{R_{11}^2} + \frac{1}{R_{12}^2}\right) w^2\right],$$

which is in agreement with $(1.71\mathrm{a})_1$.

The relations (1.71a), (1.72a) are valid in an arbitrary system of orthogonal coordinates. If the lines of the coordinates follow the lines of principal curvatures, then $1/R_{12} = 0$, and the curvatures $1/R_{11} = 1/R_1$ and $1/R_{22} = 1/R_2$ are the principal curvatures.

1.8 Constitutive equations, equations of equilibrium

Let us introduce the notion of the *internal forces* and *moments* which are the stress resultants and stress couples referred to a unit length of the middle surface. We obtain these quantities by integrating the stresses across the thickness of the wall of the shell element. Taking into account the change of length of the edge of the element with the change of ξ, we obtain

$$\begin{aligned} N^{\alpha\beta} &= \int_{-h/2}^{h/2} (\sigma^{\alpha\beta} - \underline{\xi b_\gamma^\beta \sigma^{\alpha\gamma}}) \sqrt{g/a}\, \mathrm{d}\xi, \\ M^{\alpha\beta} &= \int_{-h/2}^{h/2} (\sigma^{\alpha\beta} - \underline{\xi b_\gamma^\beta \sigma^{\alpha\gamma}}) \sqrt{g/a}\, \xi \mathrm{d}\xi, \end{aligned} \tag{1.74}$$

where $\sqrt{g/a} = 1 - 2H\xi + K\xi^2$. If the shell is thin, then $\xi/R_1 \ll 1$ and the above relations can be simplified by neglecting the underlined small terms and assuming $\sqrt{g/a} = 1$. We assume that the material of the shell is elastic, homogeneous, and isotropic. Then, using Hooke's law,

$$\sigma^{\alpha\beta} = \frac{E}{1-\nu^2}\left[\nu \varkappa_\delta^\delta a^{\alpha\beta} + (1-\nu) \mathrm{e}^{\alpha\beta}\right] + \frac{\nu}{1-\nu}\sigma_{33}\,\delta^{\alpha\beta} \tag{1.75}$$

and integrating across the thickness, we obtain the following relations:

$$\begin{aligned} M^{\alpha\beta} &= D[\nu a^{\alpha\beta}\varkappa_\delta^\delta + (1-\nu)\varkappa^{\alpha\beta}] + \frac{\nu}{1-\nu}\delta^{\alpha\beta}\int_{-h/2}^{h/2} \sigma_{33}\,\xi \mathrm{d}\xi, \\ \hat{N}^{\alpha\beta} &= \frac{Eh}{1-\nu^2}[\nu a^{\alpha\beta}\varepsilon_\delta^\delta + (1-\nu)\varepsilon^{\alpha\beta}] + \frac{\nu}{1-\nu}\delta^{\alpha\beta}\int_{-h/2}^{h/2} \sigma_{33}\,\mathrm{d}\xi. \end{aligned} \tag{1.76}$$

We sum here with respect to δ, and

$$D = \frac{Eh^3}{12(1-\nu^2)}.$$

The relations above are called the *constitutive equations.* Relations (1.74) differ in their form from the corresponding relations known from the classical theory of shells.

The application of tensor instead of physical components accounts for the difference. For example, translating the relations (1.74) to the physical components, we obtain (by $R_{12} = 0$ and $R_{11} = R_1$, $R_{22} = R_2$)

$$\sigma^{11} = \frac{\sigma_{\langle 11\rangle}}{g_{11}} = \frac{\sigma_{\langle 11\rangle}}{a_{11}}\left(1+\frac{\xi}{R_1}\right)^2, \qquad N_{\langle 11\rangle} = N^{11}g_{11}.$$

Introducing this into (1.74) and remembering that $\sqrt{g/a} = (1+\xi/R_1)(1+\xi/R_2)$, we find

$$N_{11} + \int_{-h/2}^{h/2} \sigma_{11}(1+\xi/R_2)\mathrm{d}\xi.$$

In the last relation the bracket $\langle\ \rangle$ is omitted.

We have for the bending moment

$$M_{11} = \int_{-h/2}^{h/2} \sigma_{11}(1+\xi/R_2)\xi\mathrm{d}\xi,$$

a relation well known in the theory of shells.

If we neglect the effect of σ_{33} in (1.76), we obtain the relations between the stress resultants, strains, and curvatures which are often used in the theory of shells, since they are simple and analogous to the corresponding relations in the theory of plates.

Assuming the distribution of the stresses σ_{33} in the shell the same as in the plate (1.81), we find on substitution and integration

$$\hat{N}_{11} = \frac{Eh}{1-\nu^2}(\varepsilon_{11}+\nu\varepsilon_{22}) + \frac{\nu h}{2(1-\nu)}Z,$$

$$\hat{N}_{22} = \frac{Eh}{1-\nu^2}(\varepsilon_{22}+\nu\varepsilon_{11}) + \frac{\nu h}{2(1-\nu)}Z,$$

$$\hat{N}_{12} = \hat{N}_{21} = \frac{Eh}{1+\nu}\varepsilon_{12},$$

$$M_{11} = D(\varkappa_{11}+\nu\varkappa_{22})+\frac{\nu}{1-\nu}\frac{h^2}{10}Z,$$

$$M_{22} = D(\varkappa_{22}+\nu\varkappa_{11})+\frac{\nu}{1-\nu}\frac{h^2}{10}Z,$$

$$M_{12} = M_{21} = (1-\nu)\varkappa_{12}D, \qquad D = \frac{Eh^3}{12(1-\nu^2)}, \qquad Z = X^3.$$

The positive senses of the stress resultants and couples are defined in the Fig. 7. The above relations are very often used in the theory of shells (with $Z = 0$).

However, it can be proved that these equations do not satisfy the equations of equilibrium of moments with respect to the axis perpendicular to the shell surface. In order to avoid this contradiction we can apply the constitutive equations proposed by Novozhilov. Then instead of Eq. (1.76), where $N_{12} = N_{21}$, we assume that these forces are not equal but satisfy the equation

$$N_{12}-\frac{M_{21}}{R_{22}}-\frac{M_{22}}{R_{12}} = \tilde{S} = N_{21}-\frac{M_{12}}{R_{11}}-\frac{M_{11}}{R_{12}}$$

where

$$\tilde{S} = \frac{Eh}{1+\nu}\varepsilon_{12}, \qquad M_{12}+M_{21} = 2\tilde{M} = \frac{Eh^3}{6(1+\nu)}\varkappa_{12}.$$

Then we find

$$N_{12} = \frac{Eh}{1+\nu}\varepsilon_{12}+\frac{1-\nu}{R_{22}}D\varkappa_{12}+\frac{M_{22}}{R_{12}} = \tilde{S}+\frac{M_{21}}{R_{22}}+\frac{M_{22}}{R_{12}},$$

$$N_{21} = \frac{Eh}{1+\nu}\varepsilon_{12}+\frac{1-\nu}{R_{11}}D\varkappa_{12}+\frac{M_{11}}{R_{12}} = \tilde{S}+\frac{M_{12}}{R_{11}}+\frac{M_{11}}{R_{12}}.$$

These relations can be simplified by the assumption $M_{12} = M_{21} = \tilde{M}$. It can easily be proved that the equilibrium of twisting moments in this case is satisfied. The above relations were justified by Novozhilov who based his considerations on variational theorems.

It results from the approximate character of the theory of shells that we can also base our work on different constitutive equations. For example, we can assume the relations similar to that obtained by Lure [1.6]. These relations have been obtained by the development of the integrals (1.74) in power series with respect to ξ. If we do not change the relations for the normal forces and bending moments, we obtain the relations for the internal forces:

$$N^{\alpha\beta} = \hat{N}^{\alpha\beta} - \frac{1-\nu}{2} D(b^{\alpha}_{\delta}\varkappa^{\beta\delta} - b^{\alpha}_{\delta}\varkappa^{\delta\beta}) \tag{1.77}$$

where $\hat{N}^{\alpha\beta}$ are defined by Eq. (1.76).

In the physical components we have

$$N_{12} = \frac{Eh}{1+\nu}\varepsilon_{12} - \frac{1}{2}\left(\frac{1}{R_{11}} - \frac{1}{R_{22}}\right)\tilde{M} - \frac{1-\nu}{2} D \frac{\varkappa_{11}-\varkappa_{22}}{R_{12}},$$

$$N_{21} = \frac{Eh}{1+\nu}\varepsilon_{12} + \frac{1}{2}\left(\frac{1}{R_{11}} - \frac{1}{R_{22}}\right)\tilde{M} + \frac{1-\nu}{2} D \frac{\varkappa_{11}-\varkappa_{22}}{R_{12}};$$

then

$$N_{12}+N_{21} = 2\tilde{S}.$$

The forces $N^{\alpha\beta}$ defined by relations (1.77) satisfy the equilibrium equation of the twisting moments.

The equations of equilibrium expressed in terms of the stress resultants and couples can be obtained by integration of the equations of equilibrium of an element of the three-dimensional body in the same way as by the definition of the stress resultants and couples. We obtain the following set of equations:

$$\begin{aligned}
&N^{\delta\beta}|_{\delta} - b^{\beta}_{\delta} N^{\delta 3} + X^{\beta} = 0,\\
&N^{\delta 3}|_{\delta} + \bar{b}_{\delta\gamma} N^{\delta\gamma} + X^{3} = 0,\\
&M^{\alpha\beta}|_{\alpha} - N^{\beta 3} + X^{\beta}_{M} = 0,\\
&d_{\alpha\beta}(N^{\alpha\beta} + b^{\beta}_{\gamma} M^{\gamma\alpha}) = 0, \qquad \alpha, \beta = 1, 2,
\end{aligned} \tag{1.78}$$

where $\bar{b}_{\alpha\beta} = -\varkappa_{\alpha\beta} + b_{\alpha\beta}$ takes into account the change of curvature of the middle surface of the shell produced by the applied load. Here, X^{β} are the external forces per unit area of shell and X^{β}_{M} denotes the external moments distributed over a unit area of the shell. This load is usually of small significance and will be considered only in certain cases of plates loaded by bending moments. In the first equation $(1.78)_1$ the effect of the change of the curvature is neglected, because the term $\varkappa_{\alpha} N^{\alpha 3}$ is in this equation of small importance.

Upon covariant differentiation and introduction of the physical components the above set of equations takes the form (in orthogonal coordinates α_i):

$$(A_j N_{ii})_{,i} - N_{jj} A_{j,j} + (A_i N_{ji})_{,j} + A_{i,j} N_{ij} +$$

$$+ A_i A_j \left(\frac{Q_i}{R_{ii}} - \frac{Q_j}{R_{ij}}\right) + A_i A_j X_i = 0, \tag{1.79$_1$}$$

$$-\left(\frac{1}{R_{11}} + \varkappa_{11}\right) N_{11} - \left(\frac{1}{R_{22}} + \varkappa_{22}\right) N_{22} + \left(\frac{1}{R_{12}} - \varkappa_{12}\right) N_{12} +$$

$$+\left(\frac{1}{R_{21}}-\varkappa_{21}\right)N_{21}+\frac{1}{A_1A_2}[(A_2Q_1)_{,1}+(A_1Q_2)_{,2}]+Z=0, \qquad (1.79)_2$$

$$(A_iM_{ji})_{,j}+M_{ij}A_{i,j}+(A_jM_{ii})_{,i}-M_{jj}A_{j,i}-A_iA_jQ_i+X_{Mi}=0, \qquad (1.79)_3$$

$$N_{12}-N_{21}+\frac{M_{12}}{R_{11}}-\frac{M_{21}}{R_{22}}+\frac{M_{11}}{R_{12}}-\frac{M_{22}}{R_{21}}=0 \qquad (1.79)_4$$

$(i, j$ not summed, $i \neq j)$,

where $N^{i3} = Q^i$, $X^3 = Z$.

The same equations can be obtained by considering the equilibrium of a small element cut out from the shell. Figure 7 presents such a parallelopiped

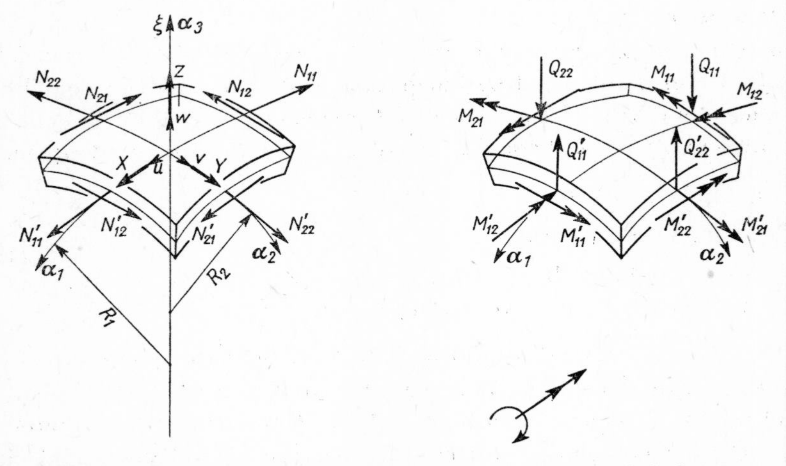

Fig 7. Positive directions of internal forces and moments acting on a shell element.

loaded by the internal forces and moments. The forces in the cross-section $\alpha_1 =$constant are the following: normal N_{11} and tangential N_{12} stress resultants, stress couples M_{11} and M_{12}, and the shear force Q_1. Similarly, in the second cross-section α_2 = constant, we have the stress resultants N_{22}, N_{21}, and the shear force Q_2. The positive senses of these quantities are indicated in Fig. 7.

The set of equilibrium equations is written in the system of coordinates connected with the actual configuration of the shell. On the assumptions of § 1.4 we assume that the coefficients of the quadratic form do not change with the deformation.

1.9 Effect of transverse shear and normal stresses

Now we shall define the values of the average angles of rotations β_α. Let us assume that the transverse shear stresses can be expressed by the formula

$$\tau_{\alpha 3} = \sigma_{\alpha 3} = \frac{3}{2}\frac{Q_\alpha}{h}\left[1-\left(\frac{2\xi}{h}\right)^2\right], \qquad \alpha = 1, 2. \tag{1.80}$$

The shear stresses $\tau_{\alpha 3}$ (1.80) satisfy the boundary conditions: for $\xi = \pm h/2$, $\tau_{\alpha 3} = 0$. The forces Q_α are transverse shear forces. In the classical theory of plates and shells the effect of transverse normal stresses is neglected. Let us take it into account and calculate the distribution of stresses σ_{33} corresponding to the assumed distribution of $\tau_{\alpha 3}$. From the equations of equilibrium (1.29) there results

$$\sigma_{33} = -\int \tau_{13}|_1\, d\xi - \int \tau_{23}|_2\, d\xi + \Psi(\alpha_1, \alpha_2).$$

Introducing $\tau_{\alpha 3}$ from Eq. (1.80), we have

$$\sigma_{33} = -\frac{3}{2}(Q_1|_1 + Q_2|_2)\int\left[1-\left(\frac{2\xi}{h}\right)^2\right]d\xi + \Psi(\alpha_1, \alpha_2).$$

Taking into account the boundary conditions:

$$\sigma_{33} = \begin{cases} Z & \text{for} \quad h/2, \\ 0 & \text{for} \quad -h/2, \end{cases}$$

and using the simplified equilibrium equation $Q_i|_i \simeq Z$, we obtain

$$\sigma_{33} \simeq \frac{1}{2}Z\left[1+3\frac{\xi}{h}-4\left(\frac{\xi}{h}\right)^3\right]. \tag{1.81}$$

The assumed distribution of shear across the thickness of the shell defines the angles β_α, according to (1.69), as functions of the displacement u_3 and the shear forces Q_α. We determine these quantities by equating the work of the resultant couples in the average rotations and the work of the stress resultants in the average displacement to the work of the corresponding stresses in the actual displacement in the same section.

We obtain

$$\begin{aligned} \int_{-h/2}^{h/2} \sigma_{\alpha\beta} u_\alpha^*\, d\xi &= N_{\alpha\beta} u_\alpha + M_{\alpha\beta}\beta_\alpha, \\ \int_{-h/2}^{h/2} \tau_{\alpha 3} u_\alpha^*\, d\xi &= Q_\alpha u_3 \end{aligned} \tag{1.82}$$

(α, β not summed, $\alpha \neq \beta$).

Here u_3 is the average displacement, normal to the middle surface.

Introducing the stress resultants (1.76) and displacement equations (1.69) into Eqs. (1.82), we see that they are satisfied identically, except the last two Eqs. (1.82). Introducing τ_{i3} from (1.80), we find the following relation between the average and actual displacement u_3^*:

$$\int_{-h/2}^{h/2} u_3^* \left[1 - \left(\frac{2\xi}{h}\right)^2\right] d\xi = u_3 .$$

Relations (1.68) can be used to define the components of the displacement vector corresponding to the assumed distribution of the shear stresses $\tau_{\alpha 3}$. Namely, from (1.68) we have

$$2e_{\alpha 3} = \frac{\partial u_\alpha^*}{\partial \xi} + u_3^* \|_\alpha = \frac{1}{G} \tau_{\alpha 3} = \frac{3Q_\alpha}{2hG} \left[1 - \left(\frac{2\xi}{h}\right)^2\right]. \tag{1.83}$$

Integrating the relation with respect to ξ, we find

$$u_\alpha^* = u_\alpha + \xi \left[-u_3^* \|_\alpha + \frac{3Q_\alpha}{2hG} \left(1 - \frac{4\xi^2}{3h^2}\right)\right].$$

On the other hand, substituting relations (1.69) into Eqs. (1.67), we find

$$\beta_\alpha^* = \frac{\partial u_\alpha^*}{\partial \xi} = -u_\alpha^* \|_\alpha + \frac{3Q_\alpha}{2hG} \left[1 - \left(\frac{2\xi}{h}\right)^2\right].$$

These results are in contradiction with our previous assumption that the angles of rotations β_α do not depend on ξ. However, this difficulty can be overcome by the calculation of the average values of β_α^*. Multiplying the above equations by $\frac{3}{2}[1-(2\xi/h)^2]d\xi/h$ and integrating between the limits $\xi = \pm h/2$, we obtain

$$\beta_\alpha = -u_3\|_\alpha + \frac{6}{5} \frac{Q_\alpha}{hG}. \tag{1.84}$$

Applying the Kirchhoff–Love hypotesis, we find

$$\beta_\alpha = -u_3\|_\alpha = -(w|_\alpha + b_\alpha^\delta u_\delta).$$

The above relations have the following form in physical components:

$$\beta_{\langle 1 \rangle} = -\frac{\partial w}{A_1 \partial \alpha_1} + \frac{u}{R_{11}} - \frac{v}{R_{12}}, \qquad \beta_{\langle 2 \rangle} = -\frac{\partial w}{A_2 \partial \alpha_2} + \frac{v}{R_{22}} - \frac{u}{R_{12}}.$$

On introduction of β_α into (1.72), we find

$$\varkappa_{\alpha\beta} = -w|_{\alpha\beta} - \frac{1}{2}[(b^\delta_\alpha u_\delta)|_\beta + (b^\delta_\beta u_\delta)|_\alpha] -$$

$$-\frac{1}{2}[b^\delta_\alpha d_{\beta\delta} + b^\delta_\beta d_{\alpha\delta}]\Omega_n + \frac{6}{10}\cdot\frac{1}{hG}(Q_\alpha|_\beta + Q_\beta|_\alpha),$$

The changes of curvature are not only functions of the normal deflection w and the shear forces Q_α, but also depend on the displacements u and v. Performing the exact calculations, we should first define these displacements from (1.71) and next calculate the changes of curvatures from (1.72). However, if the normal displacement is bigger than the tangential displacements, these calculations can be simplified neglecting the changes of curvatures produced by u and v or assuming that the expression in Eq. $(1.71)_1$* is approximately equal to

$$\tfrac{1}{2}[(b^\delta_\alpha u_\delta)|_\beta + (b^\delta_\beta u_\delta)|_\alpha + (b^\delta_\alpha d_{\beta\delta} + b^\delta_\beta d_{\alpha\delta})\Omega_n] \cong b^\delta_\alpha b_{\delta\beta} w;$$

then $\varkappa_{\alpha\beta} \simeq -w|_{\alpha\beta} - b^\delta_\alpha b_{\delta\beta} w$.

This relation results from the assumption that $\varepsilon_{\alpha\beta} \simeq 0$. This means that the changes of curvature of the shell produced by the displacements u and v are calculated approximately as for the shell deformed isometrically.

The stresses in the shell can be obtained with better accuracy if we work with formulae (1.75) instead of (1.69).

Introducing the displacements u^*_α (1.83) into (1.71) and later into (1.75), we find

$$\sigma_{\alpha\beta} = \frac{E}{1-\nu^2}\Big\{\nu\Big(u_\gamma||_\gamma\, \delta^\beta_\alpha + \frac{1}{2}\, u^*_3||_\gamma u^*_3||_\gamma\Big) +$$

$$+\frac{1}{2}(1-\nu)(u_\alpha||_\beta + u_\beta||_\alpha + u^*_3||_\alpha u^*_3||_\beta) -$$

$$-\xi[\nu u^*_3||_{\gamma\gamma}\,\delta^\beta_\alpha + (1-\nu)\, u^*_3||_{\alpha\beta}] +$$

$$+\frac{3}{2}\frac{\xi}{hG}\Big(1-\frac{4\xi^2}{3h^2}\Big)\Big[\nu Q_\gamma|_\gamma\,\delta^\beta_\alpha + \frac{1-\nu}{2}(Q_\alpha|_\beta + Q_\beta|_\alpha)\Big]\Big\} +$$

$$+\frac{\nu}{1-\nu}\sigma_{33}\,\delta_{\alpha\beta}. \tag{1.85}$$

The surface values of the stresses given by the above relations are the following

$$\sigma_{\alpha\beta}\Big(\pm\frac{h}{2}\Big) = \frac{E}{1-\nu^2}\Big\{\nu\Big(u_\gamma||_\gamma + \frac{1}{2}\, u^*_3||_\gamma u^*_3||_\gamma\Big)\delta^\beta_\alpha +$$

---

* Digit 1 behind the brackets means first equation of the group.

$$+\frac{1-\nu}{2}(u_\alpha\|_\beta+u_\beta\|_\alpha+u_3^*\|_\alpha\cdot u_3^*\|_\beta)\pm$$

$$\pm\frac{h}{2}[\nu u_3^*\|_{\gamma\gamma}\,\delta_\alpha^\beta+(1-\nu)\,u_3^*\|_{\alpha\beta}]\pm$$

$$\pm\frac{1}{2G}\left[\nu Q_\gamma|_\gamma\,\delta_\alpha^\beta+\frac{1-\nu}{2}(Q_\alpha|_\beta+Q_\beta|_\alpha)\right]\Bigg\}+\frac{\nu}{1-\nu}\sigma_{33}\,\delta_{\alpha\beta}.$$

Having Eqs. (1.85), we can obtain the expressions for bending and twisting moments introducing $\sigma_{\alpha\beta}^{\text{1}}$ into Eqs. (1.74). It should be noted that these expressions are in exact agreement with the corresponding relations (1.76) derived by means of the average values of β_α and relations (1.69). However, the surface values of the stresses $\sigma_{\alpha\beta}$ (1.85) are not equal to those resulting from formulae (1.76). E. Reissner proved this fact in 1974 [1.42] in the case of transverse bending of plates.

1.10 Compatibility equations

Between the six components of strains and changes of curvatures of the middle surface there exist three differential equations, independent of the components u, v, w of the displacement vector and called the *compatibility conditions.* The starting point in deriving these equations can be Eqs. (1.14) and the Codazzi–Gauss conditions (1.65) which express the fact that the undeformed middle surface of the shell is embedded in three-dimensional Euclidean space. Similar equations for the fundamental tensors $\bar{a}_{ij}$, $\bar{b}_{ij}$ of a deformed configuration of the middle surface can be interpreted as three compatibility conditions for the strains ε_{ij} and curvatures $\varkappa_{ij}$. We obtain in this way a set of equations

$$d^{\alpha\beta}d^{\lambda\mu}\left[\varepsilon_{\alpha\mu}|_{\beta\lambda}+b_{\alpha\mu}\varkappa_{\beta\lambda}+\frac{1}{2}\varkappa_{\alpha\mu}\varkappa_{\beta\lambda}-\frac{6(1+\nu)}{5Eh}b_{\alpha\mu}(Q_\beta|_\lambda+Q_\lambda|_\beta)\right]=0,$$

$$d^{\alpha\beta}d^{\lambda\mu}\Bigg[\left(\varkappa_{\beta\lambda}+\frac{1}{2}b_\beta^\delta\varepsilon_{\delta\lambda}+\frac{1}{2}b_\lambda^\delta\varepsilon_{\delta\beta}\right)\Bigg|_\mu-b_\lambda^\delta(\varepsilon_{\delta\beta}|_\mu+\varepsilon_{\delta\mu}|_\beta+\varepsilon_{\beta\mu}|_\delta)-\qquad(1.86)$$

$$-\frac{6(1+\nu)}{5Eh}(Q_\beta|_{\lambda\mu}+Q_\lambda|_{\beta\mu})\Bigg]=0,\qquad\alpha,\beta=1,2.$$

These equations, in the case of the classical theory, have been obtained by W. T. Koiter [1.22]. Their linearized version is in agreement with the equations of compatibility obtained by A. L. Goldenveizer [1.17], and V. V. Novozhilov [1.20]. On introducing the physical components, we obtain the following relations (in the case $1/R_{12}=0$):

$$A_2\varkappa_{22,1}+A_{2,1}(\varkappa_{22}-\varkappa_{11})-A_1\varkappa_{12,2}-2A_{1,2}\varkappa_{12}+\frac{2}{R_2}A_{1,2}\varepsilon_{12}+$$

$$+\frac{1}{R_1}[2(A_1\varepsilon_{12,})_2-A_2\varepsilon_{12,1}-A_{2,1}(\varepsilon_{22}-\varepsilon_{11})]$$

$$=\frac{12(1+\nu)}{5Eh}\left\{\left(Q_{2,1}+\frac{A_{2,1}Q_1}{A_1}\right)_{,1}-A_{2,1}\left(\frac{Q_{1,1}}{A_1}+\frac{A_{1,2}Q_2}{A_1A_2}\right)-\right.$$

$$-\frac{1}{2}\left[\frac{A_1^2}{A_2}\left(\frac{Q_1}{A_1}\right)_{,2}+A_2\left(\frac{Q_2}{A_2}\right)_{,1}\right]-$$

$$\left.-\frac{1}{2}A_{2,1}\left[\frac{A_1}{A_2}\left(\frac{Q_1}{A_1}\right)_{,2}+\frac{A_2}{A_1}\left(\frac{Q_2}{A_2}\right)_{,1}\right]\right\}. \tag{1.86a}$$

The second equation, similar to the first one, can be obtained by a simple change of indices. The third equation takes the form

$$\frac{\varkappa_{11}}{R_1}+\frac{\varkappa_{22}}{R_1}+\varkappa_{11}\varkappa_{22}-\varkappa_{12}^2+\frac{1}{A_1A_2}\times$$

$$\times\left\{\left[\left[\frac{1}{A_1}A_2\varepsilon_{22,1}+A_{2,1}(\varepsilon_{22}-\varepsilon_{11})+A_1\varepsilon_{12,2}-2A_{1,2}\varepsilon_{12}\right]\right]_{,1}+\right.$$

$$\left.+\left[\frac{1}{A_2}[A_1\varepsilon_{11,2}+A_{1,2}(\varepsilon_{11}-\varepsilon_{22})-A_2\varepsilon_{12,1}-2A_{2,1}\varepsilon_{12}]\right]_{,2}\right\}$$

$$=\frac{12}{5}\frac{1+\nu}{Eh}\cdot\frac{1}{A_1A_2}\cdot\left[\left(\frac{A_2}{A_1}\frac{Q_1}{R_2}\right)_{,1}+\left(\frac{A_1}{A_2}\frac{Q_2}{R_1}\right)_{,2}\right]. \tag{1.87}$$

The terms which appear on the right-hand side of Eqs. (1.86) and (1.87) result from the effect of transverse shear deformations.

1.11 Simplified shell equations

The field equations of the theory of thin shells may be presented in various forms. One of the possibilities is to eliminate certain of the stress resultants and stress couples from the equations of equilibrium by means of the constitutive equations. Expressing the strains and changes of curvatures in terms of displacement components and their derivatives, we obtain three equations in terms of three displacement vector components as unknowns. But the equations obtained in this way are extremely complicated and were not ever published. Also the complex equations obtained by A. L. Goldenveizer in his book [1.17] represent only a simplified version.

In the particular case of the cylindrical shell such equations were obtained by many authors. The works by W. Flügge [1.16], A. L. Goldenveizer [1.17], V. Z. Vlasov [1.12], K. Girkmann [1.23] and others [1.24] should be mentioned. The equations obtained by these authors differ between each other in a small degree. The differences lay in the limits of the errors of the assumptions of the theory of thin shells.

For the shells of revolution the shell equations in terms of the displacement components were given by S. Kaliski and others [1.25]. There are also published the equations for the spherical shell in terms of the displacement components.

An alternative possibility to present the shell field equations is to eliminate the strains and changes of the curvature from the compatibility conditions by means of the constitutive equations. W. T. Koiter obtained in this way the differential equations, this time in terms of the three stress functions as unknowns. But the resulting equations of compatibility in terms of the stress functions are equally complicated.

Novozhilov [1.20] also eliminated strains and curvature changes from the compatibility conditions by means of the constitutive equations introducing as the variables the complex internal forces and moments. He reduced in this way the set of eighth order equilibrium and compatibility equations to a set of fourth order of the three differential equations for the three unknown complex functions. This treatment required some simplifications of the equilibrium and compatibility equations.

For shallow shells very simple equations were obtained by Kh. M. Mushtari [1.13], L. H. Donnell [1.11], K. Marguerre [1.26] and V. Z. Vlasov [1.12]. The shell equations are reduced in this case to a set of two differential equations of the fourth order for two unknown functions: the normal deflection of the shell and the stress function. These equations are very useful in the numerical calculations for many engineering problems, but are not accurate enough for the non-shallow shells. It is possible to reduce the shell equations to the set of two differential equations of the fourth order also in the case of non-shallow shells. Only the condition that the curvature of the middle surface of the shell is given by the slowly varying function has to be satisfied. The derivation of these equations is presented in the next section.

1.11.1 *Shells of slowly varying curvatures*

Let us consider non-shallow shells whose curvatures are slowly varying functions of the principal coordinates α_1, α_2. The *slowly varying function* means here the function for which the ratio of its derivatives to itself is smaller than unity in the whole area except the region closed to the points where the function is equal to zero. It is possible to develop for such shells simplified equations convenient for numerical computations. The spherical and cylindrical shells whose curvatures are constant are included of course in this type of shells.

Let us assume that the shell is so thin that we can neglect terms of the order h^2/R_i^2 in comparison with unity. We shall evaluate the order of the neglected terms introducing the characteristic wave length L of the deformation pattern. Let us assume that the curvatures change so slowly that we can neglect the terms $L^2 R_{i,1}/\pi^2 R_i^2$ in comparison to unity, where R_i are the principal radii of the curvature of the middle surface of the shell.

In order to obtain simple equations convenient for engineering purposes we reduce the solution of the shell problem to the solution of the two partial differential equations. We introduce the stress function Φ in the following form [1.27]:

$$N^{\alpha\beta} = -d^{\alpha\lambda}d^{\beta\mu}\Phi|_{\lambda\mu} - Ka^{\alpha\beta}\Phi - \\ -(1-\nu)Dd^{\beta\lambda}d^{\alpha\mu}b_\lambda^\delta \varkappa_{\delta\mu} - b^{\alpha\beta}\bar{Q} + P^{\alpha\beta}, \tag{1.88}$$

where $\bar{Q}$ satisfies the equation $\bar{Q}|_\alpha = N^{\delta 3}a_{\delta\alpha}$, $N^{\alpha 3} = Q^\alpha$ are the shear forces, $K = 1/R_1 R_2$ is the Gaussian curvature, $K = b/a$, and $P^{\alpha\beta}$ is an arbitrary partial solution of the equations $(1.78)_1$. On substitution of the above expressions into the first two equations of equilibrium $(1.78)_1$ and introducing the physical notations and the coordinates, we obtain

$$(A_2 N_{11})_{,1} - N_{22}A_{2,1} + (A_1 N_{21})_{,2} + N_{12}A_{1,2} + A_1 A_2\left(\frac{Q_1}{R_{11}} - \frac{Q_2}{R_{12}}\right) \\ = O[\Phi K_{,\alpha} + 2H_{,\alpha}\bar{Q}] + O\Phi[wb^{\alpha\beta}{}_{,\alpha}] \tag{1.89}$$

where $H = \frac{1}{2}b_\alpha^\alpha = -\frac{1}{2}(1/R_1 + 1/R_2)$ is the mean curvature of the middle surface of the shell. We observe that all terms containing first, second, and third order derivatives of Φ cancel. There remains only the term containing the function Φ multiplied by the first order derivative of Gaussian curvature and a minor term resulting from the effect of the shearing forces containing the derivative of the curvature. Since it has been assumed that we consider only shells of slowly varying curvatures, these terms are small and can be neglected.

Let us examine the magnitude of the neglected terms, for example, the term $\Phi(K)_{,1}$. Let us compare them with the first term in Eq. (1.88).

Neglecting the term $\Phi(K)_{,1}$, we neglect the term of the order $\sim L^3 K_{,1}/\pi^3 = L^3(R_1 R_2)_{,1}/\pi^3 R_1^2 R_2^2$, small in comparison with unity. In order to evaluate the magnitude of the Gaussian curvature let us consider as an example an ellipsoid of revolution (Fig. 8). The Gaussian curvature of this shell is given by the equation

$$K = \frac{1}{R_1 R_2} = (a^2 \sin^2 \varphi + b^2 \cos^2 \varphi)^2 a^{-4} b^{-2}$$

where a and b are the semiaxis of the ellipse and φ is the angle between the normal to the shell surface and the axis of revolution. We have then

$$K_{,1} = 2(a^2 - b^2) a^{-4} b^{-2} (a^2 \sin^2 \varphi + b^2 \cos^2 \varphi) \sin 2\varphi .$$

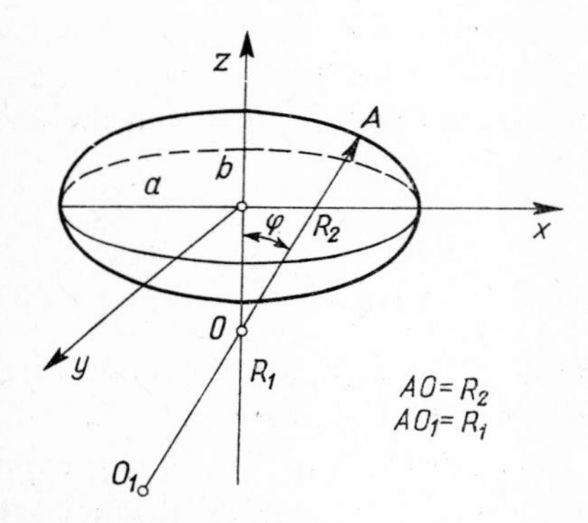

Fig. 8. Ellipsoid of revolution.

Let us calculate the value of this term for the angle $2\varphi = 90°$. Introducing $R_1 = \sqrt{8} a^2 b^2/(a^2+b^2)^{3/2}$, we find that the wave length should satisfy the condition

$$L^3 \ll \frac{\pi^3 \sqrt{8}\, a^3}{(a^4/b^4 - 1)(1 + b^2/a^2)^{3/2}} . \tag{1.90}$$

If the admissible error is of the order of one per cent, we have $L \leqslant 0.4a$ for $a = 2b$, where a is a longer semiaxis of the ellipse. This condition is very often satisfied. The errors of the same order are produced neglecting the remaining terms on the right-hand side of Eqs. (1.89). When the radii of curvature are constant, these terms vanish and the stress function (1.89) satisfies the equilibrium equations exactly.

Introducing the stress function Φ into the third equation of equilibrium (1.78) and using the identities

$$\begin{aligned} b_{\alpha\beta}a^{\alpha\beta} &= 2H = -\left(\frac{1}{R_1}+\frac{1}{R_2}\right); \\ b_{\alpha\beta}b^{\alpha\beta} &= 4H^2-2K = \frac{1}{R_{11}^2}+\frac{2}{R_{12}^2}+\frac{1}{R_{22}^2} \end{aligned} \tag{1.91}$$

and the relation $\bar{b}_{\alpha\beta} = b_{\alpha\beta}-\varkappa_{\alpha\beta}$, we obtain

$$\begin{aligned} Q^{\alpha}|_{\alpha} = {} & d^{\alpha\lambda}d^{\beta\mu}b_{\alpha\beta}\Phi|_{\lambda\mu}+2HK\Phi+(1-\nu)DK\varkappa^{\lambda}_{\lambda}- \\ & - d^{\alpha\lambda}d^{\beta\mu}\varkappa_{\alpha\beta}\Phi|_{\lambda\mu}-[4H^2-2\bar{K}]Q-b_{\alpha\beta}P^{\alpha\beta}-X^3- \\ & - \underline{Ka^{\alpha\beta}\varkappa_{\alpha\beta}\Phi}-\underline{(1-\nu)Dd^{\beta\lambda}d^{\alpha\mu}b^{\delta}_{\lambda}\varkappa_{\alpha\beta}\varkappa_{\delta\mu}}+\underline{\varkappa_{\alpha\beta}\bar{b}^{\alpha\beta}Q}+\varkappa_{\alpha\beta}P^{\alpha\beta}. \end{aligned} \tag{1.92}$$

Because of the fact that in our considerations we limit ourself to such geometrically nonlinear problems in which the normal deflection w is of the order of several times the thickness, we can reject in Eq. (1.92) the underlined non-linear terms, usually much smaller than the remaining terms in this equation. Let us see when they can be disregarded. Namely, let us compare them with the term $d^{\alpha\lambda}d^{\beta\mu}\Phi|_{\lambda\mu}$, remaining in this equation, which is of the order of the product of the second derivatives of the functions w and Φ.

$$d^{\alpha\lambda}d^{\beta\mu}\varkappa_{\alpha\beta}\Phi|_{\lambda\mu} \simeq w_{,11}\Phi_{,22} = w_0\Phi_0\frac{\pi^4}{L^4}.$$

The first underlined term $Ka^{\alpha\beta}\varkappa_{\alpha\beta}\Phi$ is of the order

$$\Phi_0 w_0 \frac{\pi}{R_1R_2L^2}.$$

This term is much smaller than the left term if $L^2 \ll \pi^2 R_1 R_2$. This condition is often satisfied because the length of the half wave of the deformation is usually smaller than the radius of the curvature of the shell. In a similar way the magnitude of the other terms can be evaluated. The term $d^{\alpha\lambda}d^{\beta\mu}\varkappa_{\alpha\beta}\Phi|_{\lambda\mu}$ can be presented in a different form using the simplified relation $\varkappa_{\alpha\beta} \sim -w|_{\alpha\beta}$.

Then Eq. (1.92) can be rewritten in the more compact form. Introducing the differential operators Δ_k and $L(\ )$, we have

$$Q^{\alpha}|_{\alpha} = -\Delta_k\Phi+L(w,\Phi)+(1-\nu)DK\varkappa^{\lambda}_{\lambda}-(4H^2-2K)\bar{Q}-X^3-(b_{\alpha\beta}-\varkappa_{\alpha\beta})P^{\alpha\beta},$$

where

$$\Delta_k \Phi = -d^{\alpha\lambda} d^{\beta\mu} b_{\alpha\beta} \Phi|_{\lambda\mu} - 2HK\Phi,$$
$$L(w, \Phi) = d^{\alpha\lambda} d^{\beta\mu} w|_{\alpha\beta} \Phi|_{\lambda\mu}.$$

Now let us introduce the constitutive equations (1.76) into the two equilibrium equations (1.78). Making use of (1.81), we obtain the following equation for the shear forces:

$$Q^\alpha = M^{\beta\alpha}|_\beta = Da^{\beta\alpha} \varkappa^\lambda_\lambda|_\beta + (1-\nu) D d^{\alpha\beta} d^{\lambda\mu} \varkappa_{\beta\lambda}|_\mu + \frac{h^2}{10} \frac{\nu}{1-\nu} a^{\alpha\beta} X^3|_\beta. \tag{1.93}$$

Using formula (1.71), we can write now

$$\varkappa^\lambda_\lambda = -\Delta w - b^\alpha_\lambda b^\lambda_\alpha w - b^{\alpha\lambda} \varepsilon_{\alpha\lambda} - \underline{b^{\alpha\lambda}|_\alpha u_\lambda} + \frac{12(1-\nu)}{5Eh} N^{\alpha 3}|_\alpha \tag{1.94}$$

where Δ denotes the two-dimensional Laplacian operator, $\Delta w = a^{\alpha\beta} w|_{\alpha\beta}$. The underlined term in (1.94) contains the derivative of the mean curvature and is neglected in further calculations. If we neglect this term, we admit an error of the same order while neglecting the terms on the right-hand side of Eq. (1.89).

The third term on the right-hand side of Eq. (1.94) can be written with the help of (1.71) and (1.88) in the following way:

$$b^{\alpha\lambda} \varepsilon_{\alpha\lambda} = b^\alpha_\lambda \varepsilon^\lambda_\alpha = \frac{-1}{2Eh} [(1-\nu) b^\alpha_\alpha \Delta\Phi - (1+\nu) \Delta_h \Phi] \tag{1.95}$$

where

$$\Delta_h \Phi = (b^\alpha_\alpha a^{\lambda\mu} - 2 b^\alpha_\alpha d^{\delta\lambda} d^{\varepsilon\mu} a_{\delta\varepsilon}) \Phi|_{\lambda\mu}. \tag{1.96}$$

The term $d^{\alpha\beta} d^{\lambda\mu} \varkappa_{\beta\lambda}|_\mu$ in Eq. (1.93) can be replaced using the compatibility equation $(1.86)_2$. We have

$$d^{\delta\beta} d^{\lambda\mu} \varkappa_{\beta\lambda}|_\mu = -d^{\delta\beta} d^{\lambda\mu} \Big[\frac{1}{2} (b^\alpha_\beta \varepsilon_{\alpha\lambda} + b^\alpha_\lambda \varepsilon_{\alpha\beta})|_\mu - \\ - b^\alpha_\lambda (\varepsilon_{\alpha\beta}|_\mu + \varepsilon_{\alpha\mu}|_\alpha) - \frac{6(1+\nu)}{5Eh} (Q_\beta|_{\lambda\mu} + Q_\lambda|_{\beta\mu}) \Big].$$

Introducing the stress function, we obtain

$$d^{\delta\beta} d^{\lambda\mu} \varkappa_{\beta\lambda}|_\mu \sim -\frac{1}{Eh} [b^\delta_\alpha a^{\lambda\mu} \Phi|_{\lambda\mu} + (1+\nu) b^\alpha_\mu d^{\delta\lambda} d^{\beta\mu} N_{\beta\lambda}]|_\alpha + \\ + \frac{6(1+\nu)}{5Eh} d^{\delta\beta} d^{\lambda\mu} (Q_\beta|_{\lambda\mu} + Q_\lambda|_{\beta\mu}). \tag{1.97}$$

This simplified equation is obtained by neglecting some small terms of the order L^2/π^2R^2 containing the derivatives of the curvatures of the shell. Introducing the above result to Eq. (1.93), we have the following equation for the shear forces:

$$Q^\alpha - \frac{h^2}{10} d^{\alpha\beta} d^{\lambda\mu} (Q_\beta|_{\lambda\mu} + Q_\lambda|_{\beta\mu}) =$$
$$-D\Big\{ a^{\alpha\beta}[\Delta w + b^\delta_\lambda b^\lambda_\delta w] - \frac{a^{\beta\alpha}}{2Eh}[(1-\nu)b^\delta_\delta \Delta\Phi - (1+\nu)\Delta_h\Phi] +$$
$$+(1-\nu)\frac{1}{Eh}[b^{\beta\alpha}\Delta\Phi + (1+\nu)b^\beta_\mu d^{\alpha\lambda} d^{\delta\mu} N_{\delta\lambda}]\Big\}\Big|_\beta +$$
$$+\frac{1}{1-\nu}\frac{h^2}{5} a^{\alpha\beta} Q^\delta|_{\delta\beta} + \frac{h^2}{10}\frac{\nu}{1-\nu} a^{\alpha\beta} X^3|_\beta . \tag{1.98}$$

With the aid of $(1.78)_2$ and (1.98) we may write, after some manipulations, the equation of equilibrium (1.92) in the form

$$D(\Delta + 4H^2 - 2K)[\Delta + 4H^2 - (3-\nu)K]w - 1 - \left(\frac{h^2}{5(1-\nu)}\Delta\right) \cdot \Delta_k \Phi$$
$$= X^3 + b_{\alpha\beta}P^{\alpha\beta} - L(w, \Phi) - \frac{h^2\Delta}{10(1-\nu)}[(2-\nu)X^3 - 2L(w, \Phi)] +$$
$$+ P^{\alpha\beta} w|_{\alpha\beta} - w\Delta_R \Phi , \tag{1.99}$$

where

$$\Delta_k \Phi = d^{\alpha\lambda} d^{\beta\mu} b_{\alpha\beta} \Phi|_{\lambda\mu} - \underline{\frac{D}{Eh}\Delta_h \Phi} - 2HK\Phi ,$$

$$\Delta_R \Phi = d^{\alpha\lambda} d^{\beta\mu} b^\delta_\alpha b_{\delta\beta} \Phi|_{\lambda\mu} .$$

The underlined term $D\Delta_h\Phi/Eh$ is of order h^2/L^2 in comparison with the remaining terms of the operator Δ_K. As a result it can usually be neglected.

The second relation between the same functions w and Φ can be obtained from the third equation of compatibility if we express the changes of curvatures in terms of w and Φ and the strains by Φ by using the relations

$$\varepsilon^{\alpha\beta} = \frac{1}{Eh}[(1+\nu)N^{\alpha\beta} - \nu a^{\alpha\beta} N^\lambda_\lambda] - \frac{1-\nu}{2}\frac{D}{Eh}[b^\beta_\gamma \varkappa^{\gamma\delta} - b^\alpha_\gamma \varkappa^{\gamma\beta}] . \tag{1.100}$$

Without giving all the details of these manipulations, since they are entirely analogous to those involved in the deduction of Eqs. (1.98), (1.99), etc., the final result can be given in the form

$$\frac{1}{Eh}(\Delta+2K)[\Delta+(1-\nu)K]\Phi+\Delta_k w= -\frac{\nu}{2}\frac{\Delta Z}{E}-$$

$$-\frac{1}{Eh}[\Delta P^{\alpha}_{\alpha}-(1+\nu)P^{\alpha\beta}|_{\alpha\beta}]+\frac{1}{2}L(w,w)+w\Delta_R w+K^2w^2. \qquad (1.101)$$

The above equations are obtained by using the constitutive equations (1.77). Similar result would be obtained by applying the equations proposed by Novozhilov. The only difference would be in the operator Δ_h where there appears a minor additional term.

It should be mentioned here that some small terms of order $(h/r)^n \ll 1$, $n \geqslant 2$, resulting from the effect of transverse normal and shear stresses, have been neglected.

The term $-\nu\Delta Z/2E$, on the right-hand side of Eq. (1.101), represents the effect of the transverse normal stress σ_{33}. This stress is usually small and only in the case of concentrated loads should be taken into account.

The set of four equations (1.98), (1.99), and (1.101) contains four unknown functions w, Φ, Q_1, Q_2. The problem of determination of the stresses and displacements in an arbitrary shell is reduced to the solution of the above equations with the corresponding boundary conditions. Neglecting in Eqs. (1.99) and (1.101) the terms containing the functions of the curvatures of the shell, we obtain the equations given by Naghdi [1.8]. The only difference consists in taking into account here the effect of normal stresses σ_{33}. In the case of concentrated loads this effect is of the same order as the effect of transverse shear stresses.

In the system of orthogonal principal coordinates (α_1, α_2) the equations for the shear forces (1.101) take the form

$$Q_1-\frac{h^2}{10}\Bigg\{\Delta Q_1-\frac{1}{A_1^2A_2^2}[(A_{1,2})^2+(A_{2,1})^2]Q_1+\frac{2A_{2,1}Q_{2,2}}{A_1A_2^2}-$$

$$-\frac{2A_{1,2}Q_{1,1}}{A_1^2A_2}-\frac{1}{A_1^2A_2}\left[\left(\frac{A_{2,1}}{A_1}\right)_{,2}-\left(\frac{A_{1,2}}{A_2}\right)_{,1}\right]Q_2\Bigg\}$$

$$=-\frac{1}{A_1}\Bigg\{D(\Delta+4H^2-2K)w-$$

$$-\frac{h^2}{24(1+\nu)}\left[H\Delta\Phi+\frac{1+\nu}{1-\nu}\Delta_h\Phi-\frac{2}{R_{11}}(\Delta\Phi-(1+\nu)N_{22})\right]+$$

$$+\frac{h^2}{10(1-\nu)}[Z+(1+\nu)[-L(w,\Phi)+\Delta_k\Phi+b_{\alpha\beta}P^{\alpha\beta}]]\Bigg\}_{,1}. \qquad (1.102)$$

The second equation (1.98) has the same form and can be obtained by the change of indices. It can be proved that the underlined terms are of the order h^2/R^2 with respect to similar terms in the above equation and can be disregarded. The operators Δ, $L(\ )$, Δ_k take the form

$$\Delta w = \frac{1}{A_1 A_2}\left[\left(\frac{A_2 w_{,1}}{A_1}\right)_{,1} + \left(\frac{A_1 w_{,2}}{A_2}\right)_{,2}\right], \tag{1.103}$$

$$\Delta_k \Phi = \frac{1}{A_1 A_2}\left[\left(\frac{A_2 \Phi_{,1}}{A_1 R_{22}}\right)_{,1} + \left(\frac{\Phi_{,1}}{R_{12}}\right)_{,2} + \left(\frac{\Phi_{,2}}{R_{12}}\right)_{,1} + \left(\frac{A_1 \Phi_{,2}}{A_2 R_{11}}\right)_{,2}\right],$$

$$L(w, \Phi) = \frac{1}{A_1 A_2}\left[\left(\frac{w_{,1}}{A_1}\right)_{,1} + \frac{A_{1,2} w_{,2}}{A_2^2}\right]\left[\left(\frac{\Phi_{,2}}{A_2}\right)_{,2} + \frac{A_{2,1}\Phi_{,1}}{A_1^2}\right] -$$

$$- \frac{2}{A_1^2 A_2^2}\left[w_{,12} - \frac{A_{2,1} w_{,2}}{A_2} - \frac{A_{1,2} w_{,1}}{A_1}\right]\left[\Phi_{,12} - \frac{A_{2,1}\Phi_{,2}}{A_2} -\right.$$

$$\left. - \frac{A_{1,2}\Phi_{,1}}{A_1}\right] + \frac{1}{A_1 A_2}\left[\left(\frac{w_{,2}}{A_2}\right)_{,2} + \frac{A_{2,1} w_{,1}}{A_1^2}\right]\left[\left(\frac{\Phi_{,1}}{A_1}\right)_{,1} + \frac{A_{1,2}\Phi_{,2}}{A_2^2}\right], \tag{1.104}$$

$$\Delta_R \Phi = \left(\frac{1}{R_{11}^2} + \frac{1}{R_{12}^2}\right)\left[\frac{1}{A_2}\left(\frac{\Phi_{,2}}{A_2}\right)_{,2} + \frac{A_{2,1}\Phi_{,1}}{A_1^2 A_2}\right] +$$

$$+ \frac{2}{R_{12}}\left(\frac{1}{R_{11}} + \frac{1}{R_{22}}\right)\left[\Phi_{,12} - \frac{A_{2,1}\Phi_{,2}}{A_2} - \frac{A_{1,2}\Phi_{,1}}{A_1}\right]\frac{1}{A_1 A_2} +$$

$$+ \left(\frac{1}{R_{22}^2} + \frac{1}{R_{12}^2}\right)\left[\frac{1}{A_2}\left(\frac{\Phi_{,1}}{A_1}\right)_{,1} + \frac{A_{1,2}\Phi_{,2}}{A_1 A_2^2}\right].$$

The bending moments in the shell are defined by Eqs. (1.76) where the changes of the curvatures should be replaced by relations (1.73) and (1.84).

$$M_{ii} = -D\left[\frac{1}{A_i}\left(\frac{w_{,i}}{A_i}\right)_{,i} + \frac{A_{i,j} w_{,j}}{A_i A_j^2} + \frac{\nu}{A_j}\left(\frac{w_{,1}}{A_j}\right)_{,j} + \frac{\nu A_{j,i} w_{,i}}{A_i^2 A_j} +\right.$$

$$\left. + \left(\frac{1}{R_{ii}^2} + \frac{1}{R_{ij}^2}\right) w + \nu\left(\frac{1}{R_{jj}^2} + \frac{1}{R_{ij}^2}\right) w\right] +$$

$$+ \frac{h^2}{5(1-\nu) A_i A_j}\left[A_j Q_{i,i} + A_{i,j} Q_j + \nu A_i Q_{j,j} + \nu A_{j,i} Q_i\right] + \frac{\nu}{1-\nu}\frac{h^2}{10} Z,$$

$$M_{12} = M_{21} = -(1-\nu)D\frac{D}{A_1 A_2}\left[w_{,12} - \frac{A_{1,2}w_{,1}}{A_1} - \frac{A_{2,1}w_{,2}}{A_2} - \right.$$
$$\left. - w\left(\frac{1}{R_{11}}+\frac{1}{R_{22}}\right)\frac{1}{R_{12}}\right] + \frac{h^2}{10}\left[\frac{A_2}{A_1}\left(\frac{Q_2}{A_2}\right)_{,1} + \frac{A_1}{A_2}\left(\frac{Q_1}{A_1}\right)_{,2}\right]$$

$(i, j = 1, 2$ not summed, $i \neq j)$.

The membrane forces are defined by the stress function Φ and displacement w

$$N_{ii} = -\frac{1}{A_j}\left(\frac{\Phi_{,j}}{A_j}\right)_{,j} - \frac{A_{j,i}\Phi_{,i}}{A_i^2 A_j} - \frac{\Phi}{R_i R_j} +$$
$$+\frac{D}{R_{ii}}[\Delta w + (4H^2 - 2K)w] + (1-\nu)D\left[\frac{\varkappa_{jj}}{R_{jj}} - \frac{\varkappa_{ij}}{R_{ij}}\right] + P_{ii},$$

$$S = \frac{1}{A_1 A_2}\left(\Phi_{,12} - \frac{A_{1,2}\Phi_{,1}}{A_1} - \frac{A_{2,1}\Phi_{,2}}{A_2}\right), \tag{1.105}$$

$$N_{ij} = S - (1-\nu)D\left[\frac{\varkappa_{ij}}{R_{ii}} - \frac{\varkappa_{jj}}{R_{ij}}\right] + \frac{D}{R_{ij}}(\varkappa_{ii} + \varkappa_{jj}) + P_{ij}$$

$(i, j = 1, 2, i \neq j)$.

Between the forces S and $\tilde{S}$ exists the relation

$$S = \tilde{S} - \frac{1}{2}(1-\nu)D\left(\frac{1}{R_{11}} + \frac{1}{R_{22}}\right)\varkappa_{12} + \frac{1}{2}(3-\nu)\frac{D}{R_{12}}(\varkappa_{11} + \varkappa_{22}).$$

If we use Novozhilov's constitutive equations, then

$$S = \tilde{S} + (1-\nu)D\left(\frac{1}{R_{11}} + \frac{1}{R_{22}}\right)\varkappa_{12} - (1-\nu)\frac{D}{R_{12}}(\varkappa_{11} + \varkappa_{22}).$$

1.11.2 *Relations in rectangular coordinates*

Concerning the cylindrical and shallow shells we usually apply a system of orthogonal coordinates (x, y) on the shell surface assuming $A_1 = A_2 = 1$ and $R_1 = R_2 =$ const. In this case the equations for the shear forces can be presented in the following form:

$$Q_x - \frac{h^2}{10}\Delta Q_x = -\frac{\partial}{\partial x}\left\{D(\Delta + 4H^2 - 2K)w + \right.$$
$$\left. + \frac{h^2}{10(1-\nu)}[Z + (1+\nu)(-L(w, \Phi) + \Delta_k \Phi + b_{\alpha\beta}P^{\alpha\beta})]\right\}, \tag{1.106$_1$}$$

$$Q_y - \frac{h^2}{10} \Delta Q_y = -\frac{\partial}{\partial y}\Big\{D(\Delta + 4H^2 - 2K)w +$$

$$+ \frac{h^2}{10(1-\nu)}[Z + (1+\nu)(-L(w, \Phi) + \Delta_k \Phi + b_{\alpha\beta}P^{\alpha\beta})]\Big\}. \quad (1.106)_2$$

Equations (1.99) and (1.101) have the same form as previously only if the differential operators have the form

$$\Delta = \frac{\partial^2}{\partial x^2} + \frac{\partial^2}{\partial y^2},$$

$$\Delta_k = \frac{\partial}{\partial x}\left(\frac{1}{R_{22}}\frac{\partial}{\partial x}\right) + \frac{\partial}{\partial y}\left(\frac{1}{R_{12}}\frac{\partial}{\partial x}\right) + \frac{\partial}{\partial x}\left(\frac{1}{R_{12}}\frac{\partial}{\partial y}\right) +$$

$$+ \frac{\partial}{\partial y}\left(\frac{1}{R_{11}}\frac{\partial}{\partial y}\right) + \left(\frac{1}{R_{11}R_{22}} - \frac{1}{R_{12}^2}\right)\left(\frac{1}{R_{11}} + \frac{1}{R_{22}}\right), \quad (1.107)$$

$$L(w, \Phi) = \frac{\partial^2 w}{\partial x^2}\frac{\partial^2 \Phi}{\partial y^2} - 2\frac{\partial^2 w}{\partial x \partial y}\frac{\partial^2 \Phi}{\partial x \partial y} + \frac{\partial^2 w}{\partial y^2}\frac{\partial^2 \Phi}{\partial x^2},$$

$$\Delta_R w = \left(\frac{1}{R_{11}^2} + \frac{1}{R_{12}^2}\right)\frac{\partial^2 w}{\partial x^2} + \frac{2}{R_{12}}\left(\frac{1}{R_{11}} + \frac{1}{R_{22}}\right)\frac{\partial^2 w}{\partial x \partial y} +$$

$$+ \left(\frac{1}{R_{22}^2} + \frac{1}{R_{12}^2}\right)\frac{\partial^2 w}{\partial y^2}.$$

The internal forces and moments are defined by the formulae

$$N_{xx} = -\frac{\partial^2 \Phi}{\partial y^2} - \frac{1-\nu}{R_{22}} D \frac{\partial^2 w}{\partial y^2} - \frac{\Phi}{R_1 R_2} + \frac{D}{R_{11}}[\Delta w + (4H^2 - 2K)w] +$$

$$+ (1-\nu) D \frac{1}{R_{12}}\frac{\partial^2 w}{\partial x \partial y} + P_{11},$$

$$N_{yy} = -\frac{\partial^2 \Phi}{\partial x^2} - \frac{1-\nu}{R_{11}} D \frac{\partial^2 w}{\partial x^2} - \frac{\Phi}{R_1 R_2} + \frac{D}{R_{22}}[\Delta w + (4H^2 - 2K)w] +$$

$$+ (1-\nu) D \frac{1}{R_{12}}\frac{\partial^2 w}{\partial x \partial y} + P_{22}, \quad (1.108)$$

$$N_{xy} = \frac{\partial^2 \Phi}{\partial x \partial y} + \frac{1-\nu}{R_{11}} D \frac{\partial^2 w}{\partial x \partial y} - \frac{D}{R_{12}}\left[\Delta w + (1-\nu)\frac{\partial^2 w}{\partial y^2}\right] + P_{12},$$

$$N_{yx} = \frac{\partial^2 \Phi}{\partial x \partial y} + \frac{1-\nu}{R_{22}} D \frac{\partial^2 w}{\partial x \partial y} - \frac{D}{R_{12}}\left[\Delta w + (1-\nu)\frac{\partial^2 w}{\partial x^2}\right] + P_{21}.$$

The bending moments are the following

$$M_{xx} = -D\left[\frac{\partial^2 w}{\partial x^2}+\nu\frac{\partial^2 w}{\partial y^2}+\left(\frac{1}{R_{11}^2}+\frac{1}{R_{22}^2}\right)w+\nu\left(\frac{1}{R_{22}^2}+\frac{1}{R_{12}^2}\right)w\right]-$$

$$-\frac{h^2}{5}\frac{\partial Q_x}{\partial x}-\frac{\nu}{1-\nu}\frac{h^2}{10}[Z+2(-L(w,\Phi)+\Delta_k\Phi+b_{\alpha\beta}P^{\alpha\beta})], \tag{1.109}$$

$$M_{yy} = -D\left[\frac{\partial^2 w}{\partial y^2}+\nu\frac{\partial^2 w}{\partial x^2}+\left(\frac{\nu}{R_{11}^2}+\frac{1}{R_{22}^2}\right)w+\nu\left(\frac{1}{R_{11}^2}+\frac{1}{R_{12}^2}\right)w\right]-$$

$$-\frac{h^2}{5}\frac{\partial Q_y}{\partial y}-\frac{\nu}{1-\nu}\frac{h^2}{10}[Z+2(-L(w,\Phi)+\Delta_k\Phi+b_{\alpha\beta}P^{\alpha\beta})],$$

$$M_{xy} = -(1-\nu)D\left[\frac{\partial^2 w}{\partial x\,\partial y}-\left(\frac{1}{R_{11}}+\frac{1}{R_{22}}\right)\frac{w}{R_{12}}\right]+\frac{h^2}{10}\left(\frac{\partial Q_x}{\partial y}+\frac{\partial Q_y}{\partial x}\right).$$

Equations (1.103), (1.106) together are a twelfth order system. However, this system should be of tenth order since at every edge we have only five boundary conditions. In the case of Cartesian coordinate and shallow shells further reduction may easily be performed. Equations (1.106) imply the second order equation

$$\frac{\partial Q_x}{\partial y}-\frac{\partial Q_y}{\partial x}=\frac{h^2}{10}\Delta\left(\frac{\partial Q_x}{\partial y}-\frac{\partial Q_y}{\partial x}\right) \tag{1.110}$$

or

$$\Delta\Psi-\frac{10}{h^2}\Psi=0 \quad \text{where} \quad \Psi=\frac{\partial Q_x}{\partial y}-\frac{\partial Q_y}{\partial x}.$$

Now, we have a tenth order system for the three dependent variables, w, Φ, and Ψ. The shearing forces Q_x and Q_y may be expressed as a combination of derivatives of these three.

$$Q_x = -D\frac{\partial}{\partial x}[\Delta w+(4H^2-2K)w]-$$

$$-\frac{h^2}{10(1-\nu)}\frac{\partial}{\partial x}[(2-\nu)Z+2(\Delta_K\varphi-L(w,\varphi)+b_{\alpha\beta}P^{\alpha\beta})]+\frac{h^2}{10}\frac{\partial\Psi}{\partial y}, \tag{1.111}$$

$$Q_y = -D\frac{\partial}{\partial y}[\Delta w+(4H^2-2K)w]-$$

$$-\frac{h^2}{10(1-\nu)}[(2-\nu)Z+2(\Delta_K\Phi-L(w,\Phi)+b_{\alpha\beta}P^{\alpha\beta})]+\frac{h^2}{10}\frac{\partial\Psi}{\partial x}.$$

1.11.3 *Polar coordinates*

Solving the problem in polar coordinates (r, ϑ), we introduce

$$A_1 = 1, \quad A_2 = r, \quad \alpha_1 = r, \quad \alpha_2 = \vartheta.$$

In this case we have the following formulae:

$$\Delta = \frac{\partial}{\partial r^2} + \frac{1}{r}\frac{\partial}{\partial r} + \frac{1}{r^2}\frac{\partial^2}{\partial \vartheta^2},$$

$$\begin{aligned}\Delta_K = \frac{1}{r}\left[\frac{\partial}{\partial r}\left(\frac{r}{R_{22}}\frac{\partial}{\partial r}\right) + \frac{\partial}{\partial r}\left(\frac{1}{R_{12}}\frac{\partial}{\partial \vartheta}\right) + \right. \\ \left. + \frac{\partial}{\partial \vartheta}\left(\frac{1}{R_{12}}\frac{\partial}{\partial r}\right) + \frac{\partial}{r\partial \vartheta}\left(\frac{1}{R_{11}}\frac{\partial}{\partial \vartheta}\right)\right] + \\ + \left(\frac{1}{R_{11}R_{22}} - \frac{1}{R_{12}^2}\right)\left(\frac{1}{R_{11}} + \frac{1}{R_{22}}\right),\end{aligned}$$

$$\begin{aligned}L(w, \Phi) = \frac{\partial^2 \Phi}{\partial r^2}\left(\frac{1}{r}\frac{\partial w}{\partial r} + \frac{1}{r^2}\frac{\partial^2 w}{\partial \vartheta^2}\right) - \\ - \frac{2}{r^2}\left(\frac{\partial^2 \Phi}{\partial r\,\partial \vartheta} - \frac{1}{r}\frac{\partial w}{\partial r}\right)\left(\frac{\partial^2 w}{\partial r\,\partial \vartheta} - \frac{1}{r}\frac{\partial w}{\partial \vartheta}\right) + \\ + \frac{\partial^2 w}{\partial r^2}\left(\frac{1}{r}\frac{\partial \Phi}{\partial r} + \frac{1}{r^2}\frac{\partial^2 \Phi}{\partial \vartheta^2}\right),\end{aligned}$$

$$\begin{aligned}\Delta_R \Phi = \left(\frac{1}{R_{11}^2} + \frac{1}{R_{12}^2}\right)\left(\frac{1}{r}\frac{\partial \Phi}{\partial r} + \frac{1}{r^2}\frac{\partial^2}{\partial \vartheta^2}\right) + \\ + \frac{2}{R_{12}}\left(\frac{1}{R_{11}} + \frac{1}{R_{22}}\right)\left(\frac{\partial^2 \Phi}{r^2 \partial r\,\partial \vartheta} - \frac{1}{r^3}\frac{\partial \Phi}{\partial \vartheta}\right) + \\ + \left(\frac{1}{R_{22}^2} + \frac{1}{R_{12}^2}\right)\frac{\partial^2 \Phi}{\partial r^2}.\end{aligned}$$

The membrane forces are:

$$\begin{aligned}N_{rr} = -\frac{1}{r^2}\frac{\partial^2 \Phi}{\partial \vartheta^2} - \frac{1}{r}\frac{\partial \Phi}{\partial r} - \frac{\Phi}{R_1 R_2} + \\ + \frac{D}{R_{11}}[\Delta w + (4H^2 - 2K)w] - \frac{1-\nu}{R_{22}} D\left(\frac{1}{r^2}\frac{\partial^2 w}{\partial \vartheta^2} + \frac{1}{r}\frac{\partial w}{\partial r}\right) + \\ + \frac{1-\nu}{R_{12}}\left(\frac{1}{r}\frac{\partial^2 w}{\partial r\,\partial \vartheta} - \frac{1}{r^2}\frac{\partial w}{\partial \vartheta}\right) + P_{rr},\end{aligned} \qquad (1.112)_1$$

$$N_{\vartheta\vartheta} = -\frac{\partial^2 \Phi}{\partial r^2} - \frac{\Phi}{R_1 R_2} + \frac{D}{R_{11}}(\Delta w + (4H^2 - 2K)w) -$$

$$-\frac{1-\nu}{R_{11}} D \frac{\partial^2 w}{\partial r^2} - \frac{1-\nu}{R_{12}} D \left(\frac{1}{r}\frac{\partial^2 w}{\partial r \partial \vartheta} - \frac{1}{r^2}\frac{\partial \omega}{\partial \vartheta}\right) + P_{\vartheta\vartheta}, \quad (1.112)_2$$

$$S = \frac{1}{r}\frac{\partial^2 \Phi}{\partial r \partial \vartheta} - \frac{1}{r^2}\frac{\partial \Phi}{\partial r},$$

$$N_{r\vartheta} = S + \frac{1-\nu}{R_{22}} D \left(\frac{1}{r}\frac{\partial^2 w}{\partial r \partial \vartheta} - \frac{1}{r^2}\frac{\partial w}{\partial \vartheta}\right) -$$

$$-(1-\nu)\frac{D}{R_{12}}\left[\frac{\partial^2 w}{\partial r^2} + (2-\nu)\left(\frac{1}{r^2}\frac{\partial^2 w}{\partial \vartheta^2} + \frac{1}{r}\frac{\partial w}{\partial r}\right)\right] + P_{r\vartheta},$$

$$N_{\vartheta r} = S + \frac{1-\nu}{R_{11}} D \left(\frac{1}{r}\frac{\partial^2 w}{\partial r \partial \vartheta} - \frac{1}{r^2}\frac{\partial w}{\partial \vartheta}\right) -$$

$$-(1-\nu)\frac{D}{R_{12}}\left[(2-\nu)\frac{1}{r^2}\frac{\partial^2 w}{\partial \vartheta^2} + \frac{1}{r^2}\frac{\partial^2 w}{\delta \vartheta^2} + \frac{1}{r}\frac{\partial w}{\partial r}\right] + P_{\vartheta r}.$$

The fundamental equations for w and Φ (1.103) remain as previously. The bending moments are

$$M_{rr} = -D\left(\frac{\partial^2 w}{\partial r^2} + \frac{\nu}{r}\frac{\partial w}{\partial r} + \frac{\nu}{r^2}\frac{\partial^2 w}{\partial \vartheta^2}\right) + \frac{h^2}{5}\frac{\partial Q_r}{\partial r} -$$

$$-\frac{h^2}{10}\frac{\nu}{1-\nu}[Z + 2(-L(w, \Phi) + \Delta_k \Phi + b_{\alpha\beta} P^{\alpha\beta})] -$$

$$-D\left[\left(\frac{1}{R_{11}^2} + \frac{1}{R_{12}^2}\right) w + \nu\left(\frac{1}{R_{22}^2} + \frac{1}{R_{12}^2}\right) w\right],$$

$$M_{\vartheta\vartheta} = -D\left(\nu\frac{\partial^2 w}{\partial r^2} + \frac{1}{r}\frac{\partial w}{\partial r} + \frac{1}{r^2}\frac{\partial^2 w}{\partial \vartheta^2}\right) - \frac{h^2}{5}\frac{\partial Q_r}{\partial r} -$$

$$-\frac{h^2}{10}\frac{2-\nu}{1-\nu} Z - \frac{h^2}{5(1-\nu)}[L(w, \Phi) - \Delta_k \Phi - b_{\alpha\beta} P^{\alpha\beta}] - \quad (1.113)$$

$$-D\left[\left(\frac{1}{R_{22}^2} + \frac{1}{R_{12}^2}\right) w + \nu\left(\frac{1}{R_{11}^2} + \frac{1}{R_{12}^2}\right) w\right],$$

$$M_{r\vartheta} = -(1-\nu) D\left[\frac{1}{r}\frac{\partial^2 w}{\partial r \partial \vartheta} - \frac{1}{r^2}\frac{\partial w}{\partial \vartheta} - \left(\frac{1}{R_{11}} + \frac{1}{R_{22}}\right)\frac{w}{R_{12}}\right] +$$

$$+\frac{h^2}{10}\left(\frac{1}{r}\frac{\partial Q_r}{\partial \vartheta} + \frac{\partial Q_\vartheta}{\partial r} - \frac{Q_\vartheta}{r}\right).$$

The following equations are for the shear forces:

$$Q_r - \frac{h^2}{10}\left(\Delta Q_r - \frac{Q_r}{r^2} - \frac{2}{r^2}\frac{\partial Q_\vartheta}{\partial \vartheta}\right) = -\frac{\partial}{\partial r}\Big\{D(\Delta + 4H^2 - 2K)w + \\ + \frac{h^2}{10(1-\nu)}\left[Z + (1+\nu)\left(-L(w,\Phi) + \Delta_k \Phi + b_{\alpha\beta}P^{\alpha\beta}\right)\right]\Big\},$$

$$Q_\vartheta - \frac{h^2}{10}\left(\Delta Q_\vartheta - \frac{Q_\vartheta}{r^2} + \frac{2}{r^2}\frac{\partial Q_r}{\partial \vartheta}\right) = -\frac{\partial}{r\partial \vartheta}\Big\{D(\Delta + 4H^2 - 2K)w + \\ + \frac{h^2}{10(1-\nu)}\left[Z + (1+\nu)\left(-L(w,\Phi) + \Delta_k \Phi + b_{\alpha\beta}P^{\alpha\beta}\right)\right]\Big\}. \tag{1.114}$$

1.12 Classical theory

1.12.1 *Shells of slowly varying curvature*

If we neglect the transverse shear deformations and limit ourselves to shells of small deflections, we obtain equations of the classical theory of shells. This reduces to neglecting in Eqs. (1.103) the terms multiplied by h^2 and the non-linear operators $L(w, w)$ and $L(w, \Phi)$. Thus, introducing a system of coordinates following the directions of the main curvatures, we have (by $1/R_{12} = 0$, $X = Y = 0$) [1.15, 1.27]

$$D\left(\Delta + \frac{1}{R_1^2} + \frac{1}{R_2^2}\right)\left(\Delta + \frac{1}{R_1^2} - \frac{1-\nu}{R_1 R_2} + \frac{1}{R_2}\right)w - \Delta_k \Phi = Z,$$
$$\frac{1}{Eh}\left(\Delta + \frac{2}{R_1 R_2}\right)\left(\Delta + \frac{1-\nu}{R_2 R_2}\right)\Phi + \Delta_k w = 0. \tag{1.115}$$

The solution of the problem of a non-shallow shell reduces to the solution of the two partial differential equations (1.115) of fourth order and to the determination of the functions w and Φ for the given boundary conditions. If the function w is known, the bending moments and the torques will be determined from Eqs. (1.104) in which we neglect the terms multiplied by h^2.

The shear forces depend now only on the deflection w

$$Q_i = -\frac{1}{A_i}\left[D\left(\Delta + \frac{1}{R_i^2} + \frac{1}{R_j^2}\right)w\right]_{,i} \qquad (j \neq i,\ i, j = 1, 2 \text{ not summed}). \tag{1.116}$$

The formulae for other internal forces are the same as previously. Disregarding in Eqs. (1.115) the terms which are the functions of the radii of curvature (except in Δ_k), we get the *Donnell–Vlasov equations for shallow shells.* It is important for equations used for numerical computation to be simple in form. Therefore Eqs. (1.115) are proposed for simplified solving engineering problems of non-shallow shells. The resulting equations differ but little from those obtained by Vlasov and are sufficiently simple in numerical computations for all types of shells.

Their accuracy is also satisfactory for engineering purposes and was proved for the case of cylindrical and spherical shell.

When the thickness or the rigidity of the shell is variable, it is also possible to reduce the shell equations to two differential equations of fourth order. The fundamental equations (1.115) for w and Φ becomes [1.32] (by $X = Y = 0$)

$$\left(\Delta + \frac{1}{R_1^2} + \frac{1}{R_2^2}\right)\left[D\left(\Delta + \frac{1}{R_1^2} - \frac{1-\nu}{R_1 R_2} + \frac{1}{R_2^2}\right)\right]w - (1-\nu)L(D, w) - \Delta_k \Phi = Z, \tag{1.117}$$

$$\left(\Delta + \frac{2}{R_1 R_2}\right)\left[\frac{1}{Eh}\left(\Delta + \frac{1-\nu}{R_1 R_2}\right)\right]\Phi - (1+\nu)L\left(\frac{1}{Eh}, \Phi\right) + \Delta_k w = 0.$$

Formulae (1.104) and (1.105) do not change.

Now we compare the derived equations (1.115) with the equations for the cylindrical and spherical shells obtained by other authors. The equations for the cylindrical shell may be achieved by assuming that $R_1 = \infty$ and $R_2 = R = \text{constant}$ and $A_1 = A_2 = \text{constant}$. Then we can reduce the two Eqs. (1.115) to one equation.

For this purpose we apply the operator Δ^2 to both sides of Eq. $(1.115)_1$. Thus we get

$$D\Delta^2\left[\Delta + \frac{1}{R^2}\right]^2 w - \Delta^2 \Delta_k \Phi = \Delta^2 Z. \tag{1.118}$$

But from Eq. $(1.115)_2$ we have

$$\Delta^2 \Phi = -Eh\Delta_k w.$$

Introduction of this relation into (1.118) yields

$$D\Delta^2\left(\Delta + \frac{1}{R^2}\right)^2 w + Eh\Delta_k^2 w = \Delta^2 Z. \tag{1.119}$$

The above equation is identical with the equation proposed by Morley [1.18].

Morley obtained his equation from Flügge's equation for a cylindrical shell by neglecting some small terms. Morley proposed to use it to replace Donnell's equation which in the case of a closed cylinder is insufficiently accurate. The accuracy of Morley's equation was also examined by comparing it with the very accurate and complex equation derived by Flügge [1.16] for cylindrical shells. The errors that occurred were very small, not exceeding 0.5 per cent. Novozhilov [1.20] emphasized that Vlasov's equations of the engineering theory of shells do not transform themselves into the equations for a circular ring if the shell becomes very short. If the cross-section is assumed as circular, the differential equations concerning this problem should be identical with those for bending of a flat curved beam. This last equation is unobtainable from Vlasov's equations. However, it can be derived from Eqs. (1.115).

For the spherical shell we have $R_1 = R_2 = R$. Introducing this into Eqs. (1.115), we obtain

$$
\begin{aligned}
&D\left(\Delta+\frac{2}{R^2}\right)\left(\Delta+\frac{1+\nu}{R^2}\right)w-\frac{1}{R}\left(\Delta+\frac{2}{R^2}\right)\Phi = Z,\\
&\frac{1}{Eh}\left(\Delta+\frac{2}{R^2}\right)\left(\Delta+\frac{1-\nu}{R^2}\right)\Phi+\frac{1}{R}\left(\Delta+\frac{2}{R^2}\right)w = 0.
\end{aligned} \tag{1.120}
$$

The above equations are identical with the equations obtained by W. T. Koiter for a spherical shell [1.19]. Equations (1.120) can easily be transformed to a single differential equation. Introducing the new function Ψ, and expressing w and Φ by means of the equations

$$
\Phi = -\frac{1}{R}\left(\Delta+\frac{2}{R^2}\right)\Psi, \qquad w = \frac{1}{Eh}\left(\Delta+\frac{1-\nu}{R^2}\right)\left(\Delta+\frac{2}{R^2}\right)\Psi, \tag{1.121}
$$

the second equation of the group (1.120) is automatically fulfilled. After the substitution into Eq. $(1.120)_1$ we have

$$
\frac{D}{Eh}\left(\Delta+\frac{2}{R^2}\right)^2\left[\left(\Delta+\frac{1}{R^2}\right)^2-\frac{\nu^2}{R^4}\right]\Psi+\frac{1}{R^2}\left(\Delta+\frac{2}{R^2}\right)^2\Psi = Z,
$$

which yields (neglecting ν^2/R^4)

$$
\left[\left(\Delta+\frac{1}{R^2}\right)^2+\frac{Eh}{DR^2}\right]\left(\Delta+\frac{2}{R^2}\right)^2\Psi = Z\frac{Eh}{D}. \tag{1.122}
$$

This fundamental equation is identical with the equation derived by Vlasov for the spherical shell [1.12].

Relations (1.121) differ little from the corresponding Vlasov's relations. But we should remember that the stress function determined by Eqs. (1.121) is not exactly the same as the corresponding stress function for the spherical shell derived by Vlasov [1.12].

1.12.2 *Shallow shells*

In the shallow shells, the radii of curvature of the middle surface are large in comparison with the other dimensions of the shell. It enables us to introduce certain simplifications into the previously derived equations. Namely, we may:

(1) Neglect the terms $A_1 A_2 Q_1/R_1$ and $A_1 A_2 Q_2/R_2$ in the first two equations of equilibrium (Eqs. $(1.79)_1$);

(2) Neglect the terms w/R_1^2 and w/R_2^2 as well as the terms containing the derivatives of the curvature $\dfrac{u}{A_1}\dfrac{\partial}{\partial \alpha_1}\left(\dfrac{1}{R_1}\right)\ldots$ in the relations between the displacements and the changes of the curvature $\varkappa_{ij}$;

(3) Apply the curvilinear system of coordinates (α_1, α_2) satisfying the condition

$$\frac{A_1 A_2}{R_1 R_2} \ll 1.$$

Then the terms which are functions of the radii of curvature vanish in the basic equations (1.115). We obtain

$$\begin{aligned} D\Delta^2 w - \Delta_k \Phi &= Z, \\ \frac{1}{Eh}\Delta^2 \Phi + \Delta_k w &= 0. \end{aligned} \tag{1.123}$$

The membrane forces are determined by the stress function in the following way:

$$\begin{aligned} N_{11} &= -\frac{1}{A_2}\frac{\partial}{\partial \alpha_2}\left(\frac{1}{A_2}\frac{\partial \Phi}{\partial \alpha_2}\right) - \frac{1}{A_1^2 A_2}\frac{\partial A_2}{\partial \alpha_1}\frac{\partial \Phi}{\partial \alpha_1}, \\ N_{22} &= -\frac{1}{A_1}\frac{\partial}{\partial \alpha_1}\left(\frac{1}{A_1}\frac{\partial \Phi}{\partial \alpha_1}\right) - \frac{1}{A_1 A_2^2}\frac{\partial A_1}{\partial \alpha_2}\frac{\partial \Phi}{\partial \alpha_2}, \\ N_{12} &= \frac{1}{A_1 A_2}\left(\frac{\partial^2 \Phi}{\partial \alpha_1 \partial \alpha_2} - \frac{1}{A_1}\frac{\partial A_1}{\partial \alpha_2}\frac{\partial \Phi}{\partial \alpha_1} - \frac{1}{A_2}\frac{\partial A_2}{\partial \alpha_1}\frac{\partial \Phi}{\partial \alpha_2}\right). \end{aligned} \tag{1.124}$$

The bending and twisting moments are given by

$$
\begin{aligned}
M_{11} &= -D\left[\frac{1}{A_1}\frac{\partial}{\partial\alpha_1}\left(\frac{1}{A_1}\frac{\partial w}{\partial\alpha_1}\right)+\frac{1}{A_1A_2^2}\frac{\partial A_1}{\partial\alpha_2}\frac{\partial w}{\partial\alpha_2}+\right.\\
&\qquad\left.+\frac{\nu}{A_2}\frac{\partial}{\partial\alpha_2}\left(\frac{1}{A_2}\frac{\partial w}{\partial\alpha_2}\right)+\frac{\nu}{A_1^2A_2}\frac{\partial A_2}{\partial\alpha_1}\frac{\partial w}{\partial\alpha_1}\right],\\
M_{22} &= -D\left[\frac{1}{A_2}\frac{\partial}{\partial\alpha_2}\left(\frac{1}{A_2}\frac{\partial w}{\partial\alpha_2}\right)+\frac{1}{A_1^2A_2}\frac{\partial A_2}{\partial\alpha_1}\frac{\partial w}{\partial\alpha_1}+\right.\\
&\qquad\left.+\frac{\nu}{A_1}\frac{\partial}{\partial\alpha_1}\left(\frac{1}{A_1}\frac{\partial w}{\partial\alpha_1}\right)+\frac{\nu}{A_1A_2^2}\frac{\partial A_1}{\partial\alpha_2}\frac{\partial w}{\partial\alpha_2}\right],\\
M_{12} &= -D(1-\nu)\left[\frac{\partial^2 w}{\partial\alpha_1\partial\alpha_2}-\frac{1}{A_1}\frac{\partial A_1}{\partial\alpha_2}\frac{\partial w}{\partial\alpha_1}-\frac{1}{A_2}\frac{\partial A_2}{\partial\alpha_1}\frac{\partial w}{\partial\alpha_2}\right].
\end{aligned}
\tag{1.125}
$$

The equations above are called the *equations of the engineering theory of shells* or the *Donnell–Vlasov equations.* These equations can be applied successfully to solve the problems of shallow shells as well as the problems in which the functions w and Φ are rapidly variable functions. The term "rapidly variable function" means here the function whose ratio to its first derivative is everywhere small in comparison to unity, except the points where this derivative is zero. The last feature of the Donnell–Vlasov equations is caused by the fact that the neglected terms contain not only the functions of the curvature of the shell but also the lower derivatives of the functions w and Φ.

Two equations (1.123) can be combined into a single equation for a single complex unknown function. Introducing

$$\tilde{\Psi} = \Phi - \frac{iEh^2}{\sqrt{12(1-\nu^2)}}\,w,$$

we obtain the equation

$$ic\Delta\Delta\tilde{\Psi} - \Delta_k\tilde{\Psi} = Z \tag{1.126}$$

where

$$c = h/\sqrt{12(1-\nu^2)}.$$

Separating from this equation the real and imaginary parts, we obtain Eqs. (1.123).

The equations for arbitrary shallow shells were developed by V. Z. Vlasov in 1944. The equations for the particular case of the cylindrical shell were known earlier (see Donnell [1.11], [1.33] 1933). The equations for slightly curved plates were given by Marguerre in 1938 [1.26] as the generalization of the non-linear Karman equations.

If the form of the shell is defined by the equation $z = z(x, y)$, where x, y are the coordinates of the projection of the point of the shell on the plane x, y, it is convenient to express the radii of curvature by relations (1.58).

We obtain then the equations

$$D\Delta^2 w - L(z, \Phi) = Z, \qquad \frac{1}{Eh}\Delta^2\Phi + L(z, w) = 0,$$

where $L(z, \Phi)$ is defined by (1.107).

The formulae for the internal forces and moments are the same as previously.

1.13 Loads tangential to middle surface

To find a solution for a shell loaded by distributed forces tangential to its middle surface, we have to find first the particular solution $P^{\alpha\beta}$ of the equilibrium equations (1.78) and then to solve the fundamental set of equations (1.99) and (1.101). Then we can simplify Eqs. $(1.78)_1$ neglecting the small terms $b^{\beta}_{\delta}N^{\delta 3}$ presenting the projections of the shear forces Q^{δ} in the directions α_1, α_2. Then the particular solution $P^{\alpha\beta}$ can be found from the equation

$$N^{\alpha\lambda}|_{\lambda} + X^{\alpha} = 0.$$

If we consider a shallow or cylindrical shell, we introduce usually on its middle surface a system of rectangular coordinates x, y. Then these equations can be satisfied by the assumption

$$P_{11} = -\int X\mathrm{d}x, \qquad P_{22} = -\int Y\mathrm{d}y, \qquad P_{12} = P_{21} = 0.$$

The third equation of equilibrium of the forces acting in the direction normal to the middle surface of the shell takes the form

$$D[\Delta + 4H^2 - 2K][\Delta + 4H^2 - (3-\nu)K]w - \Delta_k\Phi$$
$$= Z + \frac{1}{R_1}\int X\mathrm{d}x + \frac{1}{R_2}\int Y\mathrm{d}y. \qquad (1.127)_1$$

In expressing the strains in terms of the displacement w and the stress function Φ the compatibility equation reduces to

$$\frac{1}{Eh}(\Delta+2K)[\Delta+(1-\nu)K]\Phi+\Delta_k w$$
$$=\frac{1}{Eh}\left\{\frac{\partial^2}{\partial x^2}\int Y\mathrm{d}y-\nu\frac{\partial Y}{\partial y}+\frac{\partial^2}{\partial y^2}\int X\mathrm{d}x-\nu\frac{\partial X}{\partial x}\right\}. \tag{1.127}$$

The relations for the shear forces and moments remain the same as previously. In the case of a shallow shell, if we introduce the auxiliary complex function $\tilde{\psi}$, we obtain Eqs. (1.126) in the form

$$\Delta\Delta\tilde{\psi}-i\frac{\sqrt{12(1-\nu^2)}}{h}\Delta_k\tilde{\psi}$$
$$=i\frac{\sqrt{12(1-\nu^2)}}{h}(b_{\alpha\beta}P^{\alpha\beta}+Z)-\Delta P^{\alpha}_{\alpha}+(1+\nu)P^{\alpha\beta}|_{\alpha\beta}.$$

The solution for the tangential loads X, Y may be derived also in another way, provided that these forces have a potential U. Let us introduce the curvilinear coordinates α_1, α_2. Assuming that

$$X=-\frac{1}{A_1}\frac{\partial U}{\partial\alpha_1},\qquad Y=-\frac{1}{A_2}\frac{\partial U}{\partial\alpha_2},$$

the two first of Eqs. (1.79) will be satisfied identically if the forces N_{11} and N_{22} are expressed by Φ and U in the following way:

$$N_{11}=-\frac{1}{A_2}\frac{\partial}{\partial\alpha_2}\left(\frac{1}{A_2}\frac{\partial\Phi}{\partial\alpha_2}\right)-\frac{1}{A_1^2A_2}\frac{\partial A_2}{\partial\alpha_1}\frac{\partial\Phi}{\partial\alpha_1}-$$
$$-\frac{\Phi}{R_1R_2}+\frac{D\Delta w}{R_{11}}-(1-\nu)D\left[\frac{\varkappa_{22}}{R_{22}}-\frac{\varkappa_{12}}{R_{12}}\right]+U,$$
$$N_{22}=-\frac{1}{A_1}\frac{\partial}{\partial\alpha_1}\left(\frac{1}{A_1}\frac{\partial\Phi}{\partial\alpha_1}\right)-\frac{1}{A_1A_2^2}\frac{\partial A_1}{\partial\alpha_2}\frac{\partial\Phi}{\partial\alpha_2}-$$
$$-\frac{\Phi}{R_1R_2}+\frac{D\Delta w}{R_{22}}-(1-\nu)D\left[\frac{\varkappa_{11}}{R_{11}}-\frac{\varkappa_{12}}{R_{12}}\right]+U. \tag{1.128}$$

The expression for S is the same as previously. The new terms appear on the right-hand side of Eq. (1.115). We have

$$D(\Delta+4H^2-2K)[\Delta+4H^2-(3-\nu)K]w-\Delta_k\Phi$$
$$=Z-\left(\frac{1}{R_1}+\frac{1}{R_2}\right)U+U\Delta w,$$
$$\frac{1}{Eh}(\Delta+2K)[\Delta+(1-\nu)K]\Phi+\Delta_k w=(1-\nu)\Delta\left(\frac{U}{Eh}\right). \tag{1.129}$$

Equations (1.129) are valid in the arbitrary system of coordinates. The equations presented above hold true in the case of displacements small in comparison with the thickness of the shell. Taking into account the non-linear terms, Eqs. $(1.127)_1$ and $(1.127)_2$ take the form

$$D(\Delta+4H^2-2K)[\Delta+4H^2-(3-\nu)K]w-\Delta_k\Phi = Z-L(w,\Phi)+$$
$$+\left(\frac{1}{R_1}-\frac{\partial^2 w}{\partial x^2}\right)\int X\mathrm{d}x+\left(\frac{1}{R_2}-\frac{\partial^2 w}{\partial y^2}\right)\int Y\mathrm{d}y-w\Delta_R\Phi,$$
$$(1.130)$$
$$\frac{1}{Eh}(\Delta+2K)[\Delta+(1-\nu)K]\Phi+\Delta_k w = \frac{1}{2}L(w,w)+w\Delta_R w+$$
$$+K^2w^2+\frac{1}{Eh}\left\{\frac{\partial^2}{\partial x^2}\int Y\mathrm{d}y-\nu\frac{\partial Y}{\partial y}+\frac{\partial^2}{\partial y^2}\int X\mathrm{d}x-\nu\frac{\partial X}{\partial x}\right\}.$$

If we introduce a potential U, Eqs. (1.130) take the following form:

$$D[\Delta+4H^2-2K][\Delta+4H^2-(3-\nu)K]w-\Delta_k\Phi$$
$$= Z-L(w,\Phi)-\left(\frac{1}{R_1}+\frac{1}{R_2}\right)U+U\Delta w-w\Delta_R\Phi,$$
$$(1.131)$$
$$\frac{1}{Eh}(\Delta+2K)[\Delta+(1-\nu)K]\Phi+\Delta_k w$$
$$= (1-\nu)\Delta\left(\frac{U}{Eh}\right)+\frac{1}{2}L(w,w)+w\Delta_R w+K^2w^2.$$

The above equations are valid when the loads X and Y are tangential to the deformed middle surface. If these loads are tangential to the undeformed middle surface, then in the equilibrium equations of the forces in the normal direction there will appear the projections of the tangential loads X, Y. Equation $(1.131)_1$ takes then the following form:

$$D(\Delta+4H^2-2K)[\Delta+4H^2-(3-\nu)K]w-\Delta_k\Phi$$
$$= Z-L(w,\Phi)-\left(\frac{1}{R_1}+\frac{1}{R_2}\right)U+U\Delta w-$$
$$-X\frac{1}{A_1}\frac{\partial w}{\partial\alpha_1}-Y\frac{1}{A_2}\frac{\partial w}{\partial\alpha_2}-w\Delta_R\Phi.$$

The second equation remains as previously.

1.14 The strain energy

The *strain energy* in the elastic body arises as a result of the change of shape of the body under external loads. We calculate this energy as the work done by the stresses moving through the displacements in an infinitely small element dV.

In Cartesian coordinates x, y, z we have

$$U = \frac{1}{2} \iiint_V (\sigma_{xx}\varepsilon_{xx} + \sigma_{yy}\varepsilon_{yy} + \ldots) \mathrm{d}x \mathrm{d}y \mathrm{d}z, \tag{1.132}$$

where V denotes the volume of the body.

In the arbitrary system of curvilinear coordinates the volume of the infinitely small element is given by $\sqrt{g}\mathrm{d}\theta^1 \mathrm{d}\theta^2 \mathrm{d}\theta^3$. The energy stored in the body is given by the integral

$$U = \frac{1}{2} \iiint_V \sigma^{ij}\varepsilon_{ij}\sqrt{g}\mathrm{d}\theta^1 \mathrm{d}\theta^2 \mathrm{d}\theta^3, \qquad i, j = 1, 2, 3.$$

Introducing the orthogonal coordinates α_1, α_2, ξ, connected with the middle surface of the shall, and integrating across the thickness by using relations (1.83), we find

$$U = \frac{1}{2} \iint_S \Big[N_{11}\varepsilon_{11} + N_{22}\varepsilon_{22} + 2\tilde{S}\varepsilon_{12} + + Q_1 \left(\beta_1 + \frac{1}{A_1} \frac{\partial w}{\partial \alpha_1} \right) + Q_2 \left(\beta_2 + \frac{1}{A_2} \frac{\partial w}{\partial \alpha_2} \right) + + M_{11}\varkappa_{11} + M_{22}\varkappa_{22} + 2M_{12}\varkappa_{12} \Big] A_1 A_2 \mathrm{d}\alpha_1 \mathrm{d}\alpha_2, \tag{1.133}$$

where S denotes the surface of the shell.

The element of the surface is defined by $\mathrm{d}S = \sqrt{a}\mathrm{d}\alpha_1 \mathrm{d}\alpha_2$. If the coordinates are orthogonal, then $\sqrt{a} = A_1 A_2$. The first three terms in (1.133) present the strain energy produced by the strains in the middle surface of the shell.

The work done by the transverse shear forces Q_i is presented by the terms $Q_i\gamma_{i3}$ where $\gamma_{i3} = \beta_i + \frac{1}{A_1} \frac{\partial w}{\partial \alpha_1}$. The relation (1.133) can be written in the following way:

$$U = U_1 + U_2, \tag{1.134}$$

where U_1 is the membrane strain energy and U_2 is the energy of the bending.

We have

$$U_1 = \frac{1}{2}\iint_S N_{11}\varepsilon_{11}+N_{22}\varepsilon_{22}+2\tilde{S}\varepsilon_{12})A_1 A_2 \, d\alpha_1 \, d\alpha_2. \qquad (1.135$$

The bending energy is

$$U_2 = \frac{1}{2}\iint_S \Big[M_{11}\varkappa_{11}+M_{22}\varkappa_{22}+2M_{12}\varkappa_{12}+Q_1\Big(\beta_1+\frac{1}{A_1}\frac{\partial w}{\partial \alpha_1}\Big)+$$

$$+Q_2\Big(\beta_2+\frac{1}{A_2}\frac{\partial w}{\partial \alpha_2}\Big)\Big] A_1 A_2 \, d\alpha_1 \, d\alpha_2. \qquad (1.136)$$

Using the relations (1.75) between the internal forces and strains and changes of the curvatures, the strain energy can be represented by the strains or internal forces. We have

$$U_1 = \frac{Eh}{2(1-\nu^2)}\iint_S \Big[(\varepsilon_{11}+\varepsilon_{22})^2-2(1-\nu)(\varepsilon_{11}\varepsilon_{22}-\varepsilon_{12}^2)+$$

$$+(\varepsilon_{11}+\varepsilon_{22})\frac{\nu(1+\nu)Z}{2E}\Big] A_1 A_2 \, d\alpha_1 \, d\alpha_2 \qquad (1.137)$$

or

$$U_1 = \frac{1}{2Eh}\iint_S \Big[(N_{11}+N_{22})^2-2(1+\nu)(N_{11}N_{22}-\tilde{S}^2)-$$

$$-\nu(N_{11}+N_{22})\frac{Zh}{2}\Big] A_1 A_2 \, d\alpha_1 \, d\alpha_2.$$

Representing the membrane forces in terms of the stress function and in spite of neglecting the effect of the transverse normal and shear stresses we obtain rather complex relations. After simplification by neglecting small terms we obtain the following formula which is sufficiently accurate for engineering purposes:

$$U_1 = \frac{1}{2Eh}\iint_S [(\Delta\Phi+2K\Phi)^2-(1+\nu)L(\Phi,\Phi)] A_1 A_2 \, d\alpha_1 \, d\alpha_2.$$

For shallow shells the term $2K\Phi$ can be neglected in the above equation.

The bending energy can be represented by the internal forces and moments or by changes of curvatures. Using the relations (1.75) we find

$$U_2 = \frac{D}{2} \iint_S \Big\{ (\varkappa_{11}+\varkappa_{22})^2 - 2(1-\nu)(\varkappa_{11}\varkappa_{22}-\varkappa_{12}^2) +$$

$$+ (\varkappa_{11}+\varkappa_{22})\frac{6}{5}\nu(1+\nu)\frac{Z}{Eh} + \frac{5(1-\nu)}{h^2}\Big[\Big(\beta_1+\frac{\partial w}{A_1\,\partial\alpha_1{}^2}\Big)^2 +$$

$$+ \Big(\beta_2+\frac{\partial w}{A_2\,\partial\alpha_2}\Big)^2\Big]\Big\} A_1 A_2 \mathrm{d}\alpha_1 \mathrm{d}\alpha_2 . \tag{1.138}$$

Two last terms in the above integral result from the effects of the transverse normal and shear stresses. Expressing the changes of the curvatures in terms of the moments and forces, using relations (1.75), the strain energy can be represented as a function of these quantities. Neglecting the effects of the transverse shear and normal stresses, we obtain

$$U_2 = \frac{D}{2} \iint_S \{[\Delta w + (4H^2-2K)w]^2 - (1-\nu)L(w, w)\} A_1 A_2 \mathrm{d}\alpha_1 \mathrm{d}\alpha_2 . \tag{1.139}$$

The strain energy of the shell is equal to the work of the external loads. Let us assume that the shell or plate is in equilibrium under the action of the edge and surface loads. The work of these loads on the corresponding displacements over the surface and along the edge is defined by the integral

$$L = \tfrac{1}{2} \iint_S (Xu + Yv + Zw) A_1 A_2 \mathrm{d}\alpha_1 \mathrm{d}\alpha_2 +$$

$$+ \tfrac{1}{2} \oint_C (\overline{X}_b u_b + \overline{Y}_b v_b + \overline{Z}_b w_b + M_{nb}\beta_{nb} + M_{tb}\beta_{tb}) \mathrm{d}s$$

where the index b denotes the forces, moments, and displacements at the edge. The coefficient $\frac{1}{2}$ before the integral appears as the result of the progressive quasistatic increase of the load.

The formulae above are often used to approximately solve many engineering problems concerning plates and shells. The solution is based then on one of the variational methods, for example, Ritz's or Galerkin's. We shall make use of the above relations in Chapter 13, in which we shall consider the large deflections of the shell, namely the behaviour of the shell in the case where its normal deflections are large in comparison with its thicknesss. Such problems lead to non-linear differential equations whose exact solutions are not always possible. One usually adopts then approximate methods which most often are based on the calculation of the strain energy in the shell and plate.

1.15 Thermal stresses

If a shell with free edges undergoes a uniform temperature change, no thermal stresses are produced. But if the heating is non-uniform, the thermal expansion of the various parts of the shell is different and thermal stresses appear.

Also in the case where temperature change is uniform, but the edges are supported or clamped, free expansion of the shell is prevented and local stresses are set up at the edges. Knowing the thermal expansion of a shell when the edges are free, the values of reactive loads at the edges can be obtained. The solution reduces in this case to the determination of stresses in the shell with given displacement at the edges. We devote more space to discussing the first type of thermal stresses produced by non-uniform temperature. Let us assume that the temperature at any point of the shell is given by

$$T(\alpha_1, \alpha_2, \xi) = T_1(\alpha_1, \alpha_2) + \frac{\xi}{h} T_2(\alpha_1, \alpha_2), \tag{1.139}$$

where $T_1(\alpha_1, \alpha_2)$ is the average temperature over the section and $T_2(\alpha_1, \alpha_2)$ is the temperature difference of the face surfaces. The temperature produces the additional field of strain in the shell which is equal to $\alpha_t \cdot T(\alpha_1, \alpha_2, \xi)$; here, α_t is the coefficient of thermal expansion. Taking into account Hooke's law (1.31), we have

$$e_{\alpha\beta} = \frac{1+\nu}{E} \sigma_{\alpha\beta} - \frac{\nu}{E} \sigma^{\gamma}_{\gamma} \delta_{\alpha\beta} + \alpha_t T(\alpha_1, \alpha_2 \xi) \delta_{\alpha\beta}.$$

Introducing this into Eqs. (1.74) and (1.75), and neglecting transverse shear deformations, we obtain the following expressions for the bending stress resultants:

$$M_{\alpha\beta} = D[\nu \varkappa^{\delta}_{\delta} \delta_{\alpha\beta} + (1-\nu) \varkappa_{\alpha\beta}] - \frac{E\alpha_t h^2}{12(1-\nu^2)} T_2(\alpha_1, \alpha_2) \delta_{\alpha\beta}. \tag{1.140}$$

Expressing the shear forces by means of Eqs. $(1.78)_3$ and (1.140) and introducing them into the equation of equilibrium $(1.78)_2$, we obtain one of the fundamental equations of a *non-uniformly heated shell*:

$$\begin{aligned} &D(\Delta + 4H^2 - 2K)(\Delta + 4H^2 - (3-\nu)K) w - \Delta_k \Phi \\ &\quad = -(1+\nu) \frac{D\alpha_t}{h} \Delta T_2 - L(w, \Phi). \end{aligned} \tag{1.141}$$

It is assumed here that the shell is free of external loads and the material coefficients E, ν are independent of temperature. We can derive the second equation for the shell from the equation of compatibility (1.85). Expressing the strains first in terms of membrane forces and next in terms of the stress function, we obtain

$$\frac{1}{Eh}(\Delta+2K)(\Delta+(1-\nu)K)\Phi+\Delta_k w = -\alpha_t \Delta T_1+\tfrac{1}{2}L(w, w). \qquad (1.142)$$

The membrane resultant forces are determined by the formulae as previously.

We observe in the equations above that the effect of the temperature is expressed only in the terms ΔT_1 and ΔT_2. This means that when the temperature is constant or changes linearly, it does not affect the stresses. However, there can appear stresses produced by built-in or immovables edges.

Thermoelastic problems of shells are considered by W. Nowacki [1.37] and P. M. Ogibalov and V. F. Gribonov [1.31].

1.16 Orthotropic shells

The shells made of anisotropic materials find wide applications in practice. The anisotropy may result from the elastic properties of the material as well as from the structure of the shell. For example: the shell with closely placed stiffeners. In this section we derive the governing differential equations for orthotropic shells. We limit ourselves to the consideration of the most popular anisotropic shells, namely, orthotropic shells, which are characterized by the following feature. The material of the shell has various elastic properties in three mutually perpendicular directions, and has three planes of symmetry with respect to its elastic properties. Then, the stress-strain relations take the form

$$\begin{aligned}
\varepsilon_{11} &= a_{11}\sigma_{11}+a_{12}\sigma_{22}+a_{13}\sigma_{33},\\
\varepsilon_{22} &= a_{12}\sigma_{11}+a_{22}\sigma_{22}+a_{23}\sigma_{33},\\
\gamma_{12} &= a_{66}\tau_{12},\\
\gamma_{13} &= a_{55}\tau_{13},\\
\gamma_{23} &= a_{44}\tau_{23},
\end{aligned} \qquad (1.143)$$

where a_{ij} are constant coeffcients.

Let us assume that the curvilinear orthogonal coordinates α_1, α_2 follow the directions of principal curvature of the middle surface of the shell and simultaneously the directions of the orthotropy.

Solving the set of equations (1.143) with respect to σ_{ij}, we find the following relations:

$$\begin{aligned}
\sigma_{11} &= B_{11}\varepsilon_{11}+B_{12}\varepsilon_{12}+F_1\sigma_{33}, \\
\sigma_{22} &= B_{12}\varepsilon_{11}+B_{22}\varepsilon_{22}+F_2\sigma_{33}, \\
\tau_{12} &= B_{66}\gamma_{12}, \qquad \tau_{23} = B_{44}\gamma_{23}, \qquad \tau_{13} = B_{55}\gamma_{13},
\end{aligned} \tag{1.144}$$

where

$$B_{11} = \frac{a_{22}}{\Omega_a} = \frac{E_1}{1-\nu_1\nu_2},$$

$$B_{12} = -\frac{a_{12}}{\Omega_a} = \frac{\nu_2 E_1}{1-\nu_1\nu_2} = \frac{\nu_1 E_2}{1-\nu_1\nu_2},$$

$$B_{22} = \frac{a_{11}}{\Omega_a} = \frac{E_2}{1-\nu_1\nu_2},$$

$$B_{44} = \frac{1}{a_{44}}, \qquad B_{55} = \frac{1}{a_{55}}, \qquad B_{66} = \frac{1}{a_{66}},$$

$$\Omega_a = a_{11}a_{22}-a_{12}^2 = \frac{1-\nu_1\nu_2}{E_1 E_2},$$

$$F_1 = a_{13}B_{11}+a_{23}B_{12} = \frac{E_1}{E_3}\,\frac{\nu_{13}+\nu_2\nu_{23}}{1-\nu_1\nu_2},$$

$$F_2 = a_{13}B_{12}+a_{23}B_{22} = \frac{E_2}{E_3}\,\frac{\nu_{23}+\nu_1\nu_{13}}{1-\nu_1\nu_2};$$

E_1, E_2, ν_1, and ν_2 are the moduli of elasticity and the Poissons ratios in the directions α_1 and α_2, respectively. B_{66}, B_{55}, B_{44} are shear moduli in the plane of the plate and in the directions perpendicular to the plane of the plate. ν_{13} and ν_{23} are the Poisson's ratios of the material of the plate in the directions perpendicular to the middle plane of the plate. The notations above are called *engineering notations.*

Let us assume that the stresses σ_{33} acting in the direction perpendicular to the middle surface and the stresses τ_{13} and τ_{23} can be represented by Eqs. (1.80).

Relations (1.71) and (1.72) for the components of deformation do not depend on the properties of the material and are, therefore, the same as for the isotropic shell. Introducing them into the above equations and integrating along the thickness, we find the following relations for the internal forces:

$$
\begin{aligned}
N_{11} &= C_{11}\varepsilon_{11}+C_{12}\varepsilon_{22}+\tfrac{1}{2}F_1 hZ,\\
N_{22} &= C_{12}\varepsilon_{11}+C_{22}\varepsilon_{22}+\tfrac{1}{2}F_2 hZ,\\
N_{12} &= C_{66}\varepsilon_{12} = N_{21},\\
M_{11} &= D_{11}\varkappa_{11}+D_{12}\varkappa_{22}+F_1\tfrac{1}{10}h^2Z,\\
M_{22} &= D_{22}\varkappa_{22}+D_{12}\varkappa_{11}+F_2\tfrac{1}{10}h^2Z,\\
M_{12} &= D_{66}\varkappa_{12} = M_{21},
\end{aligned}
$$

where $C_{ij} = hB_{ij}$ and $D_{ij} = \frac{1}{12}h^3B_{ij}$.

The curvatures $\varkappa_{11}$, $\varkappa_{22}$, $\varkappa_{12}$ are determined by Eqs. (1.72a).

In order to obtain the relations for the average angles β_1 and β_2 we write Eqs. (1.82), (1.83) which take now the following form:

$$
\begin{aligned}
\beta_1 &= -\frac{1}{A_1}\frac{\partial w}{\partial \alpha_1}+\left(\frac{6}{5}-\frac{27h^2}{140R_1R_2}\right)\frac{Q_1}{h}a_{55},\\
\beta_2 &= -\frac{1}{A_2}\frac{\partial w}{\partial \alpha_2}+\left(\frac{6}{5}-\frac{27h^2}{140R_1R_2}\right)\frac{Q_2a_{44}}{h}.
\end{aligned}
\tag{1.146}
$$

Introducing the angles β_1 and β_2 into Eqs. (1.72a), we find

$$
\begin{aligned}
\varkappa_{11} &= -\left[\frac{1}{A_1}\frac{\partial}{\partial\alpha_1}\left(\frac{1}{A_1}\frac{\partial w}{\partial\alpha_1}\right)+\frac{1}{A_1A_2^2}\frac{\partial A_1}{\partial\alpha_2}\frac{\partial w}{\partial\alpha_2}+\frac{w}{R_1}\right]+\\
&\quad+\frac{6}{5}\frac{1}{h}\left(a_{55}\frac{\partial Q_1}{A_1\partial\alpha_1}+a_{44}\frac{Q_2}{A_1A_2}\frac{\partial A_1}{\partial\alpha_2}+\frac{w}{R_2}\right),\\
\varkappa_{22} &= -\left[\frac{1}{A_2}\frac{\partial}{\partial\alpha_2}\left(\frac{1}{A_2}\frac{\partial w}{\partial\alpha_2}\right)+\frac{1}{A_1^2A_2}\frac{\partial A_2}{\partial\alpha_1}\frac{\partial w}{\partial\alpha_1}+\frac{w}{R_2}\right]+\\
&\quad+\frac{6}{5}\frac{1}{h}\left(a_{44}\frac{\partial Q_2}{A_2\partial\alpha_2}+a_{55}\frac{Q_1}{A_1A_2}\frac{\partial A_2}{\partial\alpha_1}\right),\\
\varkappa_{12} &= \frac{1}{A_1A_2}\left[\frac{\partial^2 w}{\partial\alpha_1\partial\alpha_2}-\frac{1}{A_1}\frac{\partial w}{\partial\alpha_1}\frac{\partial A_1}{\partial\alpha_2}-\frac{1}{A_2}\frac{\partial w}{\partial\alpha_2}\frac{\partial A_2}{\partial\alpha_1}\right]+\\
&\quad+\frac{3}{5}\frac{1}{h}\left[a_{55}\frac{A_1}{A_2}\frac{\partial}{\partial\alpha_2}\left(\frac{Q_1}{A_1}\right)+a_{44}\frac{A_2}{A_1}\frac{\partial}{\partial\alpha_1}\left(\frac{Q_2}{A_2}\right)\right].
\end{aligned}
\tag{1.147}
$$

The small terms $27h^2/140R_1R_2$ in (1.146) have been neglected. The bending moments M_{11}, M_{22}, M_{12} may be written in the following way using the differential operators $I(\ )$ and $J(\ )$:

$$
\begin{aligned}
M_{11} &= -I(D_{11})w + \frac{h^2}{10}[J(D_{11})Q + ZF_1], \\
M_{22} &= -I(D_{22})w + \frac{h^2}{10}[J(D_{22})Q + ZF_2], \\
M_{12} &= -I(D_{66})w,
\end{aligned}
\tag{1.148}
$$

where

$$
\begin{aligned}
I(D_{jj}) &= \frac{D_{1j}}{A_1}\left[\frac{\partial}{\partial\alpha_1}\left(\frac{1}{A_1}\frac{\partial}{\partial\alpha_1}\right) + \frac{1}{A_2^2}\frac{\partial A_1}{\partial\alpha_2}\frac{\partial}{\partial\alpha_2} + \frac{1}{R_1^2}\right] + \\
&+ 2\frac{D_{6j}}{A_1 A_2}\left[\frac{\partial^2}{\partial\alpha_1\partial\alpha_2} - \frac{1}{A_2}\frac{\partial A_2}{\partial\alpha_1}\frac{\partial}{\partial\alpha_2} - \frac{1}{A_1}\frac{\partial A_1}{\partial\alpha_2}\frac{\partial}{\partial\alpha_1}\right] + \\
&+ \frac{D_{2j}}{A_2}\left[\frac{\partial}{\partial\alpha_2}\left(\frac{1}{A_2}\frac{\partial}{\partial\alpha_2}\right) + \frac{1}{A_1^2}\frac{\partial A_2}{\partial\alpha_1}\frac{\partial}{\partial\alpha_1} + \frac{1}{R_2^2}\right], \\
J(D_{jj})Q &= \frac{D_{1j}}{A_1}\left[a_{55}\frac{\partial Q_1}{\partial\alpha_1} + a_{44}\frac{1}{A_2}\frac{\partial A_1}{\partial\alpha_2}Q_2\right] + \\
&+ D_{6j}\left[a_{55}\frac{A_1}{A_2}\frac{\partial}{\partial\alpha_2}\left(\frac{Q_1}{A_1}\right) + a_{44}\frac{A_2}{A_1}\frac{\partial}{\partial\alpha_1}\left(\frac{Q_2}{A_2}\right)\right] + \\
&+ \frac{D_{2j}}{A_2}\left[a_{44}\frac{\partial Q_2}{\partial\alpha_2} + a_{55}\frac{1}{A_1}\frac{\partial A_2}{\partial\alpha_1}Q_1\right].
\end{aligned}
\tag{1.149}
$$

We should remember here that $D_{12} = D_{21}$ and $D_{61} = D_{62} = D_{16} = D_{26} = 0$. The shearing forces take the form

$$
\begin{aligned}
Q_i = &-\frac{1}{A_i A_j}\left\{\frac{\partial}{\partial\alpha_i}\left[A_j I(D_{ii}) - \frac{\partial A_j}{\partial\alpha_i} I(D_{jj}) + \right.\right. \\
&\left.\left. + \frac{\partial}{\partial\alpha_j}[A_i I(D_{66})]\right] + \frac{\partial A_i}{\partial\alpha_j} I(D_{66})\right\} w + \\
&+ \frac{h^2}{10}\frac{1}{A_i A_j}\left\{\frac{\partial}{\partial\alpha_i}\left[A_n J(D_{ii}) - \frac{\partial A_j}{\partial\alpha_i} J(D_{jj}) + \right.\right. \\
&\left.\left. + \frac{\partial}{\partial\alpha_j}[A_i J(D_{66})]\right] - \frac{\partial A_i}{\partial\alpha_j} J(D_{66})\right\} + \\
&+ \frac{h^2}{10}\frac{1}{A_i A_j}\left[\frac{\partial}{\partial\alpha_i}(A_j F_i Z) - F_j Z\frac{\partial A_j}{\partial\alpha_i}\right];
\end{aligned}
\tag{1.150}
$$

$i, j = 1, 2$ ($i \neq j$ not summed).

From the equation of compatibility (1.87) we infer the following equation:

$$(1/h)L_2(C_{ij})\Phi - \Delta_k w = -\tfrac{1}{2}\Delta(a_{ij}Z) + \tfrac{1}{2}L(w, w) + w\Delta_R w + K^2 w^2. \quad (1.151)$$

The differential operator $L_2(C_{ij})$ takes the form

$$L_2(C_{ij}) = \frac{1}{A_1 A_2}\frac{\partial}{\partial\alpha_1}\frac{1}{A_1}\left\{A_2\frac{\partial}{\partial\alpha_1}I_1(C_{22}) + \right.$$

$$+\frac{\partial A_2}{\partial\alpha_1}[I_1(C_{22}) - I_1(C_{11})] - \frac{A_1}{2}\frac{\partial}{\partial\alpha_2}I_1(C_{66}) - \left.\frac{\partial A_1}{\partial\alpha_2}I_1(C_{66})\right\} +$$

$$+\frac{1}{A_1 A_2}\frac{\partial}{\partial\alpha_2}\frac{1}{A_2}\left\{A_1\frac{\partial}{\partial\alpha_2}I_1(C_{11}) + \frac{\partial A_1}{\partial\alpha_2}[I_1(C_{11}) - I_1(C_{22})] - \right.$$

$$-\frac{A_2}{2}\frac{\partial}{\partial\alpha_1}I_1(C_{66}) - \left.\frac{\partial A_2}{\partial\alpha_1}I_1(C_{66})\right\} + \frac{1}{R_1 R_2}[I_1(C_{11}) + I_1(C_{22})]. \quad (1.152)$$

where $I_1(C_{ij})$ are differential operators

$$I_1(C_{11}) = \frac{C_{22}}{\Omega_c}\frac{1}{A_2}\left[\frac{\partial}{\partial\alpha_2}\left(\frac{1}{A_2}\frac{\partial}{\partial\alpha_2}\right) + \frac{1}{A_1^2}\frac{\partial A_2}{\partial\alpha_1}\frac{\partial}{\partial\alpha_1} + \frac{1}{R_1 R_2}\right] -$$

$$-\frac{C_{12}}{\Omega_c}\frac{1}{A_1}\left[\frac{\partial}{\partial\alpha_1}\left(\frac{1}{A_1}\frac{\partial}{\partial\alpha_1}\right) + \frac{1}{A_2^2}\frac{\partial A_1}{\partial\alpha_2}\frac{\partial}{\partial\alpha_2} + \frac{1}{R_1 R_2}\right];$$

$$I_1(C_{22}) = \frac{C_{11}}{\Omega_c}\frac{1}{A_1}\left[\frac{\partial}{\partial\alpha_1}\left(\frac{1}{A_1}\frac{\partial}{\partial\alpha_1}\right) + \frac{1}{A_2^2}\frac{\partial A_1}{\partial\alpha_2}\frac{\partial}{\partial\alpha_2} + \frac{1}{R_1 R_2}\right] -$$

$$-\frac{C_{12}}{\Omega_c}\frac{1}{A_2}\left[\frac{\partial}{\partial\alpha_2}\left(\frac{1}{A_2}\frac{\partial}{\partial\alpha_2}\right) + \frac{1}{A_1^2}\frac{\partial A_2}{\partial\alpha_1}\frac{\partial}{\partial\alpha_1} + \frac{1}{R_1 R_2}\right],$$

where $\Omega_c = C_{11}C_{22} - C_{12}^2$,

$$I_1(C_{66}) = -\frac{1}{C_{66}}\frac{1}{A_1 A_2}\left(\frac{\partial^2}{\partial\alpha_1\partial\alpha_2} - \frac{1}{A_1}\frac{\partial A_1}{\partial\alpha_2}\frac{\partial}{\partial\alpha_1} - \frac{1}{A_2}\frac{\partial A_2}{\partial\alpha_1}\frac{\partial}{\partial\alpha_2}\right).$$

The operator Δ_k has the same form as previously (Eq. (1.104)).

$$\Delta(a_{ij}Z) = \frac{1}{A_1 A_2}\left\{\frac{\partial}{\partial\alpha_1}\frac{1}{A_1}\left[A_2\frac{\partial Z}{\partial\alpha_1}a_{23} + (a_{23} - a_{13})\frac{\partial A_2}{\partial\alpha_1}Z + \right.\right.$$

$$\left.\left.+\frac{\partial}{\partial\alpha_2}\frac{1}{A_2}\left[A_1\frac{\partial Z}{\partial\alpha_2}a_{13} + (a_{13} - a_{23})\frac{\partial A_1}{\partial\alpha_2}Z\right]\right\}.$$

In the particular case where $a_{13} = a_{23} =$ constant,

$$\Delta(a_{ij}Z) = a_{13}\Delta Z.$$

Equation $(1.78)_2$ which may be written in the following way:

$$\frac{1}{A_1 A_2}\left[\frac{\partial}{\partial\alpha_1}(A_2 Q_1)+\frac{\partial}{\partial\alpha_2}(A_1 Q_2)\right]$$

$$= -Z+L(w,\Phi)-\Delta_k\Phi+\left[\frac{1}{R_1^2}I(D_{11})+\frac{1}{R_2^2}I(D_{22})\right]w+w\Delta_R\Phi, \tag{1.153}$$

together with Eqs. (1.150) and (1.151) make a set of four equations containing the four unknown functions Q_{11}, Q_{22}, w and Φ. This set (together with the boundary conditions) enables us to find the solution for any problem of the bending of an arbitrary orthotropic shell.

1.16.1 *Cylindrical shells. Equations of the classical theory*

In the case of a cylindrical shell we assume that $\alpha_1 = x$, $\alpha_2 = y = R\varphi$. We have also $R_1 = \infty$; $R_2 = R$.

If we neglect the effects of transverse normal and shear stresses, we find the following relations for the shear forces:

$$Q_x = -\frac{\partial}{\partial x}\left[D_{11}\frac{\partial^2 w}{\partial x^2}+(D_{12}+2D_{66})\frac{\partial^2 w}{\partial y^2}+D_{12}\frac{w}{R^2}\right],$$

$$Q_y = -\frac{\partial}{\partial y}\left[D_{22}\left(\frac{\partial^2 w}{\partial y^2}+\frac{1}{R^2}\right)+(D_{12}+2D_{66})\frac{\partial^2 w}{\partial x^2}\right]. \tag{1.154}$$

Introducing these forces into the equation of equilibrium (1.153), we find the following fundamental equation:

$$L_1(D_{ij})w-\Delta_k\Phi = Z-L(w,\Phi)-w\Delta_R\Phi \tag{1.155a}$$

where the operator $L_1(D_{ij})$ takes the form

$$L_1(D_{ij}) = D_{11}\frac{\partial^4 w}{\partial x^4}+2(D_{12}+2D_{66})\frac{\partial^4 w}{\partial x^2\,\partial y^2}+$$

$$+D_{22}\left(\frac{\partial^4 w}{\partial y^4}+\frac{2}{R^2}\frac{\partial^2 w}{\partial y^2}+\frac{w}{R^4}\right)+\frac{2}{R^2}(D_{12}+D_{66})\frac{\partial^2 w}{\partial x^2}.$$

The second equation combining the functions w and Φ takes the form

$$\frac{1}{h}L_2(C_{ij})\Phi+\Delta_k w = \frac{1}{2}L(w,w)+w\Delta_R w, \tag{1.155b}$$

where the operator $L_2(C_{ij})$ is defined in the following way:

$$L_2(C_{ij}) = \frac{C_{11}}{\Omega_c}\frac{\partial^4}{\partial x^4} + \left(\frac{1}{C_{66}} - \frac{2C_{12}}{\Omega_c}\right)\frac{\partial^4}{\partial x^2\,\partial y^2} + \frac{C_{22}}{\Omega_c}\frac{\partial^4}{\partial y^4}.$$

The operators $L(w, w)$, $L(w, \Phi)$, and Δ_k remain the same as previously. The expressions for the bending moments take the form

$$\begin{aligned} M_{xx} &= -\left[D_{11}\frac{\partial^2 w}{\partial x^2} + D_{12}\left(\frac{\partial^2 w}{\partial y^2} + \frac{w}{R^2}\right)\right], \\ M_{yy} &= -\left[D_{22}\left(\frac{\partial^2 w}{\partial y^2} + \frac{w}{R^2}\right) + D_{12}\frac{\partial^2 w}{\partial x^2}\right], \\ M_{xy} &= -2D_{66}\frac{\partial^2 w}{\partial x\,\partial y}. \end{aligned} \tag{1.156}$$

The membrane forces are defined by formulae (1.108) which are the same as for isotropic shells.

1.16.2 *Shallow shells*

We consider now the case of shallow shells. If we reduce the equations by neglecting the effect of transverse shear deformations, transverse normal stresses σ_{33}, and the terms which are functions of the radii of curvatures, we obtain the following relations between the bending and twisting moments and the displacement w:

$$\begin{aligned} M_{11} &= -I(D_{11})w, \\ M_{22} &= -I(D_{22})w, \\ M_{12} &= -I(D_{66})w, \end{aligned} \tag{1.157}$$

where the operators $I(D_{ij})$ have the same form as previously Eq. (1.149). We neglect only the terms $1/R_i^2$. The corresponding expressions for the shearing forces may readily be obtained from conditions (1.150) by neglecting the terms multiplied by $h^2/10$.

Thus we have

$$Q_i = -\frac{1}{A_i A_j}\left\{\frac{\partial}{\partial \alpha_i}\left[A_j I(D_{ii}) - \frac{A_j}{\partial \alpha_i} I(D_{jj})\right] + \right.$$
$$\left. + \frac{\partial}{\partial \alpha_j}\left[A_i I(D_{66}) + \frac{A_i}{\partial \alpha_j} I(D_{66})\right]\right\} \quad (i \neq j,\ i, j = 1, 2). \tag{1.158}$$

The equations for the membrane forces N_{11}, N_{22}, N_{12} remain the same as previously.

Introducing the shearing forces Q_1 and Q_2 into the equation of equilibrium (1.153) and the strains from Eqs. (1.72) and the changes of curvature (1.73) into the equation of compatibility (1.87), we find the following set of the governing differential equations for orthotropic shallow shells:

$$L_1(D_{ij})w - \Delta_k \Phi = Z - L(w, \Phi),$$
$$\frac{1}{h} L_2(C_{ij}) + \Delta_k w = \frac{1}{2} L(w, w) \tag{1.159}$$

where $L_1(D_{ij})$ is the differential operator

$$\begin{aligned} L_1(D_{ij}) = & \frac{1}{A_1 A_2} \frac{\partial}{\partial \alpha_1} \frac{1}{A_1} \left\{ \frac{\partial}{\partial \alpha_1} \left[A_2 I(D_{11}) - \frac{\partial A_2}{\partial \alpha_1} I(D_{22}) \right] + \right. \\ & \left. + \frac{\partial}{\partial \alpha_2} \left[A_1 I(D_{66}) + \frac{\partial A_1}{\partial \alpha_2} I(D_{66}) \right] \right\} + \\ & + \frac{1}{A_1 A_2} \frac{\partial}{\partial \alpha_2} \frac{1}{A_2} \left\{ \frac{\partial}{\partial \alpha_2} \left[A_1 I(D_{22}) - \frac{A_1}{\partial \alpha_2} I(D_{11}) \right] + \right. \\ & \left. + \frac{\partial}{\partial \alpha_1} \left[A_2 I(D_{66}) + \frac{\partial A_2}{\partial \alpha_1} I(D_{66}) \right] \right\}. \end{aligned} \tag{1.160}$$

The operators $L(w, \Phi)$, $L_2(C_{ij})$, Δa_{ij} and Δ_k have the same form as previously, Eqs. (1.104), (1.152), (1.153).

If we consider the orthotropic shallow shells in rectangular Cartesian coordinates, we find relations which have the same form as presented in the previous section for the cylindrical shell. These formulae can be simplified by the neglect of the small terms containing the curvatures of the shell.

1.16.3 *Equations in polar coordinates*

In the case of cylindrical orthotropy we often use the polar coordinates r, ϑ; then $A_1 = 1$, $A_2 = r$. The operators $L(D_{ij})$ and $L(C_{ij})$ take the form

$$\begin{aligned} L_1(D_{ij})w = & D_r \frac{1}{r} \frac{\partial^2}{\partial r^2} \left(r \frac{\partial^2 w}{\partial r^2} \right) + 2H \frac{1}{r^2} \frac{\partial^3}{\partial r \partial \vartheta^2} \left[r \frac{\partial}{\partial r} \left(\frac{w}{r} \right) \right] + \\ & + D_\vartheta \left[\frac{1}{r^4} \frac{\partial^4 w}{\partial \vartheta^4} - \frac{1}{r} \frac{\partial}{\partial r} \left(\frac{1}{r} \frac{\partial w}{\partial r} \right) + \frac{2}{r^4} \frac{\partial^2 w}{\partial \vartheta^2} \right], \end{aligned} \tag{1.161$_1$}$$

$$L_2(C_{ij}) \Phi = \frac{1}{E_\vartheta} \frac{1}{r} \frac{\partial^2}{\partial r^2} \left(r \frac{\partial^2 \Phi}{\partial r^2} \right) +$$

$$+\left(\frac{1}{G}-\frac{2\nu_r}{E_r}\right)\frac{1}{r^2}\frac{\partial^3}{\partial r\,\partial\vartheta^2}+\left[r\frac{\partial}{\partial r}\left(\frac{\Phi}{r}\right)\right]+$$

$$+\frac{1}{E_r}\left[\frac{1}{r^4}\frac{\partial^4\Phi}{\partial\vartheta^4}-\frac{1}{r}\frac{\partial}{\partial r}\left(\frac{1}{r}\frac{\partial\Phi}{\partial r}\right)+\frac{2}{r^4}\frac{\partial^2\Phi}{\partial\vartheta^2}\right], \qquad (1.161)_2$$

where $D_1 = D_r$, $D_2 = D_\vartheta$, $\nu_r = \nu_1$, $\nu_\vartheta = \nu_2$, $H = D_{12}+2D_{66}$. The bending moments are

$$M_{rr} = -\left[D_r\frac{\partial^2 w}{\partial r^2}+D_{12}\left(\frac{1}{r^2}\frac{\partial^2 w}{\partial\vartheta^2}+\frac{1}{r}\frac{\partial w}{\partial r}\right)\right],$$

$$M_{\vartheta\vartheta} = -\left[D_{12}\frac{\partial^2 w}{\partial r^2}+D_\vartheta\left(\frac{1}{r^2}\frac{\partial w}{\partial\vartheta}+\frac{1}{r}\frac{\partial w}{\partial r}\right)\right], \qquad (1.162)$$

$$M_{r\vartheta} = -2D_{66}\frac{\partial^2}{\partial r\,\partial\vartheta}\left(\frac{w}{r}\right).$$

The membrane forces are expressed by formulae (1.112). In the case of shallow shells we neglect the terms $\Phi/R_1 R_2$ in (1.112) and the terms multiplied by D/R_i, $i = 1, 2$.

The shearing forces are the following:

$$Q_r = -\left[D_r\frac{1}{r}\frac{\partial}{\partial r}\left(r\frac{\partial^2 w}{\partial r^2}\right)+H\frac{1}{r}\frac{\partial^3}{\partial r\,\partial\vartheta}\left(\frac{w}{r}\right)-\right.$$

$$\left.-D_\vartheta\left(\frac{1}{r^3}\frac{\partial^2 w}{\partial\vartheta^2}+\frac{1}{r^2}\frac{\partial w}{\partial r}\right)\right], \qquad (1.163)$$

$$Q_\vartheta = -\left[H\frac{1}{r}\frac{\partial^3 w}{\partial r^2\,\partial\vartheta}+D_\vartheta\left(\frac{1}{r^3}\frac{\partial^3 w}{\partial\vartheta^3}+\frac{1}{r^2}\frac{\partial^2 w}{\partial r\,\partial\vartheta}\right)\right].$$

The theory of anisotropic shells can be found in the book by Ambartsumian [1.29].

1.17 The boundary conditions

The solutions of the differential equations (1.102), (1.103) naturally do not yet determine completely the state of stress in a shell as long as they are not subject to boundary conditions. That means that certain number of relations between the forces, moments, displacements or functions of these quantities are specified at the edges of the shell. In the case where the shell has no boundary, for example, the spherical shell, the solution must be a periodic function of the coordinates α_1, α_2. In this case the boundary conditions are replaced by the conditions of periodicity. The boundary conditions of a shell can be expressed by means of the displacements or internal forces. In the case of conditions expressed by the displacements we require the displacements u, v, w and angles of rotations β_1, β_2 to have the values given in advance at the edge of the shell. When the forces are given at the edge, the forces obtained from the general solution should fulfil the given conditions. In the case of mixed boundary conditions we can use both the equations expressed in terms of displacements and in terms of forces with the restriction that only β_1 or M_{11}, β_2 or M_{22}, w or Q_1, u or N_{11} and v or N_{22} can be adopted as the boundary conditions.

For example, for the built-in edge we have $\beta_1 = \beta_2 = w = u = v = 0$. For the free edge:

$$M_{11} = M_{12} = Q_1 = N_{11} = N_{12} = 0.$$

At the simply supported edge we can demand, for example:

$$M_{11} = M_{12} = w = v = N_{11} = 0.$$

In the case of the classical theory, when we neglect the displacements involved by the shear forces, the differential equations of the problem are of the eight order. Then we have only four arbitrary constants for one edge and that permits the satisfaction of only four boundary conditions. In order to have the number of boundary conditions the same as the number of arbitrary constants the following treatment should be performed. We define the boundary conditions expressed by the displacements in terms of the deflections w, u, v and the angles $\dfrac{\partial w}{A_1\,\partial\alpha_1}$ or $\dfrac{\partial w}{A_2\,\partial\alpha_2}$.

For example, for the built-in edge we have

$$w_0 = 0, \qquad u = 0, \qquad v = 0, \qquad \frac{\partial w}{A_1\,\partial\alpha_1} = 0.$$

When the boundary conditions are given in terms of the internal forces, we introduce the so-called *equivalent shear force* summing up the effect of the twisting moment M_{12} and the shear force along the edge. Instead of the condition requiring the twisting moment and the shear force to be equal to the given loads at various points of the edge we require this agreement to take place for the effective shear force:

$$\bar{Q}_1 = Q_1 + \frac{\partial M_{12}}{A_2\,\partial\alpha_2}. \tag{1.164}$$

The edge shear forces equivalent to the moments M_{12} are not strictly parallel, they have the radial directions and give the components parallel to the middle surface. The tangential forces N_{12} together with these forces give the effective tangential force.

$$\bar{N}_{12} = N_{12} - \frac{1}{R_2} M_{12}. \tag{1.165}$$

It turns out from the above that the five edge forces and moments can be replaced by the four statically equivalent edge loads N_{11}, N_{12}, M_{11}, and $\bar{Q}_1$.

At the free edge these forces should be equal to zero. At the simply supported edge we can apply the following conditions:

$$N_{11} = M_{11} = w = N_{12} = 0.$$

1.18 Sandwich shells

The equations for three-layer sandwich shells are similar to those obtained previously for the homogeneous shells in case where the bending rigidity of the external layers can be neglected. Therefore all the solutions for the homogeneous shells obtained taking into account the effect of transverse shear deformations can be applied to such sandwich shells provided that the coefficients characterizing the shear rigidity are changed. Let as consider the shell made of two thin layers of equal thickness δ and the core placed between them. The dimensions of the cross-section of the shell are presented in Fig. 9a. Let as assume that the material of the core is incompressible in the direction normal to the middle surface of the shell and is very weak in the tangential direction. It means that $E_{3c} = \infty$; $E_{1c} = E_{2c} = G_{12c} = 0$. If the thickness of the external layers is small in comparison with the thickness of the core, we can assume that the bending rigidity of these layers is zero and that the external layers carry only the normal and tangential forces acting in their surfaces.

The angles of rotations of the lateral planes of the shell element are, according to Fig. 9b, defined by

$$\beta_\alpha = -u_3||_\alpha + \gamma_{\alpha_c} = \frac{u_\alpha^{(1)} - u_\alpha^{(2)}}{2h_1}, \qquad \alpha = 1, 2. \tag{1.166}$$

where γ_{α_c} is the shear angle of the core and $u_\alpha^{(1)}$ and $u_\alpha^{(2)}$ are the displacements of the upper and lower external layers, respectively. Since the tangential stresses are constant across the thickness of the core, we have

$$\gamma_{\alpha_c} = \frac{\tau_{\alpha 3}}{G_c} = \frac{Q_\alpha}{2h_1 G_c} \tag{1.167}$$

where G_c is the transverse shear modulus of the core.

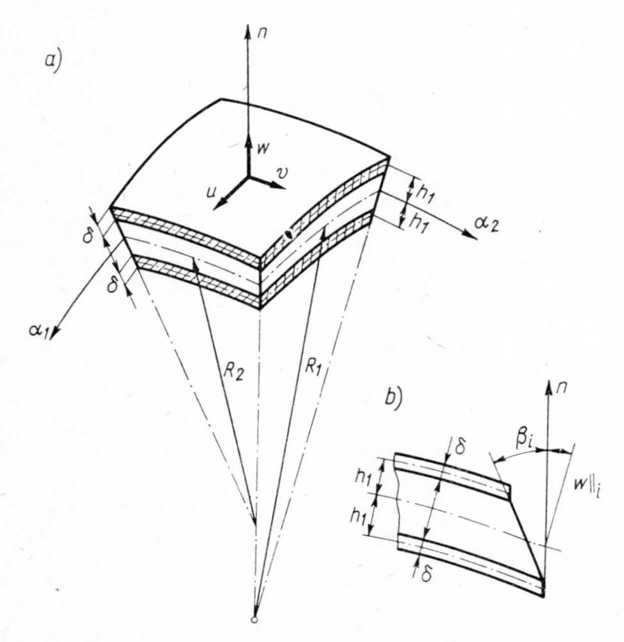

Fig 9. a) *Dimensions of an element of the sandwich shell*; b) *Deformations of a sandwich shell.*

It is very easy to notice that the above formulae are analogous to those previously obtained for homogeneous shells. The membrane forces in the external layers are defined by the following formulae:

$$N^{\alpha\beta(1)} = \frac{E\delta}{1-\nu^2}\left[\nu a^{\alpha\beta}\varepsilon_{\delta}^{\delta(1)} + (1-\nu)\varepsilon^{\alpha\beta(1)}\right] \tag{1.168}$$

where

$$\varepsilon_{\alpha\beta}^{(1)} = \frac{1}{2}\left(u_\alpha^{(1)}||_\beta + u_\beta^{(1)}||_\alpha\right);$$

$\mathbf{u}^{(1)}$ means here the displacement vector for the points of the upper layer.

A similar relation can be written for the lower layer. Calculating the bending moments from $(1.74)_2$ we find that these are defined by

$$M_s^{\alpha\beta} = (N^{\alpha\beta(1)} - N^{\alpha\beta(2)})h_1 = D_z[\nu^{\alpha\beta}\varkappa_\delta^\delta + (1+\nu)\varkappa^{\alpha\beta}] \tag{1.169}$$

where D_z is the bending rigidity of the sandwich shell

$$D_z = \frac{2Eh_1^2\,\delta}{1-\nu^2}$$

where E and ν are the modulus of elasticity and the Poisson ratio of the material of the external layers of the shell. Taking into account the equilibrium equations, we obtain the following relations for the shear forces:

$$\begin{aligned} Q^\alpha - (1-\nu)\frac{\eta}{2}\,\mathrm{d}^{\alpha\beta}\mathrm{d}^{\lambda\mu}(Q_\beta|_{\lambda\mu} + Q_\lambda|_{\beta\mu}) \\ = -D_z\{a^{\alpha\beta}[\Delta w + b_\lambda^\delta b_\delta^\lambda w]\}|_\beta + \eta a^{\alpha\beta}Q^\delta|_{\delta\beta} \end{aligned} \tag{1.170}$$

where

$$\eta = \frac{E\delta h_1}{G_c(1-\nu^2)}.$$

The effect of transverse normal stress σ_{33} is neglected here.

The equation of equilibrium in the normal direction takes the following form on introduction of the stress function from (1.88):

$$\begin{aligned} D_z(\Delta + 4H^2 - 2K)[\Delta + 4H^2 - (3-\nu)K]w - (1-\eta\Delta)\Delta_k\Phi \\ = (1-\eta\Delta)X^3 + b_{\alpha\beta}P^{\alpha\beta}. \end{aligned} \tag{1.171}$$

The second governing equation obtained from the equation of compatibility remains the same as for the homogeneous shell

$$\begin{aligned} \frac{1}{Eh}(\Delta + 2K)[\Delta + (1-\nu)K]\Phi + \Delta_k w \\ = -\frac{1}{Eh}[\Delta P_\alpha^\alpha - (1-\nu)P^{\alpha\beta}|_{\alpha\beta}] \quad \text{where} \quad h = 2\delta. \end{aligned} \tag{1.172}$$

The non-linear effects are neglected in the above equations. As we see, the equations for the three-layer sandwich shells (1.170), (1.171) and (1.172) are analogous to the equations of homogeneous shells if the effect of transverse shear deformations is taken into account. If we replace the bending rigidity D by D_z and introduce the coefficient η in place of the coefficient $h^2/5(1-\nu)$ in Eq. (1.99), we obtain the equations for sandwich shells.

The same equations are valid for homogeneous shells made of material with transverse anisotropy. If we can assume that

$$\gamma_{\alpha 3} = \frac{Q_\alpha}{k'G'h}$$

where G' is the transverse shear modulus and k' is certain constant coefficient, we obtain $\eta = D/k'G'h$. For the isotropic shell $k' = 5/6$ and $G' = E/2(1+\nu)$,

$\eta = h^2/5(1-\nu)$.

References 1

[1.1] A. Cauchy: *Bulletin de sciences a la Société Philomatique*, (1823).

[1.2] S. Poisson: Mémoire sur l'equilibre et le mouvement des corps elastiques, *Paris Mém. de l'Acad. 8* (1829).

[1.3] A. E. H. Love: *A treatise on the mathematical theory of elasticity*, Dover Publication, New York 1944.

[1.4] F. B. Hildebrand, E. Reissner and G. B. Thomas: Notes on the foundations of the theory of small displacements of orthotropic shells, *Nat. Advisory Comm. Aero. Tech. Notes, No. 1833*, 1949.

[1.5] A. E. Green and W. Zerna: *Theoretical elasticity*, Clarendon Press, Oxford 1954.

[1.6] A. I. Lure: *Statics of thin elastic shells*, Gostekhizdat, Moskva 1947 (in Russian).

[1.7] E. Reissner: Stress-strain relations in the theory of thin elastic shells, *J. Math. Phys. 31* (1952) 109–119.

[1.8] P. M. Naghdi: On the theory of thin elastic shells, *Quart. Appl. Math. 14* (1957) 369–380; *Quart. Appl. Math. 14* (1956) 331.

[1.9] N. A. Kilchevskii: Generalization of contemporary theory of shells, *P.M.M. 4.2* (1939) (in Russian).

[1.10] Kh. M. Mushtari, I. G. Teregulov: On the theory of shells of middle thickness, *Dokl. AN SSSR 6* (1959) 128 (in Russian).

[1.11] L. H. Donnell: Stability of thin-walled tubes under torsion, *N.A.C.A. Report No. 479* (1933).

[1.12] V. Z. Vlasov: *General theory of shells*, Moskva–Leningrad 1949 (in Russian).

[1.13] Kh. M. Mushtari: Some generalizations of the theory of thin shells, *Izv. Fiz. Mat. Kazansk. Univ. 3, ser. 8* (1938) (in Russian).

[1.14] S. Łukasiewicz: An improvement of the equations of the engineering theory of shells, *Rozp. Inżyn. 11.1* (1963) (in Polish).

[1.15] S. Łukasiewicz: The equations of the theory of the non-shallow shells, *Arch. Bud. Maszyn 4* (1965).

[1.16] W. Flügge: *Stresses in Shells*, Springer Verlag, Berlin-Göttingen-Heidelberg 1960.

[1.17] A. L. Goldenveizer: *Theory of elastic thin shells*, Gostekhizdat, Moskva 1953 (in Russian).

[1.18] L. S. D. Morley: An improvement on Donnell's approximation for thin-walled circular cylinders, *J. Mech. Appl. Math. 12* (1959) 89.
[1.19] W. T. Koiter: A spherical shell under point loads at its poles, *Advances in Appl. Mech., Prager Anniversary Volume*, 1963, 155–169.
[1.20] V. V. Novozhilov: *The theory of thin shells*, Groningen 1959.
[1.21] G. Kirchhoff: Über das Gleichgewicht und die Bewegung einer elastischen Scheibe, *Crelles J. 40* (1850) 51.
[1.22] W. T. Koiter: On the nonlinear theory of thin elastic shells, *Koninkl. Nederl. Akademie van Vetenschappen, ser. B, 69, no. 1*, Amsterdam 1966.
[1.23] K. Girkmann: *Flächentragwerke*, Springer, Wien 1956.
[1.24] R. Byrne: Theory of small deformations of a thin elastic shell, *University of California, Publ. Math. N.S., 2* (1944) 103–152.
[1.25] S. Kaliski: *Vibrations and waves*, PWN, Warszawa 1966 (in Polish).
[1.26] K. Marguerre: Zur Theorie der gekrümmten Platte grosser Formänderung, *Proc. 5th Inst. Congr. Appl. Mech.*, 93–101.
[1.27] S. Łukasiewicz: On the equations of the theory of shells of slowly varying curvatures, *ZAMP 22* (1971) 6.
[1.28] E. Reissner: On a variational theorem in elasticity, *J. Math. Phys. 29* (1960) 90–95.
[1.29] S. A. Ambartsumian: *Theory of anisotropic shells*, Moskva 1961, Gos. Izd. Fiz.-Mat. Lit. (in Russian).
[1.30] C. Woźniak: *Nonlinear theory of shells*, PWN, Warszawa 1966 (in Polish).
[1.31] P. M. Ogibalov, V. F. Gribonov: *Thermal stability of plates and shells*, 1968 Izd. Moskovskogo Un-ta (in Russian).
[1.32] S. Łukasiewicz: The equations of the technical theory of shells of variable rigidity, *Arch. Mech. Stosow. 13.1* (1961).
[1.33] L. H. Donnell: A discussion of thin-shell theory, *Proc. Fifth Int. Congr. Appl. Mech., Cambridge, Mass. 66* (1938).
[1.34] H. Duddeck: Biegetheorie der allgemeinen Rotationschalen mit schwacher Värenderlichkeit der Shälenkrummungen, *Ing.-Archiv. 33, B 5 H* (1964) 279–300.
[1.35] J. Kempner: Remarks on Donnell's equations, *J. Appl. Mech. Trans. ASME 77* (1955) 117–118.
[1.36] A. L. Goldenveizer: Methods of derivation and improvement of the theory of shells, *Trudy Mat. Mekh. 32.4* (1968) (in Russian).
[1.37] W. Nowacki: *Problems of thermoelasticity*, PWN, Warszawa 1960 (in Polish).
[1.38] W. Nowacki: *Theory of elasticity*, PWN, Warszawa 1960 (in Polish).
[1.39] S. Łukasiewicz: Equations of the technical theory of shells with the effect of transverse shear deformations, *Quart. Appl. Math.* (1971).
[1.40] A. I. Lure: General theory of thin elastic shells, *P.M.M. 4.2* (1940) (in Russian).
[1.41] N. A. Kilchevskii: *Foundations of analytical mechanics of shells*, Kiev, 1963 (in Russian).
[1.42] E. Reissner: On transverse bending of plates, including the effect of transverse shear deformation, *Int. J. Solids Structures 11* (1975) 569–573.
[1.43] M. Duszek: *Equations of the theory of plastic shells with large deflections*, Instytut Podst. Probl. Techn. 13, 1971.

2

Fundamental equations of plates

2.1 Differential equations of isotropic plates

A *plate* is a plane body whose one dimension, thickness, can be considered small in comparison with other dimensions. Usually the plates are loaded by the forces acting both perpendicularly to its surface and in its middle plane—the plane laying in the middle of the thickness of the plate. The plate loaded only in its middle plane is called a *sheet*.

The fundamental equations of the theory of plates can easily be obtained from the equations of the theory of shells by a simple assumption that the radii of curvature of the shell increase to infinity. In this way we obtain for the isotropic plate the following set of differential equations, taking into account the effect of moderate large normal deflection and the effect of transverse shear deformations:

$$D\Delta\Delta w = Z - L(w, \Phi) - \frac{h^2}{10(1-\nu)}[(2-\nu)Z - 2L(w, \Phi)] + \frac{\partial X_M}{\partial x} + \frac{\partial Y_M}{\partial y}, \tag{2.1}$$

$$\frac{1}{Eh}\Delta\Delta\Phi = \frac{1}{2}L(w, w) - \frac{\nu\Delta Z}{2E}$$

where $D = Eh^3/12(1-\nu^2)$. Only the effect of the normal load Z and distributed moments X_M and Y_M is taken into account in the above equations. The loads tangential to the middle surface of the plate are neglected. We solve usually the problem of a plate, introducing the orthogonal Cartesian coordinates. In this case the operators Δ and $L(w, \Phi)$ take the form

$$\Delta = \partial^2/\partial x^2 + \partial^2/\partial y^2,$$

$$L(w, \Phi) = \frac{\partial^2 w}{\partial x^2}\frac{\partial^2 \Phi}{\partial y^2} - 2\frac{\partial^2 w}{\partial x \partial y}\frac{\partial^2 \Phi}{\partial x \partial y} + \frac{\partial^2 w}{\partial y^2}\frac{\partial^2 \Phi}{\partial x^2}.$$

The relations for the bending moments are

$$M_{xx} = -D\left[\frac{\partial^2 w}{\partial x^2} + \nu\frac{\partial^2 w}{\partial y^2}\right] + \frac{h^2}{5}\frac{\partial Q_x}{\partial x} - \frac{h^2}{10}\frac{\nu}{1-\nu}[Z - 2L(w, \Phi)],$$

$$M_{yy} = -D\left[\frac{\partial^2 w}{\partial x^2} + \nu\frac{\partial^2 w}{\partial y^2}\right] + \frac{h^2}{5}\frac{\partial Q_y}{\partial y} - \frac{h^2}{10}\frac{\nu}{1-\nu}[Z - 2L(w, \Phi)], \quad (2.2)$$

$$M_{xy} = -(1-\nu)D\frac{\partial^2 w}{\partial x \partial y} + \frac{h^2}{10}\left(\frac{\partial Q_x}{\partial y} + \frac{\partial Q_y}{\partial x}\right).$$

The equations for the shear forces Q_x and Q_y take the form

$$Q_x - \frac{h^2}{10}\Delta Q_x$$

$$= -D\frac{\partial \Delta w}{\partial x} - \frac{h^2}{10(1-\nu)}\frac{\partial}{\partial x}[Z - (1+\nu)L(w, \Phi)] + X_M,$$

$$(2.3)$$

$$Q_y - \frac{h^2}{10}\Delta Q_y$$

$$= -D\frac{\partial \Delta w}{\partial y} - \frac{h^2}{10(1-\nu)}\frac{\partial}{\partial y}[Z - (1+\nu)L(w, \Phi)] + Y_M.$$

The membrane forces are defined by the stress function in the following way:

$$N_{xx} = -\frac{\partial^2 \Phi}{\partial y^2}, \quad N_{yy} = -\frac{\partial^2 \Phi}{\partial x^2}, \quad N_{xy} = \frac{\partial^2 \Phi}{\partial x \partial y}; \quad (2.4)$$

X_M and Y_M denote here the external bending moments distributed over unit area of the shell. Fig. 10.

The term $-\frac{\nu}{2}\frac{\Delta Z}{E}$ standing on the right-hand side of Eq. $(2.1)_2$ represents the effect of the stresses σ_{zz} produced by the load perpendicular to the plate surface. Usually it is small and only in the case of the concentrated loads can it be larger. The expression $\frac{1}{2}L(w, w)$ is non-linear and appears as a result of a change of geometry of the plate during deformation. Equations

(2.1) and (2.3) together with the boundary conditions determine the four functions w, Φ and Q_x, Q_y.

Having the stress function Φ, we can determine the stresses in the middle surface of the plate. Having the deflection w, the shear forces Q_x, Q_y, the stress function Φ, the bending and shear stresses can be obtained by using Eqs. (2.2) and (2.4). Thus, in general, the investigation of the plate with the effect of transverse shear deformations and large normal deflections reduces to the solution of two coupled non-linear differential equations (2.1) and later two equations (2.3).

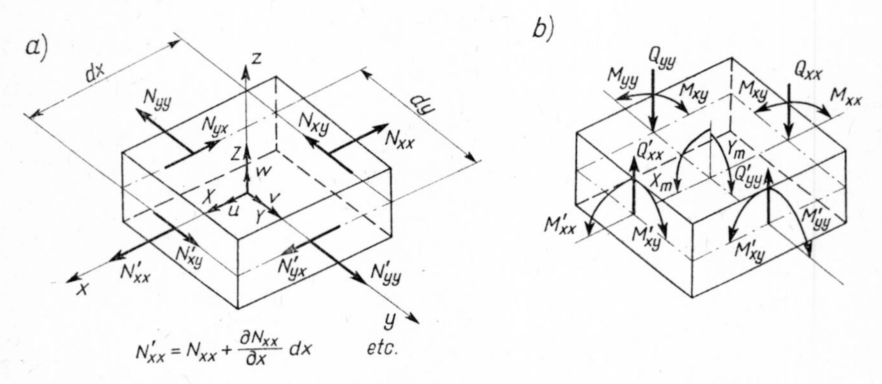

Fig. 10. a), b) *Positive values of stress resultants in an element of a plate.*

The stresses in the plate are defined by Eq. (1.85). Introducing Cartesian coordinates and $u_3^* = w$, we have, for example, the stress σ_{xx} given by the following formula:

$$\sigma_{xx} = \frac{E}{1-\nu^2}\left[\frac{\partial u}{\partial x} + \nu\frac{\partial v}{\partial y} + \frac{1}{2}\left(\frac{\partial w}{\partial x}\right)^2 + \frac{\nu}{2}\left(\frac{\partial w}{\partial y}\right)^2\right] -$$

$$-\frac{Ez}{1-\nu^2}\left[\frac{\partial^2 w}{\partial x^2} + \nu\frac{\partial^2 w}{\partial y^2}\right] +$$

$$+\frac{1}{1-\nu}\frac{z}{h}\left[3-\left(\frac{2z}{h}\right)^2\right]\left(\frac{\partial Q_x}{\partial x} + \nu\frac{\partial Q_y}{\partial y}\right) + \frac{\nu}{1-\nu}\sigma_{33}.$$

Using Eq. (1.74), we can express the stresses by the membrane forces N_{xx}, N_{yy}, N_{xy} instead of the displacements u, v. We have

$$\sigma_{xx} = \frac{N_{xx}}{h} - \frac{Ez}{1-\nu^2}\left[\frac{\partial^2 w}{\partial x^2} + \nu\frac{\partial^2 w}{\partial y^2}\right] + \frac{z}{h}\left[3-\left(\frac{2z}{h}\right)^2\right]\frac{\partial Q_x}{\partial x} +$$

$$+\frac{\nu}{2(1-\nu)}\frac{z}{h}\left[3-\left(\frac{2z}{h}\right)^2\right][-Z+2L(w,\Phi)],$$

$$\sigma_{yy}=\frac{N_{yy}}{h}-\frac{Ez}{1-\nu^2}\left[\nu\frac{\partial^2 w}{\partial x^2}+\frac{\partial^2 w}{\partial y^2}\right]+\frac{z}{h}\left[3-\left(\frac{2z}{h}\right)^2\right]\frac{\partial Q_y}{\partial y}+ \tag{2.5}$$

$$+\frac{\nu}{2(1-\nu)}\frac{z}{h}\left[3-\left(\frac{2z}{h}\right)^2\right][-Z+2L(w,\Phi)],$$

$$\tau_{xy}=\frac{N_{xy}}{h}-\frac{Ez}{1+\nu}\frac{\partial^2 w}{\partial x\,\partial y}+\frac{z}{2h}\left[3-\left(\frac{2z}{h}\right)^2\right]\left(\frac{\partial Q_x}{\partial y}+\frac{\partial Q_y}{\partial x}\right)\Bigg].$$

Let us observe that in order to find the stresses in the laterally loaded plate, taking into account the effect of transverse normal stress σ_{zz}, we need the solution of the second equation of the group (2.1), which represents the equation of compatibility of strains in the middle surface of the plate. Usually we apply this equation only if we consider the plate loaded in its plane. But if we take into account the effect of the stress σ_{zz}, there appear the stresses in the middle surface. However, these stresses are usually very small and often are neglected.

The stresses in the external, upper, and lower surfaces of the plate can be obtained by introducing $z = \pm h/2$.

$$\sigma_{xx}\left(\pm\frac{h}{2}\right)=\frac{N_{xx}}{h}\mp\frac{Eh}{2(1-\nu^2)}\left(\frac{\partial^2 w}{\partial x^2}+\nu\frac{\partial^2 w}{\partial y^2}\right)\pm\frac{\partial Q_x}{\partial x}\pm$$

$$\pm\frac{\nu}{2(1-\nu)}[-Z+2L(w,\Phi)], \tag{2.6}$$

$$\sigma_{yy}\left(\pm\frac{h}{2}\right)=\frac{N_{yy}}{h}\mp\frac{Eh}{2(1-\nu^2)}\left(\frac{\partial^2 w}{\partial y^2}+\nu\frac{\partial^2 w}{\partial x^2}\right)\pm\frac{\partial Q_y}{\partial y}\pm$$

$$\pm\frac{\nu}{2(1-\nu)}[-Z+2L(w,\Phi)],$$

$$\tau_{xy}\left(\pm\frac{h}{2}\right)=\frac{N_{xy}}{h}\mp\frac{Eh}{2(1+\nu)}\frac{\partial^2 w}{\partial x\,\partial y}\pm\frac{1}{2}\left(\frac{\partial Q_x}{\partial y}+\frac{\partial Q_y}{\partial x}\right).$$

Representing the stresses in terms of bending moments (2.2), we find

$$\sigma_{\alpha\beta}\left(\pm\frac{h}{2}\right)=\frac{N_{\alpha\beta}}{h}\pm\frac{6M_{\alpha\beta}}{h^2}\mp\frac{1}{10}(Q_{\alpha,\beta}+Q_{\beta,\alpha})\pm\frac{\nu\delta_{\alpha\beta}}{10(1-\nu)}[Z-2L(w,\Phi)],$$

where $N_{\alpha\beta}$ are the membrane forces obtained by solving the equation

$$\frac{1}{Eh}\Delta\Delta\Phi = \frac{1}{2}L(w, w) - \frac{\nu\Delta Z}{2E}.$$

Since the effect of the transverse normal stress was taken into account in the approximate manner, it is possible to simplify the calculations by neglecting in the above equation the term $\nu\Delta Z/2E$. However, in this case we shoul dadd to the forces $N_{\alpha\beta}$ the term representing the effect of the transverse normal s.ress from (1.76). Then we have

$$N_{\alpha\beta} = \overline{N}_{\alpha\beta} + \frac{\nu}{2(1-\nu)}Z$$

where $\overline{N}_{\alpha\beta}$ are the membrane forces calculated from

$$\frac{1}{Eh}\Delta\Delta\Phi = \frac{1}{2}L(w, w).$$

It is interesting to compare the contents of Eq. (2.2) with the above expressions (2.6). The noteworthy difference between the right-hand side of Eq. (2.2) and Eq. (2.6) is the absence in Eq. (2.6) of the factor 6/5 which occurs in front of the terms with Q_α if we calculate the stresses from (2.2). E. Reissner showed the significance of this difference and proved its effect in the solution of the problem of the stress concentration due to the presence of a circular hole in the pure bending of an infinite plate [1.42]. E. Reissner obtained an excellent agreement with results of Alblas exact analysis within the framework of the three-dimensional elasticity theory [2.23]. V. Panc also considered the problem of the calculation of the stresses in the plate, taking into account the effect of transverse shear deformations, and obtained similar results [2.24]. If we calculate the stresses, introducing into formulae (1.75) the strains defined by the average values of the angles β_α, we obtain less accurate values.

It is covenient in some cases to perform the calculations in polar coordinates. It particularly concerns circular plates, because in this case the edge of the plate is defined by the simple relation r = constant and which simplifies the boundary conditions. If we introduce into the previously derived equations new coordinates (r, ϑ), we find on transformations the following formulae for bending and twisting moments:

$$M_{rr} = D\left[\frac{\partial\beta_r}{\partial r} + \nu\frac{1}{r}\frac{\partial\beta_\vartheta}{\partial\vartheta} + \nu\frac{\beta_r}{r} + \frac{6\nu(1+\nu)}{5Eh}Z\right],$$

$$M_{\vartheta\vartheta} = D\left[\nu\frac{\partial\beta_r}{\partial r} + \frac{1}{r}\frac{\partial\beta_\vartheta}{\partial\vartheta} + \frac{\beta_r}{r} + \frac{6\nu(1+\nu)}{5Eh}Z\right],$$

$$M_{r\vartheta} = \frac{1-\nu}{2} D \left[\frac{1}{r} \frac{\partial \beta_r}{\partial \vartheta} + \frac{\partial \beta_\vartheta}{\partial r} - \frac{\beta_\vartheta}{r} \right].$$

The average angles of rotations take the form

$$\beta_r = -\frac{\partial w}{\partial r} + \frac{12}{5} \frac{1+\nu}{Eh} Q_r; \qquad \beta_\vartheta = -\frac{1}{r} \frac{\partial w}{\partial \vartheta} + \frac{12}{5} \frac{1+\nu}{Eh} Q_\vartheta.$$

Introducing these relations into above formulae, we find

$$M_{rr} = -D \left[\frac{\partial^2 w}{\partial r^2} + \nu \frac{1}{r} \frac{\partial w}{\partial r} + \frac{\nu}{r^2} \frac{\partial^2 w}{\partial \vartheta^2} \right] + \\ + \frac{h^2}{5} \frac{\partial Q_r}{\partial r} - \frac{h^2}{10} \frac{\nu}{1-\nu} (Z - 2L(w, \Phi)],$$

$$M_{\vartheta\vartheta} = -D \left[\nu \frac{\partial^2 w}{\partial r^2} + \frac{1}{r} \frac{\partial w}{\partial r} + \frac{1}{r^2} \frac{\partial^2 w}{\partial \vartheta^2} \right] - \\ - \frac{h^2}{5} \frac{\partial Q_r}{\partial r} - \frac{h^2}{10(1-\nu)} [(2-\nu) Z - 2L(w, \Phi)],$$

$$M_{r\vartheta} = -(1-\nu) D \left[\frac{1}{r} \frac{\partial^2 w}{\partial \vartheta \partial r} - \frac{1}{r^2} \frac{\partial w}{\partial \vartheta} \right] + \frac{h^2}{10} \left(\frac{1}{r} \frac{\partial Q_r}{\partial \vartheta} + \frac{\partial Q_\vartheta}{\partial r} - \frac{Q_\vartheta}{r} \right).$$

Equations for shear forces have the form

$$Q_r - \frac{h^2}{10} \left(\Delta Q_r - \frac{Q_r}{r^2} - \frac{2}{r^2} \frac{\partial Q_\vartheta}{\partial \vartheta} \right) \\ = -D \frac{\partial}{\partial r} (\Delta w) - \frac{h^2}{10(1-\nu)} \frac{\partial}{\partial r} [Z - (1+\nu) L(w, \Phi)],$$

$$Q_\vartheta - \frac{h^2}{10} \left(\Delta Q_\vartheta - \frac{Q_\vartheta}{r^2} + \frac{2}{r^2} \frac{\partial Q_r}{\partial \vartheta} \right) \\ = -D \frac{1}{r} \frac{\partial}{\partial \vartheta} (\Delta w) - \frac{h^2}{10(1-\nu)} \frac{1}{r} \frac{\partial}{\partial \vartheta} [Z - (1+\nu) L(w, \Phi)],$$

where the Laplace operator

$$\Delta = \frac{\partial^2}{\partial r^2} + \frac{1}{r} \frac{\partial}{\partial r} + \frac{1}{r^2} \frac{\partial^2}{\partial \vartheta^2}.$$

The operator $L(w, \Phi)$ takes the form as given in § 1.11.3.

The stresses in the plate can be calculated by using Eqs. (1.85). We obtain the following relations by introducing a system of polar coordinates (r, ϑ):

$$\sigma_{rr} = -\frac{Ez}{1-\nu^2}\left[\frac{\partial^2 w}{\partial r^2}+\frac{\nu}{r}\frac{\partial w}{\partial r}+\frac{\nu}{r^2}\frac{\partial^2 w}{\partial \vartheta^2}\right]+$$
$$+\frac{1}{1-\nu}\left[\frac{\partial Q_r}{\partial r}+\frac{\nu}{r}\left(Q_r+\frac{\partial Q_\vartheta}{\partial \vartheta}\right)\right]\frac{z}{h}\left[3-\left(\frac{2z}{h}\right)^2\right]+$$
$$+\frac{N_{rr}}{h}+\frac{\nu}{1-\nu}\frac{Z}{2}\left[3\frac{z}{h}-4\left(\frac{z}{h}\right)^3\right],$$

$$\sigma_{\vartheta\vartheta} = -\frac{Ez}{1-\nu^2}\left[\nu\frac{\partial^2 w}{\partial r^2}+\frac{1}{r}\frac{\partial w}{\partial r}+\frac{\nu}{r^2}\frac{\partial^2 w}{\partial \vartheta^2}\right]+$$
$$+\frac{1}{1-\nu}\left[\nu\frac{\partial Q_r}{\partial r}+\frac{1}{r}\frac{\partial Q_\vartheta}{\partial \vartheta}+\frac{Q_r}{r}\right]\frac{z}{h}\left[3-\left(\frac{2z}{h}\right)^2\right]+$$
$$+\frac{N_{\vartheta\vartheta}}{h}+\frac{\nu}{1-\nu}\frac{Z}{2}\left[3\frac{z}{h}-4\left(\frac{z}{h}\right)^3\right],$$

$$\sigma_{r\vartheta} = -\frac{Ez}{1+\nu}\left[\frac{1}{r}\frac{\partial^2 w}{\partial r\,\partial \vartheta}-\frac{1}{r^2}\frac{\partial w}{\partial \vartheta}\right]+$$
$$+\frac{1}{2}\left(\frac{1}{r}\frac{\partial Q_r}{\partial \vartheta}+\frac{\partial Q_\vartheta}{\partial r}-\frac{Q_\vartheta}{r}\right)\frac{z}{h}\left[3-\left(\frac{2z}{h}\right)^2\right]+\frac{N_{r\vartheta}}{h}.$$

The stresses in the external surfaces of the plate can be obtained by introducing $z = \pm h/2$ into the formulae above:

$$\sigma_{rr}\left(\pm\frac{h}{2}\right) = \frac{N_{rr}}{h}\mp\frac{Eh}{2(1-\nu^2)}\left[\frac{\partial^2 w}{\partial r^2}+\frac{\nu}{r}\frac{\partial w}{\partial r}+\frac{\nu}{r^2}\frac{\partial^2 w}{\partial \vartheta^2}\right]\pm$$
$$\pm\frac{\partial Q_r}{\partial r}\mp\frac{\nu}{2(1-\nu)}(Z-2L(w,\Phi)),$$

$$\sigma_{\vartheta\vartheta}\left(\pm\frac{h}{2}\right) = \frac{N_{\vartheta\vartheta}}{h}\mp\frac{Eh}{2(1-\nu^2)}\left[\nu\frac{\partial^2 w}{\partial r^2}+\frac{\nu}{r}+\frac{1}{r^2}\frac{\partial^2 w}{\partial \vartheta^2}\right]\mp$$
$$\mp\frac{\partial Q_r}{\partial r}\mp\frac{1}{2(1-\nu)}[(2-\nu)Z-2L(w,\Phi)], \tag{2.7}$$

$$\sigma_{\vartheta r}\left(\pm\frac{h}{2}\right) = \frac{N_{r\vartheta}}{h}\mp\frac{Eh}{1+\nu}\left[\frac{1}{r}\frac{\partial^2 w}{\partial r\,\partial \vartheta}-\frac{1}{r^2}\frac{\partial w}{\partial \vartheta}\right]\pm$$
$$\pm\frac{1}{2}\left(\frac{1}{r}\frac{\partial Q_r}{\partial \vartheta}+\frac{\partial Q_\vartheta}{\partial r}-\frac{Q_\vartheta}{r}\right).$$

Representing the stresses in terms of bending moments, we find

$$\sigma_{rr}\left(\pm\frac{h}{2}\right) = \frac{N_{rr}}{h} \pm \frac{6M_{rr}}{h^2} \mp \frac{1}{5}\frac{\partial Q_r}{\partial r} \pm \frac{\nu}{10(1-\nu)}[Z-2L(w,\Phi)],$$

$$\sigma_{\vartheta\vartheta}\left(\pm\frac{h}{2}\right) = \frac{N_{\vartheta\vartheta}}{h} \pm \frac{6M_{\vartheta\vartheta}}{h^2} \pm \frac{1}{5}\frac{\partial Q_r}{\partial r} \pm \frac{1}{10(1-\nu)}[Z-2L(w,\Phi)],$$

$$\sigma_{r\vartheta}\left(\pm\frac{h}{2}\right) = \frac{N_{r\vartheta}}{h} \pm \frac{6M_{r\vartheta}}{h^2} \mp \frac{1}{5}\left(\frac{1}{r}\frac{\partial Q_r}{\partial \vartheta} + \frac{\partial Q_\vartheta}{\partial r} - \frac{Q_\vartheta}{r}\right).$$

In the case of a very thin plate, the resistance of the plate to bending can be neglected. Setting the flexural rigidity D equal to zero, we obtain Eqs. (2.8) determining the deflection of a flexible membrane.

$$\frac{1}{Eh}\Delta\Delta\Phi = \tfrac{1}{2}L(w,w), \qquad L(w,\Phi) = Z. \tag{2.8}$$

In the case of a plate with variable rigidity, i.e. $h = h(x, y)$ and therefore $D = D(x, y)$, Eqs. (2.1) take the following form:

$$\begin{aligned} &\Delta(D\Delta w) - (1-\nu)L(D, w) = Z - L(w, \Phi), \\ &\Delta\left(\frac{1}{Eh}\Delta\Phi\right) - (1+\nu)L\left(\frac{1}{Eh}, \Phi\right) = \tfrac{1}{2}L(w, w). \end{aligned} \tag{2.9}$$

The effect of the transverse shear deformations and normal stress σ_{zz} is here neglected. The formulae for internal forces in the plate of variable rigidity are the same as previously, only the expressions for Q_x and Q_y are different.

$$Q_x = -\frac{\partial}{\partial x}(D\Delta w),$$

$$Q_y = -\frac{\partial}{\partial y}(D\Delta w).$$

The above equations are obtained by the assumption that the thickness of the plate is small as compared with the other dimensions and changes slowly, therefore the state of stress in the plate can be considered as plane.

2.2 Linear equations

In a particular case where the plate is not subjected to the action of normal forces in its middle surface or where the normal forces in the plate's plane caused by the large displacements w can be neglected, Eqs. (2.1) simplify to

$$D\Delta\Delta w = q - \frac{2-\nu}{1-\nu}\frac{h^2}{10}\Delta q + \frac{\partial X_M}{\partial x} + \frac{\partial Y_M}{\partial y} \tag{2.10}$$

and

$$\begin{aligned} Q_x - \frac{h^2}{10}\Delta Q_x &= -D\frac{\partial \Delta w}{\partial x} - \frac{h^2}{10(1-\nu)}\frac{\partial q}{\partial x} + X_M, \\ Q_y - \frac{h^2}{10}\Delta Q_y &= -D\frac{\partial \Delta w}{\partial y} - \frac{h^2}{10(1-\nu)}\frac{\partial q}{\partial y} + Y_M, \end{aligned} \tag{2.11}$$

where the often used notation $q = Z$ is introduced.

The investigation of the plate reduces here to the solution of the three partial differential equations (2.10), (2.11) of the fourth and second order and to the determination of the functions w, Q_x, and Q_y for given boundary conditions. Having the functions w, Q_x and Q_y, we can determine all internal forces and stresses in the plate.

These equations were obtained in another way by M. Schäfer [2.6] in 1952. A system of three differential equations for the three functions determining the displacements was also found by H. Hencky [2.7] by a similar, as here, assumption that the points located before deformation on straight lines normal to the middle surface of the plate lie still on straight lines after deformation, but these are no more perpendicular to the deformed surface. He assumed simultaneously that the shear stresses τ_{xz} and τ_{yz} are uniformly distributed over the thickness of the plate.

The work by A. Kromm [2.8] in which a solution approaching E. Reissner theory was obtained is also worth mentioning. E. Reissner's solutions were generalized to the case of orthotropic plates by K. Girkmann and R. Beer [2.9] in 1958 and independently by J. Mossakowski [2.10] in 1959.

Also some dynamic solutions concerning either the propagation of elastic waves in isotropic plates [2.12], [2.13] or vibration of plates made of a cristal were obtained on the basis of E. Reissner's assumptions.

The general solution of Eqs. (2.10) and (2.11) can be found by the use of the method proposed by M. Schäfer [2.6]. As we know, the general solution of a differential equation can be represented in the form of two functions

$$w = w_0 + w_1 \tag{2.12}$$

where w_0 is the particular solution of Eq. (2.10) and w_1 is the general solution of the simplified equation

$$\Delta\Delta w_1 = 0. \tag{2.13}$$

We introduce now a new function Ψ and express the shear forces in the following manner:

$$Q_x = Q_{x0} - D\frac{\partial \Delta w_1}{\partial x} + \frac{\partial \Psi}{\partial y},$$
$$Q_y = Q_{y0} - D\frac{\partial \Delta w_1}{\partial y} - \frac{\partial \Psi}{\partial x}, \tag{2.14}$$

where Q_{x0} and Q_{y0} are the particular solutions of Eqs. (2.11). Differentiating Eq. $(2.14)_1$ with respect to x and Eq. $(2.14)_2$ with respect to y and adding them each to other, we have

$$\left(1 - \frac{h^2}{10}\Delta\right)\left(\frac{\partial Q_{x0}}{\partial x} + \frac{\partial Q_{y0}}{\partial y}\right) = -\left(1 - \frac{h^2}{10}\Delta\right)q, \tag{2.15}$$

which ensures the fulfilment of Eq. (2.1) with $L(w, \Phi) = 0$.

The substitution of Eqs. (2.12) to Eq. (2.10) by taking into account Eqs. (2.11) yields

$$\frac{\partial}{\partial y}\left(\Psi - \frac{h^2}{10}\Delta\Psi\right) = -\frac{\partial}{\partial x}\left(\Psi - \frac{h^2}{10}\Delta\Psi\right) = 0, \tag{2.16}$$

which gives

$$\Delta\Psi - \frac{10}{h^2}\Psi = 0.$$

The general solution of the problem consists in finding the functions w_0, w_1, Ψ satisfying Eqs. (2.10), (2.13), and (2.16) together with boundary conditions.

2.3 The strain energy of the plate

The expression for the strain energy of the plate can easily be obtained from the formulae given in § 1.14. Introducing rectangular Cartesian coordinates x, y, we have

$$\alpha_1 = x, \qquad \alpha_2 = y, \qquad A_1 A_2 \mathrm{d}\alpha_1 \mathrm{d}\alpha_2 = \mathrm{d}x\mathrm{d}y.$$

Relations (1.136) can be rewritten in the form using the relations

$$\varkappa_{xx} = \frac{\partial \beta_x}{\partial x}, \qquad \varkappa_{yy}\frac{\partial \beta_y}{\partial y},$$
$$2\varkappa_{xy} = \frac{\partial \beta_y}{\partial x} + \frac{\partial \beta_x}{\partial y}, \tag{2.17}$$

where β_x, β_y are the angles of the rotations of the lateral sides of the element of the plate.

In the case where we neglect the effects of the transverse shear and normal stresses we have

$$\beta_x = -\frac{\partial w}{\partial x}, \qquad \beta_y = -\frac{\partial w}{\partial y},$$

which gives the bending energy of the plate

$$U_2 = \tfrac{1}{2} D \iint_S \{(\Delta w)^2 - (1-\nu) L(w, w)\} \mathrm{d}x \mathrm{d}y. \tag{2.18}$$

The membrane energy

$$U_1 = \tfrac{1}{2} \iint_S (N_{xx}\varepsilon_{xx} + N_{yy}\varepsilon_{yy} + 2N_{xy}\varepsilon_{xy}) \mathrm{d}x \mathrm{d}y \tag{2.19}$$

can be presented in the following form if we take into account the relations:

$$2U_1 = \iint_S \left[N_{xx}\frac{\partial u}{\partial x} + N_{yy}\frac{\partial v}{\partial y} + N_{xy}\left(\frac{\partial u}{\partial y} + \frac{\partial v}{\partial x}\right)\right] \mathrm{d}x \mathrm{d}y +$$
$$+ \frac{1}{2} \iint_S \left[N_{xx}\left(\frac{\partial w}{\partial x}\right)^2 + N_{yy}\left(\frac{\partial w}{\partial y}\right)^2 + 2N_{xy}\frac{\partial w}{\partial x}\frac{\partial w}{\partial y}\right] \mathrm{d}x \mathrm{d}y. \tag{2.20}$$

It can be proved, integrating by parts, that the first integral (2.20) presents the work done by the edge forces acting in the middle plane of the plate

$$\iint_S \left[N_{xx}\frac{\partial u}{\partial x} + N_{yy}\frac{\partial v}{\partial y} + N_{xy}\left(\frac{\partial u}{\partial y} + \frac{\partial v}{\partial x}\right)\right] \mathrm{d}x \mathrm{d}y$$
$$= \oint_C [uN_{xx}l + vN_{yy}m + uN_{xy}m + vN_{xy}l]\, ds -$$
$$- \iint_S \left[u\left(\frac{\partial N_{xx}}{\partial x} + \frac{\partial N_{xy}}{\partial y}\right) + v\left(\frac{\partial N_{yy}}{\partial y} + \frac{\partial N_{xy}}{\partial x}\right)\right] \mathrm{d}x \mathrm{d}y, \tag{2.21}$$

where l and m denote the sine and cosine of the angle between the normal to the edge and the direction of x axis; $l = \cos(\mathbf{n}, x)$, $m = \sin(\mathbf{n}, x)$.

The first integral on the right-hand side of Eq. (2.21) is calculated along the edge of the plate. Of course, the edge forces $\bar{X}$ and $\bar{Y}$ are in equilibrium with the internal forces in immediate vicinity of the edge. The second integral vanishes if the volume forces X and Y are zero (see Eq. $(1.78)_1$). If these forces are not equal to zero, this integral represents the work of the volume forces done by the deformation of the plate.

2.4 Variational equations

The differential equations of the plate and the boundary conditions can be derived based on the equations of the minimum of the potential energy known from the calculus of variations. These equations and the natural boundary conditions can be obtained considering the first variation of the strain energy stored in the plate.

Let us consider only the bending of the plate. Then from (1.138) and (2.17) we have

$$\begin{aligned}\delta U = D\iint\Bigg\{&\left(\frac{\partial\beta_x}{\partial x}+\frac{\partial\beta_y}{\partial y}\right)\left(\frac{\partial\delta\beta_x}{\partial x}+\frac{\partial\delta\beta_y}{\partial y}\right)-\\ &-(1-\nu)\left[\frac{\partial\delta\beta_x}{\partial x}\frac{\partial\beta_y}{\partial y}+\frac{\partial\beta_x}{\partial x}\frac{\partial\delta\beta_y}{\partial y}\right)-\\ &-\left(\frac{\partial\beta_x}{\partial y}+\frac{\partial\beta_y}{\partial x}\right)\left(\frac{\partial\delta\beta_x}{\partial y}+\frac{\partial\delta\beta_y}{\partial x}\right)\Bigg]+\\ &+\frac{5(1-\nu)}{h^2}\left[\left(\beta_x+\frac{\partial w}{\partial x}\right)\left(\delta\beta_x+\frac{\partial\delta w}{\partial x}\right)+\left(\beta_y+\frac{\partial w}{\partial y}\right)\left(\delta\beta_y+\frac{\partial\delta w}{\partial y}\right)\right]+\\ &+\frac{3\nu(1+\nu)}{5Eh}q\left[\frac{\partial\delta\beta_x}{\partial x}+\frac{\partial\delta\beta_y}{\partial y}\right]\Bigg\}\mathrm{d}x\,\mathrm{d}y.\end{aligned} \tag{2.22}$$

Applying Green's theorem, we can represent the functional δU in the different form separating the variations $\delta\beta_i$ and δw. For example, we have

$$\iint_S \frac{\partial w}{\partial x}\frac{\partial\delta w}{\partial x}\,\mathrm{d}x\,\mathrm{d}y = -\iint_S \frac{\partial^2 w}{\partial x^2}\,\delta w\,\mathrm{d}x\,\mathrm{d}y+\oint_C \frac{\partial w}{\partial x}\,\delta w\cos(\mathbf{n},x)\,\mathrm{d}s,$$

$$\iint_S q\frac{\partial\delta\beta_x}{\partial x}\,\mathrm{d}x\,\mathrm{d}y = -\iint_S \frac{\partial q}{\partial x}\,\delta\beta_x\,\mathrm{d}x\,\mathrm{d}y+\oint_C q\,\delta\beta_x\cos(\mathbf{n},x)\,\mathrm{d}s.$$

We assumed here that the function w is continuous in the area S and continuous with its first derivatives at the edge C. Here cos $(\mathbf{n}, x)$ is the cosine of the angle between the normal to the edge and the direction of the x axis.

We obtain on transformations

$$\delta U = -\frac{5(1-\nu)D}{h^2}\iint\limits_S \left(\Delta w+\frac{\partial\beta_x}{\partial x}+\frac{\partial\beta_y}{\partial y}\right)\delta w\,\mathrm{d}x\,\mathrm{d}y+$$

$$+\frac{5(1-\nu)D}{h^2}\iint\limits_S \left\{\frac{\partial w}{\partial x}+\left[1-\frac{h^2}{5(1-\nu)}\frac{\partial}{\partial x^2}-\frac{h^2}{10}\frac{\partial}{\partial y^2}\right]\beta_x-\right.$$

$$\left.-\frac{h^2}{10}\frac{1+\nu}{1-\nu}\frac{\partial^2\beta_y}{\partial x\,\partial y}-\frac{6\nu(1+\nu)h}{25E(1-\nu)}\frac{\partial q}{\partial x}\right\}\delta\beta_x\,\mathrm{d}x\,\mathrm{d}y+$$

$$+\frac{5(1-\nu)D}{h^2}\iint\limits_S \left\{\frac{\partial w}{\partial y}-\frac{h^2}{10}\frac{1+\nu}{1-\nu}\frac{\partial^2\beta_x}{\partial x\,\partial y}+\right.$$

$$\left.+\left[1-\frac{h^2}{5(1-\nu)}\frac{\partial}{\partial y^2}-\frac{h^2}{10}\frac{\partial}{\partial x^2}\right]\beta_y-\frac{6\nu(1+\nu)}{25E(1-\nu)}\frac{h\partial q}{\partial y}\right\}\delta\beta_y\,\mathrm{d}x\,\mathrm{d}y+$$

$$+\frac{5(1-\nu)D}{h^2}\int\limits_C \left[\left(\beta_x+\frac{\partial w}{\partial x}\right)\cos(\mathbf{n},x)+\left(\beta_y+\frac{\partial w}{\partial y}\right)\cos(\mathbf{n},y)\right]\delta w\,\mathrm{d}s+$$

$$+D\int\limits_C \left[\left(\frac{\partial\beta_x}{\partial x}+\nu\frac{\partial\beta_y}{\partial y}\right)\cos(\mathbf{n},x)+\right.$$

$$\left.+\frac{1-\nu}{2}\left(\frac{\partial\beta_y}{\partial x}+\frac{\partial\beta_x}{\partial y}\right)\cos(\mathbf{n},y)\right]\delta\beta_x\,\mathrm{d}s+$$

$$+D\int\limits_C \left[\frac{1-\nu}{2}\left(\frac{\partial\beta_x}{\partial y}+\frac{\partial\beta_y}{\partial x}\right)\cos(\mathbf{n},x)+\right.$$

$$\left.+\left(\frac{\partial\beta_y}{\partial y}+\nu\frac{\partial\beta_x}{\partial x}\right)\cos(\mathbf{n},y)\right]\delta\beta_y\,\mathrm{d}s+$$

$$+\frac{6\nu(1+\nu)D}{5Eh}\int\limits_C q[\delta\beta_x\cos(\mathbf{n},x)+\delta\beta_y\cos(\mathbf{n},y)]\,\mathrm{d}s. \tag{2.23}$$

The work of the load on the assumed variations of the deformations δw, $\delta\beta_x$, $\delta\beta_y$ is

$$\delta L = \iint\limits_S (q\delta w+X_M\,\delta\beta_x+Y_M\,\delta\beta_y)\,\mathrm{d}S. \tag{2.24}$$

The Lagrange equation for the considered variational problem has the form

$$\delta W = \delta U-\delta L = 0. \tag{2.25}$$

Equating to zero the expressions by the variations δw, $\delta\beta_x$, $\delta\beta_y$, we obtain the set of three differential equations for three unknown functions w, β_x, β_y

$$
\begin{aligned}
&\Delta w+\frac{\partial\beta_x}{\partial x}+\frac{\partial\beta_y}{\partial y}=-\frac{6q}{5hG},\\
&\frac{\partial w}{\partial x}+\left[1-\frac{h^2}{5(1-\nu)}\frac{\partial^2}{\partial x^2}-\frac{h^2}{10}\frac{\partial^2}{\partial y^2}\right]\beta_x-\frac{h^2}{10}\frac{1+\nu}{1-\nu}\frac{\partial^2\beta_y}{\partial x\,\partial y}\\
&\quad=\frac{6\nu(1+\nu)h}{25E(1-\nu)}\frac{\partial q}{\partial x}-\frac{6}{5hG}X_M,\\
&\frac{\partial w}{\partial y}+\left[1-\frac{h^2}{5(1-\nu)}\frac{\partial^2}{\partial y^2}-\frac{h^2}{10}\frac{\partial^2}{\partial x^2}\right]\beta_y-\frac{h^2}{10}\frac{1+\nu}{1-\nu}\frac{\partial^2\beta_x}{\partial x\,\partial y}\\
&\quad=\frac{6\nu(1+\nu)h}{25E(1-\nu)}\frac{\partial q}{\partial y}-\frac{6}{5hG}Y_m.
\end{aligned}
\tag{2.26}
$$

The effect of the external bending moments X_M, Y_M is taken into account.

The identical set of equations can be obtained from the equilibrium equations $(1.79)_{3,4}$ expressing the internal forces by the displacement w and the angles β_i. Introducing in Eq. (2.26) $\beta_i=-\dfrac{\partial w}{\partial x_i}+\dfrac{6}{5}\dfrac{Q_i}{Gh}$, we obtain Eqs. $(2.1)_1$ and (2.3).

2.5 Boundary conditions for plates

If we consider a plate of small deflections under the lateral loading perpendicular to its surface, neglecting the effect of the stress σ_{zz}, the membrane forces are zero. Then the boundary conditions concern only the forces, moments and displacements of the bending state of the plate.

The boundary conditions can be derived in a more consistent way basing on the variation of the functional (2.23). Let us take into consideration the relations

$$
\begin{aligned}
\beta_x &= \beta_n\cos(\mathbf{n},x)-\beta_t\cos(\mathbf{n},y),\\
\beta_y &= \beta_n\cos(\mathbf{n},y)+\beta_t\cos(\mathbf{n},x),
\end{aligned}
$$

and similar formulae for the edge moments and shear forces

$$
\begin{aligned}
M_{nn} &= M_{xx}\cos^2(\mathbf{n},x)+M_{yy}\cos^2(\mathbf{n},y)+2M_{xy}\cos(\mathbf{n},x)\cos(\mathbf{n},y),\\
M_{nt} &= (M_{yy}-M_{xx})\cos(\mathbf{n},x)\cos(\mathbf{n},y)+M_{xy}[\cos^2(\mathbf{n},x)-\cos^2(\mathbf{n},y)],\\
Q_n &= Q_x\cos(\mathbf{n},x)+Q_y\cos(\mathbf{n},y).
\end{aligned}
$$

Introducing it into formula (2.23) we find that the integrals along the edge C define so-called *natural boundary conditions*. The condition that the integrals along the edge C are equal to zero demands that the integral:

$$\oint_C (Q_n \delta w + M_{nn} \delta\beta_n + M_{nt} \delta\beta_t) ds = 0$$

is equal to zero. Then we find the following homogeneous boundary conditions:

$$\begin{aligned}
&1.\quad Q_n = M_{nn} = M_{nt} = 0,\\
&2.\quad Q_n = M_{nn} = \beta_t = 0,\\
&3.\quad w = M_{nn} = M_{nt} = 0,\\
&4.\quad w = M_{nn} = \beta_t = 0,\\
&5.\quad w = \beta_n = M_{nt} = 0,\\
&6.\quad w = \beta_n = \beta_t = 0,\\
&7.\quad Q_n = \beta_n = M_{nt} = 0,\\
&8.\quad Q_n = \beta_n = \beta_t = 0.
\end{aligned} \tag{2.27}$$

Conditions 1, 2 can be interpreted as the condition for the free edge. Conditions 3 and 4 correspond to the simply supported edge, 5 and 6 correspond to the clamped edge. The above conditions do not concern the forces and displacements of the membrane state of the plate.

We define the boundary conditions for plate of large deflection in the same way as for shells.

2.6 Classical theory of plates under lateral loads

If we simplify our theory by rejecting the non-linear terms and effect of transverse shear deformations, consisting in neglecting the expressions proportional to h^2 from (2.1), (2.2), and (2.3), we obtain the equations of the classical theory of plates which are valid only when displacements of the plate are small in comparison with the thickness.

The solution of the plate problem is limited then to determining of the deflection satisfying a simple equation

$$D\Delta\Delta w = q \tag{2.28}$$

together with the boundary conditions.

Knowing the deflection w, we can determine all internal forces and stresses from the following expressions:

Bending moments

$$M_{xx} = -D\left[\frac{\partial^2 w}{\partial x^2} + \nu\frac{\partial^2 w}{\partial y^2}\right],$$

$$M_{yy} = -D\left[\frac{\partial^2 w}{\partial y^2} + \nu\frac{\partial^2 w}{\partial x^2}\right], \tag{2.29}$$

$$M_{xy} = -D[1-\nu]\frac{\partial^2 w}{\partial x\,\partial y}.$$

Shear forces

$$Q_x = -D\frac{\partial \Delta w}{\partial x}; \qquad Q_y = -D\frac{\partial \Delta w}{\partial y}. \tag{2.30}$$

The stresses caused by bending in the external layer of the plate can be obtained on the basis of the moments

$$\sigma_{ij} = \frac{6M_{ij}}{h^2}, \qquad i, j = x, y.$$

In the cases where it is convenient to solve the plate in the system of polar coordinates r, ϑ the expressions (2.29), (2.30) can be transformed to

$$M_{rr} = -D\left[\frac{\partial^2 w}{\partial r^2} + \nu\left(\frac{1}{r}\frac{\partial w}{\partial r} + \frac{1}{r^2}\frac{\partial^2 w}{\partial \vartheta^2}\right)\right],$$

$$M_{\vartheta\vartheta} = -D\left[\frac{1}{r}\frac{\partial w}{\partial r} + \frac{1}{r^2}\frac{\partial^2 w}{\partial \vartheta^2} + \nu\frac{\partial^2 w}{\partial r^2}\right], \tag{2.31}$$

$$M_{r\vartheta} = -(1-\nu)D\left[\frac{1}{r}\frac{\partial^2 w}{\partial r\,\partial \vartheta} - \frac{\partial w}{r^2\partial \vartheta}\right].$$

Shear forces

$$Q_r = -D\frac{\partial}{\partial r}\Delta w, \qquad Q_\vartheta = -D\frac{\partial w}{r\partial \vartheta}\Delta w.$$

If the loading is distributed symmetrically with respect to the axis of the symmetry crossing the centre of the plate, the deflected surface is also symmetrical and the terms in expressions (2.31) containing the derivatives with respect to ϑ vanish.

The deflection and bending moments in the plate have to satisfy the boundary conditions. However, in this case Eq. (2.3) gives only two arbitrary functions for one edge and therefore enables the satisfaction of only two boundary conditions. This imperfection is the result of the assumed simplification in the derivations of the theory. In order to have a sufficient number of boundary conditions, the following treatment can be performed. The conditions in displacements can be expressed by the deflection w and angle of rotation $\partial w/\partial x$. For example, for the built-in edge we have

$$w = 0, \quad \frac{\partial w}{\partial x} = 0.$$

When the boundary conditions are given in terms of the internal forces, we introduce the equivalent shear force, summing the effect of the twisting moments M_{xy} and shear forces along the edge. Instead of the conditions requiring that the magnitudes of the twisting moment and shear force at various points of the edge are equal to an external moment and force, we demand this agreement to take place for statically equivalent, the so-called *effective shear force*

$$\bar{Q}_x = Q_x + \frac{\partial M_{xy}}{\partial y}.$$

Substituting Q_x and M_{xy} from Eqs. (2.34) and (2.35), we finally obtain for free edge

$$\bar{Q}_x = -D\left[\frac{\partial^3 w}{\partial x^3} + (2-\nu)\frac{\partial^3 w}{\partial x\,\partial y^2}\right] = 0 \qquad (2.32)$$

and

$$M_{xx} = -D\left[\frac{\partial^2 w}{\partial x^2} + \nu\frac{\partial^2 w}{\partial y^2}\right] = 0.$$

For the simply supported edge we have

$$w = 0, \; M_{xx} = 0 \quad \text{or} \quad w = 0, \; \frac{\partial^2 w}{\partial x^2} = 0.$$

2.7 Sandwich plates

The sandwich plates are widly used nowadays in many branches of the technique as ship and airplanes building, civil engineering, etc. These plates are usually build of two external layers made of the material of high strength, which are joined together with the core made of the weaker and much lighter material. The modulus of elasticity of the external layers surpasses to a high extent that of the jointing material which allows some simplifications in the mathematical description of such a structure. The theory and methods of the solution of the problems of the sandwich plates were the subject of papers by many authors ([2.15]–[2.18]). In this work we shall present the theory in its classical, linear form given by N. J. Hoff [2.14].

It has been assumed that the material of the core is infinitely rigid in the direction perpendicular to the middle plane and very flexible in the tangential directions. Then $E_z = \infty$, $E_x = E_y = 0$ and the shear modulus $G_{xy} = 0$. As the result of the above assumptions there appear only uniformly distributed shear stresses τ^c_{xz} and τ^c_{yz} in the core. It has also been assumed that external layers of the plate are equal and fulfil all assumptions of the theory of thin isotropic plates and sheets. It also results from the above assumptions that the state of stress and displacement of such a sandwich plate are antisymmetrical in respect to its middle plane.

Let us describe the behaviour of the plate by means of three displacements u, v, w, where u and v are displacements of the upper layer in the directions of x and y axis, respectively, $w(x, y)$ is the normal displacement equal for all layers. The displacements of the lower layer are $-u(x, y)$ and $-v(x, y)$, respectively.

The assumed scheme of the deformations of lateral cross-section of the plate is presented in Fig. 11b. The diagrams of the normal and tangential stresses are presented in Fig. 11a where the components and resultants of the internal forces are defined. Using Hooke's law, we write the following equations for the membrane forces in the upper layer:

$$
\begin{aligned}
N_{xx} &= \frac{E\delta}{1-\nu^2}\left(\frac{\partial u}{\partial x}+\nu\frac{\partial v}{\partial y}\right),\\
N_{yy} &= \frac{E\delta}{1-\nu^2}\left(\frac{\partial v}{\partial y}+\nu\frac{\partial u}{\partial x}\right),\\
N_{xy} &= \frac{E\delta}{2(1+\nu)}\left(\frac{\partial u}{\partial y}\quad\frac{\partial v}{\partial x}\right).
\end{aligned}
\tag{2.33}
$$

If the displacement of the external layer is u, the corresponding shear strain γ^c_{xz} of the core can be defined from the Fig. 11b; we find

$$
\gamma^c_{xz} = \frac{u}{h}+\frac{2h+\delta}{2h}\frac{\partial w}{\partial x}. \tag{2.34}
$$

Then the shear stresses in the core are

$$\tau_{\alpha z} = G_c \gamma_{\alpha z}^c = G_c \left[\frac{u_\alpha}{h} + \frac{2h+\delta}{2h} w_{,\alpha} \right], \quad \alpha = x, y, \tag{2.35}$$

where G_c is the shear modulus of the core.

Integrating along the thickness, we find

$$Q_\alpha = G_c(2h+\delta)\left[\frac{u_\alpha}{h} + \frac{2h+\delta}{2h} w_{,\alpha} \right]. \tag{2.36}$$

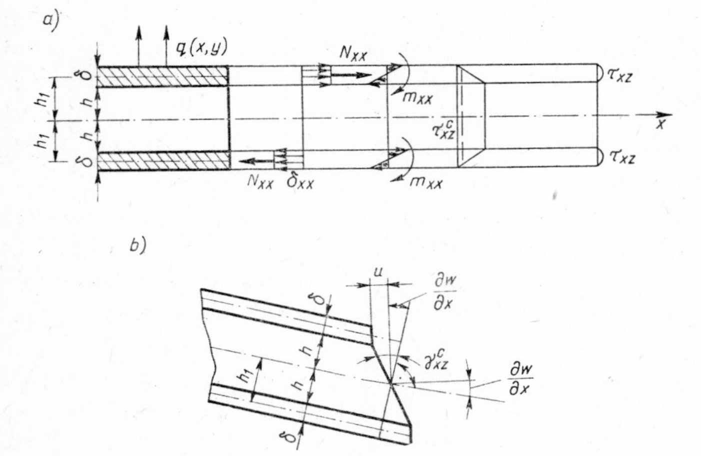

Fig. 11a). Stresses and stress resultants in a sandwich plate.
Fig. 11b). Displacements in a sandwich plate.

The bending moments in the external layers are denoted by small letters m_{ij} to differentiate them from resultant bending moments acting in the whole cross-section of the plate.

We have

$$m_{\alpha\beta} = -D[\nu\delta_\beta^\alpha w_{,\delta\delta} + (1-\nu) w_{,\alpha\beta}]. \tag{2.37}$$

The shear forces in the external layers are defined also by small letters q_α. We have

$$q_\alpha = -(D\Delta w)_{,\alpha},$$

where $D = E\delta^3/12(1-\nu^2)$ is rigidity of the external layers, E and ν are the modulus of elasticity and Poisson ratio of the external layers of the plate.

If the equations of equilibrium of the element of the upper sheet and the equation of equilibrium of the element cut out from the whole sandwich plate are expressed by the displacements u, v, w, we obtain the following set of differential equations derived by Hoff [2.14]:

$$\begin{aligned} L_{11}w+L_{12}u+L_{13}v &= \frac{1-\nu^2}{2E\delta}q, \\ L_{21}w+L_{22}u+L_{23}v &= \frac{1-\nu^2}{E\delta}n_x, \\ L_{31}w+L_{32}u+L_{33}v &= \frac{1-\nu^2}{E\delta}n_y, \end{aligned} \tag{2.38}$$

where corresponding differential operators L_{ij} are defined by the following formulae:

$$\begin{aligned} L_{11} &= \frac{D(1-\nu^2)}{E\delta}\Delta\Delta - \frac{G_c(2h+\delta)^2(1-\nu^2)}{4E\delta h}\Delta, \\ L_{12} = L_{21} &= -\frac{G_c(2h+\delta)(1-\nu^2)}{2E\delta h}\frac{\partial}{\partial x}, \\ L_{13} = L_{31} &= -\frac{G_c(2h+\delta)(1-\nu^2)}{2E\delta h}\frac{\partial}{\partial y}, \\ L_{22} &= \frac{\partial^2}{\partial x^2}+\frac{1-\nu}{2}\frac{\partial^2}{\partial y^2}-\frac{G_c(1-\nu^2)}{E\delta h}, \\ L_{23} = L_{32} &= \frac{1+\nu}{2}\frac{\partial^2}{\partial x\,\partial y}, \\ L_{33} &= \frac{\partial^2}{\partial y^2}+\frac{1-\nu}{2}\frac{\partial^2}{\partial x^2}-\frac{G_c(1-\nu^2)}{E\delta h}. \end{aligned} \tag{2.39}$$

The effect of the antisymmetric loadings n_x and n_y is taken into account in Eqs. (2.38). These loadings have the form of normal forces, uniformly distributed along the thickness of the external layers. It corresponds to the action of the bending moments X_M and Y_M where

$$X_M = 2n_x\left(h+\frac{\delta}{2}\right), \qquad Y_M = 2n_y\left(h+\frac{\delta}{2}\right).$$

The set of differential equations can be reduced to one equation for the displacement function F. This function is defined by the following operator determinant:

$$\begin{vmatrix} L_{11} & L_{12} & L_{13} \\ L_{21} & L_{22} & L_{23} \\ L_{31} & L_{32} & L_{33} \end{vmatrix} F_i = \frac{1-\nu^2}{2E}q_i, \qquad i = 1, 2, 3, \tag{2.40}$$

which gives, on transformations,

$$\left[1-\frac{1}{2}\varkappa_c(1-\nu)\Delta\right]\left[1-\varkappa_c\frac{2D}{D_z}\Delta\right]\Delta\Delta F_i = \frac{q_i}{D_z}, \tag{2.41}$$

where

$$q_1 = q(x, y), \quad q_2 = 2n_x(x, y), \quad q_3 = 2n_y(x, y),$$

and

$$D_z = \frac{E\delta(2h+\delta)^2}{2(1-\nu^2)} + 2D$$

is the rigidity of the sandwich plate.

The coefficient $\varkappa_c = Eh\delta/G_c \cdot (1-\nu^2)$ is the coefficient of the flexibility of the core. The displacements are expressed by the displacement functions F_i by means of the following determinants:

$$\begin{aligned}
w &= \begin{vmatrix} L_{22} & L_{23} \\ L_{32} & L_{33} \end{vmatrix} F_1 - \begin{vmatrix} L_{12} & L_{13} \\ L_{32} & L_{33} \end{vmatrix} F_2 + \begin{vmatrix} L_{12} & L_{13} \\ L_{22} & L_{23} \end{vmatrix} F_3, \\
u &= -\begin{vmatrix} L_{21} & L_{23} \\ L_{31} & L_{33} \end{vmatrix} F_1 + \begin{vmatrix} L_{11} & L_{13} \\ L_{31} & L_{33} \end{vmatrix} F_2 - \begin{vmatrix} L_{11} & L_{13} \\ L_{21} & L_{23} \end{vmatrix} F_3, \\
v &= \begin{vmatrix} L_{21} & L_{23} \\ L_{31} & L_{33} \end{vmatrix} F_1 - \begin{vmatrix} L_{11} & L_{12} \\ L_{31} & L_{32} \end{vmatrix} F_2 + \begin{vmatrix} L_{11} & L_{12} \\ L_{21} & L_{22} \end{vmatrix} F_3,
\end{aligned} \tag{2.42}$$

where F are the solutions of Eqs. (2.41).

Developing the determinants, we obtain the displacements given by the formulae

$$\begin{aligned}
w = {} & (1-\varkappa_c\Delta)\left[1-\frac{1-\nu}{2}\varkappa_c\Delta\right]F_1 - \frac{2h+\delta}{2}\frac{\partial}{\partial x}\left[1-\frac{1-\nu}{2}\varkappa_c\Delta\right]F_2 - \\
& -\frac{2h+\delta}{2}\frac{\partial}{\partial y}\left[1-\frac{1-\nu}{2}\varkappa_c\Delta\right]F_3,
\end{aligned} \tag{2.43$_1$}$$

$$\begin{aligned}
u = {} & -\frac{2h+\delta}{2}\frac{\partial}{\partial x}\left[1-\frac{1-\nu}{2}\varkappa_c\Delta\right]F_1 + \\
& +\frac{(2h+\delta)^2}{4}\left\{\Delta\left[1-\frac{1-\nu}{2}\varkappa_c\Delta\right]-\frac{\partial^2}{\partial y^2}\left[1+\frac{1-\nu}{2}\varkappa_c\Delta\right]\right\}F_2 + \\
& +\frac{(2h+\delta)^2}{4}\frac{\partial^2}{\partial x\,\partial y}\left[1+\frac{1-\nu}{2}\varkappa_c\Delta\right]F_3 -
\end{aligned}$$

$$-\frac{Dh}{G_c}\Delta\Delta\left[1-\varkappa_c\frac{\partial^2}{\partial y^2}-\frac{1-\nu}{2}\varkappa_c\frac{\partial^2}{\partial x^2}\right]F_2-$$

$$-\frac{Dh}{G_c}\frac{1+\nu}{2}\varkappa_c\frac{\partial^2}{\partial x\,\partial y}\Delta\Delta F_3, \tag{2.43$_2$}$$

$$v=-\frac{2h+\delta}{2}\frac{\partial}{\partial y}\left[1-\frac{1-\nu}{2}\varkappa_c\Delta\right]F_1+$$

$$+\frac{(2h+\delta)^2}{4}\frac{\partial^2}{\partial x\,\partial y}\left[1+\frac{1-\nu}{2}\varkappa_c\Delta\right]F_2+$$

$$+\frac{(2h+\delta)^2}{4}\left\{\Delta\left[1-\frac{1-\nu}{2}\varkappa_c\Delta\right]-\frac{\partial^2}{\partial x^2}\left[1+\frac{1-\nu}{2}\varkappa_c\Delta\right]\right\}F_3-$$

$$-\frac{Dh}{G_c}\frac{1+\nu}{2}\varkappa_c\frac{\partial^2}{\partial x\,\partial y}\Delta\Delta F_2-\frac{Dh}{G_c}\Delta\Delta\left[1-\varkappa_c\frac{\partial^2}{\partial x^2}-\frac{1-\nu}{2}\varkappa_c\frac{\partial^2}{\partial x^2}\right]F_3. \tag{2.43$_3$}$$

The above set of equations enables us to solve the problem of an arbitrarily loaded sandwich plate.

If the plate is loaded only by the load perpendicular to its surface, the function F_1 is a solution of the non-homogeneous equation, and F_2 and F_3 are solutions of the homogeneous equations.

Let us consider such a case and denote by u_1, v_1, w_1 the solution corresponding to the function F_1. We obtain

$$D_z\left[1-\frac{1-\nu}{2}\varkappa_c\Delta\right]\left[1-\varkappa\frac{2D}{D_z}\Delta\right]\Delta\Delta F_1=q,$$

$$w_1=(1-\varkappa_c\Delta)\left[1-\frac{1-\nu}{2}\varkappa_c\Delta\right]F_1,$$

$$u_1=-\frac{2h+\delta}{2}\frac{\partial}{\partial x}\left[1-\frac{1-\nu}{2}\varkappa_c\Delta\right]F_1, \tag{2.44}$$

$$v_1=-\frac{2h+\delta}{2}\frac{\partial}{\partial y}\left[1-\frac{1-\nu}{2}\varkappa_c\Delta\right]F_1.$$

It can easily be observed that the operator $\left[1-\frac{1-\nu}{2}\varkappa_c\Delta\right]F_1=U$ appears in all above equations. Thus we can lower the range of the differential equation. We obtain on transformation

$$D_z\left(1-\varkappa_c\frac{2D}{D_z}\Delta\right)\Delta\Delta U = q, \tag{2.45}$$

$$w_1^{(1)} = (1-\varkappa_c\Delta)\,U,$$

$$u_1^{(1)} = -\frac{2h+\delta}{2}\frac{\partial U}{\partial x},$$

$$v_1^{(1)} = -\frac{2h+\delta}{2}\frac{\partial U}{\partial y}.$$

Of course, the solution of Eq. (2.45) does not contain all solutions of Eq. (2.44) because the integral of the equation

$$\left[1-\frac{1-\nu}{2}\varkappa_c\Delta\right]F_1 = 0$$

has not been taken yet into the consideration.

The solution of the above equation can be defined as $F_1 = \theta$. This integral satisfies also the solutions of Eq. (2.44). We obtain the complete solution of the problem by adding to the previous solution $w_1^{(1)}$ the solution $w_1^{(2)}$ defined by the equation

$$\left[1-\frac{1-\nu}{2}\varkappa_c\Delta\right]\theta = 0. \tag{2.46}$$

The displacements $w_1^{(2)}$, $u_1^{(2)}$ and $v_1^{(2)}$ are expressed by the function θ in the following way:

$$w_1^{(2)} = 0,\quad u_1^{(2)} = -\frac{2h+\delta}{2}\frac{\partial\theta}{\partial y},\quad v_1^{(2)} = -\frac{2h+\delta}{2}\frac{\partial\theta}{\partial x}.$$

It can easily be proved that the homogeneous equations are also satisfied. If the shear modulus of elasticity G_c increases to infinity, according to the formulae for $\varkappa$, $\varkappa \to 0$ and θ tends to zero. This function is then the quantity characteristic of the sandwich plate.

The equations of the sandwich plate take a more simple form if we assume a more simplified model of the plate. Namely, if we assume that the external layers carry only membrane stresses in their planes and have no bending rigidity. This assumption can be applied to the plates with thin external layers.

The bending rigidity of a plate is

$$D_z = \frac{2Eh_1^2\,\delta}{1-\nu^2}$$

where ν, E are the moduli of elasticity of the external layers, and δ is their thickness.

The angle of the rotation of the lateral cross-section of such plate is

$$\beta_\alpha = \frac{u_\alpha}{h_1} = -w_{,\alpha} + \frac{Q_\alpha^c}{2h_1 G_c}.$$

These formulae are analogous to the previously developed (1.83). The bending moments in the plate are defined by relations (2.29).

Using the equilibrium equations, we find the equations for the shear forces

$$Q_{,\alpha} - (1-\nu)\frac{\eta}{2}\Delta Q_{,\alpha} = -D[\Delta w]_{,\alpha} - \eta q_{,\alpha} + X_M, \qquad \alpha = x, y. \tag{2.47}$$

The equation of equilibrium of the forces in the normal direction takes the form

$$D_Z \Delta\Delta w = [1-\eta\Delta]q + \frac{\partial X_M}{\partial x} + \frac{\partial Y_M}{\partial y}, \tag{2.48}$$

where

$$\eta = \frac{E\delta h_1}{G_c(1-\nu^2)}.$$

The above-developed set of equations is identical with that for isotropic plate (2.1) if we replace $h^2/5(1-\nu)$ by η and neglect the effect of transverse normal stresses. This effect is expressed in $(2.1)_1$ by the coefficient ν appearing in the numerator in the expression $h^2(2-\nu)q/10(1-\nu)$, and by $\nu\Delta Z/E$ in $(2.1)_2$. Replacing the first expression by $h^2 q/5(1-\nu)$ and neglecting the second, we obtain the equations for homogeneous plates with the effect of transverse shear deformations. The bending moments in such a sandwich plate are defined by the identical relations as those in the classical theory of isotropic plates in which the effects of transverse shear stresses are disregarded.

We can conclude from the above considerations that the sandwich plates satisfying the conditions that the bending rigidity of the external layers is negligible are the particular case of the Reissner plates. Thus the solutions for such plates are also valid for sandwich plates.

The numerical results can be obtained by introducing the coefficients characterizing the rigidity of the plate and neglecting certain terms resulting from the effect of the transverse normal stress σ_{33}.

In the above equations, the non-linear terms were neglected. The governing differential equations of moderate large-deflection theory for sandwich plates of the simplified model, subjected to the lateral load q and distributed bending moments X_M, Y_M, are

$$D_Z \Delta\Delta w = (1-\eta\Delta)[q-L(w,\Phi)]+\frac{\partial X_M}{\partial x}+\frac{\partial Y_M}{\partial y},$$

$$\frac{1}{Eh}\Delta\Delta\Phi = \frac{1}{2}L(w,w),$$

where

$$\eta = \frac{E\delta h_1}{G_c(1-\nu^2)}, \qquad h = 2\delta.$$

The solution of sandwich plate problems can be reduced to those for corresponding homogeneous plates. Consequently, all methods described in the following sections are applicable to layered plates.

References 2

[2.1] A. Nadai: *Elastische Platten*, Springer, Berlin, 1925.

[2.2] J. M. Michell: *Proc. Lond. Math. Soc. 34* (1902) 223.

[2.3] Th. v. Kármán: *Encyklopedie der Mathematischen Wissenchaften* IV (1910) 349.

[2.4] E. Reissner: On the theory of bending of elastic plates, *J. Math. Phys. 23* (1944) 184–191.

[2.5] E. Reissner: The effect of transverse shear deformation on the bending of elastic plates, *J. Appl. Mech. 12, A68–A77* (1945).

[2.6] M. Schäfer: Über eine Verfeinerung der klasischen Theorie dünner schwach gebogener Platten, *ZAMM 32* (1952) 161.

[2.7] H. Hencky: Über die Berücksichtigung der Schubverzerrung in ebenen Platten, *Ing.-Arch. 16* (1947) 42–76.

[2.8] A. Kromm: Verallgemeinte Theorie der Plattenstatik, *Ing.-Arch. 21* (1953) 266; *35* (1955) 231.

[2.9] K. Girkmann, R. Beer: Anwendung der verschärften Plattentheorie nach Erick Reissner auf orthotrope Platten, *Öster. Ing. Arch. 12* (1958) 101–110.

[2.10] J. Mossakowski: *Equations of Reissner's theory of orthotropic plates*, in: *Jubilee Volume of M. Wierzbicki*, PWN, Warszawa (1959) 145–155 (in Polish).

[2.11] E. Reissner: On bending of elastic plates, *Quart. Appl. Math. 5* (1947) 55.

[2.12] R. Gran Olson: *Ingr. Arch. 5* (1934) 363.

[2.13] Z. Kączkowski: Statics of non-homogenous plates and discs, non-homogenity in elasticity and plasticity, *Proc. IUTAM Symposium*, London 1958, pp. 77–82.

[2.14] N. J. Hoff: Bending and buckling of rectangular sandwich plates, *NACA T. N. No. 2225*, November 1950.

[2.15] A. Aleksandrov, L. Briuker, L. Kurtin, A. Prusakov: *Calculattions of sandwich plates*, Moskva 1960 (in Russian).

[2.16] C. Libove, S. B. Batdorf: A general small deflection theory for flat sandwich plates, *NACA No 899*, April 1948.

[2.17] S. A. Ambartsumian: *Theory of anizotropic plates*, Fizmatizdat., Moskva 1967 (in Russian).

[2.18] I. Solvey: Bibliography and summaries of sandwich construction (1939–1954), *Res. Lab. Melbourne Austral., ARL/SM, 2* (1955).

[2.19] Z. Kączkowski: *Plates, static calculations*, Arkady, Warszawa 1968 (in Polish).

[2.20] R. Ganowicz: Selected problems of Reissner's theory of plates and sandwich plates, *Mech. Teoret. Stosow. 3* (1966) 4 (in Polish).

[2.21] Z. Kączkowski: Der Einfluss der Schubverzerrungen und des Drehbeharrungsvermögens auf die Schwingungsfrequenz von anizotropen Platten, *Bull. Acad. Pol. Sci. Série Sci. Techn. 7* (1960) 8.

[2.22] D. V. Peshtamaldjan: On bending of orthotropic plates, *Dokl. Akad. Arm. SSR, Mekhanika 32, 1* (1961) 17–122 (in Russian).

[2.23] J. B. Alblas: *Theorie van de Driedimensionale Spanningstoestand in een Doorboorde Plaat*, Dissertation, Delft 1957.

[2.24] V. Panc: *Theories of elastic plates*, Noordhoff I.P., Leyden 1975.

3

Plates under lateral loads at an interior point

3.1 Methods of solution

The most evident way of defining the state of stress and displacement in plates and shells seems to be to solve the problem for a load distributed over a finite small surface of the structure. Next, calculating the limiting values when this surface tends to zero and keeping a constant value of the resultant of this load, we obtain the solution for the case of a concentrated load acting on one point of the structure. This method is convenient in the axi-symmetrical cases, in other cases it is not usually applied because of certain difficulties met by the solution of the equations of continuity at the border of the loaded and unloaded regions.

A second method often applied consists of the presentation of the discontinuous load as the sum of certain continuous forces acting on the whole area of the structure. In applying this method we express the load by means of Fourier series. However, this treatment is not very convenient in the case of concentrated loads because of slow convergence of the series defining the internal forces and moments in the structure. In the close vicinity of the point of application of the load the series are very slowly convergent. This requires the calculation of a very large number of terms and produces errors in the results. Due to this we seek at a closed solution in the case of concentrated loads. These solutions can be found more easily by applying the method of the Fourier integral or the infinite Fourier transforms.

In the mathematical analysis it is convenient to present the concentrated load by the Dirac delta function $\delta(x)$ defined by

$$\delta(x) = 0 \text{ if } x \neq 0 \quad \text{and} \quad \delta(0) = \infty .$$

While representing the concentrated load P acting at the point $x = \xi$ we write

$$q(x, \xi) = P\delta(x-\xi).$$

The delta function is not a "true" function since the definition above implies the inconsistent relations

$$\delta(x) = 0 \text{ if } x \neq 0; \qquad \int_{-\infty}^{\infty} \delta(x)\mathrm{d}x = 1.$$

The formal application of the $\delta(x)$ function provides a convenient notation permitting generalization of many mathematical relations. The common use of Dirac's function $\delta(x)$ and other non-functional quantities in physics and technical disciplines becomes the reason for a certain kind of computational formalism which found its realization in the theory of the generalized functions, called the *distributions*. The definition of the operations, such as the rules of differentiation and integration of such functions is particularly important. The ordinary derivative of $\delta(x)$ is equal to zero for any $x \neq 0$ and is undefined at the point $x = 0$. The theory of distributions gives, for example, the answer to the question of what is the derivative or the integral of this function.

Taking as our starting point the functional approach, we can define the delta function as (see § 3.4)

$$\delta(x) = \frac{1}{\pi} \int_{0}^{\infty} \cos \alpha x \,\mathrm{d}\alpha .$$

If integrating, for example, under the sign of the integral, we obtain

$$\int \delta(x)\mathrm{d}x = \frac{1}{\pi} \int_{0}^{\infty} \frac{\sin \alpha x}{\alpha} \mathrm{d}\alpha + C = I(x) + C,$$

where $I(x)$ is a unit step function defined as

$$I(x) = \begin{cases} 1 & \text{for} \quad x > 0, \\ 0 & \text{for} \quad x < 0, \end{cases}$$

and C is an arbitrary constant.

In a similar way we can define the Dirac function of two variables. We have (§ 3.4)

$$\delta(x, y) = \frac{1}{\pi^2} \int_0^\infty \int_0^\infty \cos \alpha x \cos \beta y \, d\alpha \, d\beta .$$

The third approach to the problem of concentrated loads consists of studying the singular solution of differential equations of plates and shells. Considering the equilibrium of a sector of a plate or shell containing the neighbourhood of the singular points, we come to the conclusion that this sector can be in equilibrium only if an external concentrated force or moment is applied at this point.

We usually demand from a solution for the concentrated load:

(1) satisfaction of the differential equations of the problem;
(2) satisfaction of the boundary conditions;
(3) regularity, except for the point of application of the load;
(4) satisfaction of the equilibrium of the concentrated load with the edge forces for an arbitrary region surrounding the singular point.

The above conditions are not satisfactory. In order to be sure that the solution is "pure", i.e. that it is not contaminated by self-equilibrating singularities of higher order, one must consider them as a singular part of Green's function for the concentrated load. Then integrating over an arbitrary area of the structure for a distributed load it should be shown that:

(5) the solution obtained satisfies the non-homogeneous plate or shell equations;
(6) the displacement w and $\partial w/\partial x$, $\partial w/\partial y$ are continuous everywhere;
(7) the stress resultants N_{xx}, N_{yy}, N_{xy} and M_{xx}, M_{yy}, M_{xy}, Q_x, and Q_y are continuous everywhere.

In the following sections all the above-described methods will be illustrated by several examples.

3.2 Isotropic circular plate loaded by a concentrated force at its centre

A solution of the problem of a plate loaded by a force can simply be obtained by considering at first a distributed load and then passing to the limit that a constant load acts on vanishing area.

In the beginning, let us consider the most simple case — the circular plate simply supported at the edge and loaded at its centre by a concentrated force normal to its surface. The solution can be derived by considering the solution for the plate in which the load is uniformly distributed over the inner portion bounded by a circle of radius c (Fig. 12). Assuming that the radius c becomes infinitely small whereas the total load P remains finite, we obtain the case of a concentrated force.

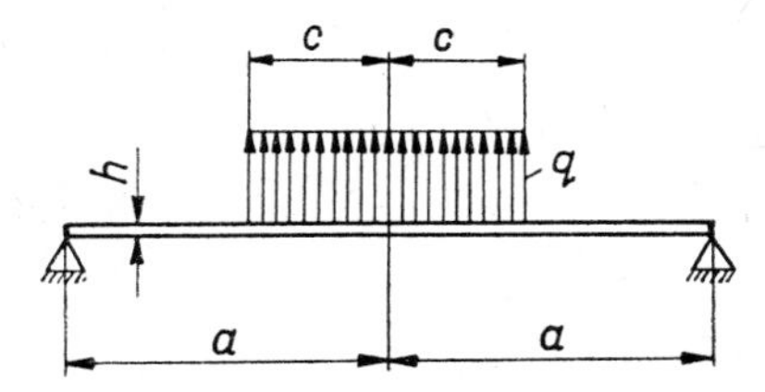

Fig. 12. Symmetrically loaded circular plate.

We obtain the deflection for the case of a uniformly distributed load on the basis of the general solution of differential equations of the plate (2.1), which in the system of polar coordinates takes the following form for the symmetrical case:

$$w = w_0 + C_1 + C_2 r^2 + C_3 r^2 \ln\frac{r}{a} + C_4 \ln\frac{r}{a}; \qquad (3.1)$$

C_i are here the constants of integration and w_0 is the particular solution of Eqs. (2.1).

Dividing the plate into two portions: loaded and unloaded, the solutions are:

$$w^{(1)} = \frac{qr^4}{64\pi D} + C_1^{(1)} + C_2^{(1)} r^2 + C_3^{(1)} r^2 \ln\frac{r}{a} + C_4^{(1)} \ln\frac{r}{a},$$

and

$$w^{(2)} = C_1^{(2)} + C_2^{(2)} r^2 + C_3^{(2)} r^2 \ln\frac{r}{a} + C_4^{(2)} \ln\frac{r}{a};$$

Owing to the complete symmetry of the plate, the shear forces in two portions of the plate are determined by the equilibrium conditions and are

$$Q_r^{(1)} = -\frac{qr}{2} \quad \text{and} \quad Q_r^{(2)} = -\frac{qc^2}{2r} \quad \text{where} \quad q = \frac{P}{\pi c^2}.$$

For the simply supported plate we have the following boundary conditions:

$$r = a, \quad w^{(2)} = 0, \quad M_{rr}^{(2)} = 0, \quad Q_r^{(2)} = -\frac{qc^2}{2a};$$
$$r = c, \quad w^{(2)} = w^{(1)}, \quad M_{rr}^{(2)} = M_{rr}^{(1)}, \quad \beta_r^{(2)} = \beta_r^{(1)}; \tag{3.2}$$
$$r = 0, \quad \beta_r^{(1)} = 0, \quad Q_r^{(1)} = 0.$$

From Eq. $(2.31)_4$ we have

$$\frac{d}{dr}(\Delta w)^{(1)} = \frac{qr}{2D}; \quad \frac{d}{dr}(\Delta w)^{(2)} = \frac{qc^2}{2Dr}.$$

After solving the above equations we find $C_3^{(1)} = C_4^{(1)} = 0$ and the deflection of the inner portion $0 \leqslant r \leqslant c$ is equal to

$$w^{(1)} = \frac{P}{16\pi D}\left\{\frac{r^4}{4c^2} + r^2\left[\frac{(1-\nu)c^2-4a^2}{2(1+\nu)a^2} + 2\ln\frac{c}{a} + \right.\right.$$
$$\left. + \frac{4}{5}\frac{h^2}{c^2}\left(\frac{\nu}{1+\nu}\frac{c^2}{a^2} - \frac{2-\nu}{2(1-\nu)}\right)\right] + \frac{4a^2(3+\nu)-(7+3\nu)c^2}{4(1+\nu)} +$$
$$\left. + c^2\ln\frac{c}{a} - \frac{4}{5}h^2\left[\frac{\nu}{1+\nu} - \frac{2-\nu}{1-\nu}\left(\frac{1}{2} - \ln\frac{c}{a}\right)\right]\right\}. \tag{3.3}$$

For the outside portion $c \leqslant r \leqslant a$ we have

$$w^{(2)} = \frac{P}{16\pi D}\left\{\left[2(3+\nu) - (1-\nu)\frac{c^2}{a^2} - \frac{8}{5}\nu\frac{h^2}{a^2}\right]\frac{a^2-r^2}{2(1+\nu)} + \right.$$
$$\left. + 2r^2\ln\frac{r}{a} + c^2\left(1 - \frac{2-\nu}{1-\nu}\frac{4}{5}\frac{h^2}{c^2}\right)\ln\frac{r}{a}\right\},$$

where $P = \pi c^2 q$.

Substituting $r = 0$ in $(3.3)_1$, we find the deflection at the centre. The terms proportional to h^2 represent the correction due to shear deformations. We see that this effect is of the order of $(h/c)^2$. When c is large in comparison with h, it can be neglected.

Only in the case where the load acts on a small area and c is small, can this effect be important and should be taken into consideration.

When $c \to 0$, we have from (3.3) the case of a concentrated force (Fig. 13)

$$w^{(2)} = \frac{P}{16\pi D}\left\{\frac{(3+\nu-\frac{4}{5}\nu h^2/a^2)(a^2-r^2)}{1+\nu} + 2\left(r^2 - \frac{2}{5}\cdot\frac{(2-\nu)}{(1-\nu)}h^2\right)\ln\frac{r}{a}\right\}. \tag{3.4}$$

For $r = 0$ the deflection of the plate is infinitely large. The term proportional to $h^2 \ln(r/a)$ increases infinitely as r approaches zero, as a consequence of our assumption that the load P is always finite. Thus, when c approaches zero, the corresponding shearing stresses and shearing strains become infinitely large.

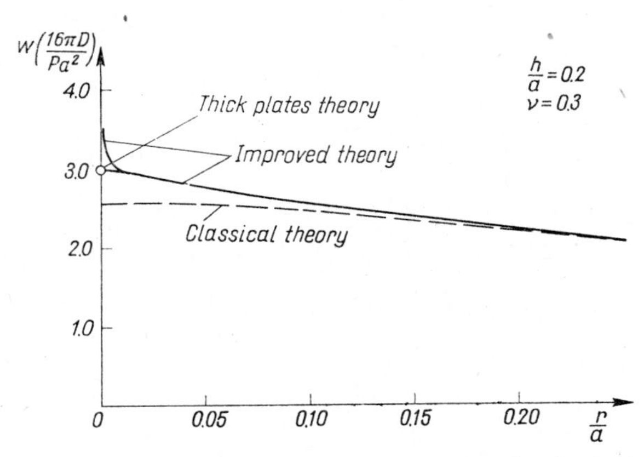

Fig. 13. Comparison of the deflection of the circular plate loaded at its centre by a concentrated force obtained by means of various plate equations.

Using the classical theory of plates, we would achieve a singular expression for the deflection w equal to $\frac{P}{8\pi D} r^2 \ln \frac{r}{a}$. For $r = 0$, the plate will have the finite deflection

$$w = \frac{3+\nu}{1+\nu} \frac{Pa^2}{16\pi D} . \tag{3.5}$$

The application of the effect of the shear deformations on the deflection of the plate results in additional singular terms proportional to $\ln(r/a)$ which for $r \to 0$ becomes infinitely large.

The only reason for appearance of singular terms is the action of the concentrated force P. The boundary forces cannot involve such an effect. It results from the above that expression (3.6) can be considered as the particular solution w_0 of the general equation of the plate (Eq. (2.1)) loaded by a concentrated force at the point $r = 0$.

$$w_0 = \frac{P}{8\pi D}\left[r^2 - \frac{2(2-\nu)}{5(1-\nu)} h^2\right] \ln \frac{r}{a} . \tag{3.6}$$

To obtain the formulae for the plate with other boundary conditions it is enough to add to the particular solution (3.6) the general solution including a certain number of constants of integration so as to fulfil the same new boundary conditions.

Let us calculate, for example, the displacement of the plate with built-in edges. Then for $r = a$ the boundary conditions yield $w = 0$ and $\beta_r = 0$. After some calculations we have

$$w = -\frac{P}{16\pi D}\left[(a^2-r^2)+2\left(r^2-\frac{2(2-\nu)}{5(1-\nu)}h^2\right)\ln\frac{r}{a}\right]. \tag{3.7}$$

It comes out from the above that the type of singularity of the functions defining the state of stresses near the point of application of the load does not depend on the boundary conditions of the plate.

The bending moments in the simply supported plate can be found by introducing displacements (3.3) into relations (1.24). For the inner portion $0 \leqslant r \leqslant c$

$$\begin{aligned} M_{rr}^{(1)} &= \frac{P}{4\pi}\left[-1(+\nu)\ln\frac{c}{a}+\right.\\ &\left.\quad +1-\frac{1-\nu}{4}\frac{c^2}{a^2}-\frac{3+\nu}{4}\frac{r^2}{c^2}-\frac{2}{5}\frac{h^2}{c^2}\nu\left(\frac{c^2}{a^2}-\frac{1}{2}\right)\right],\\ M_{\vartheta\vartheta}^{(1)} &= \frac{P}{4\pi}\left[-(1+\nu)\ln\frac{c}{a}+\right.\\ &\left.\quad +1-\frac{1-\nu}{4}\frac{c^2}{a^2}-\frac{3\nu+1}{4}\frac{r^2}{c^2}-\frac{2}{5}\frac{h^2}{c^2}\nu\left(\frac{c^2}{a^2}-\frac{1}{2}\right)\right]. \end{aligned} \tag{3.8}$$

If $c \to 0$, $M_{rr} = M_{\vartheta\vartheta} \to +\infty$. The maximum bending moments are at the centre $r = 0$. For the unloaded portion of the plate $c \leqslant r \leqslant a$

$$\begin{aligned} M_{rr}^{(2)} &= \frac{P}{4\pi}\left[-(1+\nu)\ln\frac{r}{a}+\right.\\ &\left.\quad +\frac{1-\nu}{4}\frac{c^2}{r^2}\left(1-\frac{r^2}{a^2}\right)-\frac{2}{5}\frac{h^2}{r^2}\nu\left(\frac{r^2}{a^2}-\frac{1}{2}\right)\right],\\ M_{\vartheta\vartheta}^{(2)} &= \frac{P}{4\pi}\left[-(1+\nu)\ln\frac{r}{a}+\right.\\ &\left.\quad +(1-\nu)-\frac{1-\nu}{4}\frac{c^2}{r^2}\left(1+\frac{r^2}{y^2}\right)-\frac{2}{5}\frac{h^2}{r^2}\nu\left(\frac{r^2}{a^2}+\frac{1}{2}\right)\right]. \end{aligned} \tag{3.9}$$

The maximum values of these moments are on the circle $r = c$. For $c \to 0$ $M_{rr} \to \infty$, $M_{\vartheta\vartheta} \to -\infty$.

It results from Eqs. (3.9) that the magnitude of the moments $M_{rr}^{(2)}$ and $M_{\vartheta\vartheta}^{(2)}$ is different when we come up to the radius c from the centre and from the edge of the plate. These moments differ by $a\nu Ph^2/10\pi c^2$ (Fig. 14). The moment in

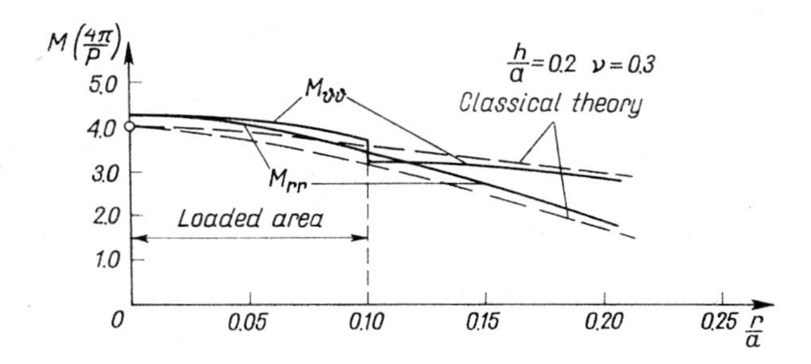

Fig. 14. Comparison of the bending moments in the circular plate loaded by a concentrated force obtained by means of the classical theory and the equations with the effect of transverse shear deformations.

the plate is hence expressed as a discontinuous function whose jump appears on the radius c in the place of the rapid change of the load. When $c \to 0$ this difference increases to infinity. Expression (3.9) is used to obtain the bending moments M_{rr} and $M_{\vartheta\vartheta}$ in the plate loaded by the concentrated force P. Neglecting the terms containing $(h/a)^2$, we find for $c = 0$

$$\begin{aligned} M_{rr} &= \frac{P}{4\pi}\left[-(1+\nu)\ln\frac{r}{a} + \frac{\nu}{5}\frac{h^2}{r^2}\right], \\ M_{\vartheta\vartheta} &= \frac{P}{4\pi}\left[-(1+\nu)\ln\frac{r}{a} + (1-\nu) - \frac{\nu}{5}\frac{h^2}{r^2}\right] \end{aligned} \tag{3.10}$$

or

$$\left.\begin{matrix} M_{xx} \\ M_{yy} \end{matrix}\right\} = \frac{P}{4\pi}\left[-(1+\nu)\ln\frac{r}{a} + (1-\nu)\frac{y^2}{r^2} \pm \frac{\nu h^2}{5r^4}(x^2-y^2)\right],$$

$$M_{xy} = -\frac{P}{4\pi}\left[1-\nu-\frac{2\nu h^2}{5}\frac{1}{r^2}\right]\frac{xy}{r^2}.$$

The moments for r approaching zero become infinitely large

$$M_{rr} \to \infty, \quad M_{\vartheta\vartheta} \to -\infty \quad \text{if} \quad r \to 0.$$

The stresses in the plate are given by formula (2.6). In order to obtain the stresses it is necessary to calculate the membrane forces in the middle surface of the plate. The solution of the second Eq. $(2.1)_2$ gives:
for $0 \leqslant r \leqslant c$

$$\Phi = \frac{\nu h P r^2}{8\pi c^2},$$

for $c \leqslant r \leqslant a$

$$\Phi = -\frac{\nu h P}{4\pi} \frac{1}{a^2 - c^2} \left(a^2 \ln \frac{r}{a} - \frac{r^2}{2} \right).$$

Then we find for $0 \leqslant r \leqslant c$

$$N_{rr} = -\frac{1}{r} \frac{\partial \Phi}{\partial r} = \frac{\nu h P}{4\pi c^2}, \qquad N_{\vartheta\vartheta} = -\frac{\partial^2 \Phi}{\partial r^2} = \frac{\nu h P}{4\pi c^2},$$

and for $c \leqslant r \leqslant a$

$$N_{rr} = \frac{\nu h P}{4\pi} \frac{1}{a^2 - c^2} \left(\frac{a^2}{r^2} - 1 \right), \qquad N_{r\vartheta} = -\frac{\nu h P}{4\pi} \frac{1}{a^2 - c^2} \left(\frac{a^2}{r^2} + 1 \right).$$

The value of the membrane force $N_{\vartheta\vartheta}$ jumps from $+\nu h P/4\pi c^2$ to $-\nu h P/4\pi c^2$ at the radius c, i.e. in the place of rapid change of the load. If $c \to 0$, the membrane forces becomes for $c \leqslant r \leqslant a$

$$N_{rr} = \frac{\nu h P(a^2 - r^2)}{4\pi a^2 r^2}, \qquad N_{\vartheta\vartheta} = -\frac{\nu h P(a^2 + r^2)}{4\pi a^2 r^2}.$$

For $r \to 0$ the membrane forces tend to $\pm\infty$. But, calculating these forces at the point $r = 0$ for the region $0 \leqslant r \leqslant c$, we find that for $c \to 0$ both the membrane forces N_{rr} and $N_{\vartheta\vartheta}$ are equal and tend to $+\infty$. However, the stresses resulting from these forces are very small in comparison with the bending stresses and except closed to the vicinity of the point of application of the load can be neglected.

Analysing the singular expression (3.10) for $c \to 0$, we can prove that it satisfies Eqs. (2.1) and (2.3) and the condition of equilibrium of the shear forces with the load P for the region containing the point $(0, 0)$

$$\int_c Q_n \, ds = P.$$

We can separate the expression w_0^{**} from the singular solution w_0 (3.6) by writing

$$w_0 = w_0^* + w_0^{**},$$

where

$$w_0^* = \frac{P}{8\pi D} r^2 \ln r - \frac{Ph^2}{10(1-\nu)\pi D} \ln r,$$

$$w_0^{**} = \frac{\nu}{20(1-\nu)} \frac{Ph^2}{\pi D} \ln r.$$

The solution w_0^{**}, which represents the deflection produced by the transverse normal stress σ_z, also satisfies the above condition separately. The solution w_0^{**} gives

$$\int_c Q_n \, \mathrm{d}s = 0.$$

Then it represents the singular self-equilibrated solution which added to an arbitrary solution does not change the equilibrium of the plate. It follows that conditions (1)-(4), § 3.1, which are usually applied in the theory of elasticity and the theory of plates to the problems concerning the action of concentrated forces, do not give a unique answer, therefore it is necessary to take into account also conditions (5), (6), (7), § 3.1.

This problem was discussed by E. Sternberg and R. A. Eubanks [3.24, 3.25]. According to these authors, the additional condition which should be satisfied is the range of singularity of the solution.

Let us analyse the solution w_0^{**} and the bending moments resulting from it. We find

$$M_{rr} = -M_{\vartheta\vartheta} = \frac{\nu h^2 P}{20\pi} \frac{1}{r^2}, \qquad M_{r\vartheta} = Q_r = 0.$$

This state of stress can be defined as the "centre of bending", analogous to the "centre of compression" known in the theory of elasticity. The above solution can be noticed in formulae (3.10). It can be proved that these expressions are the results of the effect of the transverse normal stress σ_{zz} perpendicular to the plane of the plate. According to the simplified theory, these stresses appear in the loaded region only. The load in the region of its application produces also strains and stresses in the directions perpendicular to the load, that is, in the directions tangential to the surface of the plate. The magnitude of these strains is proportional to the coefficient ν. The bending moments defined by the above formulae are the resultant values of the additional stresses. Similarly, the membrane forces defined by the solution present the membrane effect of the same stresses. This state can be defined as the "center of the compression" or "tension" depending on whether the force acts in the direction toward the plate or from the plate.

It results from all considerations above that the stresses increase to infinity in the vicinity of the point of application of the load. However, we known that bodies able to carry the infinitely large stresses and preserving simultaneously their elastic properties do not exist in nature. Practically, the force concentrated at one point cannot be realized. This notion can, however, be used as the useful static scheme for the local load acting on the structure. We know, according to de Saint-Venant's principle, that the way in which the load is distributed over a small area of the body has no influence on the stresses and displacements in the more distant points of the body.

In reality, a concentrated force is always realized as a force distributed over a certain surface. Even in the case where the load is applied through a pin-like body, the stresses resulting from this load produce a local plastic state and cause the driving of the pin in the material of the plate. If the plate is made of fragile material, the pin causes the local crumbling which always leads to the fact that the load becomes a load distributed over a certain surface. Such a load produces finite stresses.

The relations above can be used for the calculation of the deflections and stresses in the plate at all points except the points in the direct vicinity of the point of application of the load. If the radius r is smaller than the thickness of the plate, then Eqs. (3.4) and (3.8) should not be used. In order to obtain more exact results the central part of the plate should be treated as the three-dimensional body. Such a problem will be considered in § 3.6.

3.3 Circular plate with cylindrical orthotropy loaded by a concentrated force

The differential equation of the plate loaded symmetrically takes the form of the ordinary differential equation, see Eqs. (1.159) and (1.160);

$$D_r \frac{1}{r} \frac{\mathrm{d}^2}{\mathrm{d}r^2} \left(r \frac{\mathrm{d}^2 w}{\mathrm{d}r^2} \right) - D_\vartheta \frac{1}{r} \frac{\mathrm{d}}{\mathrm{d}r} \left(\frac{1}{r} \frac{\mathrm{d}w}{\mathrm{d}r} \right) = q(r). \tag{3.11}$$

Internal forces in the plate bent symmetrically can be found from Eq. (1.162) neglecting the derivatives with respect to the angle ϑ. We have

$$M_{rr} = -\left(D_r \frac{d^2w}{dr^2} + \frac{D_{12}}{r} \frac{dw}{dr}\right),$$

$$M_{\vartheta\vartheta} = -\left(D_{12} \frac{d^2w}{dr^2} + \frac{D_\vartheta}{r} \frac{dw}{dr}\right), \tag{3.12}$$

$$Q_r = -\left[D_r \frac{1}{r} \frac{d}{dr}\left(r \frac{d^2w}{dr^2}\right) - \frac{D_\vartheta}{r^2} \frac{dw}{dr}\right].$$

Let us consider the circular plate of radius a, built-in edges and of cylindrical orthotropy, loaded by a concentrated force at its centre. The boundary conditions are

$$r = 0, \qquad \frac{dw}{dr} = 0, \qquad Q_r = -\frac{P}{2\pi r},$$

$$r = a, \qquad w(a) = 0, \qquad \frac{dw}{dr} = 0.$$

We look for the solution of Eq. (3.11) in the form

$$w = r^k.$$

On substituting this function in Eq. (3.11) we have the characteristic equation

$$k(k-1)^2(k-2) - \lambda^2 k(k-2) = 0 \tag{3.13}$$

where

$$\lambda = \sqrt{\frac{D_\vartheta}{D_r}}. \tag{3.14}$$

The roots of Eq. (3.13) are

$$k_1 = 0, \qquad k_2 = 2, \qquad k_3 = 1+\lambda, \qquad k_4 = 1-\lambda. \tag{3.15}$$

In the case where $\lambda \neq 0$ and $\lambda \neq 1$ we have the solution

$$W = C_1 + C_2 r^2 + C_3 r^{1+\lambda} + C_4 r^{1-\lambda} + w_0, \tag{3.16}$$

where w_0 is the particular solution of Eq. (3.11). The case where $\lambda = 0$ concerns the plate having no bending rigidity in the circumferential direction, for example, with ribs placed only in radial direction.

When $\lambda = 0$, $k_3 = k_4 = 1$, the general solution takes the form

$$w = C_1 + C_2 r^2 + C_3 r + C_4 r \ln \frac{r}{a} + w_0. \tag{3.17}$$

We have the second particular case where $\lambda = 1$. Then $D_r = D_\vartheta$ which corresponds to the isotropic plate. The solution takes the form (3.7). On introducing the solution (3.16) into boundary conditions we obtain the following equations:

$$w = \frac{Pa^2}{4\pi(1-\lambda^2)(1+\lambda)D_r}\left[1-\lambda+(1+\lambda)\left(\frac{r}{a}\right)^2-2\left(\frac{r}{a}\right)^{1+\lambda}\right], \tag{3.18}$$

$$M_{rr} = \frac{P}{2\pi(1-\lambda^2)}\left[\left(\lambda+\frac{D_{12}}{D_r}\right)\left(\frac{r}{a}\right)^{\lambda-1}-\left(1+\frac{D_{12}}{D_r}\right)\right],$$

$$M_{\vartheta\vartheta} = \frac{P}{2\pi(1-\lambda^2)}\left[\left(1+\lambda\frac{D_{12}}{D_r}\right)\left(\frac{r}{a}\right)^{\lambda-1}-\left(1+\frac{D_{12}}{D_r}\right)\right], \tag{3.19}$$

$$Q_r = -\frac{P}{2\pi r}. \tag{3.20}$$

The above formulae lose their validity when $\lambda = 1$; then the deflection is the same as for the isotropic plate. The moments at the point of application of the load have finite value

$$M_{rr}(0) = M_{\vartheta\vartheta}(0) = \frac{P}{2\pi(\lambda^2-1)}\left(1+\frac{D_{12}}{D_r}\right) \tag{3.21}$$

if $\lambda > 1$. Otherwise when $\lambda \leqslant 1$ the bending moments have the infinite magnitudes. At the clamped edge the moment is

$$M_{rr}(a) = M_{\vartheta\vartheta}(a) = -\frac{P}{2(1+\lambda)}. \tag{3.22}$$

For the plate with simply supported edges we obtain the deflection

$$w = \frac{Pa^2}{4\pi(1-\lambda^2)(1+\lambda)D_r}\left[(1-\lambda)\left(\frac{2}{\lambda+D_{12}/D_r}+1\right)+\right.$$

$$\left.+(1+\lambda)\left(\frac{r}{a}\right)^2-2\left(\frac{1}{\lambda+D_{12}/D_r}+1\right)\left(\frac{r}{a}\right)^{1+\lambda}\right]. \tag{3.23}$$

The bending moments can be found by proper differentiation of the deflection function. We obtain, similarly as previously, finite values of the bending moments at the loading point if $\lambda > 1$.

3.4 Solutions by means of Fourier series and Fourier integrals

In the preceding sections we obtained the solution for the plate subjected to a concentrated force, using the solution for the plate subjected partly to uniformly distributed load. We assumed that the loaded area approached zero, the magnitude of the load being at the same time constant. In the case of the orthotropic plate, discussed in the previous section, we obtained the results taking into account the boundary conditions for the centre of the plate where the force was applied. The action of the concentrated force on the plate can also be considered by means of another method. The discontinuous concentrated load can be expressed by a certain continuous equivalent load $q(x, y)$ distributed over the whole surface of the plate and described mathematically by continuous functions of coordinates (x, y). This can be obtained by representing the load by Fourier series or Fourier integrals. We can write, for example, the following equation for the load $q(x, y)$:

$$q(x, y) = \sum_{m=1}^{\infty} \sum_{n=1}^{\infty} q_{mn} \sin \frac{m\pi x}{a} \sin \frac{n\pi y}{b}. \tag{3.24}$$

To calculate the coefficients q_{mn} we multiply both sides of this equation by $\sin \frac{k\pi x}{a} \sin \frac{l\pi y}{b}$ and integrate from 0 to a and from 0 to b. Observing that

$$\int_0^a \int_0^b \sin \frac{m\pi x}{a} \sin \frac{k\pi x}{a} \sin \frac{m\pi y}{b} \sin \frac{l\pi y}{b} \, dx \, dy$$

$$= \begin{cases} 0 & \text{if} \quad m \neq k,\ n \neq l, \\ ab/4 & \text{if} \quad m = k,\ n = l, \end{cases} \tag{3.25}$$

we obtain

$$q_{mn} = \frac{4}{ab} \int_0^a \int_0^b q(x, y) \sin \frac{m\pi x}{a} \sin \frac{n\pi y}{b} \, dx \, dy.$$

In the case of a concentrated force P acting at the point (x_0, y_0) on the plate of dimensions a, b (Fig. 15) we have the following formula for the coefficients q_{mn}:

$$q_{mn} = \frac{4P}{ab} \sin \frac{m\pi x_0}{a} \sin \frac{n\pi y_0}{b}. \tag{3.26}$$

It can be proved that the series (3.24) with the coefficients (3.26) for the concentrated force is divergent at the point $x = x_0$, $y = y_0$. At any other point its sum is equal to zero. In the case where the load is uniformly distributed over a certain small rectangular area $2c \times 2d$ (Fig. 16) the coefficients q_{mn} are

$$q_{mn} = \frac{4P}{\pi^2 cdmn} \sin \frac{m\pi x_0}{a} \sin \frac{n\pi y_0}{b} \sin \frac{n\pi c}{a} \sin \frac{n\pi d}{b}. \tag{3.27}$$

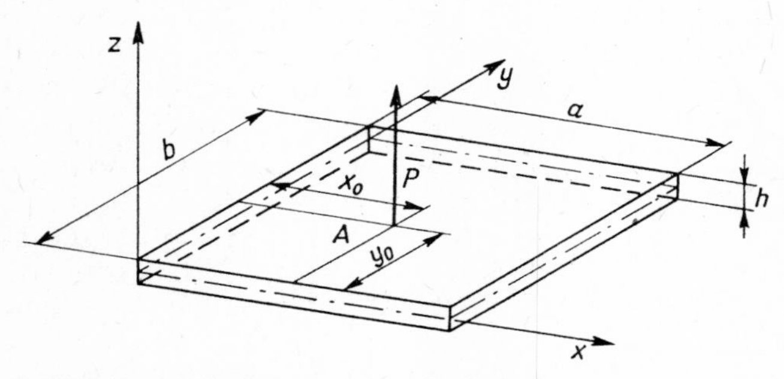

Fig. 15. Rectangular plate loaded by a concentrated force.

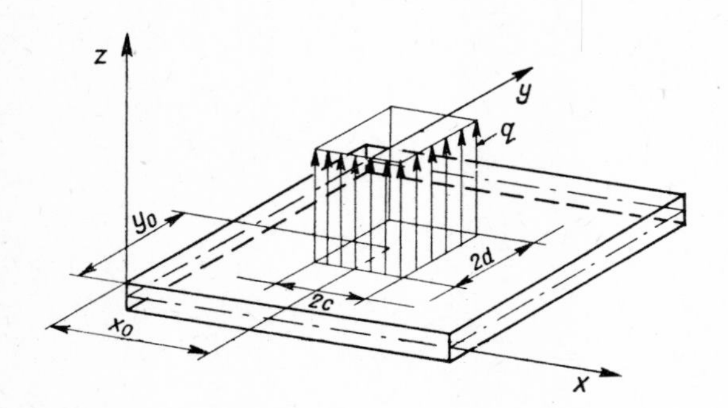

Fig. 16. Load distributed over the surface $2c \times 2d$.

Representing the load by Fourier series we assume that the deflection is also represented by a similar Fourier series. For example,

$$w = \sum_m^{\infty} \sum_n^{\infty} w_{mn} \sin \frac{m\pi x}{a} \sin \frac{n\pi y}{b}.$$

It is easy to prove that the above function satisfies the boundary conditions of the simply supported edge at all four edges. We find for $x = 0$ and $x = a$,

$w = 0$ and $M_{xx} = 0$, and for $y = 0$ and $y = b$, $w = 0$ and $M_{yy} = 0$. Sometimes when investigating the continuous plates or plates with built-in edges it is convenient to represent the deflection by the series of cosine functions. Then we write

$$q(x, y) = \sum_{m}^{\infty} \sum_{n}^{\infty} q_{mn} \cos \frac{m\pi x}{a} \cos \frac{n\pi y}{b}. \tag{3.28}$$

Observing that

$$\int_0^a \cos \frac{m\pi x}{a} \cos \frac{k\pi x}{a} \, dx = \begin{cases} 0 & \text{for} \quad k \neq m, \\ a/2 & \text{for} \quad k = m \neq 0, \\ a & \text{for} \quad k = m = 0, \end{cases}$$

and

$$\int_0^b \cos \frac{n\pi y}{b} \cos \frac{l\pi y}{b} \, dy = \begin{cases} 0 & \text{for} \quad l \neq n, \\ b/2 & \text{for} \quad l = n \neq 0, \\ b & \text{for} \quad l = n = 0, \end{cases}$$

we find, by a similar treatment, the following formulae for the coefficients q_{mn}:

$$\begin{aligned}
q_{00} &= \frac{1}{ab} \int_0^a \int_0^b q(x, y) \, dx \, dy, \\
q_{0n} &= \frac{2}{ab} \int_0^a \int_0^b q(x, y) \cos \frac{n\pi y}{b} \, dx \, dy, \\
q_{m0} &= \frac{2}{ab} \int_0^a \int_0^b q(x, y) \cos \frac{m\pi x}{a} \, dx \, dy, \\
q_{mn} &= \frac{4}{ab} \int_0^a \int_0^b q(x, y) \cos \frac{m\pi x}{a} \cos \frac{n\pi y}{b} \, dx \, dy.
\end{aligned} \tag{3.29}$$

The deflection of the plate is represented then by the series

$$w(x, y) = \sum_{m} \sum_{n} w_{mn} \cos \frac{m\pi x}{a} \cos \frac{n\pi y}{b}.$$

In this way we identically satisfy such boundary conditions at every edge of the plate as those corresponding to the axis of symmetry of a symmetrically loaded plate.

Fourier series can be replaced by Fourier integrals which are treated as a limit of a Fourier series representing the function $f(x)$ distributed along an infinitely long segment $2L \to \infty$. Let us assume that in the segment $-L < x < L$ the function $f(x)$ is represented by the series

$$f(x) = f_0 + \sum_{m=1}^{\infty} \left(f_m^{(1)} \cos \frac{m\pi x}{L} + f_m^{(2)} \sin \frac{m\pi x}{L} \right), \tag{3.30}$$

where

$$\begin{aligned} f_0 &= \frac{1}{2L} \int_{-L}^{L} f(u) \, du, \\ f_m^{(1)} &= \frac{1}{L} \int_{-L}^{L} f(u) \cos \frac{m\pi u}{L} \, du, \\ f_m^{(2)} &= \frac{1}{L} \int_{-L}^{L} f(u) \sin \frac{m\pi u}{L} \, du. \end{aligned} \tag{3.31}$$

On substitution of (3.31) into (3.30), we obtain

$$\begin{aligned} f(x) &= \frac{1}{2L} \int_{-L}^{L} f(u) \, du + \\ &\quad + \frac{1}{L} \sum_m \int_{-L}^{L} f(u) \left(\cos \frac{m\pi u}{L} \cos \frac{m\pi x}{L} + \right. \\ &\quad \left. + \sin \frac{m\pi u}{L} \sin \frac{m\pi x}{L} \right) du. \end{aligned} \tag{3.32}$$

Since the successive values of

$$\Delta \left(\frac{m\pi}{L} \right) = \frac{(m+1)\pi}{L} - \frac{m\pi}{L} = \frac{\pi}{L},$$

then introducing this into Eq. (3.32) we have

$$f(x) = \frac{1}{2L} \int_{-L}^{L} f(u) \, du +$$

$$+\frac{1}{\pi}\sum_{m}\Delta\left(\frac{m\pi}{L}\right)\int_{-L}^{L} f(u)\left(\cos\frac{m\pi u}{L}\cos\frac{m\pi x}{L}+\right.$$

$$\left.+\sin\frac{m\pi u}{L}\sin\frac{m\pi x}{L}\right)du. \tag{3.33}$$

We obtain the finite value of the function $f(x)$ for $L \to \infty$, assuming that the integral

$$\int_{-L}^{L} f(u)\,du$$

has a finite value. Then the first term of Eq. (3.33) vanishes. Letting $\pi m/L = \alpha_m$, we find that

$$\Delta\left(\frac{m\pi}{L}\right) = \Delta(\alpha_m) \to d\alpha\,,$$

when $L \to \infty$ and the quantity α_m takes all values between 0 and ∞, the limit of the function $f(x)$ as $L \to \infty$ has the form

$$f(x) = \frac{1}{\pi}\int_{0}^{\infty} d\alpha \int_{-\infty}^{\infty} f(u)(\cos\alpha u\cos\alpha x+\sin\alpha u\sin\alpha x)\,du$$

$$= \int_{0}^{\infty} (f^{(1)}(\alpha)\cos\alpha x+f^{(2)}(\alpha)\sin\alpha x)\,d\alpha, \tag{3.34}$$

where

$$f^{(1)}(\alpha) = \frac{1}{\pi}\int_{-\infty}^{\infty} f(x)\cos\alpha x\,dx,$$

$$f^{(2)}(\alpha) = \frac{1}{\pi}\int_{-\infty}^{\infty} f(x)\sin\alpha x\,dx.$$

In the case where the function $f(x)$ is symmetrical with respect to the origin we have the following representation:

$$f(x) = \int_{0}^{\infty} f(\alpha)\cos\alpha x\,d\alpha, \tag{3.35}$$

where

$$f(\alpha) = \frac{2}{\pi} \int_0^\infty f(x) \cos \alpha x \, \mathrm{d}x .$$

When the function $f(x)$ is antisymmetrical with respect to the origin we have

$$f(x) = \int_0^\infty f(\alpha) \sin \alpha x \, \mathrm{d}\alpha , \tag{3.36}$$

where

$$f(\alpha) = \frac{2}{\pi} \int_0^\infty f(x) \sin \alpha x \, \mathrm{d}x .$$

For the concentrated force we find the following representation:

$$q(x) = \frac{P}{\pi} \int_0^\infty \cos \alpha x \, \mathrm{d}\alpha . \tag{3.37}$$

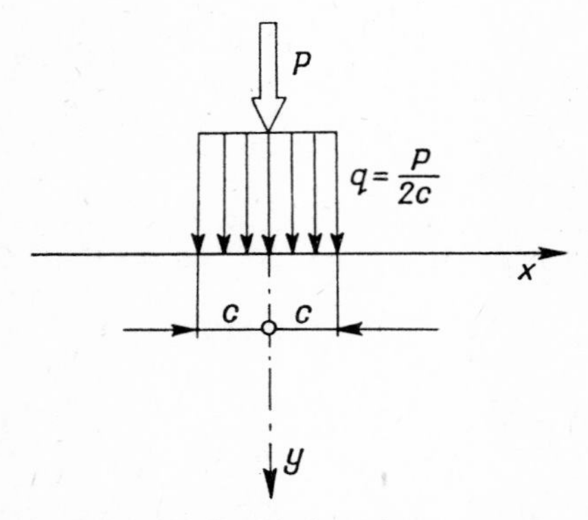

Fig. 17. Load distributed along the segment $2c$.

When the force is distributed along a certain distance $2c$ (Fig. 17) we have

$$q(x) = \frac{P}{\pi} \int_0^\infty \frac{\sin \alpha c}{\alpha c} \cos \alpha x \, \mathrm{d}\alpha . \tag{3.38}$$

In a similar way we can represent the function of two coordinates $f(x, y)$ by the double Fourier integral

$$f(x, y) = \int_0^\infty \int_0^\infty (f_{\alpha\beta}^{(1)} \cos \alpha x \cos \beta y + f_{\alpha\beta}^{(2)} \sin \alpha x \sin \beta y +$$

$$+ f_{\alpha\beta}^{(3)} \cos \alpha x \sin \beta y + f_{\alpha\beta}^{(4)} \sin \alpha x \cos \beta y) \mathrm{d}\alpha \, \mathrm{d}\beta, \tag{3.39}$$

where

$$f_{\alpha\beta}^{(1)} = \frac{1}{\pi^2} \int_{-\infty}^{\infty} \int_{-\infty}^{\infty} f(x, y) \cos \alpha x \cos \beta y \, \mathrm{d}x \, \mathrm{d}y,$$

$$f_{\alpha\beta}^{(2)} = \frac{1}{\pi^2} \int_{-\infty}^{\infty} \int_{-\infty}^{\infty} f(x, y) \sin \alpha x \sin \beta y \, \mathrm{d}x \, \mathrm{d}y,$$

$$f_{\alpha\beta}^{(3)} = \frac{1}{\pi^2} \int_{-\infty}^{\infty} \int_{-\infty}^{\infty} f(x, y) \cos \alpha x \sin \beta y \, \mathrm{d}x \, \mathrm{d}y,$$

$$f_{\alpha\beta}^{(4)} = \frac{1}{\pi^2} \int_{-\infty}^{\infty} \int_{-\infty}^{\infty} f(x, y) \sin \alpha x \cos \beta y \, \mathrm{d}x \, \mathrm{d}y.$$

Also in this case the integral

$$\int_{-\infty}^{\infty} \int_{-\infty}^{\infty} f(x, y) \mathrm{d}x \, \mathrm{d}y$$

must have a finite value.

The above operation is called the *double exponential Fourier transform* and can also be defined in the following way:

$$\tilde{f}_j(\alpha, \beta) = \frac{1}{2\pi} \int_{-\infty}^{\infty} \int_{-\infty}^{\infty} f_j(x, y) \mathrm{e}^{i(\alpha x + \beta y)} \mathrm{d}x \, \mathrm{d}y.$$

The *inverse transform* is given by the formula

$$f_j(x, y) = \frac{1}{2\pi} \int_{-\infty}^{\infty} \int_{-\infty}^{\infty} \tilde{f}_j(\alpha, \beta) \mathrm{e}^{-i(\alpha x + \beta y)} \mathrm{d}x \, \mathrm{d}y.$$

The transform of the derivatives of the function $f(x, y)$ can be found from the formulae

$$\frac{1}{2\pi} \int_{-\infty}^{\infty} \int_{-\infty}^{\infty} \frac{\partial f_j(x, y)}{\partial x} \mathrm{e}^{i(\alpha x + \beta y)} \mathrm{d}x \, \mathrm{d}y = -i\alpha \tilde{f}_j(\alpha, \beta),$$

$$\frac{1}{2\pi}\int_{-\infty}^{\infty}\int_{-\infty}^{\infty}\frac{\partial f_j(x,y)}{\partial y}\,e^{i(\alpha x+\beta y)}dx\,dy = -i\beta\tilde{f}_j(\alpha,\beta).$$

The Fourier transform is applied to the solution of the plate and shell equations in the following way. We multiply both sides of the equations by $e^{i(\alpha x+\beta y)}$ and integrate on the plane x, y. Using the above formulae, we reduce the differential equations to the linear algebraic equations which can be solved with respect to the functions $\tilde{f}_j(\alpha,\beta)$. Performing the inverse transform, we find the functions $f_j(x,y)$. It occurs in the calculation of the Fourier integrals. The performance of the Fourier transform on the right-hand sides of the equations is then equivalent to the presentation of the load

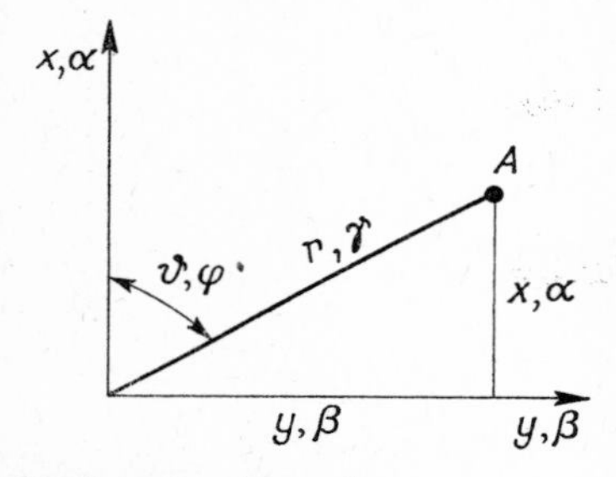

Fig. 18. Systems of coordinates.

in the form of a Fourier integral. The concentrated force P acting at the point $x = x_0$, $y = y_0$ can be expressed by the following double integral:

$$q(x,y) = \frac{P}{\pi^2 l^2}\int_0^{\infty}\int_0^{\infty}\cos\alpha\left(\frac{x-x_0}{l}\right)\cos\beta\left(\frac{y-y_0}{l}\right)d\alpha\,d\beta, \qquad (3.40)$$

where l is a certain arbitrary length. In a similar way we can represent the load distributed over a certain small area $2c\times 2d$ (Fig. 16). We get then

$$q(x,y) = \frac{P}{\pi^2 l^2}\int_0^{\infty}\int_0^{\infty}\frac{\sin\alpha c/l}{\alpha c/l}\cos\alpha\left(\frac{x-x_0}{l}\right)\frac{\sin\beta d/l}{\beta d/l}\cos\beta\left(\frac{y-y_0}{l}\right)d\alpha\,d\beta; \qquad (3.41)$$

when $c\to 0$, $d\to 0$

$$\frac{\sin\alpha c/l}{\alpha c/l}\to 1 \quad \text{and} \quad \frac{\sin\beta d/l}{\beta d/l}\to 1$$

and the formula is reduced to (3.40).

Formally, integral (3.40) does not exist and, in conclusion, all calculations should be made by means of formula (3.41). Really, the expressions for concentrated loads can be achieved by using formula (3.41) and substitution of $c \to 0$ and $d \to 0$ in the obtained results. However, as Eq. (3.40) gives the same results it can be considered as the representation of the concentrated force in the form of an integral. Integral (3.40) has features similar to those of a previously described series (3.26); at the point $x = x_0$, $y = y_0$ it is divergent and at any other point of the area of the plate it is equal to zero. In the case the calculations are performed in polar coordinates we can transform expressions (3.40). Taking the origin of new coordinates (γ, φ) connected with (α, β) by the expressions $\alpha = \gamma \cos\varphi$, $\beta = \gamma \sin\varphi$, at the point (x_0, y_0) we can reduce expression (3.40) to

$$q(x, y) = \frac{P}{\pi^2 l^2} \int_0^\infty \int_0^{\pi/2} \cos(\gamma x \cos\varphi) \cos(\gamma y \sin\varphi) \gamma \, d\gamma \, d\varphi; \tag{3.42}$$

here x, y are the non-dimensional coordinates, defined by $x = \bar{x}/l$, $y = \bar{y}/l$. Integration with respect to the variable φ yields

$$q(r) = \frac{P}{2\pi l^2} \int_0^\infty \gamma J_0(\gamma r) \, d\gamma$$

where

$$r = \sqrt{x^2 + y^2} \quad \text{and} \quad \gamma = \sqrt{\alpha^2 + \beta^2}. \tag{3.43}$$

When the force P is distributed uniformly over a small surface of radius $c = \bar{c}/l$ whose centre is at the origin of coordinates (r, ϑ), we can derive the following formula for the load $q(x, y)$ by integration of expression (3.43) over the circular surface $r \leqslant c$:

$$q(x, y) = \frac{2P}{\pi^2 l^2} \int_0^\infty \int_0^\infty \frac{J_1(\gamma r)}{\gamma c} \cos\alpha x \cos\beta y \, d\alpha \, d\beta. \tag{3.44}$$

On evaluation with respect to φ in the polar coordinates γ, φ, we have

$$q(r) = \frac{P}{\pi l^2} \int_0^\infty J_1(\gamma c) J_0(\gamma r) \, d\gamma. \tag{3.45}$$

r is here the non-dimensional coordinate $r = \dfrac{\bar{r}}{l} = \dfrac{1}{l}\sqrt{\bar{x}^2 + \bar{y}^2}$. The above integral expresses the discontinous function satisfying the conditions

$$\text{for } r < c, \quad q = \frac{P}{\pi c^2 l^2}, \quad c = \bar{c}/l,$$

$$\text{for } r = c, \quad q = \frac{P}{2\pi c^2 l^2},$$

$$\text{for } r > c, \quad q = 0.$$

When $c \to 0$, $J_1(\gamma c)/\gamma c \to \frac{1}{2}$, formula (3.45) becomes identical with (3.43). When the area of the function $q(x, y)$ is considered as unlimited in the x-axis direction, but is finite in the y-axis direction and has the dimension b, the load can be represented by the series of Fourier integrals.

For example, if the function $q(x, y)$ is symmetrical with respect to the x and y-axis, we have

$$q(x, y) = \int_0^\infty \left(q_0 + \sum_{n=1}^\infty q_{n\alpha} \cos \frac{n\pi y}{b}\right) \cos \alpha x \, d\alpha,$$

where

$$\begin{aligned} q_{0\alpha} &= \frac{2}{\pi b} \int_0^\infty dx \int_0^b q(x, y) \, dy \cos \alpha x, \\ q_n &= \frac{4}{\pi b} \int_0^\infty dx \int_0^b q(x, y) \cos \frac{n\pi y}{b} \, dy \cos \alpha x. \end{aligned} \tag{3.46}$$

It is convenient in mathematical analysis to represent the concentrated load by the *Dirac delta function* $\delta(x)$. This symmetrical unit-impulse function of the real variable x is defined by the formula

$$\int_a^b f(\xi)\,\delta(x-\xi)\,d\xi = \begin{cases} 0 & \text{if} \quad x > a \text{ or } x > b, \\ \frac{1}{2} f(x) & \text{if} \quad x = a \text{ or } x = b \\ f(x) & \text{if} \quad a < x < b, \; a < b, \end{cases}$$

where $f(x)$ is an arbitrary continuous function of x. The delta function is not a "true" function since definition (3.49) implies the inconsistent relations:

$$\delta(x) = 0 \quad (x \neq 0),$$

$$\int_{-\infty}^{\infty} \delta(\xi)\,d\xi = 1.$$

Representing the Dirac function by means of a Fourier integral, we have

$$\delta(x) = \frac{1}{\pi} \int_0^\infty \cos \alpha x \, d\alpha .$$

The formal use of the $\delta(x)$ function furnishes a convenient notation permitting generalization of many mathematical relations. We can also define the delta function of two variables (x, y) in a similar way; we have, for example,

$$\delta(x, y) = \frac{1}{\pi^2} \int_0^\infty \int_0^\infty \cos \alpha x \cos \beta y \, d\alpha \, d\beta .$$

The delta function $\delta(x, y)$ satisfies the following relation:

$$\delta(x-\xi, y-\eta) = \delta(x-\xi)\,\delta(y-\eta),$$

which enables the separation of the variables in the solution.

3.4.1 *Introduction to the theory of distributions*

The common use of Dirac's function $\delta(x)$ and other non-functional quantities in physics and engineering disciplines is the reason for a certain kind of computational formalism, for example concerning the rules of differentiation of such quantities. Further formal generalization of the classical notion of the function resulted in the theory of the generalized functions, called the *theory of distributions.*

We know that a concentrated force can be presented by $\delta(x)$. The definition of the operations, the rules of differentiation and integration of such functions is particularly important. The ordinary derivative of $\delta(x)$ is equal to zero for any $x \neq 0$. The theory of distribution gives, for example, the answer to the question of what is the derivative of this function. As the theory of distribution is connected with the subject of the present work, a short presentation of its elements seems useful. The readers interested in this discipline can enlarge their knowledge studying the papers and books such as [3.90], [3.91], [3.92].

Distributions are most often defined as:

(1) a class of equivalent sequences (Mikusiński's approach),

(2) linear functionals in appropriate linear spaces (Schwartz's approach).

3.4.2 *Sequence approach*

Defining the distribution as the class of sequences, we introduce a notion of a *fundamental sequence* $\{f_n(x)\}$. Let the functions $f_n(x)$, $n = 1, 2, \ldots$, be continuous in the interval $(-\infty, +\infty)$. A sequence is considered fundamental if for any interval $[a, b]$ there exists a sequence $\{F_n(x)\}$ such that the equation

$$\frac{d^k F_n(x)}{dx^k} = f_n(x)$$

is valid in the interval $[a, b]$. We require also that the sequence $\{F_n(x)\}$ to be uniformly convergent in the limited interval $[a, b]$ and almost uniformly convergent in the interval $(-\infty, +\infty)$.

Two fundamental sequences $\{f_n(x)\}$ and $\{g_n(x)\}$ are called *equivalent* if for each interval $[a, b]$ the conditions

$$F_n^k(x) = f_n(x), \quad G_n^k(x) = g_n(x),$$
$$F_n(x) \Rightarrow F(x), \quad G_n(x) \Rightarrow F(x)$$

are satisfied. This means that for $n \to \infty$ the sequences are convergent to the same function $F(x)$. We denote by

$$\langle f_n(x) \rangle$$

the distribution determined by the class of the sequences equiponderant to the fundamental sequence $\{f_n(x)\}$. The Dirac function

$$\delta(x) = \left\langle \frac{1}{\pi} \frac{n}{n^2x^2+1} \right\rangle$$

is here the example of the distribution defined in this way, as for any $x \neq 0$ we have

$$\lim_{n\to\infty} \frac{1}{\pi} \frac{n}{n^2x^2+1} = 0.$$

This function is constant and equal to zero in the interval $-\infty < x < 0$ and $0 < x < +\infty$ For $x = 0$

$$\lim_{n\to\infty} \frac{1}{\pi} \frac{n}{n^2x^2+1} = \infty.$$

The derivative of the distribution. Let us consider the distribution $f(x) = \langle f_n(x) \rangle$ and assume the functions $f_n(x)$ to have continuous derivatives

$f_n'(x)$. The sequence $\langle f_n'(x)\rangle$ is also the fundamental sequence, so there exists a distribution $\langle f_n'(x)\rangle$ which is called the *derivative* of the distribution $\langle f_n(x)\rangle$

$$f'(x) = \langle f'(x)\rangle .$$

Let us take as an example the *unit-step function* defined as

$$I(x) = \begin{cases} 1 & \text{for} \quad x > 0, \\ 0 & \text{for} \quad x < 0. \end{cases}$$

This function is the distribution which can be defined by the sequence

$$I(x) = \left\langle \frac{1}{\pi} \arctan nx + \frac{1}{2} \right\rangle .$$

This function has its ordinary derivative equal to zero for any $x \neq 0$. This derivative is not identical with that distributional, as according to the definition

$$I'(x) = \left\langle \frac{1}{\pi} \arctan nx + \frac{1}{2} \right\rangle' = \left\langle \frac{1}{\pi} \frac{n}{n^2x^2+1} \right\rangle = \delta(x)$$

it is equal to Dirac's function. The distributional derivative cannot be in this case identical with the ordinary derivative as the function under differentiation has a point of discontinuity. By differentiation of $\delta(x)$, we find

$$\delta'(x) = \left\langle -\frac{2}{\pi} \frac{n^3 x}{(n^2x^2+1)^2} \right\rangle, \qquad \text{etc.}$$

The distribution $\varphi(x)$ is called the *indefinite integral of* $f(x)$ if $\varphi'(x) = f(x)$.

For example, if $f(x) = I(x)$, then integrating we have

$$\begin{aligned} F(x) &= \int I(x)\,\mathrm{d}x \\ &= \left\langle \frac{x}{2} + \frac{x}{\pi} \arctan nx - \frac{1}{2\pi n} \ln(1+n^2x^2) \right\rangle \end{aligned}$$

where

$$F(x) = \begin{cases} 0 & \text{for} \quad x < 0, \\ x & \text{for} \quad x \geqslant 0. \end{cases}$$

The number

$$\int_a^b f(x)\,\mathrm{d}x = \varphi(b) - \varphi(a)$$

is called a *definite integral of the distribution* $f(x)$. In accordance with the above definition we have, for example,

$$\int_{-\infty}^{\infty} \delta(x-x_0)\,\mathrm{d}x = I(\infty)-I(-\infty) = 1 .$$

3.4.3 *Functional approach*

Defining the distribution as the functional, we introduce the so-called *fundamental function* $\Phi(x)$ satifying the following conditions:

(1) *$\Phi(x)$ is determined in the interval $(-\infty, +\infty)$ and has all derivatives in this interval.*

(2) *There exists a limited interval $[a, b]$ such that $\Phi(x) = 0$ for $x < a$ and $x > b$.*

The set of all fundamental functions is called a *fundamental space* and is denoted by $\mathscr{D}$. The space $\mathscr{D}$ of fundamental functions is linear as if $\Phi_1(x) \in \mathscr{D}$ and $\Phi_2(x) \in \mathscr{D}$, then $k_1\Phi_1(x)+k_2\Phi_2(x) \in \mathscr{D}$ for arbitrary numbers k_1 and k_2. We can notice that if $\Phi(x) \in \mathscr{D}$, then $\Phi'(x) \in \mathscr{D}$. It means that the arbitrary derivative of the fundamental function is the fundamental function. Following Schwartz, any linear functional defined in the space $\mathscr{D}$ is called the *distribution.* Denote this functional by the symbol $\langle f, \Phi\rangle$. The conditions demanding that it be the linear continuous functional, i.e. the distribution, are the following

(1) *If $\Phi_1, \Phi_2 \in \mathscr{D}$ and k_1, k_2 are arbitrary real numbers, then $\langle f, k_1\Phi_1 + +k_2\Phi_2\rangle = k_1 \langle f, \Phi_1\rangle+k_2\langle f, \Phi_2\rangle$.*

(2) *If the sequence of the fundamental functions $\{\Phi_n(x)\}$ is convergent to zero in the space $\mathscr{D}$, it means that if all sequences $\{\Phi_n^{(k)}(x)\}$ $(k = 0, 1, 2, \ldots)$ are uniformly convergent to zero in the interval $[a, b]$, then $\langle f, \Phi\rangle \to 0$ for $n \to \infty$.*

The second condition is the condition of the continuity of the functional $\langle f, \Phi\rangle$.

If the function $f(x)$ is locally integrable, then for every fundamental function $\Phi(x) \in \mathscr{D}$ there exists a linear continuous functional $\langle f, \Phi\rangle$ in the space $\mathscr{D}$, i.e. the distribution;

$$\langle f, \Phi\rangle = \int_{-\infty}^{\infty} f(x)\Phi(x)\,\mathrm{d}x .$$

The derivative of the distribution. Let us assume that the function $f(x)$ has a continuous derivative $f'(x)$. Then the distributions

$$\langle f, \Phi\rangle = \int_{-\infty}^{\infty} f(x)\Phi(x)\,\mathrm{d}x, \qquad \langle f', \Phi\rangle = \int_{-\infty}^{\infty} f'(x)\Phi(x)\,\mathrm{d}x$$

are regular. Integrating by parts and taking into account the assumed properties of the fundamental functions, we find

$$\langle f', \Phi\rangle = \int_{-\infty}^{\infty} f'(x)\Phi(x)\,\mathrm{d}x = f(x)\Phi(x)\Big|_{-\infty}^{\infty} - \int_{-\infty}^{\infty} f(x)\Phi'(x)\,\mathrm{d}x$$

$$= -\int_{-\infty}^{\infty} f(x)\Phi'(x)\,\mathrm{d}x = \langle f, -\Phi'\rangle.$$

Let us calculate, for example, the derivative of the *unit function* $I(x)$. Applying the functional approach, we have

$$\langle I'(x), \Phi(x)\rangle = \langle I(x), -\Phi'(x)\rangle$$

$$= -\int_{-\infty}^{\infty} I(x)\Phi'(x)\,\mathrm{d}x = -\Phi(x)\Big|_{0}^{\infty} = \Phi(0).$$

Since

$$\langle \delta(x), \Phi(x)\rangle = \int_{-\infty}^{\infty} \delta(x)\Phi(x)\,\mathrm{d}x = \Phi(0),$$

we find that the derivative of the unit function is the function $\delta(x)$.

It can easily be noticed that the definition of the distribution given by Schwartz differs from that proposed by Mikusiński. However, in spite of these differences, both definitions are equivalent. The functional approach is more general. The operations on distributions are defined in such a way that in the case the distribution is regular, i.e. identical to the ordinary function, these operations are reduced to the known operations on the ordinary functions. Below some fundamental operations are given:

Addition of distributions:

$$\langle f+g, \Phi\rangle = \langle f, \Phi\rangle + \langle g, \Phi\rangle.$$

Multiplication of a distribution by a number:

$$\langle kf, \Phi\rangle = \langle f, k\Phi\rangle = k\langle f, \Phi\rangle.$$

Multiplication of a distribution by a function:

$$\langle \lambda(x)f(x), \Phi\rangle = \langle f(x), \lambda(x)\,\Phi\rangle.$$

The product of two distributions in the general case is not the distribution. This means that in the field of distributions the expressions such as $\delta^2(x)$, $1/\delta(x)$ make no sense. The distributions are widely applied in the theory of differential equations. The fundamental solution of the differential equation

$$P\left(\frac{\partial^k}{\partial x^k}\right) y(x) = f(x) \tag{3.47}$$

is called the *distribution* $h(x)$ satisfying the equation

$$P\left(\frac{\partial^k}{\partial x^k}\right) h(x) = \delta(x).$$

If we know the fundamental solution $h(x)$, and $f(x)$ is the distribution for which exists the *convolution*

$$y(x) = \int_{-\infty}^{\infty} h(x-\xi)f(\xi)\,\mathrm{d}\xi,$$

then this convolution is the solution of the differential equation (3.47).

All solutions obtained in the theory of plates and shells corresponding to the concentrated force are the fundamental solutions.

3.4.4 *Methods of integration*

The calculus of residues is a useful tool for the calculation of Fourier integrals. We know from the theory of complex functions that the integral along the closed circumference is equal to the sum of the residues of the integral contained in the area of the circumference (Cauchy's theorem). We have

$$\oint f(z)\,\mathrm{d}z = 2\pi i \sum_{j=1}^{m} \operatorname{Res} f(z)_{z=z_j}.$$

This theorem can be used for the calculation of the integral of the function of the real variable. Let us consider the most frequently met case where the function $f(z)$ is given by the quotient of two functions

$$f(z) = \frac{F(z)}{G(z)}.$$

The function $G(z)$ is zero at the point $z = z_0$.

If $z = z_0$ is a single pole of $f(z)$, then the residual is

$$\operatorname*{Res}_{z=z_0} f(z) = \lim_{z\to z_0} [(z-z_0)f(z)] = \frac{F(z_0)}{G'(z_0)}, \tag{3.48}$$

where $'$ denotes differentiation with respect to z.

If $z = z_0$ is a multiple pole of the range m, we have

$$\operatorname{Res} f(z) = \frac{1}{(m-1)!}\frac{\mathrm{d}^{m-1}}{\mathrm{d}z^{m-1}}[(z-z_0)^m f(z)]_{z=z_0}.$$

Let us now take advantage of the above theorem to calculate the integral

$$\int_0^\infty \frac{\cos\alpha x}{\alpha^2+\beta^2}\,\mathrm{d}\alpha.$$

This integral can be replaced by the integral of the complex function

$$\int_0^\infty \frac{\cos\alpha x}{\alpha^2+\beta^2}\,\mathrm{d}\alpha = \mathrm{re}\int_0^\infty \frac{\mathrm{e}^{izx}}{z^2+\beta^2}\,\mathrm{d}z.$$

Let us integrate the above function along the closed circumference (Fig. 19). We obtain

$$\int_{-R}^{R} \frac{\mathrm{e}^{izx}}{\beta^2+z^2}\,\mathrm{d}z + \int_C \frac{\mathrm{e}^{izx}}{\beta^2+z^2}\,\mathrm{d}z = 2\pi i \sum \operatorname{Res} f(z).$$

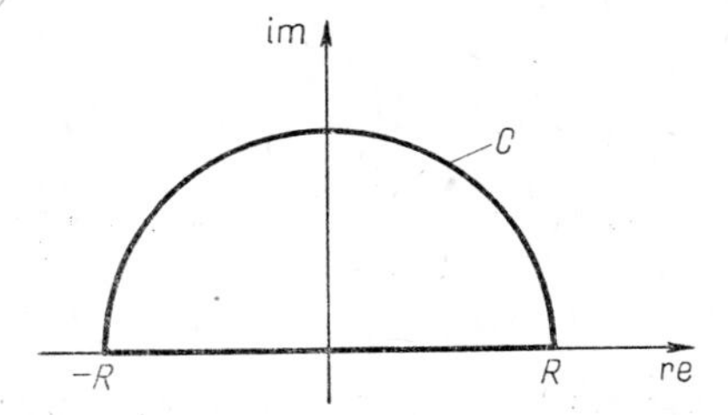

Fig. 19. Path of integration.

The only pole of the integrand in the upper halfplane iy is $i\beta$, i.e.

$$\operatorname{Res} f(z) = \mathrm{e}^{-\beta x}/2\beta i.$$

Let us assume that $R \to \infty$. If $f(z) \to 0$ for $z \to \infty$ (*Jordan conditions*), the integral along the circumference C vanishes and only the integral between the limits $-\infty$ and ∞ remains. As the function $f(z)$ is symmetrical with respect to the iy-axis, we obtain

$$\int_0^\infty \frac{\cos\alpha x\,\mathrm{d}\alpha}{\alpha^2+\beta^2} = \frac{\pi \mathrm{e}^{-\beta x}}{2\beta}\,.$$

The Fourier integral cannot always be calculated analytically. In some cases a numerical method is unavailable. Dividing the whole surface between the limits of integration in small rectangles of dimensions $\Delta\alpha$ and $\Delta\beta$, we calculate the value of the integrand $F(x, y, \alpha, \beta)$

$$f(x, y) = \int_0^\infty \int_0^\infty F(x, y, \alpha, \beta)\,\mathrm{d}\alpha\,\mathrm{d}\beta$$

at points of the net $F_{j,k} = F(j\Delta\alpha, k\Delta\beta)$. Then, using trapezoid approximation we replace the integral by the sum

$$f(x, y) = \frac{\Delta\alpha\Delta\beta}{4}\left[F_{0,0}+2\left(\sum_{j=1}^{\infty} F_{j,0}+\sum_{k=1}^{\infty} F_{0,k}\right)+4\sum_{j=1}^{\infty}\sum_{k=1}^{\infty} F_{j,k}\right].$$

Taking the parabolic approximation, we find the formula

$$f(x, y) = \frac{\Delta\alpha\Delta\beta}{9}\left[F_{0,0}+2\left(\sum_{j=1}^{\infty} F_{2j,0}+\sum_{k=1}^{\infty} F_{0,2k}\right)+\right.$$

$$+4\left(\sum_{j=1}^{\infty} F_{2j-1,0}+\sum_{k=1}^{\infty} F_{0,2k-1}+\sum_{j=1}^{\infty}\sum_{k=1}^{\infty} F_{2j,2k}\right)+$$

$$\left.+8\sum_{j=1}^{\infty}\sum_{k=1}^{\infty}(F_{2j,2k-1}+F_{2j-1,2k})+16\sum_{j=1}^{\infty}\sum_{k=1}^{\infty} F_{2j-1,2k-1}\right]. \tag{3.49}$$

Here the area of integration is unlimited. But since $\lim\limits_{\alpha,\beta\to\infty} F(\alpha, \beta) = 0$, the summation of the series (3.49) can often be limited to the first few terms.

3.5 Navier solution for the rectangular plate subjected to a concentrated load

Let us consider a rectangular simply supported isotropic plate of size $a \times b$, subjected to a concentrated load acting at the point with coordinates $\bar{x} = \bar{x}_0$, $\bar{y} = \bar{y}_0$ (Fig. 15). We solve this problem by means of the classical theory.

Expressing the force as a Fourier series, we have formula (3.24). The deflection of the plate can also be found in the form of a Fourier series [3.3]:

$$w = \sum_{m}^{\infty} \sum_{n}^{\infty} w_{mn} \sin \frac{m\pi \bar{x}}{a} \sin \frac{n\pi \bar{y}}{b}. \tag{3.50}$$

On substituting this into the plate equation (2.28), we conclude that the deflection (3.50) satisfies Eq. (2.28) if

$$w_{mn} = \frac{4P}{\pi^4 abD} \frac{\sin (m\pi \bar{x}_0/a) \sin (n\pi \bar{y}_0/b)}{(m^2/a^2+n^2/b^2)^2}. \tag{3.51}$$

Series (3.50) with w_{mn} from (3.51) converges rapidly and we can calculate the deflection at any point of the plate with sufficient accuracy by taking into account only the first few terms. For example, when the force acts at the centre of the plate, the deflection at the same point is

$$w_{\max} = \frac{4P}{\pi^4 abD} \sum_{m}^{\infty} \sum_{n}^{\infty} \frac{1}{(m^2/a^2+n^2/b^2)^2}. \tag{3.52}$$

Taking into account only the first four terms, we obtain the deflection of a square plate:

$$w_{\max} = 0.01121 \frac{Pa^2}{D}. \tag{3.53}$$

This result is about 3.5 per cent in error (see Table 1). The very simple *Navier solution* converges rapidly enough for the calculation of the deflection w, but it cannot be used for the calculation of the stresses. The reason for this fact is a very slow convergence of the series which makes it necessary to use a great number of terms. At the point of application of the load the series for the bending moments and shear forces are not convergent. Thus to obtain the stresses in the plate it is necessary to use a more effective method of calculations: In some cases it is possible to sum the series and write the solution in the form of a single series.

Below we give the example of such a treatment. Let us write solution (3.50) in the following form:

$$w = \frac{4Pb^3}{\pi^4 a} \sum_{m=1}^{\infty} S_m \sin \frac{m\pi x_0}{a} \sin \frac{m\pi \bar{x}}{a}, \tag{3.54}$$

where the coefficient S_m is expressed by the sum

$$S_m = \frac{1}{2} \sum_{n=1}^{\infty} \frac{\cos \frac{n\pi}{b} (\bar{y}-\bar{y}_0) - \cos \frac{n\pi}{b} (\bar{y}+\bar{y}_0)}{(m^2 b^2/a^2 + n^2)^2} . \tag{3.55}$$

In order to calculate the sum of series (3.55) we can use a series whose sum is known:

$$\sum_{n=1}^{\infty} \frac{\cos nz}{\alpha^2 + n^2} = -\frac{1}{2\alpha^2} + \frac{\pi}{2\alpha} \frac{\cosh \alpha(\pi - z)}{\sinh \pi\alpha} . \tag{3.56}$$

Differentiating the above equation with respect to α, we find, after some manipulation, the following formula for the deflection:

$$w = \frac{Pa^2}{\pi^3 D} \sum_{m=1}^{\infty} \left(1 + \alpha_m \coth \alpha_m - \frac{\alpha_m \bar{y}_1}{b} \coth \frac{\alpha_m \bar{y}_1}{b} - \frac{\alpha_m \bar{y}_0}{b} \coth \frac{\alpha_m y_0}{b}\right) \times$$

$$\times \frac{\sinh\left(\frac{\alpha_m \bar{y}_0}{b}\right) \sin \frac{m\pi \bar{x}_0}{a} \sin \frac{m\pi \bar{x}}{a} \sinh \frac{\alpha_m \bar{y}_1}{b}}{m^3 \sinh \alpha_m} , \tag{3.57}$$

where

$$\alpha_m = \frac{m\pi b}{a} , \quad \bar{y}_1 = b - \bar{y} .$$

When the force P acts at the centre $\bar{y} = 0$, $\bar{x} = a/2$, the deflection is maximum. The result is as follows:

$$w_{\max} = \frac{Pa^2}{2\pi^3 D} \sum_{m=1}^{\infty} \frac{1}{m^3} \left(\tanh \alpha_m - \frac{\alpha_m}{\cosh^2 \alpha_m}\right) = \alpha_1 \frac{Pa^2}{D} \tag{3.58}$$

where α_1 is given for various ratios b/a in Table 1.

Table 1.

b/a	1.0	1.1	1.2	1.4	1.6	1.8	2.0	3.0	∞
α_1	0.01160	0.01265	0.01358	0.01484	0.01670	0.01620	0.01651	0.0169	0.01695

In order to obtain the bending moments and stresses in the plate, the solution (3.57) must be differentiated twice according to Eqs. (2.2). We have then for $y = 0$

$$M_{xx} = \frac{Pa}{2\pi} \sum_{m=1}^{\infty} \frac{1}{m} \sin \frac{m\pi x_0}{a} \left[(1+\nu) \tan \alpha_m - \frac{(1-\nu)\,\alpha_m}{\cosh^2 \alpha_m} \right] \cdot \sin \frac{m\pi \bar{x}}{a} \tag{3.59}$$

and a similar expression for M_{yy}. The above series also does not converge rapidly enough near the loading point and cannot be used for the calculation of the stresses. We shall, therefore, have to sum them up once more with respect to m. We know from the results for the circular plate that shear forces and bending moments become infinitely large at the loading point. The distribution of the stresses in the plate with arbitrary boundary conditions should be near this point the same as in the circular plate. The solution can be derived by superposition of two solutions: the first, singular and the second, representing the effect of the boundary conditions.

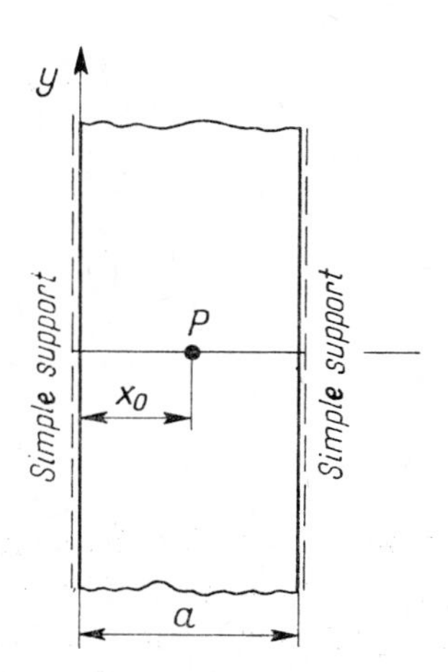

Fig. 20. Infinitely long strip plate simply supported at the edges $x = 0$ and $x = a$.

Let us transform expression (3.57) in such a way as to separate the singular part from that produced by the boundary conditions. To simplify the calculations, let us consider an infinitely long plate (Fig. 20), $b \to \infty$. Then

$$\tanh \alpha_m \simeq 1, \qquad \cosh \alpha_m \simeq \tfrac{1}{2} e^{\alpha_m},$$

$$\sinh \frac{\alpha_m}{b}(b-2\bar{y}) \approx \cosh \frac{\alpha_m}{b}(b-2\bar{y}) \approx \frac{1}{2} e^{\alpha_m \frac{b-2\bar{y}}{b}}. \tag{3.60}$$

On substitution in Eq. (3.57) we obtain the deflection of the strip plate loaded at the point with the coordinates $x = x_0$, $y = 0$:

$$w = \frac{Pa^2}{2\pi^3 D} \sum_{m=1}^{\infty} \frac{1}{m^3} \sin\frac{m\pi \bar{x}_0}{a} \sin\frac{m\pi \bar{x}}{a} \left(1+\frac{m\pi y}{a}\right) e^{-\frac{m\pi y}{a}}, \tag{3.61}$$

and the bending moment:

$$M_{xx} = \frac{P}{2\pi} \sum_{m=1}^{\infty} \frac{1}{m} \sin\frac{m\pi \bar{x}_0}{a} \sin\frac{m\pi \bar{x}}{a} \left[1+\nu+(1-\nu)\frac{m\pi \bar{y}}{a}\right] e^{-\frac{m\pi \bar{y}}{a}}, \tag{3.62}$$

and similar formulae for M_{yy} and M_{xy}.

The above series can be summed. For a small r, we get the following equations for the bending moments:

$$\left.\begin{matrix} M_{xx} \\ M_{yy} \end{matrix}\right\} = -\frac{1+\nu}{4\pi} P \ln\frac{r}{a_0} \pm \frac{1-\nu}{4\pi} P \frac{\bar{y}^2}{\bar{r}^2}, \tag{3.63}$$

where

$$\bar{r} = \sqrt{(\bar{x}-\bar{x}_0)^2+\bar{y}^2}, \qquad a_0 = \frac{2a}{\pi} \sin\frac{\pi \bar{x}_0}{a}.$$

The twisting moment is

$$M_{xy} = -(1-\nu)\frac{P}{4\pi}\frac{\bar{x}\bar{y}}{\bar{r}^2}.$$

The comparison of the expressions obtained with those for the circular plate may be interesting. We see that the first terms of formulae (3.10) and (3.63) would be identical if the radius of the circular plate were equal to $a = a_0$. Substituting $\bar{x}^2 = \bar{r}^2-\bar{y}^2$, we observe that only constant values differ in one solution from another. These differences arise from the different boundary conditions of the plate. The conclusion is that the stresses in the strip plate can be calculated by superposition of the stresses in the circular plate of radius $a_0 = \frac{2a}{\pi} \sin\frac{\pi \bar{x}_0}{a}$ and the simple bending caused by the constant moment

$$M_{yy} = -(1-\nu)\frac{P}{4\pi}.$$

Analogously, we can write the formula for the bending moment in the strip plate loaded by a force distributed over a small circle of radius c. Neglecting the small term of order c^2/a^2, we have for $\bar{x} = \bar{x}_0$, $\bar{y} = 0$

$$\left.\begin{matrix} M_{xx} \\ M_{yy} \end{matrix}\right\} = \frac{P}{4\pi}\left[-(1+\nu)\ln\frac{c}{a_0}+1\right]-(1\pm 1)\frac{1-\nu}{8\pi}P. \tag{3.64}$$

The solution for the plate of definite length can be derived by adding the boundary moments which will satisfy the new boundary conditions. We have then

$$x = \pm b/2, \quad w = 0, \quad M_{xx} = 0.$$

We obtain in this way the following solution by superposition of the previous solution with a new one

$$\left.\begin{matrix} M_{xx} \\ M_{yy} \end{matrix}\right\} = -\frac{1+\nu}{4\pi}P\ln\frac{r_0}{a_0}+\gamma_{\frac{1}{2}}\frac{P}{4\pi}+\frac{1-\nu}{4\pi}P\frac{y^2}{r^2},$$

where γ_1 and γ_2 are the coefficients which depend on the dimensions of the plate.

Table 2.

b/a	1.0	1.2	1.4	1.6	1.8	2.0	∞
γ_1	−0.565	−0.350	−0.211	−0.125	−0.073	−0.042	0
γ_2	0.135	0.115	0.085	0.057	0.037	0.023	0

Summing directly series (3.50) with the coefficients (3.51), we can find a closed form formula for the bending moments in the plate of definite length (see the solution due to G. J. Djanolidze in 1939 [3.4]).

In the case where the load P is distributed over the rectangular surface of the sides $c \times d$ the maximum bending moments can be represented in the form

$$M_{xx\,\max} = \beta_1 P, \qquad M_{yy\,\max} = \beta_2 P.$$

The values of the coefficients β_1 and β_2 for the square plate and plates with $b/a = 1.4$ and $b/a = 2$ for various sizes of the loaded rectangle are given in Table 10 on p. 539.

3.6 Particular solution accomplished by means of Fourier integral

3.6.1 *Infinite plate*

Let us consider an infinite plate loaded by a normal force P. Starting our analysis with the concentrated force expressed as a Fourier integral according to formula (3.41), we obtain a singular solution identical to that found previously. Let us assume that the origin of the system of the non-dimensional coordinates (x, y) $(x = \bar{x}/l,\ y = \bar{y}/l)$ is at the loading point, l being an arbitrary characteristic length. This expression then takes the form

$$q(x, y) = \frac{P}{\pi^2 l^2} \int_0^\infty \int_0^\infty \cos\alpha x \cos\beta y \, d\alpha \, d\beta. \tag{3.65}$$

The substitution of the deflection of the plate expressed by a similar integral in Eq. (2.10) yields

$$w(x, y) = \frac{Pl^2}{\pi^2 D} \int_0^\infty \int_0^\infty \frac{[1+(\eta-\varepsilon)(\alpha^2+\beta^2)]}{(\alpha^2+\beta^2)^2} \cos\alpha x \cos\beta y \, d\alpha \, d\beta, \tag{3.66}$$

where in the case of an isotropic plate

$$\eta = \frac{h^2}{5(1-\nu)l^2}, \qquad \varepsilon = \frac{\nu h^2}{10(1-\nu)l^2},$$

η represents the effect of transverse shear deformations, ε is the effect of transverse normal stress σ_{33}.

Formally this integral is divergent and does not exist. However, let us consider the results which can be obtained by the use of it.

The integral (3.66) can be evaluted with respect to α

$$w(x, y) = \frac{Pl^2}{4\pi D} \int_0^\infty \left[\frac{1}{\beta^3}(1+\beta x)+2(\eta-\varepsilon)\frac{1}{\beta}\right] e^{-\beta x} \cos\beta y \, d\beta. \tag{3.67}$$

The above integral is of course also divergent. This fact can be physically justified, as in the assumed conditions nothing was said about the way the plate is supported. The deflection of the plate defined by integral (3.66) can be interpreted as the deflection of the plate whose supports are at infinity. The deflection of such a plate is of course infinitely large. Let us, however, separate from this solution a local part expressing the action of the concentrated force. In order to do this we evaluate integral (3.67) from B to ∞. This yields (see Table 16, p. 554)

$$w_0 = \frac{Pl^2}{8\pi D}\left\{(r^2-4(\eta-\varepsilon))\left(\ln r+\gamma_0+\ln B+\sum_{k=1}^{\infty}(-B)^k\frac{r^k\cos k\vartheta}{kk!}\right)+\right.$$

$$\left.+\frac{e^{-Bx}}{B}\left(\frac{\cos By}{B}+x\cos By-y\sin By\right)\right\}, \qquad (3.68)$$

where $\vartheta = \arccos\frac{x}{r}$ (Fig. 19). B is here an arbitrary value, $r = \sqrt{x^2+y^2}$, and $\gamma_0 = 0.5772$ is Euler's constant.

Let us treat the above result as the solution of the differential equation consisting of the sum of the particular solution and the general one including a certain number of arbitrary constants. Let us omit the terms containing the arbitrary parameter B. We have then the singular part of the plate deflection which represents the action of the concentrated force and is identical with that calculated previously. The neglected terms which tend to infinity express the part of the plate deflection caused by the undefined boundary conditions. The bending moments as well as other internal forces can also be obtained from solution (3.67).

$$M_{xx} = \frac{P}{4\pi}\int_0^{\infty}\left[(1+\nu)\frac{1}{\beta}-\right.$$

$$\left.-(1-\nu)(x+2(\eta-\varepsilon)\beta)\right]e^{-\beta x}\cos\beta y\,d\beta+\frac{h^2}{5}\frac{\partial Q_x}{\partial x}, \qquad (3.69)$$

and similar expressions for the bending moments M_{yy} and M_{xy}. The integrals in this formula are divergent. This fact can also be justified by the undefined boundary conditions of the plate whose deflection is represented in formula (3.66). However, integrating from a certain parameter B to ∞, we have convergent integrals and can obtain the expression for the bending moment M_{xx} as

$$M_{xx} = \frac{P}{4\pi}\left\{-(1+\nu)\left(\ln r+\gamma_0+\ln B+\sum_{k=1}^{\infty}(-B)^k\frac{r^k\cos k\vartheta}{kk!}\right)-\right.$$

$$-(1-\nu)\left[\left(\frac{x}{r}+2(\eta-\varepsilon)\frac{B}{r}\right)\cos(By+\vartheta)-\right.$$

$$\left.\left.-2(\eta-\varepsilon)\frac{1}{r^2}\cos(By+2\vartheta)\right]e^{-Bx}\right\}+\frac{h^2}{5}\frac{\partial Q_x}{\partial x}. \qquad (3.70)$$

Besides the functions of coordinates x and y, the above formulae also include the term $\ln B$ which for $B \to 0$ decreases to $-\infty$. This term does not depend on the coordinates x, y and is constant over the whole area of the plate. It can appear only as the result of the infinite dimensions of the plate. Namely, it represents the moments caused by the fact that the supports of the plate are at infinity. Omitting this term and substituting $B = 0$, we obtain formulae (3.72) defining the bending moments near the loading point. In order to obtain the complete expressions for the bending moments, it is necessary to calculate the expressions $\partial Q_x/\partial x$ and $\partial Q_y/\partial y$.

Solving Eq. (2.3), we obtain

$$Q_x = -\frac{P}{\pi^2 l}\int_0^\infty\int_0^\infty \frac{\alpha \sin\alpha x \cos\beta y \, d\alpha \, d\beta}{\alpha^2+\beta^2}$$

$$= -\frac{P}{2\pi l}\frac{x}{x^2+y^2}, \tag{3.71}$$

$$Q_y = -\frac{P}{2\pi l}\frac{y}{x^2+y^2}.$$

On differentiation and substitution in $(3.40)_2$ we obtain the bending moments in the plate, $(B = 0)$:

$$\left.\begin{matrix} M_{xx} \\ M_{yy} \end{matrix}\right\} = \left\{\frac{P}{4\pi}\left[[-(1+\nu)\ln r \mp (1-\nu)\frac{x^2}{r^2} \pm 2\varepsilon(1-\nu)\frac{(x^2-y^2)}{r^4}\right]\right\} \begin{matrix} +C_1 \\ +C_2 \end{matrix}. \tag{3.72}$$

C_1 and C_2 are here arbitrary constants. Since $x^2 = r^2-y^2$, $r = \bar{r}/l$, the moment M_{xx} can be rewritten in the form identical with the previous form (3.10).

For the circular plate loaded at its centre by concentrated forces, the distance should be equal to the radius of the plate. The value of constants C_1 and C_2 depends on the edge conditions. For example, for the simply supported circular plate, the constants are

$$C_1 = (1-\nu)\frac{P}{4\pi}, \quad C_2 = 0.$$

For the clamped circular plate we have

$$C_1 = \frac{\nu P}{4\pi}, \quad C_2 = \frac{P}{4\pi}.$$

Of course, the same solution can be derived in the usual way. Then we use only the singular part of solution (3.66) and add the solution of the homogeneous equation of the plate.

It should be noticed here that the results obtained by rejecting the terms containing the indeterminate parameter B are identical with those which would be obtained by separating the so-called *finite part* from divergent integrals (3.66) and (3.69). This notion is due to J. Hadamard [3.26]. In the Polish literature it was used by W. Nowacki [3.81] and H. Zorski [3.74]. Nowacki defined the finite part of the divergent integral

$$\int_a^b f(\alpha)\,d\alpha$$

in the following way:

Let us assume that the integral

$$\int_{a+\varepsilon}^b f(\alpha)\,d\alpha$$

for $\varepsilon > 0$ is a convergent integral. Assuming $f(\alpha) = g(\alpha)+h(\alpha)$ and $G'(\alpha) = g(\alpha)$ and convergence of the integral

$$\int_a^b h(\alpha)\,d\alpha,$$

we have

$$\text{f.p.} \int_a^b f(\alpha)\,d\alpha = G(\alpha) + \int_a^b h(\alpha)\,d\alpha.$$

The finite part of the divergent integral can be calculated by multiple integration by parts. We have, for example,

$$\text{f.p.} \int_0^\infty \frac{\varphi(\alpha)}{\alpha}\,d\alpha = -\int_0^\infty \varphi'(\alpha)\ln\alpha\,d\alpha.$$

3.6.2 *Semi-infinite plate and wedge plate*

In the case where the semi-infinite plate is simply supported along the x axis and loaded symmetrically with respect to the y axis (Fig. 21c), the load can be represented by the integral ($x = \bar{x}/l$, $y = \bar{y}/l$)

$$q(x, y) = \frac{1}{l^2} \int_0^\infty \int_0^\infty q(\alpha, \beta) \cos \alpha x \sin \beta y \, d\alpha \, d\beta, \tag{3.73}$$

where

$$q(\alpha, \beta) = \frac{4l^2}{\pi^2} \int_0^\infty \int_0^\infty q(x, y) \cos \alpha x \sin \beta y \, dx \, dy.$$

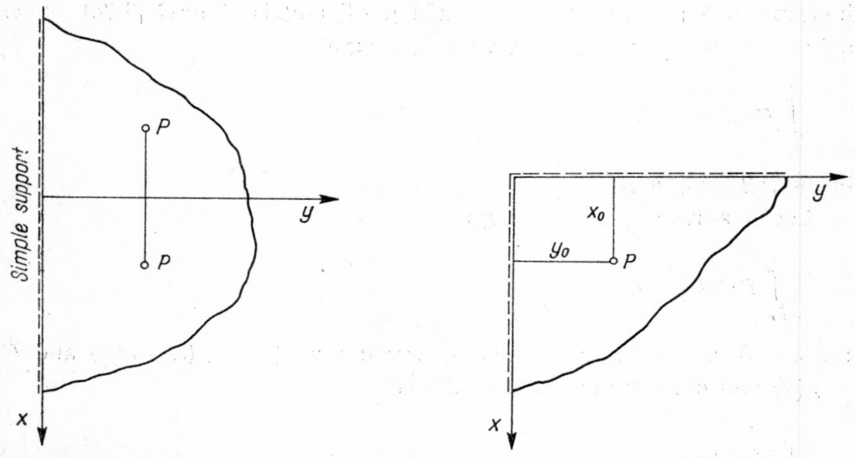

Fig. 21a), b). Simply supported semi-infinite and wedge plate.

We can express the deflection by a similar integral

$$w(x, y) = \frac{1}{l^2} \int_0^\infty \int_0^\infty w(\alpha, \beta) \cos \alpha x \sin \beta y \, d\alpha \, d\beta.$$

This function satisfies the boundary condition of simply supported edge at $y = 0$.

The case of an infinite plate simply supported at the two edges $x = 0$ and $y = 0$ can be solved by means of the integrals (Fig. 21b)

$$q(x, y) = \frac{1}{l^2} \int_0^\infty \int_0^\infty q(\alpha, \beta) \sin \alpha x \sin \alpha y \, d\alpha \, d\beta, \tag{3.74}$$

where

$$q(\alpha, \beta) = \frac{4l^2}{\pi^2} \int_0^\infty \int_0^\infty q(x, y) \sin \alpha x \sin \beta y \, dx \, dy.$$

Let us consider a case of a concentrated force acting at the point (x_0, y_0). Introducing it into Eq. (3.74) and neglecting the effect of transverse shear deformations, we obtain

$$q(\alpha, \beta) = \frac{4Pl^2}{\pi^2} \sin \alpha x_0 \sin \beta y_0 .$$

The deflection of the plate takes the form

$$w(x, y) = \frac{4Pl^2}{\pi^2 D} \int_0^\infty \int_0^\infty \frac{\sin \alpha x_0 \sin \beta y_0}{(\alpha^2+\beta^2)^2} \sin \alpha x \sin \beta y \, d\alpha \, d\beta . \tag{3.75}$$

Now, the above integral is convergent and can be evaluated analytically; we find

$$\begin{aligned} w = \frac{Pl^2}{16\pi D} \{ & [(x-x_0)^2+(y-y_0)^2] \ln [(x-x_0)^2+(y-y_0)^2] - \\ & - [(x-x_0)^2+(y+y_0)^2] \ln [(x-x_0)^2+(y+y_0)^2] + \\ & + [(x+x_0)^2+(y+y_0)^2] \ln [(x+x_0)^2+(y+y_0)^2] - \\ & - [x+x_0)^2+(y-y_0)^2] \ln [(x+x_0)^2+(y-y_0)^2] \} . \end{aligned} \tag{3.76}$$

As we have seen from the previous calculations the deflection of the infinite plate is also infinite; it can be finite only in the case where the load is in equilibrium with the supporting forces. It occurs, for example, for the plate on an elastic foundation.

3.7 Thick circular plate

3.7.1 *Equations of the theory of elasticity of the symmetrical body*

If we want to define exactly the stresses in the plate loaded over a very small area, for example over the surface of the circle of radius smaller than the thickness of the plate ($c < h$), we can procede in the following way:

Let us separate from the plate a circular cylinder surrounding the loaded place and clamped at the circumference. Let us calculate the stresses in this part of the plate based on the equations of the theory of elasticity of three-dimensional bodies. This problem presented in Fig. 22 had been considered in 1925 by A. Nadai [2.1] and in 1933 by S. Woinowsky-Krieger [3.2].

Let us put together the equations of the theory of elasticity of the symmetrically loaded three-dimensional body and assume that its displacements are also symmetrical. The displacement of the arbitrary point of the body can be defined by two components: the radial displacement u and axial displacement w. These displacements are functions of the variables r and z. The stresses: radial σ_{rr}, circumferential $\sigma_{\vartheta\vartheta}$, axial normal σ_{zz}, and tangential τ_{rz} act on the infinitesimal element of the body. The state of strain is defined by the

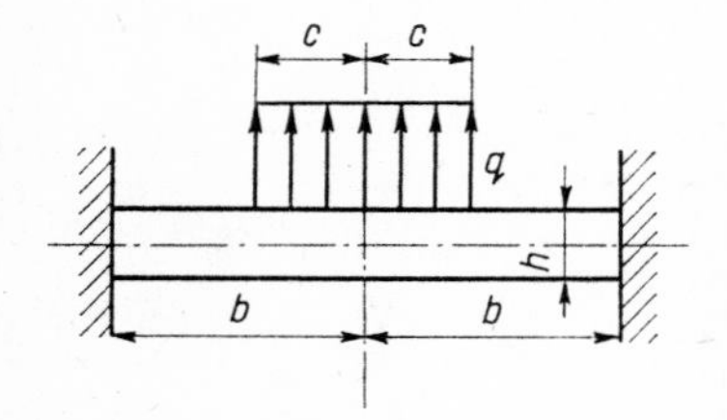

Fig. 22. Thick circular plate loaded symmetrically.

components ε_{rr}, $\varepsilon_{\vartheta\vartheta}$, ε_{zz}, γ_{rz} related to the displacements by the following formulae:

$$\varepsilon_{rr} = \frac{\partial u}{\partial r}, \qquad \varepsilon_{\vartheta\vartheta} = \frac{u}{r}, \qquad \varepsilon_{zz} = \frac{\partial w}{\partial z},$$
$$\gamma_{rz} = \frac{\partial u}{\partial z} + \frac{\partial w}{\partial r}. \tag{3.77}$$

Using Hooke's law, we find the relations

$$\sigma_{rr} = 2G\left(\frac{\partial u}{\partial r} + \frac{\nu e}{1-2\nu}\right), \qquad \sigma_{\vartheta\vartheta} = 2G\left(\frac{u}{r} + \frac{\nu e}{1-2\nu}\right),$$
$$\sigma_{zz} = 2G\left(\frac{\partial w}{\partial z} + \frac{\nu e}{1-2\nu}\right), \qquad \tau_{rz} = G\left(\frac{\partial u}{\partial z} + \frac{\partial w}{\partial r}\right), \tag{3.78}$$

where e is a volume strain:

$$e = \varepsilon_{rr} + \varepsilon_{\vartheta\vartheta} + \varepsilon_{zz} = \frac{u}{r} + \frac{\partial u}{\partial r} + \frac{\partial w}{\partial z}, \tag{3.79}$$

and G is the shear modulus and ν Poisson's ratio.

Introducing the above relations into the equilibrium equations

$$\frac{\partial \sigma_{rr}}{\partial r} + \frac{\sigma_{rr} - \sigma_{\vartheta\vartheta}}{r} + \frac{\partial \tau_{rz}}{\partial z} = 0,$$
$$\frac{\partial \sigma_{zz}}{\partial z} + \frac{\partial \tau_{rz}}{\partial r} + \frac{\tau_{rz}}{r} = 0, \tag{3.80}$$

we find the following set of equations:

$$(1-2\nu)\Delta w+\frac{\partial e}{\partial z}=0, \quad (1-2\nu)\left(\Delta u-\frac{u}{r^2}\right)+\frac{\partial e}{\partial r}=0 \tag{3.81}$$

for the functions u and w. Here Δ is the Laplace operator which takes the following form in the cylindrical coordinates r, z:

$$\Delta=\frac{\partial^2}{\partial z^2}+\frac{\partial^2}{\partial r^2}+\frac{1}{r}\frac{\partial}{\partial r}.$$

As the volume strain e in the elastic body satisfies the equation $\Delta e=0$, the set of two Eqs. (3.81) takes the form

$$\Delta^2 w=0, \qquad \left(\Delta-\frac{1}{r^2}\right)^2 u=0. \tag{3.82}$$

The above-given equations describe the axi-symmetrical state of stress and displacement of the elastic body. The solution of the above equations can be presented as a combination of an exponential function of the variable z and Bessel function of the variable r.

$$\begin{aligned} w &= \mathrm{e}^{\pm\alpha z}J_0(\alpha r) \quad \text{and} \quad w = z\mathrm{e}^{\pm\alpha z}J_0(\alpha r),\\ u &= \mathrm{e}^{\pm\beta z}J_1(\beta r) \quad \text{and} \quad u = z\mathrm{e}^{\pm\beta r}J_1(\beta r), \end{aligned} \tag{3.83}$$

where α and β are arbitrary parameters.

3.7.2 *The stresses in a semi-infinite elastic body*

Before we start to solve the problem of the thick circular plate, let us solve first a simple case, namely the problem concerning the semi-infinite elastic body bounded by the plane $z=0$ and loaded symmetrically with respect to the $r=0$ axis. This problem was solved by Boussinesq [3.89] and Hertz [3.88].

If in the plane $z=0$ there acts only the normal load $(\tau_{rz}=0)_{z=0}$ the solution of Eqs. (3.82) can be presented in the form:

$$\begin{aligned} w &= -A[2(1-\nu)+\alpha z]\mathrm{e}^{-\alpha z}J_0(\alpha r),\\ u &= A[1-2\nu-\alpha z]\mathrm{e}^{-\alpha z}J_1(\alpha r), \end{aligned} \tag{3.84}$$

where A is a constant value indefinite at the moment. Let us consider this value as a function of the parameter α:

$$A=A(\alpha).$$

The arbitrary axially symmetrical load acting on the plane $z = 0$ can be presented in the form of an integral of the Bessel function

$$p(r) = \int_0^\infty p(\alpha) J_0(\alpha r)\, \alpha\, \mathrm{d}\alpha,$$

where

$$p(\alpha) = \int_0^\infty p(r) J_0(\alpha r)\, r\, \mathrm{d}r. \tag{3.85}$$

The quantity $A(\alpha)$ can be defined from the boundary condition: for $z = 0$

$$\sigma_{zz} = -p(r).$$

In order to define the state of stress and displacement by similar functions let us multiply both sides of Eq. (3.84) by $\mathrm{d}\alpha$ and integrate between the limits $0; \infty$. In this way we find the following formulae

$$u = \int_0^\infty A(\alpha)[1-2\nu-\alpha z]\mathrm{e}^{-\alpha z} J_1(\alpha r)\, \mathrm{d}\alpha,$$

$$w = -\int_0^\infty A(\alpha)[2(1-\nu)+\alpha z]\mathrm{e}^{-\alpha z} J_0(\alpha r)\, \mathrm{d}\alpha, \tag{3.86}$$

$$e = 2(1-2\nu)\int_0^\infty A(\alpha)\, \alpha \mathrm{e}^{-\alpha z} J_0(\alpha r)\, \mathrm{d}\alpha.$$

Introducing into Eqs. $(3.78)_3$ the above relations, we find the condition for $z = 0$

$$\sigma_{zz} = 2G\int_0^\infty A(\alpha) J_0(\alpha r)\, \alpha\, \mathrm{d}\alpha;$$

then (3.87)

$$2GA(\alpha) = -p(\alpha) = -\int_0^\infty p(r)\, r J_0(\alpha r)\, \mathrm{d}r.$$

In order to define the distribution of the stress σ_{zz} resulting from the given pressure $p(r)$ we have to calculate the displacements w and u and the volume strain e from Eqs. (3.79). We have the following formula for the displacement u in the plane $z = 0$:

$$u = \frac{1-2\nu}{2G} \int_0^\infty p(\varrho)\varrho \,\mathrm{d}\varrho \int_0^\infty J_1(\alpha r) J_0(\alpha\varrho)\,\mathrm{d}\alpha.$$

The second integral on the right-hand side of the above equation, identical with the integral (3.45) presents the discontinous function

$$\int_0^\infty J_1(\alpha r) J_0(\alpha\varrho)\,\mathrm{d}(\alpha r) = \begin{cases} 1 & \text{if} \quad 0 < \varrho < r, \\ 0 & \text{if} \quad r < \varrho < \infty. \end{cases} \tag{3.88}$$

Then the radial displacement is given by the integral

$$u = -\frac{1-2\nu}{2G} \frac{1}{r} \int_0^r p(\varrho)\varrho \,\mathrm{d}\varrho. \tag{3.89}$$

The limit ∞ can be replaced here by r because the integral (3.88) vanishes for $\varrho > r$.

The last integral multiplied by 2π represents the force resultant of the pressure p inside the circle $r = c$. Introducing

$$r^2 p_m = 2 \int_0^r p(\varrho)\varrho \,\mathrm{d}\varrho,$$

we have

$$u = -\frac{1-2\nu}{4G} p_m r.$$

The stresses at the plane $z = 0$ are

$$\sigma_{zz} = -p, \qquad \sigma_{rr} = -p + \frac{1-2\nu}{2} p_m,$$

$$\sigma_{\vartheta\vartheta} = -2\nu p - \frac{1-2\nu}{2} p_m; \qquad \tau_{rz} = 0.$$

If the load acting on the plane $z = 0$ is applied over the surface of the circle $r = c$ and for $r > c$, $p = 0$, then the stresses outside the circle are

$$\sigma_{rr} = -\sigma_{\vartheta\vartheta} = \frac{1-2\nu}{2\pi r^2} P,$$

where P is the resultant of the pressure $p(r)$. These stresses are independent of the way in which the force P is distributed over the surface of the circle $r = c$. When the pressure is uniformly distributed, we have

$$p(r) = \frac{P}{\pi c^2}, \qquad 0 \leqslant r \leqslant c, \quad p(r) = 0, \qquad r > c.$$

then

$$p(\alpha) = \frac{P}{\pi c^2} \int_0^\infty r J_0(\alpha r) \mathrm{d}r = \frac{P}{\pi c} J_1(\alpha c).$$

If we consider the case of the concentrated force

$$p(\alpha) = \frac{P}{\pi} \lim_{c \to 0} \frac{J_1(\alpha c)}{c} = \frac{P\alpha}{2\pi},$$

then

$$A(\alpha) = -\frac{P\alpha}{2G\pi}.$$

Introducing it into formulae (3.78), we find on integration*

$$\begin{aligned}
\sigma_{zz} &= -\frac{P}{2\pi} \frac{3z^3}{(r^2+z^2)^{5/2}}, \qquad \tau_{rz} = -\frac{P}{2\pi} \frac{3rz^2}{(r^2+z^2)^{5/2}}, \\
\sigma_{rr} &= -\frac{P}{2\pi} \left\{ \frac{3r^2 z}{(r^2+z^2)^{5/2}} - (1-2\nu) \left[\frac{1}{r^2} - \frac{z}{r^2(r^2+z^2)^{1/2}} \right] \right\}, \\
\sigma_{\vartheta\vartheta} &= \frac{(1-2\nu)P}{2\pi} \left[\frac{z(2r^2+z^2)}{r^2(r^2+z^2)^{3/2}} - \frac{1}{r^2} \right].
\end{aligned} \tag{3.90}$$

The displacements take the form

$$\begin{aligned}
u &= \frac{P}{4\pi G} \left\{ \frac{rz}{(r^2+z^2)^{3/2}} - (1-2\nu) \left[\frac{1}{r} - \frac{z}{r(r^2+z^2)^{1/2}} \right] \right\}, \\
w &= \frac{P}{4\pi G} \left[\frac{z^2}{(r^2+z^2)^{3/2}} + 2(1-\nu) \frac{1}{(r^2+z^2)^{1/2}} \right].
\end{aligned} \tag{3.91}$$

At the edge $z = 0$

$$u = -\frac{(1-2\nu)P}{4G\pi r}, \qquad w = \frac{(1-\nu)P}{2G\pi r}.$$

It can be observed that if $r \to 0$, the displacements tend to infinity. However, we known that the case of a concentrated force has no physical meaning.

---

* r and z are here the dimensional coordinates.

3.7.3 *Axi-symmetrically loaded thick plate*

Let us consider a thick plate bounded by the planes $z = 0$ and $z = h$ and the cylindrical surface $r = a$. The plate is loaded by the pressure $-p(r)$ distributed over the surface of the circle of the radius $r = c$. Let us assume that the plate is supported along the circumference $r = a$. The problem was solved by A. Nadai [2.1] and his results will be cited below. Let us modify the solution (3.84) and represent the displacements of the plate in the form of the following series:

$$u_1 = \sum_{\lambda_i} A_{\lambda_i}\left[1-2\nu-\lambda_i\frac{z}{a}\right] e^{-\lambda_i z/a} J_1\left(\lambda_i\frac{r}{a}\right),$$

$$w_1 = -\sum_{\lambda_i} A_{\lambda_i}\left[2(1-\nu)+\frac{\lambda_i z}{a}\right] e^{-\lambda_i z/a} J_0\left(\lambda_i\frac{r}{a}\right). \tag{3.92}$$

These formulae differ from formulae (3.84) only by the notations (α is replaced by λ_i and the quantity A by A_{λ_i}), λ_i here denotes the sequence of real numbers and the sign of the sum denotes the summation of the all terms with respect to λ_i. The value of λ_i can be chosen in such a way that the condition that the displacement w is zero at the edge is satisfied. We have $w = 0$ for $r = a$ which gives $J_0(\lambda_i) = 0$; λ_i is then the root of the Bessel function of the zero order.

$$\lambda_1 = 2.4048, \qquad \lambda_2 = 5.5201, \qquad \lambda_3 = 8.6537, \ \ldots \tag{3.93}$$

Expressions (3.92) with λ_i (3.93) satisfy the condition $\tau_{rz} = 0$ for $z = 0$. The constant values A_{λ_i} can be defined by the equation

$$\sigma_{zz} = -p(r) \quad \text{for} \quad z = 0.$$

We obtain from Eqs. (3.92) and (3.78)

$$\sigma_{zz} = 2G\sum_{i=1}^{\infty} A_{\lambda_i}\frac{\lambda_i}{a}\left(1+\lambda_i\frac{z}{a}\right) e^{-\lambda_i z/a} J_0\left(\lambda_i\frac{tr}{a}\right) = -p(tr). \tag{3.94}$$

In order to calculate the values A_{λ_i} we multiply both sides of Eq. (3.94) by $\frac{r}{a}J_0\left(\frac{r}{a}\right)\mathrm{d}\left(\frac{r}{a}\right)$ and integrate between 0 and ∞. Introducing $x = r/a$ and $z = 0$, we have

$$A_{\lambda_i} = -\frac{\int_0^1 p(x)xJ_0(\lambda_i x)\mathrm{d}x}{2G\lambda_i \int_0^1 xJ_0^2(\lambda_i x)\mathrm{d}x}. \tag{3.95}$$

If the load is distributed uniformly over the surface of the circle of radius $r = c$, $p = P/\pi c^2$, and vanishes for $r > c$, we find

$$A_{\lambda_i} = -\frac{P}{\pi Gc}\frac{J_1(\lambda_i c/a)}{\lambda^2 J_1^2(\lambda_i)}.$$

If

$$c \to 0, \qquad \frac{J_1(c)}{c} \to \frac{1}{2},$$

then

$$A_{\lambda_i} = \frac{P}{2\pi aG}\frac{1}{\lambda_i J_1^2(\lambda_i)}.$$

We obtain at the moment the solution which satisfies the boundary conditions on the surfaces $z = 0$ and $r = a$. On the second free plane $z = h$, the boundary conditions are not satisfied. There appear the stresses σ_{zz} and τ_{rz} which can be removed by the addition of the solution of the homogeneous equations (3.82) in the form

$$\begin{aligned} u_2 &= \sum_\lambda \left[A\cosh\frac{\lambda z}{a} + B\frac{\lambda z}{a}\cosh\frac{\lambda z}{a} - \right. \\ &\quad \left. - C\sinh\frac{\lambda z}{a} + D\frac{\lambda z}{a}\sinh\frac{\lambda z}{a}\right] J_1\left(\frac{\lambda r}{a}\right), \\ w_2 &= \sum_\lambda \left\{-A\sinh\frac{\lambda z}{a} + B\left[(3-4\nu)\cosh\frac{\lambda z}{a} - \frac{\lambda z}{a}\sinh\frac{\lambda z}{a}\right] - \right. \\ &\quad \left. - C\cosh\frac{\lambda z}{a} + D\left[(3-4\nu)\sinh\frac{\lambda z}{a} - \frac{\lambda z}{a}\cosh\frac{\lambda z}{a}\right]\right\} J_0\left(\frac{\lambda r}{a}\right), \end{aligned} \tag{3.96}$$

where $\lambda = \lambda_i$.

The constant values A, B, C, D can be defined from the following boundary conditions:

$$\begin{aligned} z &= 0, \quad \sigma_{zz_2} = 0, \qquad \tau_{rz_2} = 0, \\ z &= h, \quad \sigma_{zz_1} + \sigma_{zz_2} = 0, \quad \tau_{rz_1} + \tau_{rz_2} = 0. \end{aligned} \tag{3.97}$$

Solving the above set of equations with respect to A, B, C, D, we satisfy the boundary conditions at both free planes of the plate. At the cylindrical surface we have $r = a$, $u_2 = 0$, and the complete state of displacement of the plate is defined by

$$w = w_1 + w_2, \qquad u = u_1 + u_2.$$

Analysing the solution obtained, we can prove that the displacement w_2 for small values of the ratios h/a and z/a tends in the limit to the displacement which would be obtained from the classical theory of plates.

We have for $h/a \to 0$ and $z/a \to 0$

$$w_2 = \frac{6(1-\nu)Pa^2}{\pi G h^3} \sum_{\lambda_i} \frac{J_0(\lambda_i r/a)}{\lambda_i^4 J_1^2(\lambda_i)}$$

$$\to \frac{3(1-\nu)Pa^2}{4\pi G h^3}\left[\frac{r^2}{a^2}\ln\frac{r}{a} + \frac{1}{2}\left(1-\frac{r^2}{a^2}\right)\right],$$

which is in agreement with (3.7).

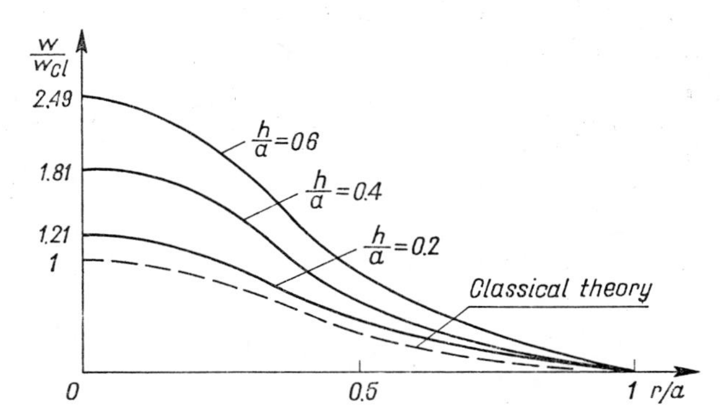

Fig. 23. Deflection of a thick circular plate loaded at its centre by a concentrated force. w_{cl}—deflection obtained by means of the classical theory of thin plates (S. Woinowsky-Krieger [3.7]).

Based on the above formulae we obtain the following results:

The total deflection has a finite magnitude at the point $z = h/2$, $r = 0$. The deflection of the central portion due to a local disturbance in stress distribution near the point of application of the load becomes of practical importance when the thickness of the plate is not very small as compared with its radius. This deflection is affected very little by the conditions at the edge and can be evaluated approximately by means of the curves calculated for the circular plate with built-in edges. The curves for different ratios of thickness to the radius a ($h/a = 0.2, 0.4, 0.6$) are shown in Fig. 23. The dotted-line curve in this figure is obtained by using Eq. (3.7) and neglecting the effect of transverse shear deformations. The additional deflections due to local stress distribution are equal to the differences between the coordinates of full lines and those of dotted lines and can be expressed by means of the approximate formula ($r = 0$), $h/a \leqslant 0.4$.

$$w_c \approx \frac{5Ph^2}{16\pi D}. \tag{3.98}$$

It can be proved, comparing formulae (3.98) and (3.7), that in order to find the correct value of the maximum deflection of the plate at $r = 0$ loaded by a concentrated force at its centre, we calculate the deflection resulting from the effect of transverse shear deformations at a small distance $r = 0.076a$ from the centre of the plate.

The maximum tensile stress at the centre of the upper face of the plate is different than calculated from formula (3.8), and for simply supported edges can be expressed by the approximate formula

$$\sigma_{rr} = \sigma_{\vartheta\vartheta} = \frac{6P}{4\pi h^2}\left[(1+\nu)\ln\frac{a}{c}+1-\frac{1-\nu}{4}\frac{c^2}{a^2}\right]+\frac{P}{\pi c^2}\left[\frac{1+2\nu}{2}-(1+\nu)\alpha_0\right], \tag{3.99}$$

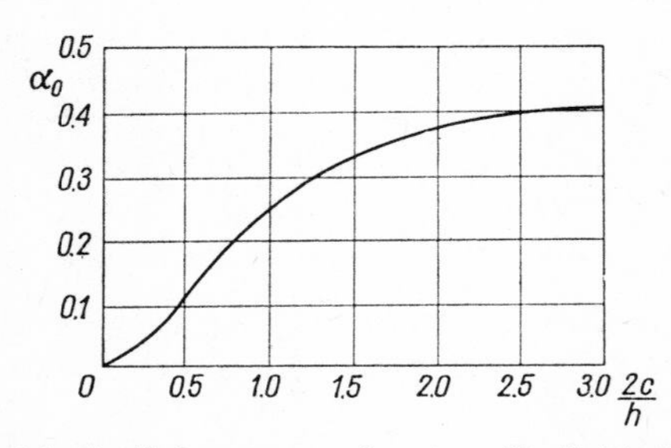

Fig. 24. Coefficient α_0 as a function of $2c/h$ (S. Woinowsky-Krieger [3.7]).

where α_0 is the coefficient depending on $2c/h$, changing from $\alpha_0 = 0.0106$ for $2c/h = 0.1$ to $\alpha_0 = 0.398$ for $2c/h = 2.5$ (see Fig. 24). For $c \to 0$ the stress increases infinitely.

We obtain the maximum compressive stress for a small c, practically independent of $2c/h$. For the simply supported plate it can be calculated approximately from the expression

$$\sigma_{rr} = \sigma_{\vartheta\vartheta} = -\frac{P}{h^2}\left[(1+\nu)\left(0.485\ln\frac{a}{h}+0.52\right)+0.48\right]. \tag{3.100}$$

These stresses occur at the centre of the lower surface of the plate.

We have for a clamped plate the maximum tensile stress at the centre of the upper surface given by the following formulae

$$\sigma_{rr} = \sigma_{\vartheta\vartheta} = \frac{6P}{4\pi h^2}\left[(1+\nu)\ln\frac{a}{c} - \frac{1-\nu}{4}\frac{c^2}{a^2}\right] +$$

$$+\frac{P}{\pi c^2}\left[\frac{1+2\nu}{2}-(1+\nu)\alpha_0\right]. \tag{3.101}$$

The maximum compressive stresses at the centre of the lower surface of the clamped plate are expressed by

$$\sigma_{rr} = \sigma_{\vartheta\vartheta} = -(1+\nu)\frac{P}{h^2}\left(0.485\ln\frac{a}{h}+0.52\right). \tag{3.102}$$

The stresses obtained from formulae (3.99), (3.100) are compared with the stresses obtained from the classical theory in Fig. 25. We see that only in the case where c is smaller than h the above formulae should be used. The dotted line represents the stresses resulting from the improved theory. They were obtained from (2.7) by the assumption that $N_{rr} = N_{\vartheta\vartheta} = P/2(1-\nu)h$. We observe that the maximum stresses at the upper surface of the plate calculated in this way are much closer to those resulting from the thick plate theory. The stress distribution across the thickness of the clamped circular plate $h/a = 0.4$; $c/a = 0.1$ and $\nu = 0.3$ is given in Fig. 26.

At the centre, the stresses σ_{rr} and $\sigma_{\vartheta\vartheta}$ are not simply proportional to the distance z. Also the distribution of tensile stress σ_{zz} normal to the surface of the plate does not exactly conform at $r = 0$ with the distribution given by (1.81). The dotted lines show the value calculated from (3.8) by adding the moment $M = -P/4$ caused by the different boundary conditions (built-in edges). We observe also the difference between the assumed and calculated distribution of tangential stresses τ_{rz}, calculated for the radius $r = c$. However, already at a small distance from this place the distribution of the stresses τ_{rz} becomes similar to the assumed.

The comparison of the stresses at the upper surface of the thick plate calculated exactly and from the classical theory of plates is presented in Fig. 27.

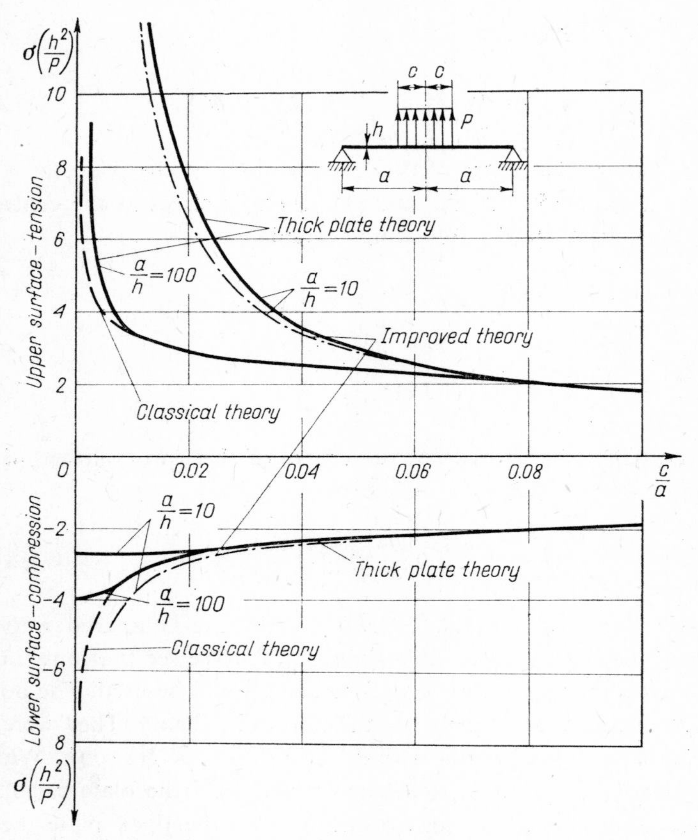

Fig. 25. Stresses $\sigma_{rr} = \sigma_{\vartheta\vartheta} = \sigma$ on upper and lower surface at the centre of plate for various c/a.

We observe the jump of the calculated stresses $\sigma_{\vartheta\vartheta}$ at the radius $r = c$. Figure 28 presents the comparison of the distribution of the stresses at the upper surface of the thick plate $h/a = 0.4$, $\nu = 0.3$ with clamped edges and loaded at the centre by a concentrated normal force.

The differences between the results of the classical and improved theory of plates and the calculations based on the three-dimensional theory of elasticity appear at the closed vicinity of the loaded region as well as at the edge of the plate.

The compatibility conditions on the thickness of the plate not being fulfilled is the main reason for the differences. It follows from the above com-

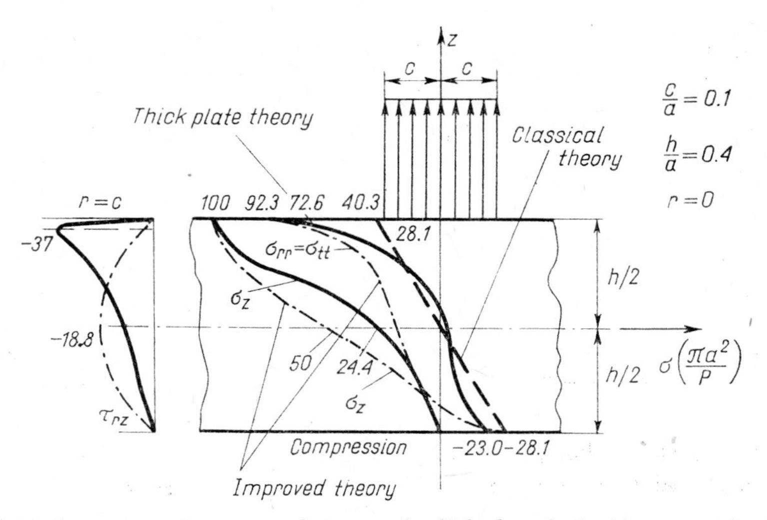

Fig. 26. Comparison of stresses at the centre of a thick plate obtained by means of various plate equations; improved theory means here the equations taking into account the effect of transverse shear deformations.

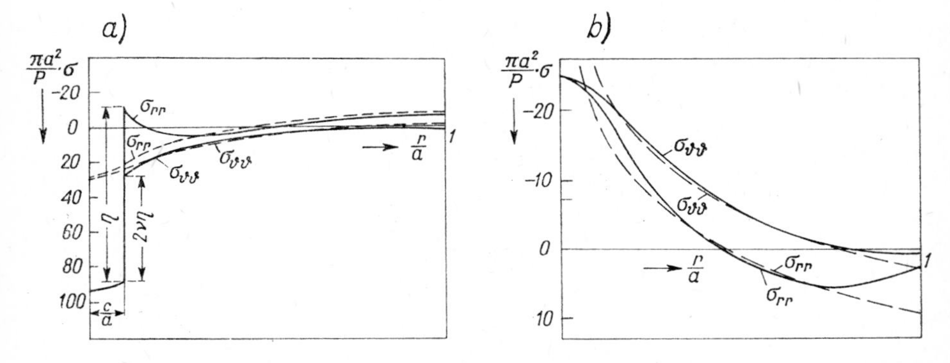

Fig. 27. Distribution of the stresses σ_{rr} and $\sigma_{\vartheta\vartheta}$ at upper and lower surface of a thick plate loaded by af orce distributed uniformly over the surface πc^2 (S. Woinowsky-Krieger [3.7]).

parison that taking into account the effects of transverse shear deformations and transverse normal stresses we obtain results closer to the results of exact calculations based on the three-dimensional equations of the theory of elasticity. However, if we consider the close vicinity of the point of application of the concentrated force ($r < 0.1h$) these improvements are not satisfactory because they do not ensure even the correct sign of the singular terms.

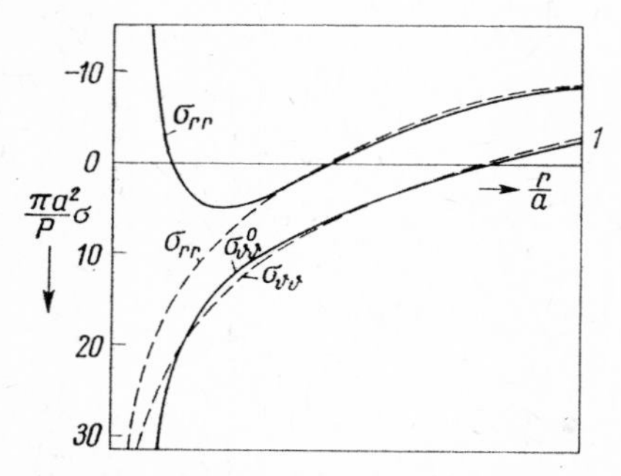

Fig. 28. Distribution of the stresses σ_{rr} and $\sigma_{\vartheta\vartheta}$ in a thick circular plate loaded by a concentrated force at its centre (S. Woinowsky-Krieger [3.7]).

This is caused also by the approximate expressions assumed for the transverse normal and shear stresses which are exact only in the case where the load $q(x, y)$ is constant in the whole area of the plate. Therefore we obtain more exact results when the load is distributed over a small but finite surface (Fig. 26). The effects of transverse shear and normal stresses are more important in the case of sandwich and anisotropic plates.

3.8 Plate on an elastic foundation

Let us consider an infinitely large plate resting on an elastic foundation and subjected to a concentrated force normal to the plate surface. The load can also be expressed with the aid of a Fourier integral. The equation of the plate resting on the elastic foundation can be derived by transforming Eq. $(2.1)_1$.

Let us assume that the elastic foundation reacts on the plate proportionally to its deflections. Then $p(x, y) = kw$, where k is a coefficient of proportionality. Equation $(2.1)_1$ takes the form

$$D\Delta\Delta w = [1-l^2(\eta-\varepsilon)\Delta](q-p), \tag{3.103}$$

where $q(x, y)$ is the active load and $p(x, y)$ is the reaction of the foundation.

Using the non-dimensional system of coordinates x, y $(x = \bar{x}/y,\ y = \bar{y}/l)$ with a characteristics length $l = \sqrt[4]{D/k}$, and introducing the relation $p = kw$, we have

$$\Delta\Delta w+(1-(\eta-\varepsilon)\Delta)w = (1-(\eta-\varepsilon)\Delta)\frac{ql^4}{\Delta}. \tag{3.104}$$

The displacement of the plate is also expected to have the form of a Fourier integral. After some manipulation the following solution for the displacement is found:

$$w = \frac{Pl^2}{\pi^2 D} \int_0^\infty \int_0^\infty \frac{1+(\eta-\varepsilon)(\alpha^2+\beta^2)}{(\alpha^2+\beta^2)^2+(\eta-\varepsilon)(\alpha^2+\beta^2)+1} \cos\alpha x \cos\beta y \, d\alpha \, d\beta. \tag{3.105}$$

In order to evaluate the above integral, let us introduce the system of polar coordinates (r, ϑ) $(r = \bar{r}/l)$ connected with the system (x, y) by the relations (Fig. 19) $x^2+y^2 = r^2$, $\alpha^2+\beta^2 = \gamma^2$ and $\alpha = \gamma\cos\varphi$, $\beta = \gamma\sin\varphi$. As

$$\int_0^{\pi/2} \cos(\gamma x \cos\varphi)\cos(\gamma y \sin\varphi)\, d\varphi = \frac{\pi}{2} J_0(\gamma r),$$

where $J_0(r)$ is Bessel's function of zero order, we obtain

$$w = \frac{Pl^2}{2\pi D} \int_0^\infty [1+(\eta-\varepsilon)\gamma^2] \frac{\gamma J_0(\gamma r)\, d\gamma}{D(\gamma)}. \tag{3.106}$$

The second term in the denominator proportional to $(\eta-\varepsilon)$ makes the analytical evaluation of this integral difficult. Therefore we evaluate it approximately, transforming the denominator as follows:

$$D(\gamma) = \gamma^4+(\eta-\varepsilon)\gamma^2+1 = (\gamma^4+1)(1+R), \quad \text{where} \quad R = (\eta-\varepsilon)\frac{\gamma^2}{\gamma^4+1}.$$

Since R is usually much smaller than 1 $(R \ll 1)$, in the case of the isotropic plate

$$R = \frac{2-\nu}{1-\nu} \frac{\sqrt{3(1-\nu^2)}}{5} \sqrt{\frac{kh}{E}} \frac{\gamma^2}{\gamma^4+1},$$

we can express the fraction with a good accuracy as

$$\frac{1}{(\gamma^4+1)(R+1)} = \frac{1-R+R^2+\dots}{\gamma^4+1} \approx \frac{1-R}{\gamma^4+1}.$$

The deflection of the plate takes then the form of the following integral:

$$w = \frac{Pl^2}{2\pi D} \int_0^\infty \frac{1+(\eta-\varepsilon)\gamma^2}{\gamma^4+1} \left(1-(\eta-\varepsilon)\frac{\gamma^2}{\gamma^4+1}+\dots\right) J_0(\gamma r)\gamma \, d\gamma. \tag{3.107}$$

This integral can be evaluated in a closed form which yields the expression

$$w = \frac{Pl^2}{2\pi D}\left[-\operatorname{kei} r + (\eta-\varepsilon)\left(\operatorname{ker} r + \frac{r}{4}\operatorname{ker}' r - (\eta-\varepsilon)\operatorname{kei} r + \ldots\right)\right] \tag{3.108}$$

where kei r and ker r are called *Thomson's (Kelvin's) functions*. The functions ker r and kei r are determined by the following series:

$$\operatorname{ker} r = \left(\ln\frac{2}{r} - \gamma_0\right)\operatorname{ber} r + \frac{\pi}{4}\operatorname{bei} r + \sum_{k=1}^{\infty}\frac{(-1)^k r^{4k}}{2^{4k}[(2k)!]^2}\sum_{s=1}^{2k}\frac{1}{s},$$

$$\operatorname{kei} r = \left(\ln\frac{2}{r} - \gamma_0\right)\operatorname{bei} r - \frac{\pi}{4}\operatorname{ber} r - \sum_{k=1}^{\infty}\frac{(-1)^k r^{4k-2}}{2^{4k-2}[(2k-1)!]^2}\sum_{s=1}^{2k-1}\frac{1}{s}, \tag{3.109}$$

where ber r and bei r are also called the *Thomson functions* and are expressed by the series

$$\operatorname{ber} r = \sum_{k=0}^{\infty}\frac{(-1)^k r^{4k}}{2^{4k}[(2k)!]^2},$$

$$\operatorname{bei} r = \sum_{k=0}^{\infty}\frac{(-1)^k r^{4k+2}}{2^{4k+2}[(2k+1)!]^2}. \tag{3.101}$$

The values of the Thomson functions can be found in tables of special functions. These functions for small magnitudes of r are expressed by the power series

$$\operatorname{ker} r = -\ln\frac{r}{2} - \gamma_0 + \pi\frac{r^2}{16} + \ldots,$$

$$\operatorname{kei} r = -\frac{r^2}{4}\ln\frac{r}{2} - \frac{\pi}{4} + (1-\gamma_0)\frac{r^2}{4} + \ldots,$$

$$\operatorname{ker}' r = -\frac{1}{r} + \frac{\pi r}{8} + \frac{r^3}{16}\left(\ln\frac{r}{2} + \gamma_0 - \frac{5}{4}\right) + \ldots,$$

$$\operatorname{kei}' r = -\frac{1}{2} r\ln\frac{r}{2} + \left(\frac{1}{2} - \gamma_0\right)\frac{r}{2} + \frac{\pi}{64} r^3 + \ldots,$$

where $\operatorname{ker}' r = \dfrac{d}{dr}(\operatorname{ker} r)$, and $\gamma_0 = 0.5772$ is Euler's constant. The functions ker r and kei r and their first derivatives are presented graphically in Fig. 29 and given in Table 12 on p. 544.

The second derivatives of the functions ker r, kei r and ber r, bei r can be obtained from the following relations:

$$
\begin{aligned}
\mathrm{ker}'' r &= -\mathrm{kei}\, r - \frac{1}{r}\,\mathrm{ker}' r, \\
\mathrm{ber}'' r &= -\mathrm{bei}\, r - \frac{1}{r}\,\mathrm{ber}' r, \\
\mathrm{kei}'' r &= \mathrm{ker}\, r - \frac{1}{r}\,\mathrm{kei}' r, \\
\mathrm{bei}'' r &= \mathrm{ber}\, r - \frac{1}{r}\,\mathrm{bei}' r.
\end{aligned}
\tag{3.111}
$$

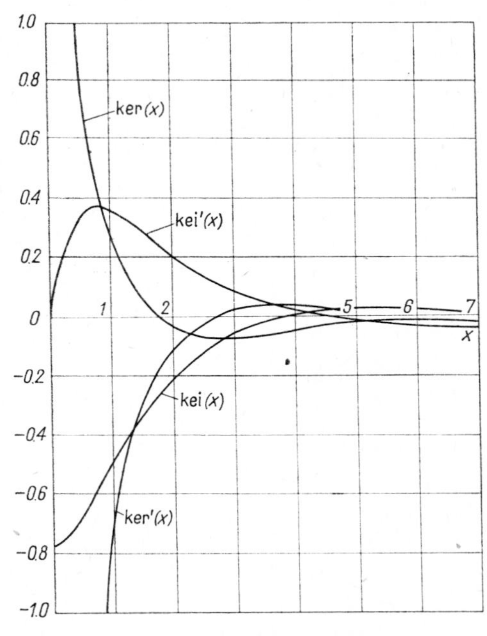

Fig. 29. Functions ker x, kei x, *and their derivatives.*

For small $r \ll 1$, a simple approximate formula for the deflection of the plate can be given.

$$
w = \frac{Pl^2}{8\pi D}\left[\pi + (r^2 - 4(\eta - \varepsilon))\left(\ln\frac{r}{2} + \gamma_0 + \frac{1}{4}\right) - r^2\right]. \tag{3.112}
$$

We observe that near the loading point the deflection of the plate resting on an elastic foundation is defined by similar expressions to those obtained previously for the simply supported plate. Taking into account the displacements produced by tangential force, we find the expression containing $\ln r$ approaching infinity when r approaches zero. But from the theory of thick plates (see (3.98)) we know that the deflection is finite. To calculate the maximum deflection, we can proceed as previously, namely calculate the deflection from formula (3.108) for $r = b \approx 2.5\,h$ and add the deflection of the central portion of the plate. This deflection can be estimated from the approximate expression

$$w = \frac{Pb^2}{16\pi D}\left(1+5\frac{h^2}{b^2}-\frac{\pi}{4}\frac{b^2}{l^2}\right). \tag{3.113}$$

Expression (3.113) is obtained by the addition of the deflection caused by the reaction of the elastic foundation to that defined by formula (3.98). From Eq. (1.114) we obtain the shear force Q_r in the following form:

$$Q_r = -\frac{P}{2\pi l}\int_0^{\infty}\frac{\gamma^4 J_1(\gamma r)\,d\gamma}{D(\gamma)}. \tag{3.114}$$

If we take only the first two terms of the series (3.114), we have on integration

$$Q_r = \frac{P}{2\pi l}\left[\ker' r-(\eta-\varepsilon)\left(\frac{3}{4}\operatorname{kei}' r+\frac{r}{4}\ker r\right)\right].$$

For $r \ll 1$

$$Q_r = -\frac{P}{2\pi l}\left[\frac{1}{r}+(\eta-\varepsilon)\frac{5}{8}r\left(\ln\frac{r}{2}+\gamma_0\right)\right].$$

On the basis of formulae (1.113), (3.108), and (3.114) we can calculate the bending moments:

$$M_{rr} = -\frac{D}{l^2}\left[\frac{d^2w}{dr^2}+\frac{\nu}{r}\frac{dw}{dr}\right]+\frac{h^2}{5l}\frac{dQ_r}{dr}-\varepsilon l^2(q-p)$$

$$= \frac{P}{2\pi}\left\{\ker r-(1-\nu)\frac{1}{r}\operatorname{kei}' r-\right.$$

$$\left.-\varepsilon(1-\nu)\frac{1}{r}\ker' r+\frac{\eta-\varepsilon}{4}\left((1+\nu)\operatorname{kei} r+r\operatorname{kei}' r\right)\right\}, \tag{3.115$_1$}$$

$$M_{\vartheta\vartheta} = -\frac{D}{l^2}\left[\nu\frac{d^2w}{dr^2}+\frac{1}{r}\frac{dw}{dr}\right]-\frac{h^2}{5l}\frac{dQ_r}{dr}-(\eta-\varepsilon)l^2(q-P)$$

$$= \frac{P}{2\pi}\left\{\nu\ker r+(1-\nu)\frac{1}{r}\operatorname{kei}'r+\varepsilon(1-\nu)\left(\frac{1}{r}\ker'r+\operatorname{kei}r\right)+\right.$$

$$\left.+\frac{\eta-\varepsilon}{4}\left((1+\nu)\operatorname{kei}r+\nu r\operatorname{kei}'r\right)\right\}. \qquad (3\ 115)_2$$

Taking into account only the first terms of the series, we have

$$\left.\begin{matrix}M_{rr}\\M_{\vartheta\vartheta}\end{matrix}\right\} = \frac{P}{4\pi}\left\{-(1+\nu)\left(\ln\frac{r}{2}+\gamma_0\right)\mp\frac{1-\nu}{2}\pm2\varepsilon(1-\nu)\frac{1}{r^2}\right\}. \qquad (3.116)$$

The bending moments, the displacement w, and the reaction of the elastic foundation are graphically presented in Fig. 30.

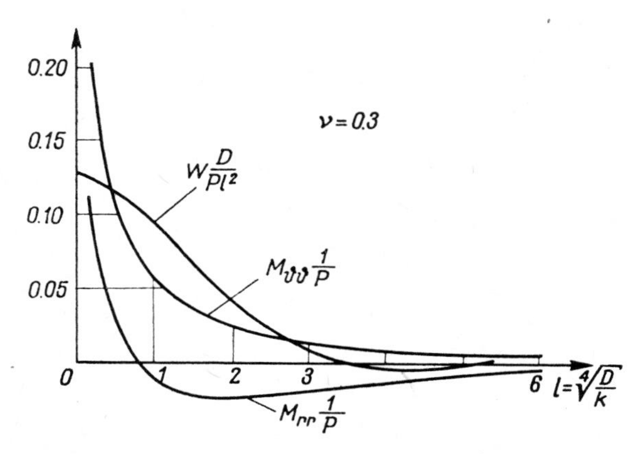

Fig. 30. Distribution of bending moments and deflection in an infinite plate resting on elastic foundation.

The moments M_{rr} and $M_{\vartheta\vartheta}$ in the plate differ one from another by a constant value $-\frac{P}{8\pi}(1-\nu)$ and a term $\frac{P}{2\pi}\varepsilon(1-\nu)\frac{1}{r^2}$ which increases to infinity as $r\to 0$. It may be interesting to compare these results with the formula obtained previously for the circular simply supported plate. We see that the solutions are almost identical when the radius of the plate is $a = 2le^{-\gamma_0} = 1.123l$.

Let us consider now the case where the load P is distributed over the area of a circle with a radius c, small in comparison with 1. The bending moments at the centre of a circular plate carrying such a load are

$$M_{rr} = M_{\vartheta\vartheta} = \frac{P}{4\pi}\left[-(1+\nu)\ln\frac{c}{a}+1+4\varepsilon(1-\nu)\frac{l^2}{c^2}\right). \tag{3.117}$$

The above formula results from Eq. (3.8) if we neglect the terms $-(1-\nu)c^2/4a^2$ and $-2\nu h^2/5a^2$ compared to unity. By substitution $a = 2le^{-\gamma_0} = 1.123\,l$ into Eq. (3.117) and addition of the moment $-P/8\pi(1-\nu)$, we obtain at the centre of the loaded circle of the infinitely large plate on the elastic foundation the following bending moment:

$$M_{rr} = M_{\vartheta\vartheta} = \frac{1+\nu}{4\pi}P\left(-\ln\frac{c}{2l}-\gamma_0+\frac{1}{2}+\frac{4}{1+\nu}\varepsilon(1-\nu)\frac{l^2}{c^2}\right). \tag{3.118}$$

In the case of a highly concentrated load the stresses resulting from Eq. (3.118) must be corrected by the addition of the results obtained by means of the thick plate theory. The following formula for the calculation of the maximum tensile stress at the bottom of the plate, under the point of application of the load, has been established by Westergaard [3.27].

$$\sigma_{rr\,\max} = 0.275(1+\nu)\frac{P}{h^2}\lg_{10}\left(\frac{Eh^3}{kb^4}\right)$$

and

$$b = \sqrt{1.6c^2+h^2}-0.675h \tag{3.119}$$

when $c < 1.724h$, and

$$b = c$$

when $c > 1.724h$, where c is the radius of the circular area over which the load P is distributed.

3.9 Infinite plate loaded by a concentrated bending moment

3.9.1 *Moment of two normal forces*

Action of the bending moment concentrated at one point of the plate surface can also be obtained by the use of Fourier integrals. We can present the moment as the resultant of two oppositely directed normal forces of infinite magnitude acting at points a vanishingly small distance $2\Delta xl$ apart (Fig. 31). The moment they produce is $M = 2l(\Delta xP)$. Each force P can be expressed as a Fourier integral. The load $q(x, y)$ caused by both forces together is

$$q(x, y) = \frac{P}{\pi^2 l^2} \int_0^\infty \int_0^\infty [\cos\alpha(x+\Delta x) - \cos(x-\Delta x)] \cos\beta y \, d\alpha \, d\beta$$

where l is an arbitrary length, and x, y are non-dimensional coordinates referred to the length l.

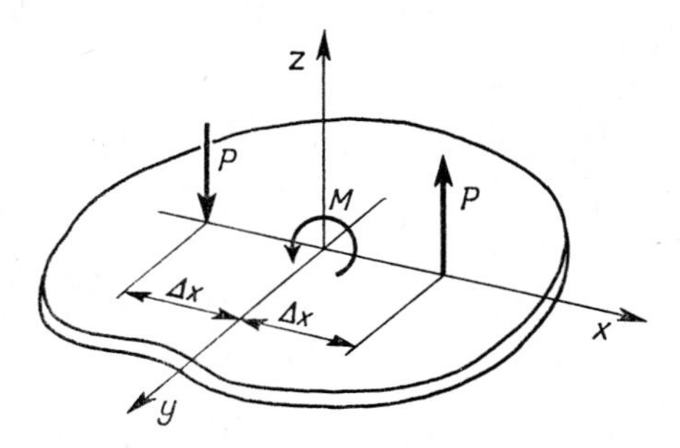

Fig. 31. Representation of bending moment by two concentrated forces acting at points vanishingly small distance apart.

Transforming the term in square brackets, we have

$$q(x, y) = -\frac{2P}{\pi^2 l^2} \int_0^\infty \int_0^\infty \sin\alpha x \sin(\alpha\Delta x) \cos\beta y \, d\beta \, d\alpha .$$

Now, let us bring both forces P toward one another, increasing simultaneously their values, so as to maintain the constant moment M. When $\Delta x \to 0$ and at the same time $2P\Delta x l = M$, we obtain the following formula describing the intensity of the load $q(x, y)$ caused by the concentrated bending moment:

$$q(x, y) = \frac{-M}{\pi^2 l^3} \int_0^\infty \int_0^\infty \alpha \sin\alpha x \cos\beta y \, d\alpha \, d\beta = \frac{M}{l} \frac{\partial}{\partial x} [\delta(x, y)]. \tag{3.120}$$

The deflection of the plate will also take the form of a Fourier integral. On substitution in the plate equation, we have

$$w(x, y) = -\frac{Ml}{\pi^2 D} \int_0^\infty \int_0^\infty \frac{\alpha[1+(\eta-\varepsilon)(\alpha^2+\beta^2)]}{(\alpha^2+\beta^2)^2} \sin\alpha x \cos\beta y \, d\alpha \, d\beta .$$

Integration with respect to the variable α yields

$$w(x, y) = -\frac{Ml}{4\pi D}\int\limits_0^\infty \left(\frac{x}{\beta} + 2(\eta - \varepsilon)\right) e^{-\beta x} \cos\beta y \, d\beta. \tag{3.121}$$

The above integral is divergent, so the deflection of the plate is infinitely large in the whole area axcept along the line $x = 0$. This results from the fact that the plate is infinitely large and its supports are at infinity. The effective solution cannot be derived until boundary conditions are imposed. Let us assume, for example, that the deflection of the plate on the straight lines $x = 0$ and $x = a$ is equal to zero. The fulfilment of this condition is possible by addition to the previously obtained particular solution of a solution of the homogeneous equation of the plate including a certain number of constants of integration. It might be expected that this solution has also form of a Fouiier integral

$$w_1(x, y) = \int\limits_0^\infty w_{(\beta)} e^{ikx} \cos\beta y \, d\beta \tag{3.122}$$

when k is a parameter indefinite at the beginning.

The substitution in the homogeneous equation $\Delta\Delta w = 0$ yields

$$(k^2 + \beta^2)^2 = 0.$$

Solving this, we obtain two double roots $k_{1,2} = \pm\beta i$. The solution of the homogeneous equation therefore takes the form

$$w_1(x, y) = \int\limits_0^\infty (C_1 e^{-\beta x} + C_2 \beta x e^{-\beta x} + C_3 e^{\beta x} + C_4 x e^{\beta x}) \cos\beta y \, d\beta. \tag{3.123}$$

Changing the constants, we can write it in the other form

$$w_1 = \int\limits_0^\infty (\overline{C_1} \cosh\beta x + \overline{C_2} x \sinh\beta x + \overline{C_3} \sinh\beta x + \overline{C_4} x \cosh\beta x) \cos\beta y \, d\beta \tag{3.124}$$

where $C_1 \div C_4$ and $\overline{C_1} \div \overline{C_4}$ are functions of the variable β.

Turning now to the problem under consideration we have to fulfil the condition demanding that the deflection of the plate for $x = 0$ and $x = a$ is equal to zero. The above condition can be fulfilled when we assume for example that $C_1 = C_2 = C_3 = 0$ and

$$C_4 = -\left(\frac{1}{\beta} - 2(\eta - \varepsilon)\frac{1}{a}\right)\frac{1}{\beta} e^{-2\beta a}.$$

we obtain then the deflection of the plate in the form of an integral adding the solutions (3.121) and (3.123). This integral can be evaluated and we find

$$w = -\frac{Ml}{4\pi D}\left\{\frac{1}{2}x\ln\frac{y^2+(2a-x)^2}{x^2+y^2} - \right.$$

$$\left. -2(\eta-\varepsilon)\left[\frac{x}{x^2+y^2}-\frac{x}{a}\frac{2a-x}{(2a-x)^2+y^2}\right]\right\}. \tag{3.125}$$

For other boundary conditions we can in the same way obtain the solution for other plates.

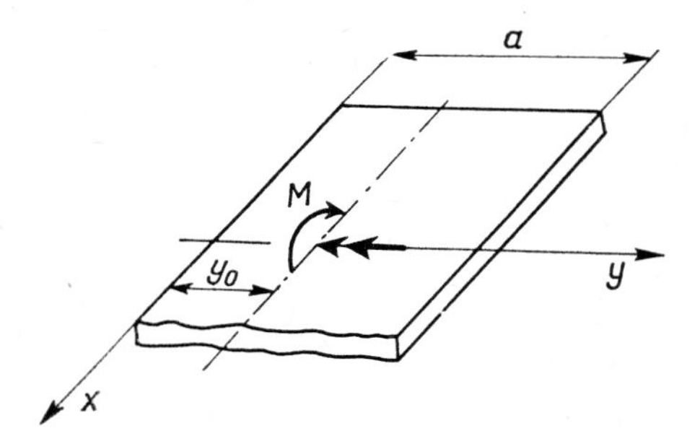

Fig. 32. Strip plate under a concentrated moment.

For example, for the strip plate (Fig. 32) we have (neglecting the effect of the transverse shear deformation)

$$w(x,y) = \frac{Ml}{8\pi D}x\ln\frac{\cosh\dfrac{\pi x}{a}-\cos\dfrac{\pi(y+y_0)}{a}}{\cosh\dfrac{\pi x}{a}-\cos\dfrac{\pi(y-y_0)}{a}}.$$

The stresses near the point of application of the load can be calculated on the basis of the particular solution (3.121). However, since the integral defining the deflection is divergent, the integrals for the bending moments are convergent and can be evaluated. The possibility of a plate rotating has no influence on the distribution of the stresses near the loading point. After a two-fold differentiation according to equations (2.2) we obtain the following bending moments:

$$\left.\begin{matrix} M_{xx} \\ M_{yy} \end{matrix}\right\} = -\frac{M}{4\pi l}x\left[\frac{1+\nu}{r^2} \pm 2(1-\nu)\frac{y^2}{r^4} \pm 4(1-\nu)\varepsilon\frac{x^2-3y^2}{r^6}\right], \tag{3.126}$$

where $r^2 = x^2+y^2$,

$$M_{xy} = (1-\nu)\frac{M}{4\pi l}\left\{\frac{y}{r^2}\left(1-\frac{2y^2}{r^2}\right)+4\varepsilon\frac{y(3x^2-y^2)}{r^6}\right\}.$$

Introducing polar coordinates $(r;\vartheta)$ (Fig. 18) and expressing the moments by the components M_{rr}, $M_{\vartheta\vartheta}$, $M_{r\vartheta}$, the following simple equations are obtained (Fig. 33):

$$\left.\begin{matrix} M_{rr} \\ M_{\vartheta\vartheta}\end{matrix}\right\} = -(1+\nu)\frac{M}{4\pi l}\frac{\cos\vartheta}{r}\left[1\pm 4\varepsilon\frac{1}{r^2}\right],$$

$$M_{r\vartheta} = (1-\nu)\frac{M}{4\pi l}\frac{\sin\vartheta}{r}\left[1-4\varepsilon\frac{1}{r^2}\right]. \tag{3.127}$$

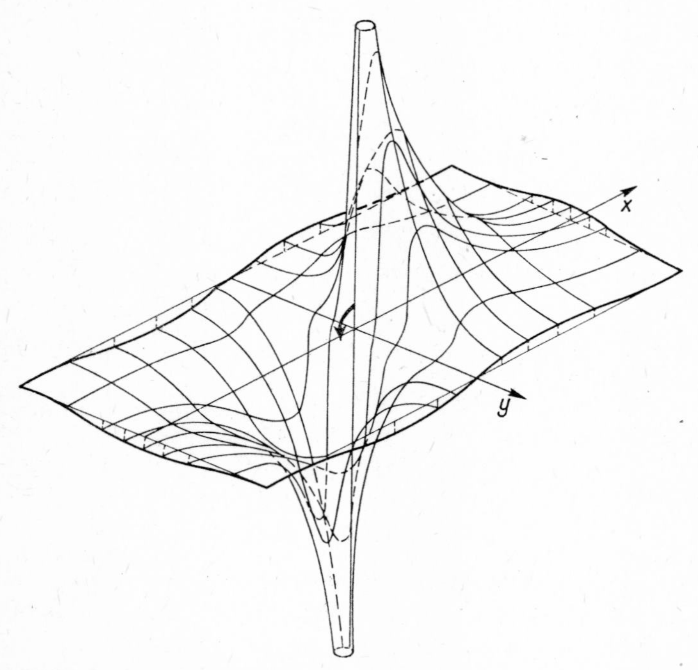

Fig. 33. Distribution of the stresses σ_{rr} and $\sigma_{\vartheta\vartheta}$ in an infinite plate loaded by a concentrated bending moment.

The stresses σ_{rr}, $\sigma_{\vartheta\vartheta}$, and $\tau_{r\vartheta}$ can be derived from the following formula. In the case of classical theory we have

$$\sigma_{i,j} = 6M_{i,j}/h^2, \quad i,j = r,\vartheta.$$

Let us now define the lines along which the stresses σ_{rr}, $\sigma_{\vartheta\vartheta}$ have a constant value. The stresses σ_{rr}, $\sigma_{\vartheta\vartheta}$ are constant (when neglecting the terms proportional to h^2)

$$\frac{\cos\vartheta}{r} = \text{const.}$$

This equation is fulfilled by each circle tangential to the y-axis at the loading point $x = 0$, $y = 0$. The tangential stresses are constant when

$$\frac{\sin\vartheta}{r} = \text{const.}$$

The circles orthogonal to the previous one and tangential to the x-axis at the point $x = 0$, $y = 0$ fulfil the condition above. These circles indicate simultaneously the places of the largest tangential stresses (Fig. 34).

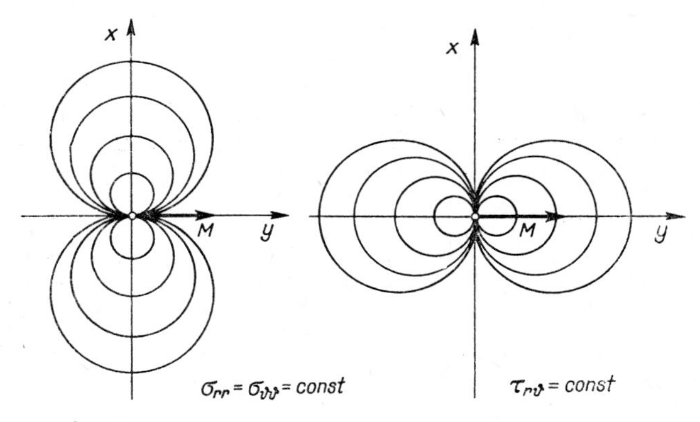

Fig. 34. Lines of equal stresses in a plate under concentrated moment (results of the classica theory).

The formulae obtained result from the superposition of two loadings P. Identical results can be obtained in another way. Let us start our considerations based on the deflection caused by a single normal load. Applying the theorem of superposition, we can obtain the deflection due to the concentrated moment by the addition of the two deflections due to two oppositely directed concentrated forces acting a small distance apart. Let us calculate the limit of the deflection as the distance between these forces tends to zero. Let us assume the deflection in the form of the singular part of (3.4).

If we replace the moment M by two forces $M/\Delta\overline{x}$ acting at the points

$(-\Delta\bar{x}, 0)$ and $(+\Delta\bar{x}, 0)$, the deflection (3.6) produced by the forces $M/\Delta x$ and $-M/\Delta x$ is

$$w = \frac{Ml}{8\pi D}\left\{[(x+\Delta x)^2+y^2-4(\eta-\varepsilon)]\ln\frac{[(x+\Delta x)^2+y^2]^{1/2}}{2\Delta\bar{x}} - \right.$$

$$\left. -[(x-\Delta x)^2+y^2-4(\eta-\varepsilon)]\ln\frac{[(x-\Delta x)^2+y^2]^{1/2}}{2\Delta x}\right\}_{\Delta x\to 0}.$$

If $\Delta\bar{x} \to 0$, we obtain the case of the concentrated moment acting at the origin of coordinates x, y. Then we have

$$w = \lim_{\Delta x\to 0}[w(P)] = \frac{M}{Pl}\frac{\partial w(P)}{\partial x},$$

where $w(P)$ is the deflection produced by a single concentrated force. On differentiation we find

$$w = \frac{Ml}{8\pi D}r\left(2\ln r+1+4(\eta-\varepsilon)\frac{1}{r^2}\right)\cos\vartheta$$

where $r = \sqrt{x^2+y^2}$. If we neglect the second term in the brackets which has no influence on the stresses, we find the singular part of the solution for the deflection of a plate loaded by the concentrated bending moment.

An important conclusion results from the above consideration: The deflection of the plate and also the bending moments and stresses can easily be obtained by the differentiation of the results derived for the normal concentrated force. It can be observed, for example, that we obtain relation (3.121) by the differentiation with respect to x in formula (3.66) and multiplying by M/Pl.

Let us consider now the case where two oppositely directed bending moments act on the surface of the plate at two points of vanishingly small distance $\Delta\bar{x}$ apart. The vectors of these moments are parallel to each other. Assuming that $M\Delta\bar{x} = H$ preserves a constant value as $\Delta\bar{x} \to 0$, we obtain the deflection of the plate under the action of this load to be equal to

$$w(H) = \frac{H}{Ml}\frac{\partial w(M)}{\partial x} = \frac{H}{Pl^2}\frac{\partial^2 w(P)}{\partial x^2}.$$

Then we obtain the relation

$$w(H) = \frac{H}{8\pi D}(2\ln r+2+\cos 2\vartheta)$$

in which only the expressions resulting from the classical theory are preserved.

The load H represents the moment acting along the length Δx and can be used for the calculation of the deflection of the plate under an arbitrarily distributed bending moment. It comes out from our considerations that the stresses in the plate are singular at the point of application of the moment. It is the result of the manner the loading is applied to the plate, i.e. from the assumption that the bending moment acts only at one point of the plate surface. In reality, however, structures are never loaded at one point. The concentrated load is always distributed over a certain surface or along the length of the plate. Therefore, it is interesting to solve the equations when the bending moment is introduced along a certain length $2c$, which is small in comparison with the other dimensions of the plate. We solve this problem and define the stresses in the plate produced by the load presented graphically in Fig. 35. The load $q(x, y)$ equal to a bending moment distributed uniformly

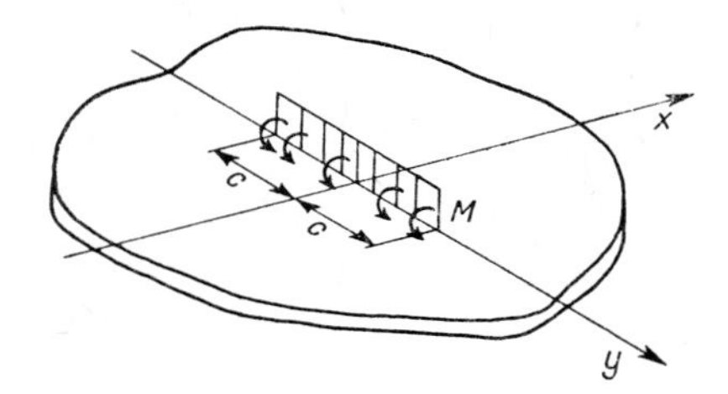

Fig. 35. Plate loaded by a moment distributed along the segment $2c$.

along the length $2c$ can be expressed by means of Fourier integral in the form ($c = \bar{c}/l$)

$$q(x, y) = -\frac{M}{\pi^2 l^3}\int_0^\infty\int_0^\infty \frac{\sin\beta c}{\beta c}\,\alpha\sin\alpha x\cos\beta y\,\mathrm{d}\alpha\,\mathrm{d}\beta. \tag{3.128}$$

The displacement then takes the form

$$w = -\frac{Ml}{\pi^2 D}\int_0^\infty\int_0^\infty \frac{\alpha[1+(\eta-\varepsilon)(\alpha^2+\beta^2)]}{(\alpha^2+\beta^2)^2}\,\frac{\sin\beta c}{\beta c}\,\sin\alpha x\cos\beta y\,\mathrm{d}\alpha\,\mathrm{d}\beta.$$

The integration with respect to α yields

$$w = -\frac{Ml}{4\pi D}\int_0^\infty \left[\frac{x}{\beta}+2(\eta-\varepsilon)\right]\mathrm{e}^{-\beta x}\frac{\sin\beta c}{\beta c}\cos\beta y\,\mathrm{d}\beta. \tag{3.129}$$

The bending moments near the loading point can be calculated from integral (3.129).

After differentiation according to (1.24) and integration, we have

$$\left.\begin{matrix} M_{xx} \\ M_{yy} \end{matrix}\right\} = \frac{M}{4\pi cl}\left[\frac{1}{2}\,[1+\nu\pm(1-\nu)](\vartheta_1-\vartheta_2)\mp\right.$$
$$\left.\mp\frac{1-\nu}{4}\,(\sin 2\vartheta_2-\sin 2\vartheta_1)\pm(1-\nu)\varepsilon\left(\frac{\sin 2\vartheta_2}{r_2^2}-\frac{\sin 2\vartheta_1}{r_1^2}\right)\right], \tag{3.130}$$

$$M_{xy} = -(1-\nu)\frac{M}{4\pi cl}\left[\frac{1}{2}\ln\frac{r_2}{r_1}+\frac{1}{4}\,(\cos 2\vartheta_2-\cos 2\vartheta_1)+\right.$$
$$\left.+\varepsilon\left(\frac{\cos 2\vartheta_2}{r_2^2}-\frac{\cos 2\vartheta_1}{r_1^2}\right)\right].$$

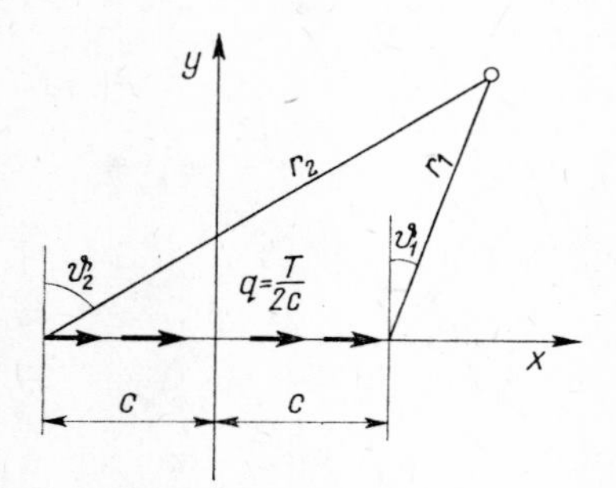

Fig. 36. Definition of the radii r_1, r_2 and the angles ϑ_1, ϑ_2, $r_1 = \sqrt{x^2+(y-c)^2}$, $r_2 = \sqrt{x^2+(y+c)^2}$.

The angles ϑ_1, ϑ_2 are defined in Fig. 36. In the whole area of the plate the moments M_{xx} and M_{yy} calculated on the basis of formulae (3.130) have finite values. For $x = 0$ and $-c < y < +c$ angles $\vartheta_1 = -\pi/2$; $\vartheta_2 = \pi/2$. Thus the moments M_{xx} and M_{yy} are constant in this portion and equal to $M_{xx} = M/4c$, $M_{yy} = M/4c$. The above yields the stresses

$$\sigma_{xx} = 3M/2h^2cl; \qquad \sigma_{yy} = 3M\nu/2h^2cl.$$

The moment M_{xy} is singular at the point $x = 0$, $y = \pm c$ as there $\ln r_2/r_1 \to \pm\infty$. The distribution of the moments M_{xx}, M_{yy} for $y = 0$ is presented graphically in Fig. 38.

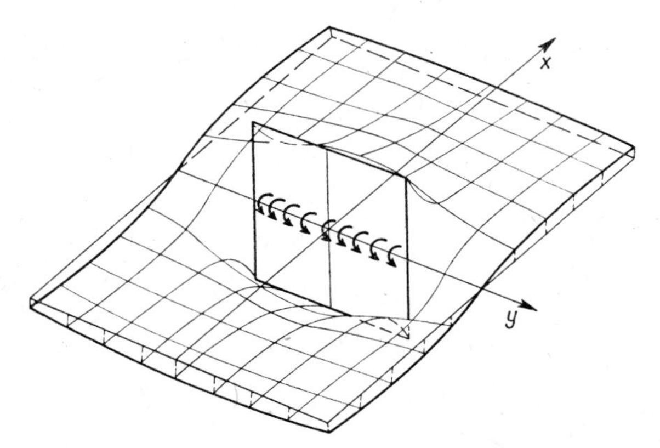

Fig. 37. Variation of stresses σ_{xx} in a plate loaded by a moment distributed uniformly along the segment 2c.

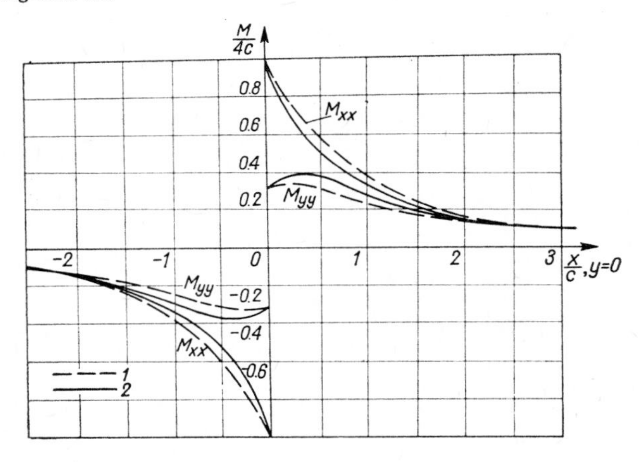

Fig. 38. Variation of the moments M_{xx} and M_{yy} for $y = 0$; 1—classical theory, 2—theory with the effect of shear deformation $h/c = 0.5$.

3.9.2 *Concentrated bending moment*

We defined in the previous section the concentrated bending moment as the resultant moment of two oppositely directed normal concentrated forces of infinite magnitude acting at points of vanishingly small distance apart. However, this moment can be defined also in another way. We took into account in the equilibrium equations (1.79) the bending loadings X_M and Y_M; then we can define the concenrated bending moment M as $X_M = M\delta(x, y)$ or $Y_M = M\delta(x, y)$ where $\delta(x, y)$ is the Dirac function.

This new definition does not introduce any difference if we use the classical theory of plates. However, in the case where we take into account the effects of transverse shear deformations it leads to new solutions. Notice that, calculating the work of the moment M realized as the couple of two normal forces, we have $M\partial w/\partial x$, but the work of the moment X_M is $X_M \beta_x$. We multiply here by a different physical quantity. The moment X_M has in the case of sandwich plates a simple physical interpretation. It is a couple of two forces tangential to the plate surface and applied to the external layers of the plate.

The singular solution obtained for the moment defined as $X_M = M\delta(x, y)$ has the form

$$w_0 = \frac{Ml}{8\pi D} x[\ln r^2 + 1],$$

which is identical with the expression obtained from the classical theory of plates. However, we obtain different relations for the shear forces [2.20]:

$$Q_x = \frac{M}{2\pi l^2}\left\{\frac{\cos 2\vartheta}{r^2}\left[1-\sqrt{10}\frac{\bar{r}}{h}K_1\left(\sqrt{10}\frac{\bar{r}}{h}\right)\right]+\frac{10}{h^2}\sin^2\vartheta K_0\left(\sqrt{10}\frac{\bar{r}}{h}\right)\right\}, \tag{3.131}$$

$$Q_y = \frac{M}{2\pi}\left\{\frac{\sin 2\vartheta}{\bar{r}^2}\left[1-\sqrt{10}\frac{\bar{r}}{h}K_1\left(\sqrt{10}\frac{\bar{r}}{h}\right)\right]-\frac{5}{h^2}\sin 2\vartheta K_0\left(\sqrt{10}\frac{\bar{r}}{h}\right)\right\}.$$

The bending moments in the plate are defined by

$$\left.\begin{matrix} M_{xx} \\ M_{yy} \end{matrix}\right\} = -\frac{M}{4\pi}\frac{\cos\vartheta}{\bar{r}}\left\{[(1+\nu)\pm 2(1-\nu)\sin^2\vartheta]+\right.$$

$$\pm\frac{2}{5}\frac{h^2}{\bar{r}^2}(\cos^2\vartheta-3\sin^2\vartheta)\left[2-2\sqrt{10}\frac{\bar{r}}{h}K_1\left(\sqrt{10}\frac{\bar{r}}{h}\right)-\right.$$

$$\left.\left.-10\frac{\bar{r}^2}{h^2}K_0\left(\sqrt{10}\frac{\bar{r}}{h}\right)\right]+\frac{\bar{r}^3 10\sqrt{10}}{h^3}\sin^2\vartheta K_1\left(\sqrt{10}\frac{\bar{r}}{h}\right)\right\},$$

$$M_{xy} = \frac{M}{4\pi}\frac{\sin\vartheta}{\bar{r}}\left\{(1-\nu)\cos 2\vartheta+\frac{2}{5}\frac{h^2}{\bar{r}^2}(\sin^2\vartheta-3\cos^2\vartheta)\times\right. \tag{3.132}$$

$$\times\left[2-2\sqrt{10}\frac{\bar{r}}{h}K_1\left(\sqrt{10}\frac{\bar{r}}{h}\right)-10\frac{\bar{r}^2}{h^2}K_0\left(\sqrt{10}\frac{\bar{r}}{h}\right)\right]+$$

$$\left.+\frac{5\bar{r}^3\sqrt{10}}{h^3}\cos 2\vartheta K_1\left(\sqrt{10}\frac{\bar{r}}{h}\right)\right\},$$

where $r^2 = x^2+y^2$, $\sin\vartheta = y/r$, $\cos\vartheta = x/r$, and $K_0(\bar{r})$ and $K_1(\bar{r})$ are the Bessel functions of the second kind. The above solutions differs from solution (3.126) only by the terms in brackets multiplied by $\bar{r}^2/h^2$ and r^3/h^3. For small argument x the functions are $K_0(x) = -\ln\frac{1}{2}x$ and $K_1(x) = \frac{1}{2}x\ln\frac{1}{2}x$, and the singularities of the bending moments are $1/r$, $1/r^3$ and $r^{-1}\ln r$.

3.10 Bending moment introduced through a rigid insert

Let us consider still one more example—a case of a circular plate of radius a loaded by a concentrated bending moment applied to the plate through a rigid central insert of radius c. We solve this case by satisfying the boundary conditions at two edge of the plate $r = c$ and $r = a$. We assume that the plate is clamped in the insert along the inner boundary which rotates under action of the moment M. At the second edge the plate is simply supported.

The general solution of the homogeneous equation $\Delta^2 w = 0$ in the polar coordinates (r, ϑ) takes the form

$$\begin{aligned} w = {} & C_{10}+C_{20}r^2+C_{30}\ln r+C_{40}r^2\ln r+ \\ & +(C_{11}r+C_{21}r^3+C_{31}r^{-1}+C_{41}r\ln r)\cos\vartheta+ \\ & +\sum_{n=2}^{\infty}(C_{1n}r^n+C_{2n}r^{n+2}+C_{3n}r^{-n}+C_{4n}r^{-n+2})\cos n\vartheta+ \\ & +(D_{11}r+D_{21}r^3+D_{31}r^{-1}+D_{41}r\ln r)\sin\vartheta+ \\ & +\sum_{n=2}^{\infty}(D_{1n}r^n+D_{2n}r^{n+2}+D_{3n}r^{-n}+D_{4n}r^{-n+2})\sin n\vartheta+ \\ & +E_1\vartheta+E_2r^2\vartheta+E_3r\vartheta\sin\vartheta+E_4r\vartheta\cos\vartheta. \end{aligned} \tag{3.133}$$

Due to antisymmetry of loading we take only the terms for $n = 1$. Then we have

$$w = (C_1r+C_2r^3+C_3r^{-1}+C_4r\ln r)\cos\vartheta.$$

To simplify the record we omit the double indexes at the constants C_{ij}. The boundary conditions are: At the inner boundary the edge of the plate is clamped, then

$$\left(\frac{\partial w}{\partial r}\right)_{r=c} = \left(\frac{w}{r}\right)_{r=c}.$$

The resultant forces must balance the moment M, then

$$-2c\int_0^{\pi}(M_{r\vartheta})_{r=c}\sin\vartheta\,d\vartheta+2c^2\int_0^{\pi}(Q_r)_{r=c}\cos\vartheta\,d\vartheta+$$

$$+2c\int_0^{\pi}(M_{rr})_{r=c}\cos\vartheta\,d\vartheta+M=0.$$

Since we consider the plate with the simply supported edge $r=a$, we have $M_{rr}=0$ and $w=0$.

Solving this set of equations, we find the deflection

$$w=\frac{Mal}{8\pi D[(3+\nu)+(1-\nu)k^4]}\left\{-[(1+\nu)+(1-\nu)k^4]\left(\frac{r}{a}\right)^3+\right.$$

$$+(1+\nu)(1-k^2)^2\left(\frac{r}{a}\right)+2[(3+\nu)-(1-\nu)k^4]\left(\frac{r}{a}\right)\ln\left(\frac{r}{a}\right)+$$

$$\left.-k^2[(1+\nu)k^2-(3+\nu)]\frac{a}{r}\right\}\cos\vartheta \tag{3.134}$$

where $k=c/a$. When $k=0$, Eq. (3.134) reduces to the equation

$$w=\frac{Mrl}{8\pi D}\left\{\frac{1+\nu}{3+\nu}\left[1-\left(\frac{r}{a}\right)^2\right]+2\ln\frac{r}{a}\right\}\cos\vartheta, \tag{3.135}$$

which corresponds to the action of the concentrated moment M. The bending moments obtained from (3.134) are:

$$M_{rr}=-\frac{M}{4\pi[3+\nu+(1-\nu)k^4]al}\left\{-(3+\nu)[1+\nu+(1-\nu)k^4]\frac{r}{a}+\right.$$

$$\left.+(1+\nu)[3+\nu+(1-\nu)k^4]\frac{a}{r}-(1-\nu)[(1+\nu)k^2-(3+\nu)]\frac{c^2a}{r^3}\right\}\cos\vartheta,$$

$$M_{\vartheta\vartheta}=-\frac{M}{4\pi[3+\nu+(1-\nu)k^4]al}\left\{-(3\nu+1)[1+\nu+(1-\nu)k^4]\frac{r}{a}+\right.$$

(3.136)

$$\left.+(1+\nu)[3+\nu+(1-\nu)k^4]\frac{a}{r}+(1-\nu)[(1+\nu)k^2-(3+\nu)]\frac{c^2a}{r^3}\right\}\cos\vartheta,$$

$$M_{r\vartheta}=(1-\nu)\frac{M}{4\pi[3+\nu+(1-\nu)k^4]al}\left\{-[1+\nu+(1-\nu)k^4]\frac{r}{a}+\right.$$

$$\left.+[3+\nu+(1-\nu)k^4]\frac{a}{r}+[(1+\nu)k^2-(3+\nu)]\frac{c^2a}{r^3}\right\}\sin\vartheta;$$

assuming that $a \to \infty$, $k \to 0$, we obtain the bending moments in the infinitely large plate

$$\left.\begin{matrix} M_{rr} \\ M_{r\vartheta} \end{matrix}\right\} = -\frac{M}{4\pi c l}\left[\frac{c}{r} \pm \frac{c^3}{r^3} + \nu\left(\frac{c}{r} \mp \frac{c^3}{r^3}\right)\right]\cos\vartheta, \tag{3.137}$$

$$M_{r\vartheta} = (1-\nu)\frac{M}{4\pi c l}\left(\frac{c}{r} - \frac{c^3}{r^3}\right)\sin\vartheta.$$

If $r = c$, the moments reach the maximum values

$$M_{rr} = \frac{M}{2\pi c l}\cos\vartheta, \qquad M_{\vartheta\vartheta} = \frac{\nu M}{2\pi c l}\cos\vartheta, \qquad M_{r\vartheta} = 0.$$

The bending moment M_{rr} for $r = c$ does not depend on Poisson's ratio. If $c \to 0$ and $r > 0$, we find

$$M_{rr} = M_{\vartheta\vartheta} = -(1+\nu)\frac{M}{4\pi r l}\cos\vartheta,$$

$$M_{r\vartheta} = (1-\nu)\frac{M}{4\pi r l}\sin\vartheta,$$

which is in agreement with the values calculated previously. If the external edge of the plate is clamped, the solution can be obtained by changing the boundary conditions.

3.11 Thermal singularities

Let us consider the stresses which appear in the plate as a result of a rapid, local change of temperature. We assume that the plate is subjected to a non-uniform heating over its surface. Moreover, the temperature $T(x, y, z)$ varies through the thickness. The temperature is the cause of the strains in the plate. In the case of plane state of stress we have

$$e_{ij} = \frac{1}{E}\left[(1+\nu)\sigma_{ij} - \nu\sigma_{\alpha}^{\alpha}\delta_{ij}\right] + \alpha_t T(x, y, z)\,\delta_{ij}, \qquad i, j = 1, 2, \tag{3.138}$$

where α_t is the linear coefficient of thermal expansion of the material of the plate, and

$$\sigma_{ij} = \frac{E}{1+\nu}\left[e_{ij} + \frac{\nu}{1-\nu}e_{\alpha}^{\alpha}\delta_{ij}\right] + \frac{E}{1-\nu}\alpha_t T\delta_{ij}. \tag{3.139}$$

We neglect in this case of thermal stresses the effect of transverse shear deformation due to its insignificance and assume Kirchhoff's hypothesis. Then

$$\beta_x = -\frac{\partial w}{\partial x}, \qquad \beta_y = -\frac{\partial w}{\partial y},$$

and

$$u_z = u - z\frac{\partial w}{\partial x}, \qquad v_z = v - z\frac{\partial w}{\partial y}.$$

Taking into account relations (1.70), (1.72a) and introducing Eqs. (3.139) into Eqs. (1.74), we obtain the following expressions for the bending moments:

$$M_{xx} = -D\left(\frac{\partial^2 w}{\partial x^2} + \nu\frac{\partial^2 w}{\partial y^2}\right) - \frac{E}{1-\nu}\int_{-h/2}^{h/2} \alpha_t T_z \, dz,$$

$$M_{yy} = -D\left(\frac{\partial^2 w}{\partial y^2} + \nu\frac{\partial^2 w}{\partial x^2}\right) - \frac{E}{1-\nu}\int_{-h/2}^{h/2} \alpha_t T_z \, dz,$$

$$M_{xy} = -(1-\nu)D\frac{\partial^2 w}{\partial x \, \partial y}.$$

It is here assumed that E, ν are independent of the temperature. Expressing the shearing forces by Eqs. $(1.79)_4$ and introducing them into the equation of equilibrium $(1.79)_3$, we have

$$D\Delta\Delta w = q - L(w, \Phi) - \frac{E}{1-\nu}\int_{-h/2}^{h/2} \alpha_t \Delta T z \, dz. \tag{3.140}$$

The second governing equation for the plate can be derived from the equation of compatibility (1.87). Introducing in it the strains from (3.138) expressed by Φ we find

$$\frac{1}{Eh}\Delta\Delta\Phi = -\int_{-h/2}^{h/2} \alpha_t \Delta T \, dz + \frac{1}{2}L(w, w). \tag{3.141}$$

Let us assume that the temperature at any point of the plate is given by

$$T(x, y, z) = T_1(x, y) + \frac{z}{h}T_2(x, y), \tag{3.142}$$

where $T_1(x, y)$ is the average temperature over the section and $T_2(x, y)$ is the temperature difference of the face surfaces. Introducing it into Eqs. (3.140) and (3.141), we find

$$
\begin{aligned}
D\Delta\Delta w &= q - \frac{1+\nu}{h} D\Delta(\alpha_t T_2) - L(w, \Phi), \\
\frac{1}{Eh}\Delta\Delta\Phi &= \frac{1}{2} L(w, w) - \Delta(\alpha_t T_1).
\end{aligned}
\tag{3.143}
$$

If we consider a plate heated uniformly over a small area S, plane and bending hot spots of intensity T_{s1} and T_{s2}, respectively are defined as

$$
\begin{aligned}
T_{s1} &= \lim_{\substack{S\to 0 \\ T_1\to\infty}} (\alpha_t T_1 S), \\
T_{s2} &= \lim_{\substack{S\to 0 \\ T_2\to\infty}} (\alpha_t T_2 S).
\end{aligned}
\tag{3.144}
$$

In the case of a bending hot spot Eq. (3.140) reduces to

$$
\Delta\Delta w = -\frac{1+\nu}{h} T_{s2}\Delta[\delta(x, y, 0, 0)] \tag{3.145}
$$

where $\delta(\;)$ is the delta function.

The functions $T(x, y)$ can be represented in the form of a Fourier integral (3.39). Solving Eq. (3.145), we find the deflection of the plate in the form

$$
w(x, y) = \frac{(1+\nu)T_{s2}}{\pi^2 h} \int_0^\infty \int_0^\infty \frac{\cos\alpha x \cos\beta y \, d\alpha \, d\beta}{\alpha^2+\beta^2}.
$$

The above integral can be evaluated with respect to α. We obtain

$$
w = -\frac{(1+\nu)T_{s2}}{2\pi h} \int_0^\infty \frac{e^{-\beta x}}{\beta} \cos\beta y \, d\beta.
$$

We see that the deflection of the plate is given by the divergent integral. It results from the infinite dimension of the plate. Integrating from B to ∞ and later assuming that $B \to 0$, we find the deflection in the form

$$
w = -\frac{(1+\nu)T_{s2}}{2\pi h} \ln r, \qquad r = \frac{\bar{r}}{l}, \tag{3.146}
$$

which is the particular solution of Eq. (3.145).

We obtain the bending moments in the form of convergent integrals

$$M_{xx} = \frac{(1-\nu^2)T_{s2}}{2\pi h l^2} D \int_0^\infty \beta e^{-\beta x} \cos \beta y \, d\beta = -\frac{ET_{s2}}{24\pi} \frac{h^2}{l^2} \frac{\cos 2\vartheta}{r^2},$$

$$M_{yy} = \frac{ET_{s2}}{24\pi} \frac{h^2}{l^2} \frac{\cos 2\vartheta}{r^2}, \qquad M_{xy} = -\frac{ET_{s2}}{24\pi} \frac{h^2}{l^2} \frac{\sin 2\vartheta}{r^2}.$$

Expressing the moments by components M_{rr}, $M_{\vartheta\vartheta}$, and $M_{r\vartheta}$, we have

$$M_{rr} = -M_{\vartheta\vartheta} = \frac{ET_{s2}}{24} \frac{h^2}{l^2} \frac{1}{r^2}, \qquad M_{r\vartheta} = 0.$$

We have called this singularity the "bending hot spot". When the same plate is heated uniformly over its thickness at one point, a plane stress system results. We arrive then at the plane "hot spot". As Eq. $(3.143)_1$ has the same form as previously (3.140), we obtain the following results (with $L(w, w) = 0$):

$$\Phi = -\frac{EhT_{s1}}{2\pi} \ln r. \tag{3.147}$$

The membrane forces are

$$N_{rr} = -N_{\vartheta\vartheta} = \frac{Eh}{2\pi} \frac{T_{s1}}{l^2} \cdot \frac{1}{r^2}.$$

The results obtained can be used to form the solution for any distribution of temperature. Every distribution of heating and cooling can be represented as an integral over infinitesimal contributions. It is possible to find the solution for a case of arbitrary heating by using the corresponding singular solution as the basis of a Green function of the problem, (see next § 3.12).

3.12 Influence surfaces and the Green functions

3.12.1 *Influence surfaces for the deflection*

The deflection of an elastic structure caused by a single concentrated force can be used for the calculation of the deflections and stresses produced by an arbitrary load $q(x, y)$ if the theorem of superposition can be applied. Let us consider, for example, the deflection of a simply supported plate loaded by a unit force $P = 1$ acting at the point A of coordinates $x = \xi$, $y = \eta$. The deflection is the function of both coordinates x, y and the parameters

ξ, η (Fig. 15). Let us denote this deflection by $K(x, y, \xi, \eta)$. Regarding x and y as variables, $w = K(x, y, \xi, \eta)$ represents the deflected surface of the plate loaded by a unit force at the point $x = \xi$, $y = \eta$. Let us consider now the coordinates ξ, η as variables; then the above equation defines the "influence surface" for the deflection of the plate at the point $C(x, y)$. Therefore if some load intensity $q(\xi, \eta)$ is distributed over an area F, the deflection of an arbitrary point of the plate can easily be obtained by integration of the results for the concentrated force.

Applying an elementary force $q(\xi, \eta)\mathrm{d}\xi\mathrm{d}\eta$ and using the principle of superposition, we find the deflection of the plate as the double integral

$$w = \iint_F q(\xi, \eta) K(x, y, \xi, \eta)\mathrm{d}\xi\mathrm{d}\eta. \tag{3.148}$$

This integral is extended over a loaded area F. The function K defined by Eq. (3.148) is called *Green's function of the plate*. By virtue of the reciprocal theorem of Maxwell this function satisfies the condition

$$K(x, y, \xi, \eta) = K(\xi, \eta, x, y)$$

and in general depends on the boundary conditions of the body. In the case under consideration it pertains to the simply supported rectangular plate. Similar functions may be developed for any other boundary conditions for plates of any shape. We may also graphically represent the influence surface $K(x, y, \xi, \eta)$ for the deflection at some fixed point (x, y) by means of contour lines.

Green's function for deflection can be used to define the influence surface for stresses. This surface can be obtained on the basis of the deflection by corresponding differentiation. In the case of the line load actions we use the "influence line" technique. The influence line gives the variation of some selected effect $E(x, y)$ at any point $C(x, y)$ of a structure as a unit action travels along the path s of the loading applied to the structure. The total effect at the point C due to the loading $q(s)$ applied along S is

$$E(x, y) = \int_0^S K(x, y, s) q(s)\mathrm{d}s.$$

This integral can be evaluated numerically, graphically or analytically whichever is found convenient. The construction of an appropriate influence line becomes quite simple since a condition of reciprocal symmetry exists between the point C and any point F on the path s. This condition may be stated as

$$K(C, F) = K(F, C)$$

where $K(C, F)$ is the effect at the point C due to the unit action applied at F and $K(F, C)$ is the same effect at F due to the unit action applied at C. Then the influence line required for C may be obtained directly by applying the unit action at C to the otherwise unloaded structure and evaluating its selected effect along the path S. To be able to carry out this evaluation it is necessary to know the effect of a unit action. Obviously, the effects of any more complex loads may be built up as a combination of basic unit actions, namely, normal and tangential forces and bending and twisting moments. A most extensive set of influence surfaces for rectangular plates with various boundary conditions was worked out by A. Pucher [3.40]. Pucher built the influence surfaces $K(x, y, \xi, \eta)$ according to the equation

$$K(x, y, \xi, \eta) = K_0(x, y, \xi, \eta)+K_1(x, y, \xi, \eta),$$

where K_0 contains a singular part whereas K_1 means the regular function for the whole surface of the plate. The singular part of the solution can be taken from formula (2.6) for the deflection of the circular plate loaded by a concentrated force $P = 1$ at the centre. In this solution the term

$$K_0 = \frac{1}{8D\pi} [r^2-4(\eta-\varepsilon)l^2]\ln \frac{r}{a}$$

makes the curvature d^2w/dr^2 at the point $r = 0$ infinitely large.

The function K_0 is independent of the shape of the edge and support. The complete influence surface $K = K_0+K_1$, as the surface of the deflection caused by the unit load acting at the point ξ, η, must fulfil the defined boundary conditions. The singular part of the solution K_0 gives, however, the boundary values which cannot correspond to the considered ones. The boundary conditions can be satisfied due to the regular part of the solution K_1 which must fulfil the homogeneous equation of the plate. Usually, the expression in the form of series built from the integrals of the biharmonic equation is used to obtain the solution K_1. The coefficients of this solution are determined in such a way that the complete solution corresponds to the boundary conditions. As we can use a limited number of terms, the boundary conditions can only be satisfied approximately. Sometimes the biharmonic algebraic expressions are used for the calculations of the regular part K_1. The coefficients are determined in such a way as to achieve the smallest sum of the square of errors. Sometimes sufficiently accurate results are obtained by the collocation method, consisting in the satisfaction of the boundary conditions only at points of edges. Also the method of developing the edge values of the singular function in Fourier series can be used. For rectangular plates supported on edges, the method worked out by Koepcke [3.41] is convenient. Besides, numerical methods based on finite differences and finite elements can be employed.

3.12.2 *Influence surfaces for the internal moments*

The singularity of the bending moments and shear forces can be determined from the function K_0. Let us represent K_0 in the system of Cartesian coordinates. We have

$$r = \sqrt{(x-\xi)^2+(y-\eta)^2},$$

$$K_0 = \frac{1}{16\pi D}\,[(x-\xi)^2+(y-\eta)^2-4(\eta-\varepsilon)]\ln\frac{(x-\xi)^2+(y-\eta)^2}{a^2}.$$

For the bending moment we have

$$M_{xx} = D(\varkappa_{xx}+\nu\varkappa_{yy})+\frac{h^2}{5}\frac{\partial Q_x}{\partial x},$$

where

$$\varkappa_{xx} = -\frac{\partial^2 w}{\partial x^2}, \qquad \varkappa_{yy} = -\frac{\partial^2 w}{\partial y^2}.$$

As a result of the symmetry of the function K_0 the curvatures $\varkappa_{xx}$, $\varkappa_{yy}$ have identical form, they are only rotated in plane by 90° one from another. Differentiating K_0, we find

$$D\varkappa_{xx} = \frac{1}{8\pi}\left(2\ln\frac{r}{a}+2\cos^2\vartheta+1+4(\eta-\varepsilon)\cdot\frac{l^2}{r^2}\cos 2\vartheta\right).$$

3.12.3 *Singularity and the influence surface for the edge moment of a clamped plate*

In this case the singularities of the influence surface of the support moment cannot be obtained from the classical singularity K_0. These singularities contain the edge values which are also partly singular. They can, however, be obtained by the transformation of the solution for the semi-infinite cantilever plate derived by Michell [3.42] (Fig. 39). The deflection of such a plate is defined by the formula

$$w = \frac{P}{8\pi D}\left\{2\bar{x}\bar{\xi}-\frac{1}{2}\,[(\bar{x}-\bar{\xi})^2+\bar{y}^2]\ln\frac{(\bar{x}+\bar{y})^2+\bar{y}^2}{(\bar{x}-\bar{\xi})^2+\bar{y}^2}\right\}. \tag{3.149}$$

Differentiating with respect ξ, we have the following expressions for the moment M_{xx} at the point $(0, \eta)$ (for $\nu = 0$):

$$(M_{xx})_{\xi,\eta} = \lim \left(-D \frac{\partial^2 w}{\partial \bar{\xi}^2} \right) = -\frac{P}{\pi} \frac{\bar{x}^2}{\bar{x}^2 + (\bar{y} - \bar{\eta})^2} = -\frac{P\cos^2\vartheta}{\pi} .$$

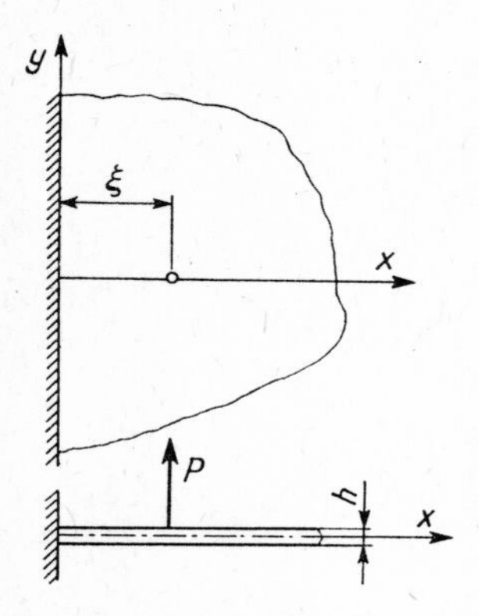

Fig. 39. Semi-infinite plate with built-in edge $x = 0$.

The above expression does not contain singular terms but the function $M_{xx}(\xi, \eta)$ is discontinuous at the origin. This discontinuity is finite in the whole area of the plate and is presented in Fig. 40 for the square plate. The influence surface of the deflection of the built-in plate can be expressed by means of the closed formulae [3.43]. If the clamped plate is loaded by a force P acting at the point $A(r_0, \vartheta)$ (Fig. 41), the deflection is

$$w = \frac{PR^2}{16\pi D} \left\{ \left[\varrho_0^2 + \varrho^2 - 2\varrho\varrho_0 \cos(\vartheta - \vartheta_0) \ln \frac{\varrho_0^2 + \varrho^2 - 2\varrho\varrho_0 \cos(\vartheta - \vartheta_0)}{1 + \varrho_0^2\varrho^2 - 2\varrho\varrho_0 \cos(\vartheta - \vartheta_0)} + \right.\right.$$

$$\left.\left. + (1 - \varrho_0^2)(1 - \varrho^2) \right] \right\},$$

where $\varrho = r/R$; $\varrho_0 = r_0/R$. The influence surface of the bending moment M_{rr} is obtained by differentiation of Eq. (3.150) according to formula (2.31) with respect to ϱ_0 and ϑ_0.

$$M_{rr}(\varrho_0, \vartheta_0, \varrho, \vartheta) = -\frac{D}{R^2} \left[\frac{\partial^2 w}{\partial \varrho_0^2} + \nu \left(\frac{1}{\varrho_0^2} \frac{\partial^2 w}{\partial \vartheta_0^2} + \frac{1}{\varrho_0} \frac{\partial w}{\partial \varrho_0} \right) \right].$$

At the edge for $\varrho_0 = 1$ we have

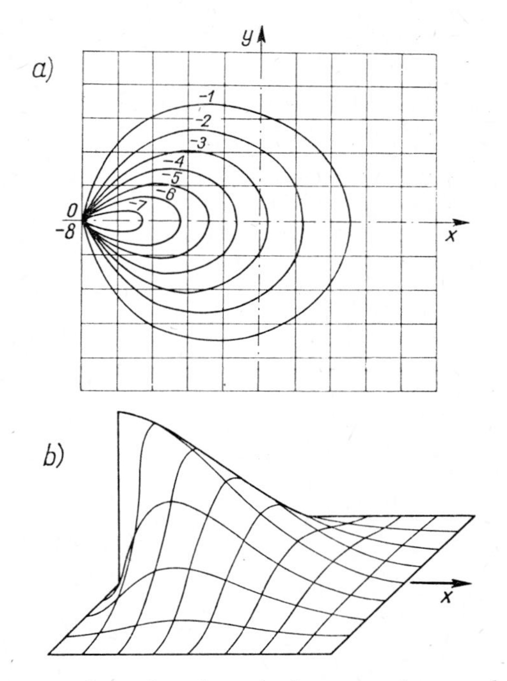

Fig. 40. Influence lines for a simply supported square plate loaded at the edge by a concentrated bending moment.

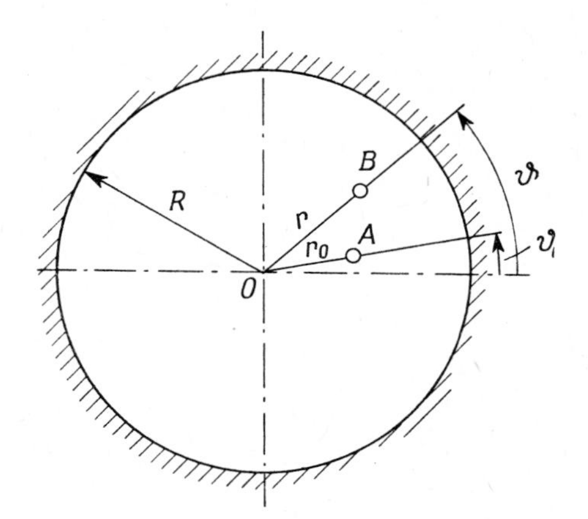

Fig. 41. Circular plate with built-in edge.

$$M_{rr}(\varrho_0, \vartheta_0, \varrho, \vartheta) = -\frac{P}{4\pi} \frac{(1-\varrho^2)^2}{1+\varrho^2-2\varrho\cos(\vartheta-\vartheta_0)}.$$

The above influence surface is presented in Fig. 42. Green's functions were the starting point of the works by S. Kaliski and W. Nowacki [3.36], H. Zorski [3.74, 3.75], and A. Kacner [3.78] to accomplish the solutions for the plates

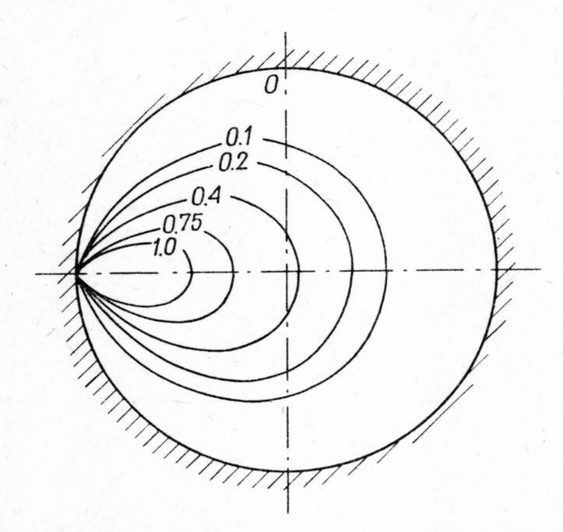

Fig. 42. Influence lines for a circular plate with built-in edge loaded at the edge by a concentrated bending moment.

of discontinuous boundary conditions. The solution of the plate reduces here to the solution of the set of Fredholm's integral equations of first and second order. In this way the solution was found for a strip plate and also a rectangular plate of discontinuous boundary conditions, for example, a plate clamped along a certain part of its edge. This solution has been widened for orthotropic and anisotropic plates [3.33, 3.34, 3.35]. The influence surfaces for continuous plates were given by G. Hoeland [3.45].

3.13 Plates of various shapes under lateral loads

Rectangular plates loaded by concentrated forces as well as by forces distributed over a small surface have been studied by many authors. The action of a concentrated force on a strip plate was worked out in particular by A. Nadai [2.1] in 1925. The same problem with regard to a rectangular plate with various boundary conditions was solved by S. Woinowsky–Krieger [3.6, 3.7]. A solution for a clamped rectangular plate loaded by a concentrated force at its centre was presented by Dana Young [3.9, 3.10]. It can also be found in the works by H. Marcus [3.11], I. Barta [3.12], G. Pickett [3.13], C. J. Thorne and J. V. Atanasoff [3.14], Y. S. Uflyand [3.15]. The same problem was solved by the method of finite differences by E. G. Odley [3.20]. A square plate resting on corner points and loaded at its centre by a force distributed over a small rectangular surface was discussed by H. Marcus [3.11]. The action of concentrated forces on many different plates with various boundary conditions was also treated in detail by S. Timoshenko and Woinowsky-Krieger in book *Theory of plates and shells* [3.1].

Closed form solutions for a strip plate loaded by a concentrated force can also be found in the works by Z. Kączkowski [3.22, 3.23], A. Kacner [3.23, 3.78], and others [3.30, 3.75, 3.77]. These solutions were widened by J. Mossakowski [3.34, 3.35] for orthotropic and anisotropic plates.

The circular plate under a concentrated load applied at a point at some distance from the centre was solved by A. Clebsch [3.58], A. Föppl [3.59], W. Flügge [3.60], H. Schmidt [3.61], and W. Müller [3.62] (see also [3.1]).

The case of a circular plate supported at several points along the boundary was discussed by W. Reissner [3.63], A. Nadai [3.69], and W. A. Bassali [3.65]. Circular plates with a linearly varying thickness submitted to a concentrated force at the centre were discussed by R. Gran Olsen [3.66] and H. D. Conway [3.67] and W. Gittleman [3.68].

3.13.1 *Method of images*

The solution for various plates with simply supported edges can easily be found by the use of the *method of images*. This method was first used by A. Nadai [3.69] and M. T. Huber [3.70]. Since it is also very useful for determination of the stresses in shells, we cite here the main idea of this method.

For example, we want to find the deflection of a rectangular simply supported plate with sides a and b loaded at the point (x_0, y_0) (Fig. 15). But we have found previously the deflection of an infinitely long plate loaded by the force P applied at the point $(x, 0)$. Let us imagine that the rectangular plate is extended in both the positive and negative y-directions and loaded with a series of negative and positive forces P applied at the distance $2y_0$ one from another (Fig. 43). The deflections of such an infinitely long plate are equal to zero along the lines AB, CD, EF, etc. The bending moments along the same lines are also zero. The given plate $ABCD$ can be treated as a portion of an infinitely long plate as shown in Fig. 43. The deflection of the given plate can be calculated by the superposition of the deflections produced by all applied forces. As a result we obtain the deflection of the simply supported rectangular plate in the form of a very rapidly convergent series. Let us consider, for example, the case where $y_0 = b/2$. The deflection at the

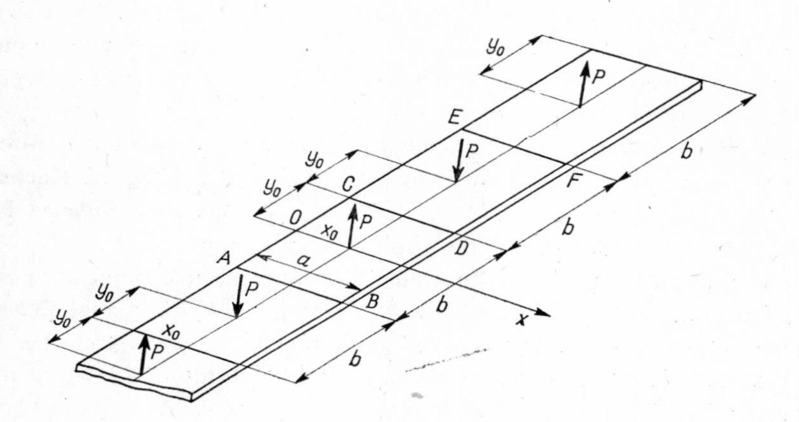

Fig. 43. Application of the method of images to simply supported strip plate.

x-axis of an infinitely long plate produced by the load P at the point O ($y = 0$) is expressed by Eq. (3.61).

$$w_1 = \frac{Pa^2}{2\pi^3 D} \sum_{m=1,3,5,\ldots}^{\infty} \frac{1}{m^3} \sin\frac{m\pi x_0}{a} \sin\frac{m\pi x}{a}.$$

The two forces acting at the distance b from the point O produce the following deflection at the x-axis:

$$w_2 = -\frac{Pa^2}{\pi^3 D} \sum_{m=1,3,5,\ldots}^{\infty} \frac{1}{m^3} \sin\frac{m\pi x_0}{a} (1+2\alpha_m)e^{-2\alpha_m} \sin\frac{m\pi x}{a},$$

where $\alpha_m = m\pi b/2a$. We obtain the total deflection at the x-axis by the summation of the resultants of all forces in the form of the series

$$w = \frac{Pa^2}{2\pi^3 D} \sum_{m=1,3,5,\ldots}^{\infty} \frac{1}{m^3} \sin\frac{m\pi x_0}{a} [1-2(1+2\alpha_m)e^{-2\alpha_m}+$$

$$+2(1+4\alpha_m)e^{-4\alpha_m}\ldots]\sin\frac{m\pi x}{a}.$$

The sum of this series can be evaluated by means of the known series. And we obtain the final expression in coincidence with expression (3.57) of § 3.4.

3.13.2 *Singular solutions for semi-infinite and wedge plates*

In the previous section (§ 3.10) the singular solution corresponding to the action of the concentrated force on an infinite plate has been considered. This solution does not satisfy any definite boundary conditions. However, based on it, it is easy to derive the singular solutions for plates with certain boundary conditions along one or two straight lines, for example, the solution for the semi-infinite plate simply supported along the edge. The conditions of simple support are identical with those existing at the axis of antisymmetry In order to obtain the solution in question we load the plate by two oppositely

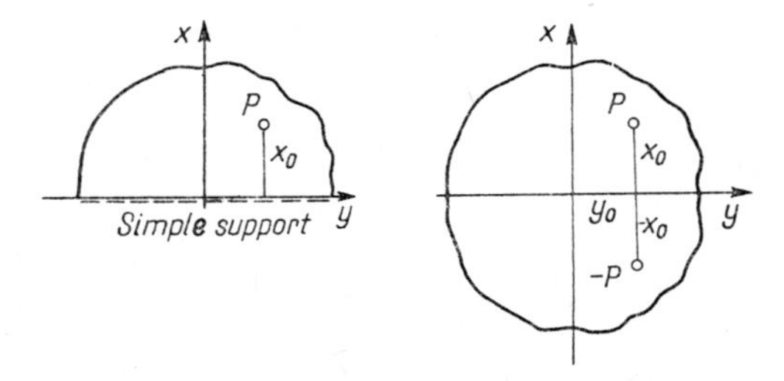

Fig. 44. Semi-infinite plate with simply supported edge $x = 0$.

directed forces acting antisymmetrically with respect to the y-axis (Fig. 44). Then we find

$$w = \frac{Pl^2}{16\pi D}\{[(x-x_0)^2+(y-y_0)^2]\ln[(x-x_0)^2+(y-y_0)^2]+ \\ -[(x+x_0)^2+(y-y_0)^2]\ln[(x+x_0)^2+(y-y_0)^2]\}. \tag{3.151}$$

Bending moments in plate can be found from (3.151) by differentiation. It should be noticed that the plate considered is not in equilibrium. It can freely rotate with respect to the y-axis or we can assume that the supports are at infinity.

Loading the plate by the four forces acting antisymmetrically with respect to the x and y-axes, we find the solution for the plate resting on simple supports on the x and y axes (see Fig. 45). This solution is identical with that obtained previously by means of the Fourier integral (3.76). The solution for the plate simply supported along the x-axis and clamped at $y = 0$ can be obtained from the solution (3.149) for the semi-infinite plate. Summing up the deflections resulting from two concentrated forces acting antisymmetrically with respect to the x-axis, we have (Fig. 46).

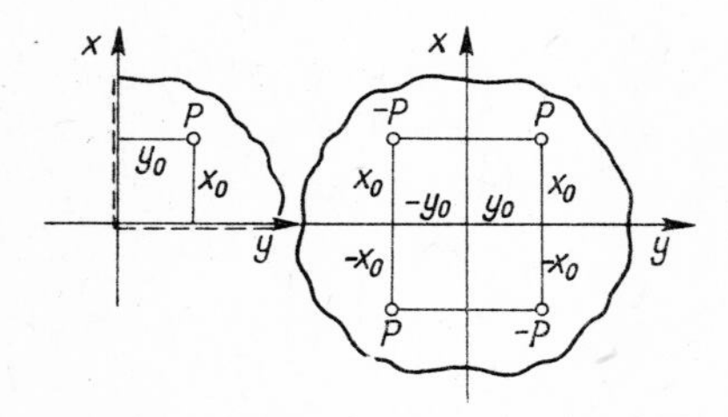

Fig. 45. Wedge plate simply supported at the edges $x = 0$ and $y = 0$.

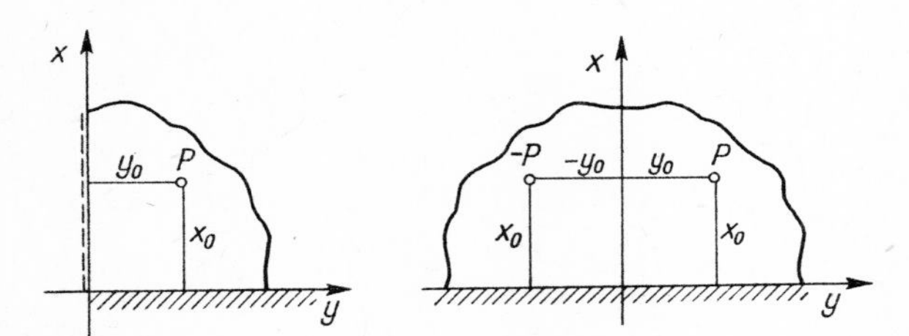

Fig. 46. Wedge plate with simply supported edge $y = 0$ and clamped edge $x = 0$.

$$w = \frac{Pl^2}{16\pi D}\left\{[(x-x_0)^2+(y-y_0)^2]\ln\frac{(x-x_0)^2+(y-y_0)^2}{(x+x_0)^2+(y-y_0)^2} - \right.$$

$$\left. - [(x-x_0)^2+(y+y_0)^2]\ln\frac{(x-x_0)^2+(y+y_0)^2}{(x+x_0)^2+(y+y_0)^2}\right\}. \qquad (3.152)$$

The case of the infinite wedge plate simply supported at two edges (Fig. 47) can also be solved by the method of images. Adding the deflections of the infinite plate loaded by a system of $2n$ forces P antisymmetrically applied with respect to the axis of symmetry, we find the deflection for the wedge plate with the angle π/n at the apex.

$$w = \frac{Pl^2}{16\pi D}\sum_{k=1}^{n}\left\{\left[r^2+r_0^2-2rr_0\cos\left(\vartheta_0-\vartheta+\frac{2k\pi}{n}\right)\right]\times\right.$$

$$\times \ln\left[r^2+r_0^2-2rr_0\cos\left(\vartheta_0-\vartheta+\frac{2k\pi}{n}\right)\right]-$$

$$-\left[r^2+r_0^2-2rr_0\cos\left(\vartheta_0+\vartheta+\frac{2k\pi}{n}\right)\right]\times$$

$$\times \ln\left[r^2+r_0^2-2rr_0\cos\left(\vartheta_0+\vartheta+\frac{2k\pi}{n}\right)\right]\Bigg\}. \tag{3.153}$$

The equilateral triangular plate and plates in the form of an isoceles right triangle with simply supported edges can also be solved by using the method of images (see also A. Nadai [2.1], S. Timoshenko [3.1]).

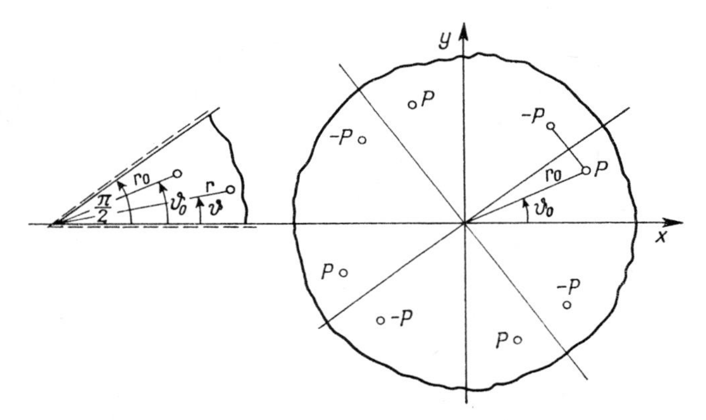

Fig. 47. Wedge plate with two simply supported edges.

3.14 Singularities in the orthotropic plate

3.14.1 *Solution by means of Fourier integral*

Let us consider now an infinite plate loaded by a concentrated force P at the point $x = 0$, $y = 0$. Taking the axes of coordinates x, y following the direction of orthotropy, we have the equation for bending of the plate (see Eq. (1.155))

$$D_x \frac{\partial^4 w}{\partial x^4} + 2H \frac{\partial^4 w}{\partial x^2 \partial y^2} + D_y \frac{\partial^4 w}{\partial y^4} = q(x, y). \tag{3.154}$$

Solving the problem by means of Fourier integrals, we obtain the deflection in the following form:

$$w = \frac{Pl^2}{\pi^2} \int_0^\infty \int_0^\infty \frac{\cos \alpha x \cos \beta y \, d\alpha \, d\beta}{D_x \alpha^4 + 2H\alpha^2\beta^2 + D_y \beta^4} \cdot$$

The above integral can be evaluated with respect to α by means of calculus of residues. We have

$$w = \frac{Pl^2}{2\pi D_x(\alpha_1^2 - \alpha_2^2)} \int_0^\infty \left[\frac{i e^{i\alpha_1 x\beta}}{\alpha_1} - \frac{i e^{i\alpha_2 x\beta}}{\alpha_2} \right] \frac{\cos \beta y}{\beta^3} \, d\beta, \tag{3.155}$$

where α_1, α_2 are the roots of the characteristic equation

$$D_x \alpha^4 + 2H\alpha^2 + D_y = 0. \tag{3.156}$$

Introducing the notations $\lambda = \sqrt[4]{D_y/D_x}$ and $\mu_1 = H/\sqrt{D_x D_y}$, we have

$$\left.\begin{matrix}\alpha_1^2 \\ \alpha_2^2\end{matrix}\right\} = -\lambda^2 [\mu_1 \pm \sqrt{\mu_1^2 - 1}]. \tag{3.157}$$

Now we have to consider the three following cases:

$$\begin{aligned} &(1) \quad H^2 > D_x D_y; \ \mu_1 > 1, \\ &(2) \quad H^2 = D_x D_y; \ \mu_1 = 1, \\ &(3) \quad H^2 < D_x D_y; \ \mu_1 < 1. \end{aligned} \tag{3.158}$$

(1) In the first case, α_1^2 and α_2^2 have real values

$$\alpha_1^2 = -k_1^2, \quad \alpha_2^2 = -k_2^2 \quad \text{where} \quad k_{1,2}^2 = \lambda^2(\mu_1 \pm \sqrt{\mu_1^2 - 1}) > 0,$$

but the roots are imaginary

$$\alpha_1 = ik_1, \quad \alpha_2 = ik_2.$$

Integral (3.155) is divergent. But we can, similarly as for isotropic plates, eliminate from it the singular part of the deflection w corresponding to the action of the concentrated force. Integrating between the limits B, ∞ and taking into account singular terms only, we find

$$\begin{aligned} w = \frac{Pl^2}{8\pi D_x(k_1^2 - k_2^2)} \bigg\{ &- \frac{k_2^2 x^2 - y^2}{k_2} \ln(k_2^2 x^2 + y^2) + \\ &+ \frac{k_1^2 x^2 - y^2}{k_1} \ln(k_1^2 x^2 + y^2) + 4xy(k_2 \vartheta_2 - k_1 \vartheta_1) \bigg\}, \end{aligned} \tag{3.159}$$

where $\vartheta_1 = \arctan \dfrac{y}{k_1 x}$, $\vartheta_2 = \arctan \dfrac{y}{k_2 x}$. This expression can still be simplified. Introducing the polar coordinates (r, ϑ) and preserving singular terms only, we find

$$w = \frac{Pl^2}{8\pi D_x \lambda^3}\sqrt{\frac{2}{1+\mu_1}}(\lambda^2\cos^2\vartheta+\sin^2\vartheta)r^2\ln r.$$

For the isotropic plate $\mu_1 = 1$ the above deflection is identical with that obtained previously.

(2) In the second case we have double roots,

$$\alpha_1^2 = \alpha_2^2 = k^2 = \lambda^2 = \sqrt{\frac{D_x}{D_y}}$$

and the solution takes the form

$$w = \frac{Pl^2}{16\pi D_x \lambda^3}(\lambda^2x^2+y^2)\ln(\lambda^2x^2+y^2). \tag{3.160}$$

If we introduce new coordinates $\bar{x} = x\lambda$, $\bar{y} = y$, the above solution coincides with that obtained previously for an isotropic plate. In polar coordinates the singular expression for the deflection is

$$w = \frac{Pl^2}{8\pi D_x \lambda^3}(\lambda^2\cos^2\vartheta+\sin^2\vartheta)r^2\ln r.$$

(3) In the third case, $H^2 < D_xD_y$, k has a complex value and the roots of Eq. (3.156) are

$$\alpha_{1,2} = i\lambda[\mu_1 \pm i\sqrt{1-\mu_1^2}]^{1/2} = i\lambda\left[\sqrt{\frac{1+\mu_1}{2}} \pm i\sqrt{\frac{1-\mu_1}{2}}\right] = i(a \pm ib). \tag{3.161}$$

On substituting it into (3.155) and integrating, we obtain the singular solution

$$\begin{aligned} w = \frac{Pl^2}{16\pi D_x\sqrt{1-\mu_1^2}}\Big\{&[b(x^2+y^2)+2xy]\ln[x^2+y^2+2bxy]+ \\ &+[b(x^2+y^2)-2xy]\ln[x^2+y^2-2bxy]- \\ &-2a(x^2-y^2)(\vartheta_1-\vartheta_2)\Big\}, \end{aligned} \tag{3.162}$$

where

$$\vartheta_1 = \arctan\frac{bx-y}{ax}, \qquad \vartheta_2 = \arctan\frac{bx+y}{ax}.$$

Introducing polar coordinates, we obtain the singular term

$$w = \frac{Pl^2\lambda}{8\pi D_x} \sqrt{\frac{2}{1+\mu_1}}\, r^2 \ln r$$

which also coincides with the result for an isotropic plate if $\lambda = 1$ and $\mu_1 = 1$. In order to calculate the stresses we should know the values of the curvatures of the deflected surface of the plate. In the first case we find from (3.155) the following integral for the curvature $\partial^2 w/\partial x^2$

$$\frac{\partial^2 w}{\partial x^2} = \frac{P}{2\pi D_x(k_1^2-k_2^2)} \int_0^\infty (k_2 e^{-k_2\beta x} - k_1 e^{-k_1\beta x}) \frac{\cos\beta y}{\beta}\, d\beta. \tag{3.163}$$

Upon integration and taking only singular terms, we find

$\mu_1 > 1$,

$$\frac{\partial^2 w}{\partial x^2} = \frac{P}{4\pi D_x(k_1^2-k_2^2)} [k_1 \ln(k_1^2 x^2+y^2) - k_2 \ln(k_2^2 x^2+y^2)\}. \tag{3.164}$$

Introducing $x = r\cos\vartheta$, $y = r\sin\vartheta$ and preserving only singular terms, we find

$$\frac{\partial^2 w}{\partial x^2} = \frac{\partial^2 w}{\partial y^2} = -\frac{P}{4\pi D_x \lambda} \sqrt{\frac{2}{1+\mu_1}} \ln r.$$

In the second case the curvatures are

$$\mu_1 = 1, \qquad \frac{\partial^2 w}{\partial x^2} = \frac{\partial^2 w}{\partial y^2} = -\frac{P}{4\pi D_x \lambda} \ln r \tag{3.165}$$

and in the third case we have

$$\mu_1 < 1, \qquad \frac{\partial^2 w}{\partial x^2} = \frac{\partial^2 w}{\partial y^2} = -\frac{P\lambda}{4\pi D_x} \sqrt{\frac{2}{1+\mu_1}} \ln r. \tag{3.166}$$

We observe that in all three cases the curvatures and also the stresses are infinitely large at the point of application of the load.

3.14.2 *Solution in a double Fourier series*

When the rectangular orthotropic plate is simply supported on all edges, the deflection can be found by the method used for the isotropic plate in § 3.4.

Representing the load in the form of a double trigonometric series, we find the solution

$$w = \frac{4P}{\pi^4 ab} \sum_{m=1}^{\infty} \sum_{n=1}^{\infty} \frac{\sin \dfrac{m\pi x_0}{a} \sin \dfrac{n\pi y_0}{b} \sin \dfrac{m\pi x}{a} \sin \dfrac{n\pi y}{b}}{\dfrac{m^4}{a^4} D_x + \dfrac{2m^2n^2}{a^2b^2} H + \dfrac{n^4}{b^4} D_y}. \tag{3.167}$$

When the plate is made of an isotropic material, $D_x = D_y = H = D$, and this solution coincides with that given in § 3.4.

In the particular case of $H = D_x D_y$ the deflection at the centre of the orthotropic plate may be obtained from the results calculated for the isotropic plate. Comparing the results (3.50) and (3.167), we conclude that the deflection of the orthotropic plate is the same as that for an isotropic plate having the rigidity D and sides $a_0 = a\sqrt[4]{D/D_x}$ and $b_0 = b\sqrt[4]{D/D_y}$. In a similar manner the curvatures of orthotropic plates may be expressed as those of the isotropic plates.

3.14.3 *Infinitely long strip plate. Solution by means of single trigonometric series*

In this case let us consider an infinitely long plate with simply supported edges $x = 0$ and $x = l$. We assume that the load is expanded along the x-axis in the series

$$q = \sum_{m=1}^{\infty} q_{mn} \sin \frac{m\pi x}{l}.$$

The solution satisfying the boundary conditions at the edges parallel to the y-axis can be assumed in the similar form

$$w = \sum_{m=1}^{\infty} w_m \sin \frac{m\pi x}{l},$$

where $w_m = w_m(y)$ is a function of y only. Substituting this in (3.154), we obtain the following ordinary differential equation:

$$D_y w_m^{\text{IV}} - 2H \frac{m^2\pi^2}{l^2} w_m'' + D_x \frac{m^4\pi^4}{l^4} w_m = 0. \tag{3.168}$$

Assuming that $w_m = Ae^{my/k}$, we find the value of k

$$k_{1,2,3,4} = \pm \frac{l\lambda}{\pi} \sqrt{\mu_1 \pm \sqrt{\mu_1^2 - 1}}, \tag{3.169}$$

where

$$\lambda^4 = \frac{D_y}{D_x}, \qquad \mu_1 = \frac{H}{D_x D_y}.$$

If we neglect the positive roots, the integral of Eq. (3.154) becomes

$$w = \sum_{m=1}^{\infty} (A_m e^{-my/k_1} + B_m e^{-my/k_2}) \sin \frac{m\pi x}{l}. \tag{3.170}$$

From symmetry we conclude that if along the y-axis $(\partial w/\partial x)_{y=0} = 0$, then we have $B_m = -k_2 Am/k_1$. The second boundary condition for the x-axis gives the relation between the shearing forces Q_y and the load.

$$\overline{Q}_y = -\frac{\partial}{\partial y}\left(D_y \frac{\partial^2 w}{\partial y^2} + H \frac{\partial^2 w}{\partial x^2}\right) = -\frac{1}{2}\sum_{m=1}^{\infty} q_m \sin\frac{m\pi x}{l}. \tag{3.171}$$

Solving the above equations, we find the deflection of the plate. For the concentrated force P acting at the point $x = x_0$, $y = 0$ we have the following expressions:

Case 1: $\mu_1 > 1$;

$$w = \frac{Pl^3}{\pi^4 D_x (k_1^2 - k_2^2)} \sum_{m=1}^{\infty} \frac{1}{m^3} (k_1 e^{-my/k_1} -$$

$$- k_2 e^{-my/k_2}) \sin\frac{m\pi x_0}{l} \sin\frac{m\pi x}{l}. \tag{3.172}$$

Case 2: $\mu_1 = 1$;

$$w = \frac{Pl^2}{2\pi^3 D_x \lambda} \sum_{m=1}^{\infty} \frac{1}{m^3}\left(1 + \frac{m\pi y}{l\lambda}\right) e^{-my/b} \sin\frac{m\pi x_0}{l} \sin\frac{m\pi x}{l}.$$

Case 3: $\mu_1 < 1$;

$$w = \frac{Pl}{2\pi^2 \sqrt{D_x D_y}} \times$$

$$\times \sum_{m=1}^{\infty} \frac{1}{m^3}\left(a' \sin\frac{my}{a'} + b' \cos\frac{my}{a'}\right) e^{-my/b'} \sin\frac{m\pi x_0}{l} \sin\frac{m\pi x}{l},$$

where

$$a' = \frac{l\lambda}{\pi}\sqrt{\frac{2}{1-\mu_1}}, \qquad b' = \frac{l\lambda}{\pi}\sqrt{\frac{2}{1+\mu_1}}.$$

In order to obtain the bending stresses we must first calculate the curvatures of the deflected surface of the plate. In the first case the curvatures are given by the following series:

$$\varkappa_{xx} = -\frac{\partial^2 w}{\partial x^2}$$

$$= \frac{Pl}{\pi^2 D_x(k_1^2 - k_2^2)} \sum_{m=1}^{\infty} \frac{1}{m} (k_1 \mathrm{e}^{-my/k_1} - k_2 \mathrm{e}^{-my/k_2}) \sin \frac{m\pi x_0}{l} \sin \frac{m\pi x}{l}, \tag{3.173}$$

$$\varkappa_{yy} = -\frac{\partial^2 w}{\partial y^2}$$

$$= \frac{Pl}{\pi^2 D_x(k_1^2 - k_2^2)\lambda^2} \sum_{m=1}^{\infty} \frac{1}{m} (k_1 \mathrm{e}^{-my/k_2} - k_2 \mathrm{e}^{-my/k_1}) \sin \frac{m\pi x_0}{l} \sin \frac{m\pi x}{l}.$$

To evaluate the sum we can use the series whose sum is known

$$\sum_{m}^{\infty} \frac{1}{m} \mathrm{e}^{-my/k} \sin \frac{m\pi x_0}{l} \sin \frac{m\pi x}{l}$$

$$= \frac{1}{4} \ln \frac{\cosh \dfrac{y}{k} - \cos \dfrac{\pi(x+x_0)}{l}}{\cosh \dfrac{y}{k} - \cos \dfrac{\pi(x-x_0)}{l}}.$$

Then we have for $\mu_1 > 1$

$$\varkappa_{xx} = \frac{Pl}{4\pi^2 D_x(k_2^2 - k_1^2)} [k_2 L(k_2) - k_1 L(k_1)],$$

$$\varkappa_{yy} = \frac{Pl}{4\pi^2 \sqrt{D_x D_y}(k_2^2 - k_1^2)} [k_2 L(k_1) - k_1 L(k_2)], \tag{3.174}$$

where

$$L\left(\frac{y}{k}\right) = \ln \frac{\cosh \dfrac{y}{k} - \cos \dfrac{\pi(x+x_0)}{l}}{\cosh \dfrac{y}{k} - \cos \dfrac{\pi(x-x_0)}{l}}.$$

In the second case we have the following closed-form expressions for the curvatures:

$$\varkappa_{xx} = \frac{P}{8\pi D_x \lambda} L_1 + \frac{P\lambda y}{8D_y l} L_2,$$
$$\varkappa_{yy} = \frac{P\lambda}{8\pi D_y} L_1 - \frac{P\lambda y}{8D_y l} L_2. \tag{3.175}$$

where

$$L_1 = \ln \frac{\cosh \dfrac{\pi y}{\lambda l} - \cos \dfrac{\pi(x+x_0)}{l}}{\cosh \dfrac{\pi y}{\lambda l} - \cos \dfrac{\pi(x-x_0)}{l}},$$

$$L_2\left(\frac{\pi y}{\lambda l}\right)$$
$$= \frac{\sinh \dfrac{\pi y}{l\lambda}}{\cosh \dfrac{\pi y}{l\lambda} - \cos \dfrac{\pi(x-x_0)}{l}} - \frac{\sinh \dfrac{\pi y}{l\lambda}}{\cosh \dfrac{\pi y}{l\lambda} - \cos \dfrac{\pi(x+x_0)}{l}}.$$

In the third case we have

$$\mu_1 < 1,$$
$$\varkappa_{xx} = \frac{P}{8lD_y}\left[a' \ln \frac{\varrho_1}{\varrho_2} + b'(\vartheta_1 - \vartheta_2)\right],$$
$$\varkappa_{yy} = \frac{P}{8l\sqrt{D_x D_y}}\left[a' \ln \frac{\varrho_1}{\varrho_2} - b'(\vartheta_1 - \vartheta_2)\right], \tag{3.176}$$

where

$$\left.\begin{matrix}\varrho_1\\ \varrho_2\end{matrix}\right\} = \left[\cosh \frac{y}{a'} \cos \frac{y}{b'} - \cos \frac{(x \pm x_0)}{l}\right]^2 + \sinh^2 \frac{y}{a'} \sin^2 \frac{y}{b'},$$

$$\left.\begin{matrix}\vartheta_1\\ \vartheta_2\end{matrix}\right\} = \arctan \frac{\sinh \dfrac{y}{a'} \sin \dfrac{y}{b'}}{\cosh \dfrac{y}{a'} \cos \dfrac{y}{b'} - \cos \dfrac{\pi(x \pm x_0)}{l}}.$$

We observe that in all three cases the bending moments are infinitely large at the point of application of the load.

The above solutions have been obtained by W. Nowacki [3.82] and S. Woinowsky-Krieger [3.83]. More particular results regarding influence surfaces of orthotropic rectangular plates may be found in papers by H. Olsen and F. Reinitz-Huber [3.85], H. Homberg and J. Weinmeister [3.86], and by S. G. Lekhnitzkii [3.87].

3.15 Singular solutions for sandwich plates

3.15.1 *Normal force*

The solution for the infinite sandwich plate on the basis of the simplified model, i.e. by the assumption that the bending rigidity of the external layers is negligible in comparison with the rigidity of the plate, can be obtained immediately from (3.66) and (3.4) by using the analogy of the governing differential equations. The effect of the normal stress σ_{zz} should be neglected here because it has in this case no physical meaning. Then the distribution of the moments in the sandwich plate is the same as in the isotropic plate. The only difference consists of the relations for the deflection. However, we should remember here that the assumption of the theory of sandwich plates presented here contains also the condition that the material of the core is underformable in the direction perpendicular to the middle surface of the plate and the normal displacements of the external layers are equal. In order to satisfy this condition approximately, in the case of the plate loaded by a lateral concentrated force the plate should be strengthened in this place by a lateral element which can withstand the normal stresses σ_{zz}. The singular solutions for the sandwich, three-layer plate have been obtained by R. Ganowicz [3.73] based on the complete equations (2.45). Using the Fourier transform, the following solution can be obtained:

$$U(x, y) = \frac{Pl^2}{4\pi^2 D_z} \int_{-\infty}^{\infty}\int_{-\infty}^{\infty} \frac{e^{i(\alpha x+\beta y)}\,d\alpha\,d\beta}{\left[1+\varkappa_c \dfrac{2D}{D_z l^2}(\alpha^2+\beta^2)\right](\alpha^2+\beta^2)^2}$$

where $\varkappa_c$, D, D_z are defined by relations (2.41), $x = x/l$, $y = y/l$, $r = r/l$ are the non-dimensional coordinates referred to the arbitrary length l. Let us assume that $l^2 = 2\varkappa_c D/D_z$. This integral can be presented in the form of the sum of simple integrals. Separating the finite part we obtain the following singular solution for the displacement function:

$$U(x, y) = \frac{Pl^2}{8\pi D_z}\left[\frac{r^2}{2}\ln r^2 + 2\ln r^2 + 4K_0(r)\right],$$

where $K_0(r)$ is the modified Bessel function of the second kind. The singular solution for the displacements can be found using the above solution and Eqs. (2.44). We find

$$w = \frac{Pl^2}{8\pi D_z}\left\{r^2 \ln r - \frac{D_z}{D}(\ln r^2+2)+4\left[\ln r+\left(1-\frac{D_z}{2D}\right)K_0(r)\right]\right\},$$

$$u = -\frac{Pl(2h+\delta)}{16\pi D_z}\left\{x(\ln r^2+1)-4\frac{\partial}{\partial x}[\ln r+K_0(r)]\right\},$$

$$v = -\frac{Pl(2h+\delta)}{16\pi D_z}\left\{y(\ln r^2+1)-4\frac{\partial}{\partial y}[\ln r+K_0(r)]\right\}.$$

The internal forces in the external layers can be obtained using relations (2.33) and (2.36). It develops that the resultant internal moments, that is, the moments calculated for the whole sandwich plate are identical with (3.10) obtained previously (by means of the classical theory) for the isotropic thin plate. Let us consider the immediate vicinity of the point of application of the load and let us examine the solution obtained.

Since for $r \to 0$ $K_0(r) \simeq -\ln r$, for $r \to 0$ the deflection of the plate w has the finite value

$$w(0,0) = -\frac{P}{2\pi D_z}\varkappa_c = -\frac{Pl^2}{4\pi D}$$

similarly as in the classical theory of thin plates. This result is interesting because in spite of taking into account the effect of transverse shear stresses, the displacement at the point of application of the load is finite. Let us notice that if $r \to 0$, the concentrated force P is equilibrated only by the shear forces in the external layers:

$$2q_r = Q_r = -\frac{P}{2\pi r l}.$$

This means that the concentrated force acts first of all on the external layers and next is transmitted to the core. Because of the assumed antisymmetry of the solution, this loading should be considered as two concentrated forces $\frac{1}{2}P$ each acting on the upper and lower surface of the plate.

3.15.2 *Concentrated bending moment*

The concentrated bending moment can be defined as the resultant of two concentrated normal forces acting on the plate separated by a vanishingly small distance $\Delta x \to 0$. Thus:

$$M = \lim_{\Delta x \to 0} P \cdot \Delta x.$$

The solution for this case can be obtained by the differentiation of the solution for the concentrated normal force and because of it will not be considered separately. The case where the concentrated moment is defined as the resultant of two concentrated tangential forces acting on the external layers is more justified in the case of sandwich plates. A singular solution for this case has been obtained based on the complete equations (2.41). We can assume that the forces acting on the external layers are uniformly distributed across its thickness. This type of loading has been taken into consideration in Eqs. (2.41). Then the concentrated bending moment is given by

$$M = n_x(2h+\delta) = N\delta(x, y)(2h+\delta).$$

Applying equation $(2.41)_2$ we obtain the deflection function F_2 in the form of the integral

$$F_2 = \frac{Nl^2}{2\pi^2 D_z} \times$$

$$\times \int_{-\infty}^{\infty}\int_{-\infty}^{\infty} \frac{e^{-i(\alpha x+\beta y)} d\alpha\, d\beta}{(\alpha^2+\beta^2)^2\left[1+\varkappa_w \frac{2D}{D_z}(\alpha^2+\beta^2)\right]\left[1+\frac{1}{2}(1-\nu)\varkappa_w(\alpha^2+\beta^2)\right]},$$

which can be evaluated by separating it into simple integrals. The integrals can be evaluated by the separation of the "finite parts". As the result we obtain the following expression for the singular part of the function F_2:

$$F_2 = \frac{Nl^2}{2\pi D_z}\left\{\frac{r^2}{2}\ln r+2\frac{1+\gamma^2}{\gamma^2}\ln r+\right.$$

$$\left.+\frac{2}{\gamma^2-1}\left[\gamma^2 K_0(r)-\frac{1}{\gamma^2}K_0(r\gamma)\right]\right\},$$

where

$$\gamma = \sqrt{2/\varkappa_c(1-\nu)}.$$

We obtain the deflection w and displacements u and v introducing F_2 into Eq. (2.43). We find for the deflection w the following singular solution:

$$w = -\frac{N(2h+\delta)l}{4\pi D_z}\left\{\frac{x}{2}(\ln r^2+1)+2\frac{x}{r^2}(1-rK_1(r))\right\},$$

and similar expression for the displacements u and v. The internal forces in the plate can be defined from Eqs. (2.33) and (2.36). It develops that the resultant internal forces and moments calculated for the complete cross-section of the plate do not depend on the bending stiffness of the external layers, D. They are identical with those which would be obtained from the equations of sandwich plates based on the assumption that the bending rigidity of the external layers is negligible. We know, however, that this solution can be found from the analogy between the equations of sandwich plates and the equations of isotropic plates with the consideration of the effect of transverse shear deformations.

References 3

[3.1] S. TIMOSHENKO, S. WOINOWSKY-KRIEGER: *Theory of plates and shells*, McGraw-Hill, New York 1959.

[3.2] S. WOINOWSKY-KRIEGER, *Ing.-Arch. 4* (1933) 305.

[3.3] C. L. M. H. NAVIER: *Bull. Soc. Philomath. 96* (1823).

[3.4] G. YU. DJAMELIDZE: Summing of Navier's solution for rectangular plates, *Sb. Prikl. Mat. Mekh. 3.4* (1939) (in Russian).

[3.5] W. FLÜGGE: *Die Strenge Berechnung von Kreisplatten unter Einzellasten*, Springer, Berlin 1928.

[3.6] S. WOINOWSKY-KRIEGER: *Ing.-Arch. 3* (1932) 340.

[3.7] S. WOINOWSKY-KRIEGER: *Ing.-Arch. 21* (1953) 336, 337.

[3.8] A. NADAI: Über die Spannungsverteilung in einer durch eine Einzelkraft belasteten rechteckigen Platte, *Bauing.*, *2* (1921) 11.

[3.9] DANA YOUNG: *J. Appl. Mech. 6, A-114* (1939).

[3.10] DANA YOUNG: *J. Appl. Mech. A-139* (1940).

[3.11] H. MARCUS: *Die Theorie elastischer Gewebe*, 2d. ed., 155 pp., Berlin 1932.

[3.12] I. BARTA: *Z. Angew. Math. Mech. 17* (1937) 184.

[3.13] G. PICKETT: *J. Appl. Mech. 6, A-168*, (1937).

[3.14] C. J. THORNE and J. V. ATANASOFF: *Iowa State Coll. J. Se. 14* (1940) 333.

[3.15] YA. S. UFLYAND: *Doklady Ak. Nauk SSSR 72* (1950) 665.

[3.16] C. P. SIESS, N. M. NEWMARK: *Univ. Illinois Bull. 47* (1950) 95.

[3.17] R. OHLIG: Zwei- und vierseitig aufgelagerte Rechteckplatten unter Einzelkraftbelastung, *Ing. Arch. 16* (1947) 51.

[3.18] E. MELAN: Die Berechnung einer exzentrisch durch eine Einzellast belasteten Kreisplatte, *Eisenbau 17* (1926) 190.

[3.19] E. BITTNER: *Momententafeln und Einflussflächen für Krenzweise bewehrte Eisenbetonplatten*, Springer, Wien 1938.

[3.20] E. G. ODLEY: *J. Appl. Mech. 14, A-289* (1947).

[3.21] S. TIMOSHENKO: Über die Biegung der allseitig unterstützten Platten unter der Wirkung einer Einzellast, *Bauing. 3* (1922) 51.

[3.22] Z. KĄCZKOWSKI: Certain closed solution for the deflection of strip plates, *Arch. Mech. Stos. 5* (1953) 4 (in Polish).

[3.23] A. Kacner, Z. Kączkowski: Application of tabular function for the calculation of the deflection and internal forces in orthotropic strip plates, *Rozpr. Inż. 7* (1959) 41–91 (in Polish).

[3.24] E. Sternberg, R. A. Eubanks: On the concept of concentrated loads and an extension of the uniqueness theorem in the linear theory of elasticity, *J. Rat. Mech. 4* (1955).

[3.25] E. Sternberg: On some recent developments in the linear theory of elasticity, *Structural Mechanics — Proc. 1st Symposium on Naval Struct. Mech.*, Pergamon Press 1960.

[3.26] J. Hadamard: *Lectures on Cauchy's problem in partial differential equations*, Yale Univer. Press., 1923.

[3.27] H. M. Westergaard: *Ingeniøren 32* (1929) 513; *Public Roads 7* (1926) 25; *10* (1929) 65; *14* (1933) 185.

[3.28] M. Suchar: Calculation of the influence surfaces for equilateral plates, *Rozpr. Inż. 7* (1959) 237–260 (in Polish).

[3.29] M. Suchar: Certain problems of the theory of elasticity of anisotropic shells, *Rozpr. Inż. 12.2* (1964) (in Polish).

[3.30] M. Suchar: An anisotropic elastically restrained half-plane loaded by a concentrated force, *Bull. Acad. Polon. Sci., Série Sci. Techn. 9* (1961) 617.

[3.31] I. Hanuška: The bending of wedge-shaped plate, *Arch. Mech. Stos. 15,2* (1963) 209–224.

[3.32] M. Suchar: Certain problem of the theory of elasticity of anisotropic plates, *Rozpr. Inż. 12* (1964) (in Polish).

[3.33] M. Suchar: On singular solution in the theory of anisotropic plates, *Bull. Acad. Polon. Sci., Série. Sci. techn. 12* (1964) 29–38.

[3.34] J. Mossakowski: Singular solutions in the theory of orthotropic plates, *Arch. Mech. Stos. 3* (1954) (in Polish).

[3.35] J. Mossakowski: Singular solutions for anisotropic plates, *Arch. Mech. Stos. 1* (1955) (in Polish).

[3.36] S. Kaliski, W. Nowacki: Some problems of structural analysis of plates with mixed boundary conditions, *Arch. Mech. Stos. 8.4* (1956) 413.

[3.37] V. K. Prokopov: Deformation of the circular plate with three rigid supports under concentrated loads, *Trudy Leningr. Polit. Un-ta 252* (1965) 107–113 (in Russian).

[3.38] G. Nitsiotas: Die Singularitäten der frei drehbar gelagerten orthotropen Halfplattenstreifen, *Ing.-Arch. 29.3* (1960) 223.

[3.39] S. Woinowsky-Krieger: Über die Biegung des orthotropen Plattenstreifens durch Einzellasten, *Ing.-Arch. 25.2* (1957) 226.

[3.40] A. Pucher: *Die Einflussfelder elastischer Platten*, Springer, Wien 1951.

[3.41] W. Koepcke: *Umfangsgelagerte Rechteckplatten mit drehbaren und eingespannten Rändern*, Dis. Berlin 1940, Printed by R. Noske. Borna–Leipzig 1940; Zur Ermittlung der Einflussflächen und inneren Kräfte umfangsgelagerter Rechteckplatten, *Ing.-Arch. 18* (1950) 106.

[3.42] J. H. Michell: *Proc. Lond. Math. Soc. 34* (1902), 223.

[3.43] A. Pucher: *Ing.-Arch. 12* (1941) 76.

[3.44] A. Pucher: *Die Einflussfelder des Plattenstreifen mit zwei eingespannten Rändern*, Federhofer-Girkmann Festschrift, Wien 1950.

[3.45] G. Hoeland: *Ing. Arch. 24* (1956) 124.

[3.46] M. SOKOLOVSKY: Über einige Probleme der Elastizitätstheorie orthotroper Körper, *Bull. Acad. Polon. Sci., Série Cl. 4, 4* (1954).

[3.47] H. D. CONWAY: Some problems of orthotropic plane stress, *J. Appl. Mech. 20* (1953) 72; Note on the orthotropic half plane subjected to concentrated loads, *J. Appl. Mech. 22* (1955) 130.

[3.48] K. MARGUERRE: Über den Träger auf elastischer Unterlage, *ZAMM 17* (1937) 224.

[3.49] K. GIRKMANN: Stegblechbeulung unter ortlichen Lastangriff, *Sitz. Ber. Österr. Akad. d. Wiss. Wien. Math. Nat. Kl., 145* (1936).

[3.50] M. A. BIOT: Bending of an infinite beam on an elastic foundation, *J. Appl. Mech.* (1937).

[3.51] E. REISSNER: *On the theory of beams on an elastic foundation*, Federhofer-Girkmann Festschrift, Wien 1950.

[3.52] H. BOROWICKA: Druckverteilung unter elastischen Platten, *Ing. Arch. 10* (1939) 113.

[3.53] G. SCHUBERT: Zur Frage der Druckverteilung unter elastisch gelagerten Tragwerken, *Ing. Arch. 13* (1942) 132.

[3.54] M. SADOWSKI: Zweidimensionale Probleme der Elastizitätstheorie, *ZAMM 8* (1928) 507.

[3.55] S. D. CAROTHERS: *Proc. Roy. Soc. Lond. 97* (1920) 110.

[3.56] HUNG-GAU WU and C. W. NIELSON: The stresses in a flat curved bar resulting from concentrated tangential boundary loads, *J. Appl. Mech. Trans. ASME 21* (1954) 151–159.

[3.57] YI-YUAN-YU: Bending of isotropic thin plates by concentrated edge couples and forces, *J. Appl. Mech. Trans. ASME 21* (1954).

[3.58] A. CLEBSCH: *Theorie der Elastizität fester Körper*, 1862.

[3.59] A. FÖPPL: *Sitzber. Bayer. Akad. Wiss. Jahrg. 155* (1912).

[3.60] W. FLÜGGE: *Die Strenge Berechnung von Kreisplatten unter Einzellasten*, Berlin 1928.

[3.61] H. SCHMIDT: *Ing.-Arch. 1* (1930) 147.

[3.62] W. MÜLLER: *Ing.-Arch. 13* (1943) 355.

[3.63] E. REISSNER: *Math. Ann. 111* (1935) 777.

[3.64] A. NADAI: *Z. Phys. 23* (1922) 366.

[3.65] W. A. BASSALI: *Proc. Cambridge Phil. Soc. 53* (1957) 726.

[3.66] R. GRAN OLSEN: *Ing.-Arch. 10* (1939) 14.

[3.67] H. D. CONWAY: *J. Appl. Mech. 15* (1948) 1.

[3.68] W. GITTLEMAN: *Aircraft Eng. 22* (1950) 224.

[3.69] A. NADAI: *Z. Angew. Math. Mech. ZAMM 2* (1922) 1.

[3.70] M. T. HUBER: *ZAMM 6* (1926) 228.

[3.71] R. GANOWICZ: A note on sandwich plate under the action of a concentrated torque, *Arch. Mech. Stos. 23* (1971) 329–338.

[3.72] R. GANOWICZ, On certain solution for sandwich plates, *Rozpr. Inżyn. 14, 3* (1966) (in Polish).

[3.73] R. GANOWICZ: Singular solutions in the general theory of sandwich plates, *Mech. Teoret. Stos. 3* (1967) 5 (in Polish).

[3.74] H. ZORSKI: Plates with discontinuous supports, *Arch. Mech. Stos. 10.3* (1958) 271.

[3.75] H. ZORSKI: Some case of bending of anisotropic plates, *Arch. Mech. Stos. 12.1* (1959) 71.

[3.76] H. ZORSKI: A semi-infinite strip with discontinuous boundary conditions, *Arch. Mech. Stos. 10.3* (1958) 371.

[3.77] M. SOKOŁOWSKI: The bending nonhomogeneous plates, *Arch. Mech. Stos. 10.3* (1958) 376.

[3.78] A. KACNER: I. A closed solution in the case of semi-infinite plate with discontinuous boundary conditions, *Arch. Mech. Stos. 4* (1957); II. A closed solution in the case of semi-infinite plate with discontinuous boundary conditions, *Arch. Mech. Stos. 10.4* (1958)

[3.79] A. KACNER: A method of gradual approximations applied to the problems of plates with discontinuous boundary conditions, *Arch. Inżyn. Ląd. 4,3* (1958) 397–408 (in Polish).

[3.80] Z. KĄCZKOWSKI: Rectangular orthotropic thin plates with arbitrary boundary conditions, *Arch. Mech. Stos. 10* (1958) 525.

[3.81] W. NOWACKI, J. MOSSAKOWSKI: Influence surfaces for ring plates *Arch. Mech. Stos. 5.2* (1953) 237–272.

[3.82] W. NOWACKI: *Acta Techn. Sci. Hung. 8* (1954) 109.

[3.83] S. WOINOWSKY-KRIEGER, *Ing.-Arch. 25* (1957) 90.

[3.84] S. G. LEKHNITSKY: On the theory of anisotropic thick plates, *Izv. Ak. Nauk SSSR, Otd. Tekhn. Nauk, Mekhanika i Mashinostroenie 2* (1959) 142–145 (in Russian).

[3.85] H. OLSEN, F. REINITZ-HUBER: *Die zweiseitig gelagerte Platte*, Berlin 1950.

[3.86] H. HOMBERG, J. WEINMEISTER: *Einflussflächen für Kreuzwerke*, 2nd ed., Berlin 1956.

[3.87] S. G. LEKHNITZKII, *Anisotropic plates*, Gostekhizdat, Moskva 1947 (in Russian).

[3.88] H. HERTZ: Über die Beruhrung fester elastischer Korper, *J. reine und angewandte Mathematik* (Crelle) Bb. 92, 1882.

[3.89] J. V. BOUSSINESQ: *Application des potentieles à l'etude d'équilibre et du mouvement des solides elastiques*, Gauthier-Villars, Paris 1885, p. 63.

[3.90] J. MIKUSIŃSKI, R. SIKORSKI: The elementary theory of distributions I, *Rozprawy Matematyczne 12* (1957).

[3.91] L. SCHWARTZ: *Théorie des distributions*, Hermann, Paris 1966.

[3.92] J. OSIOWSKI: *Elements of the theory of distributions*, Chapter V of the book: *Elements of modern mathematics for engineers*, PWN, Warszawa–Wrocław 1964 (in Polish).

4

Concentrated lateral loads at the edge of a plate

4.1 Cantilever plate loaded at its free edge by a lateral concentrated force

We discuss now the case of a cantilever plate loaded by a force P at its free edge. This problem was considered by many authors by means of the classical theory of plates (see C. W. Mac Gregor [4.1], D. L. Holl [4.2], T. J. Jaramillo [4.3], K. Girkman [1.23], W.T. Koiter and J. B. Alblas [4.4], and E. Bittner [4.5]). Let us find a solution resulting from the improved theory taking into account the effect of shear deformations.

In order to define the type of singularities of the stresses at the loading point, we consider at first a simplified problem, namely a semi-infinite plate spreading indefinitely in the x and y directions and loaded at its free edge $x = 0$ by a concentrated force (Fig. 48). Let us introduce a system of non-dimensional coordinates defined by $x = \bar{x}/l$; $y = \bar{y}/l$; l is an arbitrary length. Because the load is applied to the edge, the displacements of the plate $w(x, y)$ fulfil the homogenous equation. The solution can be found in the form (3.123). We have to satisfy the following boundary conditions: at the free edge $x = 0$ the moments $M_{xx} = 0$ and $M_{xy} = 0$, and the shear force $Q_x = P\delta(y, 0)$.

For $x \to \infty$ the displacement w should vanish. We have therefore for $x = 0$ (Fig. 48):

$$-\frac{D}{l^2}\left[\frac{\partial^2 w}{\partial x^2}+\nu\frac{\partial^2 w}{\partial y^2}\right]+\frac{h^2}{5l}\frac{\partial Q_x}{\partial x}=0, \qquad (4.1)_1$$

$$-(1-\nu)\frac{D}{l^2}\frac{\partial^2 w}{\partial x\,\partial y}+\frac{h^2}{10l}\left(\frac{\partial Q_x}{\partial y}+\frac{\partial Q_y}{\partial x}\right)=0, \tag{4.1$_2$}$$

$$\bar{Q}_x=\frac{P}{\pi l}\int_0^\infty \cos\beta y\,\mathrm{d}\beta \tag{4.1$_3$}$$

for

$$x\to\infty,\quad w\to 0,\quad \frac{\partial w}{\partial x}\to 0,\quad Q_x\to 0. \tag{4.2}$$

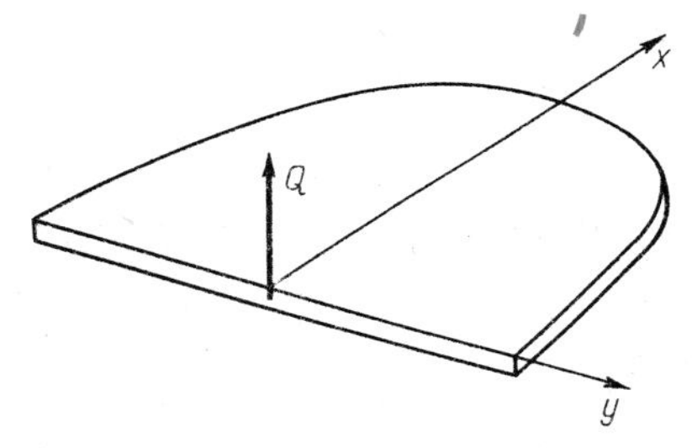

Fig. 48. Plate loaded at its free edge by a concentrated shear force.

The condition demanding that with the growing distance from the loading point, the disturbances caused by the applied load vanish, can be satisfied when $C_3 = C_4 = 0$ in Eq. (3.123). Forming the equation for the boundary conditions, we must calculate first the shear forces Q_x and Q_y. Solving Eq. (2.16) by means of Fourier integrals, we find

$$\Psi=\int_0^\infty (E_1\mathrm{e}^{-\beta_1 x}+E_2\mathrm{e}^{\beta_1 x})\sin\beta y\,\mathrm{d}\beta, \tag{4.3}$$

where $\beta_1 = \sqrt{\beta^2+10l^2/h^2}$, E_1 and E_2 are arbitrary constants.

$E_2 = 0$ results from conditions (4.2); then the shear forces take the following form:

$$Q_x=\int_0^\infty [E_1\beta\mathrm{e}^{-\beta_1 x}-2C_2 D\beta^3\mathrm{e}^{-\beta x}]\cos\beta y\,\mathrm{d}\beta,$$

$$Q_y=\int_0^\infty [E_1\beta_1\mathrm{e}^{-\beta_1 x}-2C_2 D\beta^3\mathrm{e}^{-\beta x}]\sin\beta y\,\mathrm{d}\beta. \tag{4.4}$$

Introducing relations (4.3) and (4.4) in (4.2) and solving the equations, we find values of the constants E_1, E_2 and C_1, C_2. Then the displacement w takes the form

$$w = -\frac{Pl^2}{\pi D(1-\nu)} \int_0^\infty \Big\{ 2 -$$

$$-\left(1-\nu+\frac{2h^2}{5l^2}\beta^2\right)\frac{h^2}{5l^2}\beta(\beta_1-\beta)+\frac{h^2}{5l^2}\beta^2(3+\nu)+$$

$$+(1-\nu)\left[1-\frac{h^2}{5l^2}\beta(\beta_1-\beta)\right]x\beta\Big\}\frac{e^{-\beta x}}{D_{(\beta)}\beta^3}\cos\beta y \, d\beta, \tag{4.5}$$

where

$$D_{(\beta)} = 3+\nu-\frac{2h^2}{5l^2}\beta(\beta_1-\beta).$$

We see that the above integral is divergent and has no solution. Physically this is due to the fact that the plate is not in equilibrium. We can imagine only that its supports are at an infinite distance from the edge $x = 0$ ($w \to 0$ as $x \to \infty$). That is the reason for the infinite displacement. However, based on the solution, we can define the stresses near the loading point. Differentiating expression (4.5) according to formulae (2.2), we obtain the moment M_{xx} to be equal to

$$M_{xx} = \frac{P}{\pi}\int_0^\infty \Big\{(1-\nu)\beta x\left[1-\frac{h^2}{5l^2}\beta(\beta_1-\beta)\right]-$$

$$-(1+\nu)\frac{h^2}{5l^2}\beta\beta_1\cdot[e^{-(\beta_1-\beta)x}-1]\Big\}\cdot\frac{e^{-\beta x}}{D_{(\beta)}}\cos\beta y \, d\beta, \tag{4.6}$$

where

$$\beta_1 = \sqrt{\beta^2+\frac{10l^2}{h^2}}.$$

The above integral has a complex form and can only be evaluated numerically. The complicated denominator $D_{(\beta)}$ is the main trouble. However, it can be evaluated approximately if we notice that the denominator changes slowly from $(3+\nu)$ for $\beta = 0$ to $(1+\nu)$ as $\beta \to \infty$. Since $e^{-\beta x}$ is a very rapidly damped function, except in the region for small x, the exact value of the denominator is important only for small β. Then we may neglect the second term in the denominator. If the value of $10l^2/h^2$ is large for small β, we may neglect β^2 in comparison with $10l^2/h^2$, then $\sqrt{\beta^2+10l^2/h^2}-\beta \simeq \sqrt{10}\,l/h$. As a result, on evaluation, we obtain an approximate expression

$$M_{xx} \approx \frac{P}{(3+\nu)\pi}\Bigg\{(1-\nu)\cos^2\vartheta -$$

$$-\frac{\sqrt{10}h}{5l}\left[(1-\nu)\cos 2\vartheta-(1+\nu)(e^{\frac{\sqrt{10}lx}{h}}-1)\right]\frac{\cos\vartheta}{r}\Bigg\}. \tag{4.7}$$

Neglecting the terms proportional to $\sqrt{10}h/5l$ in (4.7), we obtain the same results as if we used the classical theory of plates and equivalent boundary conditions (2.32). We observe that the terms appearing as a result of taking into account the effects of shear deformations include the singularities of higher order than those resulting from the classical theory. Limiting ourselves to the classical theory, we find the deflection

$$w = -\frac{Pl^2}{(3+\nu)\pi D}\int_0^\infty\left(\frac{2}{1-\nu}+x\beta\right)\frac{e^{-\beta x}}{\beta^3}\cos\beta y\,d\beta. \tag{4.8}$$

In a similar way as previously we can eliminate from the above integral the singular expression for the deflection oι the plate.

$$w = -\frac{Pl^2}{2(1-\nu)(3+\nu)\pi D}\{(y^2-\nu x^2)\ln r^2+2(1+\nu)xy\vartheta+y^2-x^2\}.$$

The above formula was found by A. Nadai [2.1, p. 204] in the other way.

The moments M_{xx} and M_{xy} are

$$M_{xx} = \frac{1-\nu}{3+\nu}\frac{P}{\pi}\cos^2\vartheta,$$

$$M_{xy} = \frac{1-\nu}{3+\nu}\frac{P}{\pi}\left[\frac{1+\nu}{1-\nu}\vartheta+\frac{xy}{r^2}\right]. \tag{4.9}$$

It develops from formulae (4.9) that the moments M_{xx} and M_{xy}, according to the classical theory, have no infinitely large values at the point of application of the load. The moment M_{xx} is constant along the x-axis and its value is $(1-\nu)P/(3+\nu)$. We obtain the following integral for the moment M_{yy}:

$$M_{yy} = -\frac{P}{(3+\nu)\pi}\int_0^\infty[2(1+\nu)+(1-\nu)\beta x]\frac{e^{-\beta x}}{\beta}\cos\beta y\,d\beta. \tag{4.10}$$

We see that the moment M_{yy} is expressed by a divergent integral. The assumption that the supports of the plate are at an infinite distance from the loading point accounts for this result. Then the singular part of the solution consists not only of the local effect of the force P, but also of an infinitely large moment resulting from the infinitely large size of the plate. We can, however, get an idea of the kind of singularity of the moment M_{yy} at the point ($x = 0$, $y = 0$) by evaluation of integral (4.10) from B to ∞ and next by the assumption that $B = 0$. We have (see table of integrals)

$$M_{yy} = \frac{2(1+\nu)P}{(3+\nu)\pi}\left[[\ln r + \ln B + \gamma_0 + \sum_{k=1}^{\infty} (-B)^k \frac{r^k \cos k\vartheta}{kk!} - \right.$$
$$\left. - \frac{1-\nu}{2(1+\nu)} \frac{x}{r} e^{-Bx} \cos(By+\vartheta) \right]. \tag{4.11}$$

The term $\ln B$ appearing in the solution and for $B \to 0$ increasing to infinity does not depend on the coordinates x, y and, therefore, is constant in the area of the plate. It can appear only as a result of infinite dimensions of the plate. In rejecting this term and then substituting $B = 0$, we obtain a formula representing the singular expression for the moment M_{yy} valid near the loading point with the accuracy to constant values:

$$M_{yy} = \frac{2(1+\nu)}{3+\nu} \frac{P}{\pi} \ln r - \frac{1-\nu}{3+\nu} \frac{P}{\pi} \frac{x^2}{r^2} + C_0, \tag{4.12}$$

where C_0 is a constant value.

Calculating the bending moment M_{yy} by the direct differentiation of the deflection w (4.8), we obtain the value of the constant, $C_0 = -2(1+\nu)/(3+\nu)\pi$.

By Eq. (4.12), the moment M_{yy} for $x = 0$, $y = 0$ (contrary to the moments M_{xx} and M_{xy}) increases to infinity and gives at this point infinitely large stresses.

The above idealized problem was solved with the assumption of simplified boundary conditions. Namely, we have assumed that the plate is infinitely large. As a result the solution obtained describes exactly the state of stresses only near the loading point.

Let us consider now a more real problem applying the classical theory. Namely, let us calculate the displacements and the stresses in the strip plate of width a loaded by the force P at the free edge. At the other edge $x = a$ the plate is clamped (Fig. 49). Thus two new boundary conditions must be satisfied

$$x = 0; \quad M_{xx} = 0, \quad \bar{Q}_x = P\delta(y, 0),$$
$$x = a, \quad w = 0, \quad \frac{\partial w}{\partial x} = 0.$$

We assume that the solution takes the form (3.124) and the length $l = a$. After solving the equations for the boundary conditions, we obtain the following constants of integration:

$$
\begin{aligned}
C_1' &= -\frac{Pl^2}{\pi D}\frac{1}{\beta D(\beta)}(\sinh 2\beta a - 2\beta a),\\
C_2' &= -\frac{Pl^2}{\pi D}\frac{1-\nu}{2\beta D(\beta)}(\sinh 2\beta a - 2\beta a),\\
C_3' &= -\frac{Pl^2}{\pi D}\frac{1}{\beta D(\beta)}[(1-\nu)\beta^2 a^2 + 2\cosh^2\beta a],\\
C_4' &= -\frac{Pl^2}{\pi D}\frac{1}{\beta D(\beta)}[2 + (1-\nu)\sinh^2\beta a],
\end{aligned}
\tag{4.13}
$$

where

$$D(\beta) = (1-\nu)(3+\nu)\sinh^2\beta a + (1-\nu)^2\beta^2 a^2 + 4$$

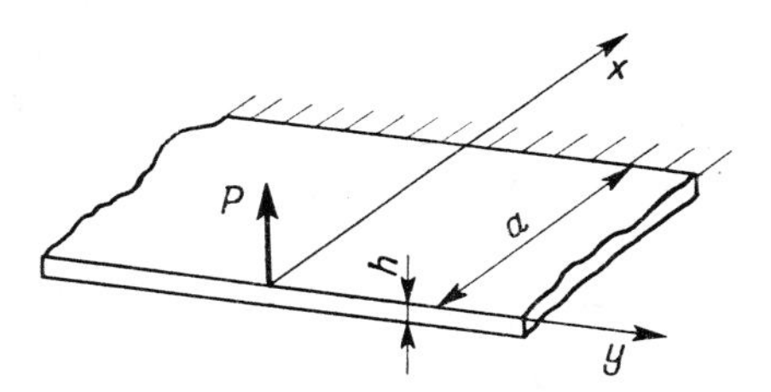

Fig. 49. Strip plate loaded at its free edge.

The deflection along the free edge obtained by means of numerical computations is presented in Fig. 50. The deflection rapidly diminishes with the distance from the point of application of the load. The bending moments can be obtained according to (2.29).

Assuming $a \to \infty$, we should obtain the solution found previously. It is worth noting that in the case where the clamped edge is at a finite distance, the value of the moment $M_{xx} = 0.465P$ at the point $x = a$, $a = 0$ and for $\nu = 0$ is obtained by numerical integration, while from Eq. (7.9) we obtain, for $\nu = 0$, $M_{xx} = 0.106P$, i.e. about four times less. For the semi-infinite clamped plate (2.126) we found for $\nu = 0$ the value $M_{xx} = 0.318P$ at the same point. This contradiction can be explained in the following way. Due to the clamped edge additional terms appear as a result of the increase of the rigidity of the plate. When this edge is at a very large distance, its effect on the points lying near the loading point disappears. Thus the solution (3.9) is valid only at a small distance from the loading point.

The bending moment M_{yy} is

$$M_{yy} = -\frac{D}{l^2}\int_0^\infty \{[C_1'(1-\nu)-2\nu C_2']\cosh\beta x + \\ + C_2'(1-\nu)\beta x\sinh\beta x + [C_3'(1-\nu)-2\nu C_4']\sinh\beta x + \\ + C_4'(1-\nu)\beta x\cosh\beta x\}\cos\beta y\,d\beta. \tag{4.14}$$

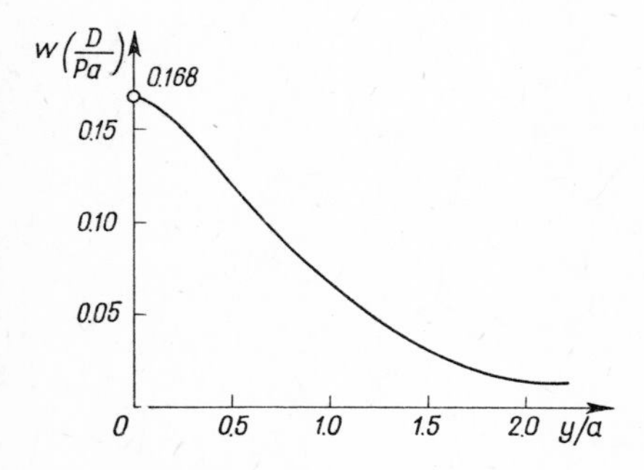

Fig. 50. Deflection at the edge of a strip plate loaded at the free edge by a concentrated shear force.

This formula can also be integrated only in a numerical way. It converges, however, very slowly. In order to facilitate the calculations we divide integral (4.14) into two parts and evaluate it in two stages: from zero to a certain arbitrary value B and then from B to ∞. The evaluation from B to ∞ can be done analytically by replacing the complicated integrand by its simple asymptotic expression which for large β represents the integrand with a sufficient accuracy. We assume that, for $\beta > B$, $\sinh\beta a$ and $\cosh\beta a$ can be replaced by $\sinh\beta a = \cosh\beta a = \frac{1}{2}e^{\beta a}$.

Neglecting all terms small in comparison to $e^{\beta a}$, we find the following formula for the moment:

$$M_{yy}^B = -\frac{P}{\pi(3+\nu)}\int_B^\infty [2(1+\nu)+(1-\nu)\beta x]\frac{e^{-\beta x}}{\beta}\cos\beta y\,d\beta.$$

This integral can be evaluated analytically (see (4.11)). The integral between the limits 0 and B cannot be simplified and should be evaluated numerically.

We see that the numerical integration between the infinite limits can be reduced to an integration between the finite limits.

Also, this example proved once more that the type of singularity does not depend on the shape of the plate. For the free edge $x = 0$, Eq. (4.14) for the moment M_{yy} simplifies. We obtain then the value independent of the width of the plate a

$$M_{yy} = -2(1-\nu^2)\frac{P}{\pi}\int_0^\infty \frac{(\frac{1}{2}\sinh 2\beta - 1)\cos\beta y \, d\beta}{4+(1-\nu)^2\beta^2+(1-\nu)(3+\nu)\sinh^2\beta;} \tag{4.15}$$

and for $\beta > \beta\lim = B$

$$M_{yy}^B = -\frac{2(1+\nu)P}{(3+\nu)\pi}\int_B^\infty \frac{\cos\beta y}{\beta}\, d\beta = \frac{2P(1+\nu)}{(3+\nu)\pi}\,\mathrm{Ci}(By). \tag{4.16}$$

In Fig. 51 there is plotted the exact value of the integrand (4.15) and its approximation (4.16) valid for large β. We see that for $\beta > B$ the difference between

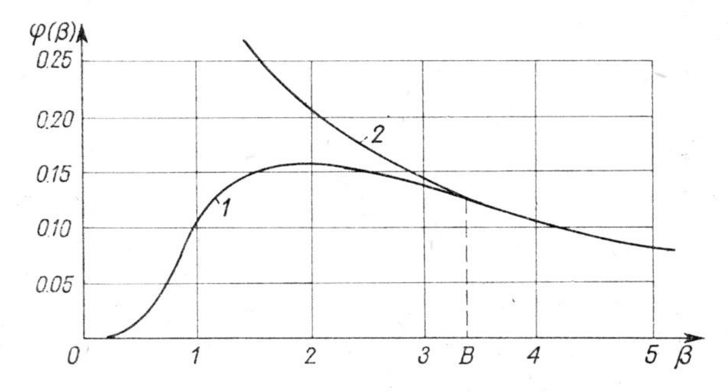

Fig. 51. Comparison of the simplified and exact value of integrand (4.15). (*1—Eq.* (4.15), *2—Eq.* (4.16).)

the exact and approximate value is very small. The error for $B = 3$ amounts to about 5.8 per cent and for $B = 4$ only about 1.1 per cent. Let us find the influence of the value of B on the final result. Assuming, for example, $B = 3$, we have for $y = 0.2a$ the moment

$$M_{yy} = -2(1-\nu)^2\frac{P}{\pi}\left[0.309 - \frac{\mathrm{Ci}(0.6)}{(3+\nu)(1-\nu)}\right] = -0.1848P.$$

Assuming $B = 4$, we find $M_{yy} = -0.1830P$.

Values 0.309 for $B = 3$ and 0.402 for $B = 4$ were obtained by a numerical evaluation of integral (4.15) between the limits 0 and B. We observe that the difference between these solutions is less than about one per cent. The diagram of the moment M_{yy} at the free edge, calculated by the method described, is given in Fig. 52. The constant C_0 (see Eq. (4.12)) calculated numerically is $C = -0.25P$ ($\nu = 0.3$).

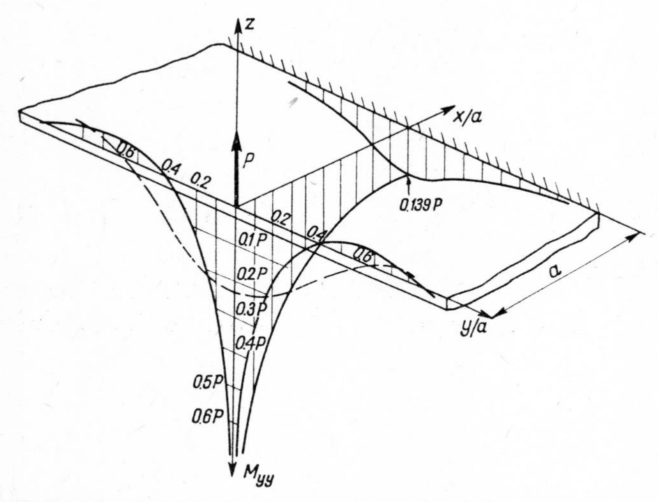

Fig. 52. Distribution of the bending moment M_{yy} in a plate loaded at its free edge by a concentrated shear force.

If the plate is loaded at its edge by the force P uniformly distributed along a certain small length $2c$ (Fig. 53), we achieve a solution which does not have the above-described singularities. Fulfilling the same simplified boundary conditions (4.13), we obtain the deflection of the plate in the form of the integral

$$w = -\frac{Pl^2}{(3+\nu)\pi D}\int_0^\infty \frac{\sin\beta c}{\beta c}\left(\frac{2}{1-\nu}+x\beta\right)\frac{e^{-\beta x}}{\beta^3}\cos\beta y\,d\beta. \tag{4.17}$$

Differentiating the displacement w under the sign of the integral (4.17), we find the bending moments presented by the integrals, some of which can be evaluated directly. We find, see Fig. 53,

$$M_{xx} = \frac{1-\nu}{2(3+\nu)}\frac{P}{\pi c}x(\vartheta_2-\vartheta_1), \qquad (4.18)_1$$

$$M_{xy} = \frac{1-\nu}{2(3-\nu)} \frac{P}{\pi c} \left\{ \frac{1+\nu}{1-\nu} [(y+c)\vartheta_2 - (y-c)\vartheta_1] - \frac{2\nu}{1-\nu} x \ln \frac{r_2}{r_1} \right] . \quad (4.18)_2$$

The moment M_{xy} is singular at the points $y = \pm c$, as $\ln(r_2/r_1)$ increases there infinitely. The integral for the bending moment M_{yy} is divergent. In order to calculate the local part of the moment M_{yy} we use the same treatment as previously. Integrating from B to ∞, we have for small $r \ll 1$

$$M_{yy}^B = -\frac{1}{3+\nu} \frac{P}{\pi c} \left\{ (1+\nu) \frac{e^{-Bx}}{B} [\sin B(y+c) - \sin B(y-c)] - \right.$$

$$- (1+\nu)[(y+c)\ln r_2 - (y-c)\ln r_1] - \frac{1-3\nu}{2} x(\vartheta_2 - \vartheta_1) -$$

$$\left. - \frac{2(1+\nu)}{3+\nu} \frac{P}{\pi} (\ln B + \gamma_0) \right\} . \quad (4.19)$$

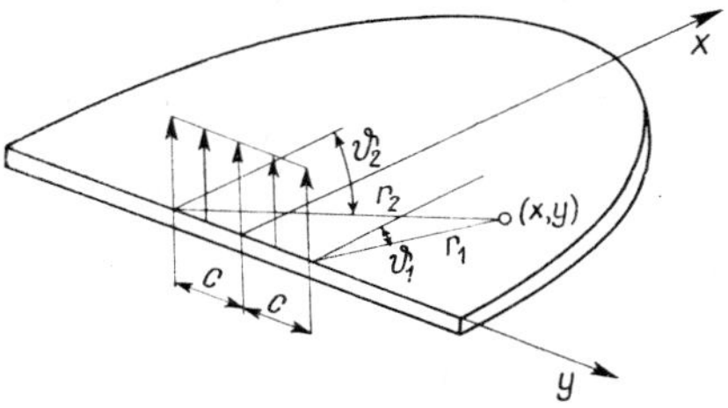

Fig. 53. Plate loaded at its free edge by a force distributed along the segment $2c$.

By neglecting the last term, which is constant in the whole area of the plate, we obtain the approximate formula for the moment M_{yy}^B, valid near the loading point. If c is small, $c \ll 1$, $\dfrac{\sin \beta c}{\beta c} \simeq 1$, we can easily evaluate the moment M_{yy} based on the previous results. The integrals evaluated between the limits 0 and B do not undergo a change; only the integral evaluated between B and ∞ is different. For $x = 0$ and $y = 0$ it yields

$$M_{yy} \approx 2 \frac{1+\nu}{3+\nu} \frac{P}{\pi} [\ln c - 1] - 0.25P .$$

We observe that in the case where the load is distributed along a certain length, the moment M_{yy} is not singular. In the case of a semi-infinite plate with the free edge loaded at the point $x = x_0$, $y = y_0$ (Fig. 54), the singular part of the deflection is

$$w = \frac{Pl^2}{16\pi D}\Bigg([(x-x_0)^2+(y-y_0)^2]\ln\frac{(x-x_0)^2+(y-y_0)^2}{(x+x_0)^2+(y-y_0)^2}+$$
$$+\frac{8}{(1-\nu)(3+\nu)}\Big\{[(x-x_0)^2+(y-y_0)^2-(1+\nu)(x^2+x_0^2)]\ln[(x+x_0)^2+$$
$$+(y-y_0)^2]+2(1+\nu)(x+x_0)(y-y_0)\arctan\frac{y-y_0}{x+x_0}+x^2+(y-y_0)^2\Big\}\Bigg). \tag{4.20}$$

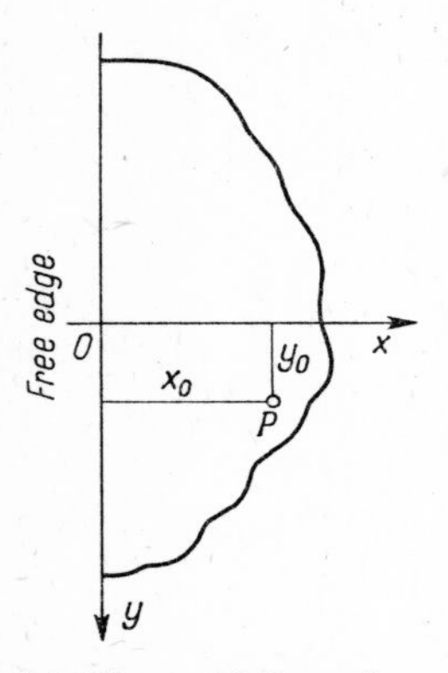

Fig. 54. Plate with free edge x =).

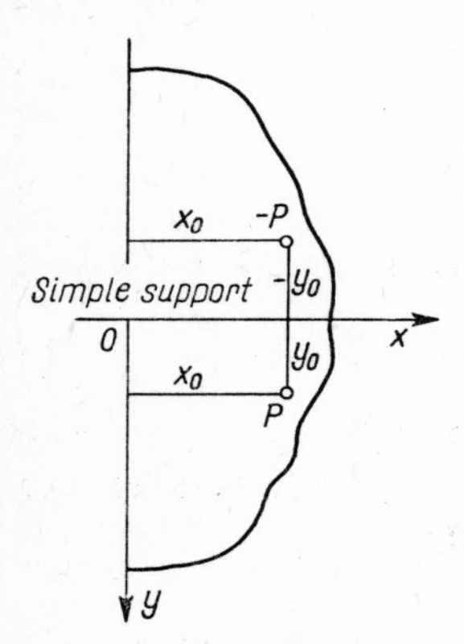

Fig. 55. Plate simply supported along the line $y = 0$ with free edge $x = 0$.

The above formula was obtained by Z. Kączkowski [4.6]. Based on the solution (4.20), we can easily find the deflection of an infinite plate simply supported at the edge $y = 0$ and free at the edge $x = 0$, loaded by the concentrated force P applied at the point $x = x_0$, $y = y_0$ (Fig. 55):

$$w = \frac{Pl^2}{16\pi D}\left([(x-x_0)^2+(y-y_0)^2]\ln\frac{(x-x_0)^2+(y-y_0)^2}{(x+x_0)^2+(y-y_0)^2} -\right.$$

$$-\,[(x-x_0)^2+(y+y_0)^2]\ln\frac{(x-x_0)^2+(y+y_0)^2}{(x+x_0)^2+(y+y_0)^2}+$$

$$+\frac{8}{(1-\nu)(3+\nu)}\Big\{[(x-x_0)^2+(y-y_0)^2-(1+\nu)(x^2+x_0^2)]\ln[(x+x_0)^2+$$

$$+(y-y_0)^2]-[(x-x_0)^2+(x+x_0)^2-(1+\nu)(x^2+x_0^2)\ln[(x+x_0)^2+$$

$$+(y+y_0)^2]+2(1+\nu)(x+x_0)\left[[(y-y_0)\operatorname{arc\,tan}\frac{y-y_0}{x+x_0}-\right.$$

$$\left.\left.-\,(y+y_0)\operatorname{arc\,tan}\frac{y+y_0}{x+x_0}\right]-4yy_0\Big\}\right).$$

The case where the force acts on a cantilever plate not at the edge but at a certain distance from it was considered by T. J. Jaramillo [4.3] and H. Jung who treated several problems of this kind [4.7] (Fig. 56). The distribution of

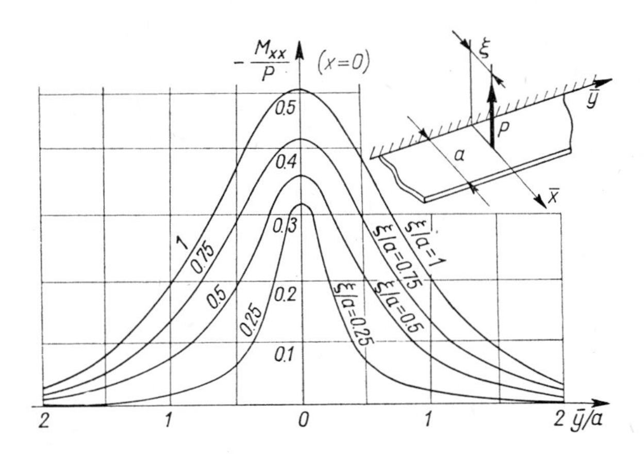

Fig. 56. Strip plate loaded by a concentrated force at the interior point $x = \xi$ (T. J. Jaramillo [4.3]).

the bending moments along the built-in edge, computed by means of Fourier integrals, for various positions ξ of the concentrated force P, for $\nu = 0.3$ is shown in Fig. 56. The case of the wedge plate (Fig. 57) was discussed by

S. Woinowsky–Krieger [4.8] and W. T. Koiter [4.9] by means of Mellin transform. The plate with two clamped edges was considered by Y. S. Ufliand [4.10] as well as W. T. Koiter and J. B. Alblas [4.4].

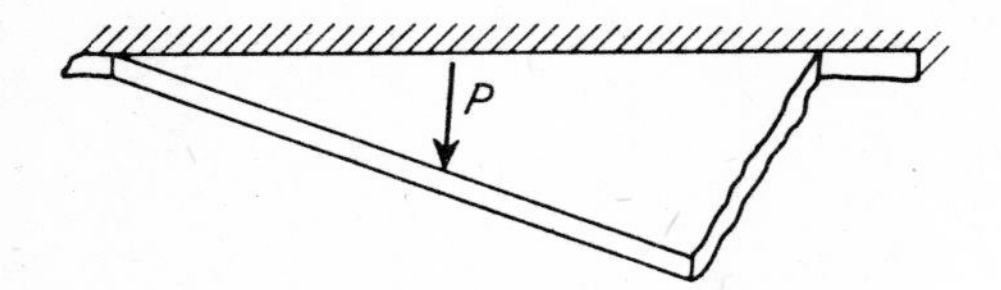

Fig. 57. Wedge plate loaded at its free edge.

4.2 Plate loaded by a concentrated moment at its free edge

Let us consider now a semi-infinite isotropic plate loaded at its free edge by a concentrated moment M (Fig. 58). The boundary conditions for the free edge are (for $x = 0$)

$$M_{xx} = -\frac{D}{l^2}\left[\frac{\partial^2 w}{\partial x^2} + \nu \frac{\partial^2 w}{\partial y^2}\right] + \frac{h^2}{5l}\frac{\partial Q_x}{\partial x} = M\delta(y, 0),$$

$$M_{xy} = -(1-\nu)\frac{D}{l^2}\frac{\partial^2 w}{\partial x\,\partial y} + \frac{h^2}{10l}\left(\frac{\partial Q_y}{\partial x} + \frac{\partial Q_x}{\partial y}\right) = 0,$$

$$\overline{Q}_x = 0.$$

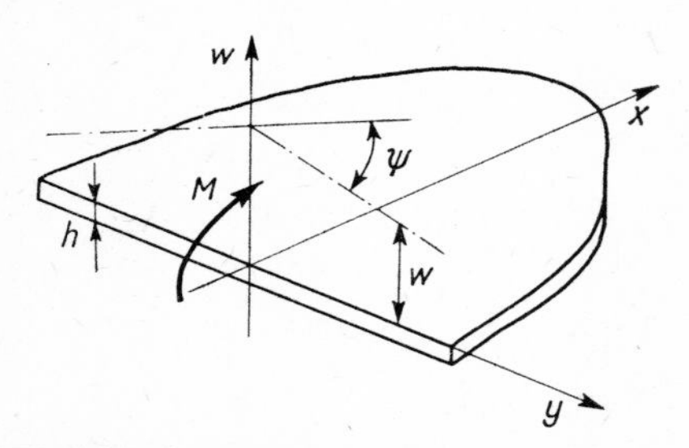

Fig. 58. Plate loaded at free edge by a concentrated bending moment.

Solving the above equations, we obtain the following values for the constants C_i, E_i (for E_i see formulae (4.3) and (4.4)):

$$C_1 = -\frac{1+\nu}{1-\nu}C_2, \quad C_2 = \frac{Ml}{\pi^2 \beta D \cdot D_{(\beta)}}, \quad E_1 = 2C_2\beta^2 D,$$

$$C_3 = C_4 = E_2 = 0.$$

The denominator $D_{(\beta)}$ is given by formula (4.5).

The deflection of the plate takes the form

$$w = -\frac{Ml}{\pi D}\int_0^\infty \frac{\left[\frac{1+\nu}{1-\nu} - \beta x\right] e^{-\beta x}\cos\beta y\, d\beta}{\left[3+\nu - \frac{2}{5}\frac{h^2}{l^2}\beta(\beta_1-\beta)\right]\beta^2} \tag{4.21}$$

where β_1 is given by (4.3).

In the case of the classical theory we have the following integral for the deflection:

$$w = -\frac{1}{3+\nu}\frac{Ml}{\pi D}\int_0^\infty \left(\frac{1+\nu}{1-\nu} - \beta x\right)\frac{e^{-\beta x}}{\beta^2}\cos\beta y\, d\beta.$$

The above integral is divergent, but eliminating the singular terms we find

$$w = -\frac{Ml}{(1-\nu)(3+\nu)\pi D}\left[x\ln r^2 - (1+\nu)y\arctan\frac{y}{x}\right]. \tag{4.22}$$

The above formula (4.22) was obtained by A. Nadai [2.1]. At $x = 0$ we have

$$w = \frac{Ml(1+\nu)}{2(1-\nu)(3+\nu)D}|y| + \text{const}.$$

The deflection of the plate at the edge has the form of two straight lines crossing each other at the loading point intersecting at an angle ψ, where

$$\psi = \frac{1+\nu}{(3+\nu)(1-\nu)}\frac{Ml}{D}. \tag{4.23}$$

The bending moments evaluated from expression (4.21) are

$$M_{xx} = \frac{M}{\pi l}\int_0^\infty \left[3+\nu-(1-\nu)\beta x + \frac{2h^2}{5l^2}\beta\beta_1(e^{-(\beta_1-\beta)x}-1)\right]\times$$

$$\times\frac{e^{-\beta x}}{D_{(\beta)}}\cos\beta y\, d\beta.$$

The exact analytical evaluation of the above integral is impossible because of the complex function appearing in the denominator $D_{(\beta)}$. This integral can be evaluated numerically, however, we can realize the nature of the correction resulting from the shear deformations by simplifying the integrals as before.

$$M_{xx} = \frac{M}{\pi l}\left\{\frac{\cos\vartheta}{r}\left(1-\frac{1-\nu}{3+\nu}\cos 2\vartheta\right) - \right.$$

$$\left. - \frac{2h\sqrt{10}\cos 2\vartheta}{5l(3+\nu)r^2}\left(e^{-\sqrt{10}lx/h}-1\right)\right\}.$$

We observe that the terms resulting from the effect of transverse shear deformations contain stronger singularities. Neglecting them, we obtain the expressions which would be derived from the classical theory and equivalent boundary conditions. We have then

$$M_{xx} = \frac{2M}{(3+\nu)\pi l}\frac{\cos\vartheta}{r}\left[1+\nu+(1-\nu)\sin^2\vartheta\right],$$

$$M_{yy} = -\frac{2M(1-\nu)}{(3+\nu)\pi l}\frac{\cos\vartheta\sin^2\vartheta}{r},$$

$$M_{xy} = \frac{2M}{(3+\nu)\pi l}\frac{\sin\vartheta}{r}\left[1-(1-\nu)\cos^2\vartheta\right].$$

The above expressions may also be obtained by direct differentiation of Eq. (4.22). Expressing the moments by the components M_{rr}, $M_{\vartheta\vartheta}$, $M_{r\vartheta}$, we have

$$M_{rr} = \frac{2(1+\nu)}{3+\nu}\frac{M}{\pi l}\frac{\cos\vartheta}{r},$$

$$M_{\vartheta\vartheta} = 0, \qquad (4.24)$$

$$M_{r\vartheta} = -\frac{2M}{(3+\nu)\pi l}\frac{\sin\vartheta}{r}.$$

We see that M_{rr} has a constant value on circles whose centres are situated on the normal line to the edge and cross the loading point.

The circles orthogonal to the above, crossing the point at which the moment is applied, presents the lines of constant twisting moment M_{xy}. Their centres lie at the edge of the plate (Fig. 59).

Also in this case we obtain infinitely large stresses at the point where the moment acts. This singularity results, of course, from the manner in which the load is applied. However, real structures are never subjected to a bending moment acting at one point. The moments are always distributed over a certain length or surface. Solving such problems, we obtain finite values of stresses. When the moment is distributed along a small length $2c$, then displacement of the plate takes the form of the integral (classical theory)

$$w = -\frac{1}{3+\nu}\frac{Ml}{\pi D}\int_0^\infty \frac{\sin\beta c}{\beta c}\left[\frac{1+\nu}{1-\nu}-\beta x\right]\frac{e^{-\beta x}}{\beta^2}\cos\beta y\,d\beta. \tag{4.25}$$

The bending moments calculated from Eq. (4.25) take the form

$$\left.\begin{matrix} M_{xx} \\ M_{yy}\end{matrix}\right\} = \pm\frac{M}{2\pi c(3+\nu)}\left\{[2\pm(1+\nu)](\vartheta_2-\vartheta_1)-\frac{1-\nu}{2}(\sin 2\vartheta_2-\sin 2\vartheta_1)\right\},$$

$$M_{xy} = \frac{1-\nu}{3+\nu}\frac{M}{2\pi c}\left[\frac{2}{1-\nu}\ln\frac{r_2}{r_1}+\frac{1}{2}(\cos 2\vartheta_2-\cos 2\vartheta_1)\right]. \tag{4.26}$$

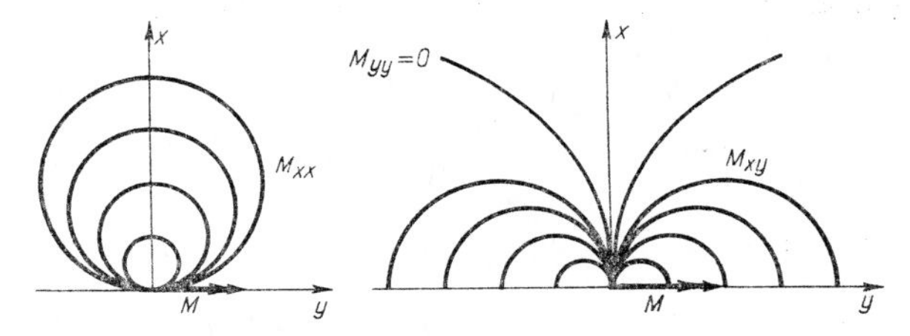

Fig. 59. Distribution of the moments M_{xx}, M_{yy}, and M_{xy} in a plate loaded at a free edge.

The angles ϑ_1 and ϑ_2 and the radii r_1 and r_2 are presented in Fig. 53. The moments M_{xx} and M_{yy} have finite values in the entire area of the plate. For the edge $x = 0$ we find $\vartheta_1 = -\pi/2$; $\vartheta_2 = +\pi/2$ and

$$M_{xx} = \frac{M}{2c}, \qquad M_{yy} = -\frac{M}{2c}\frac{1-\nu}{3+\nu}, \qquad -c < y < +c.$$

The moment M_{xy} has singularities at points $y = \pm c$ as $\ln(r_2/r_1) \to \pm\infty$ and the tangential stresses are there infinitely large. The deflection at the free edge may easily be found also in the case where the moment is distributed according to the arbitrary function $m_{xx}(y)$. We have from Eq. (4.22) the following relation:

$$\frac{d^2}{dy^2}[w_{0,y}] = -\frac{m_{xx}(y)}{D}\frac{1+\nu}{(1-\nu)(3+\nu)}.$$

On integration we find the deflection at the edge.

4.3 Plate on an elastic foundation loaded at its free edge by a concentrated force

Now, we shall investigate a semi-infinite plate on an elastic foundation loaded at its free edge by a concentrated force P. We limit ourselves to the classical theory of plates.

Assuming that the deflection of the plate has the form (3.122) and substituting in $\Delta\Delta w = 0$, we obtain

$$\int_0^\infty [(k^2+\beta^2)+4\varkappa^4]\mathrm{e}^{ikx}\cos\beta y\,\mathrm{d}\beta = 0. \tag{4.27}$$

This equation is satisfied only when the term in brackets is equal to zero. Then we receive an algebraic characteristic equation which enables us to define the parameter k

$$(k^2+\beta^2)^2+4\varkappa^4 = 0 \quad \text{where} \quad 4\varkappa^4 = \frac{l^4 k}{D}.$$

This equation has four roots

$$\left.\begin{matrix} k_1 \\ k_2 \end{matrix}\right\} = \pm a+bi, \qquad \left.\begin{matrix} k_3 \\ k_4 \end{matrix}\right\} = \pm a-bi. \tag{4.28}$$

Defining l in such a way that $4\varkappa^4 = 1$, we find that $l = \sqrt[4]{D/k}$ and we obtain the following values of a and b depending only on β:

$$\left.\begin{matrix} a \\ b \end{matrix}\right\} = \sqrt{\tfrac{1}{2}\left(\sqrt{\beta^4+1}\mp\beta^2\right)}. \tag{4.29}$$

Equation (3.122) can be transformed to

$$w = \int_0^\infty [C_1\mathrm{e}^{-bx}\sin ax+C_2\mathrm{e}^{-bx}\cos bx+C_3\mathrm{e}^{bx}\sin ax+$$
$$+C_4\mathrm{e}^{bx}\cos ax]\cos\beta y\,\mathrm{d}\beta. \tag{4.30}$$

Now it is necessary to satisfy the boundary conditions. We have for the free edge $x = 0$,

$$\bar{Q}_x = -\frac{D}{l^3}\left[\frac{\partial^3 w}{\partial x^3}+(2-\nu)\frac{\partial^3 w}{\partial x\,\partial y^2}\right] = P\delta(y,0),$$

$$M_{xx} = 0$$

and for $x \to \infty$,

$$w \to 0, \qquad \frac{\partial w}{\partial x} \to 0.$$

The conditions for $x \to \infty$ are satisfied when $C_3 = C_4 = 0$. Forming the equations of the boundary conditions for the free edge, we obtain

$$C_2 = -\frac{Pl^2}{\pi D}\frac{2b}{D'(\beta)}, \qquad C_1 = -\frac{Pl^2}{\pi D}\frac{2b(1-\nu)\beta^2}{D'(\beta)}, \tag{4.31}$$

where

$$D'(\beta) = 2(1-\nu)\beta^2\sqrt{\beta^4+1}+(1-\nu^2)\beta^4+1.$$

In order to calculate the deflection w we must first integrate Eq. (4.30) with the substituted values of the constants C_1 and C_2 and replace a and b by expressions (4.29).

The bending moments in the plate have the form

$$M_{xx} = \frac{P}{\pi}\int_0^\infty 2b[1+(1-\nu)^2\beta^4]\sin ax\frac{e^{-bx}}{D'(\beta)}\cos\beta y\,d\beta,$$

$$M_{yy} = -\frac{P}{\pi}\int_0^\infty 2b\{(1-\nu^2)\beta^2\cos ax+$$
$$+[(1-\nu)^2\beta^4-\nu)]\sin ax\}\frac{e^{-bx}}{D'(\beta)}\cos\beta y\,d\beta, \tag{4.32}$$

$$M_{xy} = (1-\nu)\frac{P}{\pi}\int_0^\infty\{[\sqrt{\beta^4+1}+\nu\beta^2]\cos ax+$$
$$+[1+(1-\nu)(\beta^4+\beta^2\sqrt{\beta^4+1})]\sin ax\}\frac{\beta e^{-bx}}{D'(\beta)}\sin\beta y\,d\beta.$$

The exact analytical values of the integrals (4.32) are unknown. We can evaluate them only numerically. They converge, however, very slowly (at the point $x = 0$, $y = 0$ they diverge). For large β the function $\cos\beta$ becomes a rapidly changing function and the usual calculations would become tedious.

We can however, calculate it approximately. Let us divide the integrals in two parts, integrating from 0 to B and from B to ∞. (B is here a certain arbitrary value.) For $\beta > B$ we can simplify the integrands so as to obtain simple integrals which are easy to evaluate. The roots a and b of the characteristic equation can be expressed as a power series with respect to $1/\beta$. Taking into consideration only the first terms we have $a = 1/2\beta$, $b = \beta$. Neglecting unity in $\sqrt{\beta^4+1}$ we obtain

$$D'(\beta) = (1-\nu)(3+\nu)\beta^4+1. \tag{4.33}$$

Neglecting unity in the denominator $D'(\beta)$, we obtain integrals similar to those calculated in Section 3.1 for the semi-infinite plate loaded at its free edge by a concentrated force. It develops that in the vicinity of the loading point the stresses in the free plate and in the plate resting on an elastic foundation have the same type of singularity. When the plate on the elastic foundation is loaded by the force uniformly distributed along a certain small length $2c$ it is adequate to multiply the integrands by $(\sin\beta c)/\beta c$. The evaluation of these integrals can be next done by means of Table 16 given on the end of the book. For the bending moments M_{yy}^B we obtain the identical expression as previously (3.19) for the bending moment M_{yy}^B in the cantilever plate.

The case of the loads P applied along the edge of the semi-infinite thick plate resting on an elastic foundation were considered by Westergaard [4.11]. The final expression for the maximum tensile strength at the bottom of the plate under the load is

$$(\sigma_{xx})_{\max} = 0.529(1+0.54\nu)\frac{P}{h^2}\left[\log_{10}\left(\frac{Eh^3}{kb^4}\right)-0.71\right],$$

where b is defined by formulae (3.119) and c is the radius of the semicircle surface on which the load P is uniformly distributed. The above formula proved to be very useful in the design of concrete roads.

The formulae developed in previous sections make it possible to separate the singular expressions in the solutions for various plates subjected to concentrated lateral loads. The division of the load into a local and general parts is worth doing when the solution is reduced to the summation of a slowly converging series or to numerical evaluation of the integrals singular at the loading point. The solutions for many particular problems can be found in the references cited.

References 4

[4.1] C. W. Mac Gregor: *Mech. Eng.* 57 (1935) 225.

[4.2] D. L. Holl: Cantilever plate with concentrated edge load, *J. Appl. Mech.* 4 (1937).

[4.3] T. J. JARAMILLO: *J. Appl. Mech. 7* (1950) 67.
[4.4] W. T. KOITER, J. B. ALBLAS: *Proc. Koninkl. Ned. Akad. Wetenschap, ser. B, 57.2* (1954) 259.
[4.5] E. BITTNER, *Momententafeln und Einflussflächen für Kreuzweise bewahrte Eisenbetonplatten*, Springer, Wien 1938.
[4.6] Z. KĄCZKOWSKI, Certain closed solution for the deflection of a strip plate, *Arch. Mech. Stos. 5.4* (1953) 591–628. (in Polish).
[4.7] H. JUNG, *Math. Nachr. 6* (1952) 343.
[4.8] S. WOINOWSKY-KRIEGER, *Ing.-Arch. 20* (1952) 391.
[4.9] W. T. KOITER, *Ing.-Arch. 21* (1953) 381.
[4.10] Y. S. UFLIAND, *Doklady AN SSSR 84* (1952) 463.
[4.11] H. M. WESTERGAARD, *Ingeniøren 32* (1923) 513; *Public Roads 7* (1926) 25; *10* (1929) 65; *14* (1933) 185.
[4.12] D. L. HOLL, Cantilever plate with concentrated edge load, *J. Appl. Mech. 4.1, A-8* (1937).

5

Plates loaded only in their middle plane

5.1 General equations

Now we shall consider several basic cases of the plate loaded only in its plane. The problem consists in determining the stresses in the flat plate when the stresses are constant across its thickness. This is also known as the *two-dimensional problem of the theory of elasticity.* The solution of this case is limited to finding the stress function from Eq. $(2.1)_2$ since by assumption $q = 0$, $w = 0$, and the second fundamental equation $(2.1)_1$ is identically satisfied. At present we shall limit ourselves to determining the distribution of the stresses and displacements near the loading point. We neglect here the problem of stability connected with bending of the plate caused by the load acting in its plane. These problems will be discussed in Chapter 6.

Equation $(2.1)_2$ is derived on the assumption that no external surface or volume loads except Z act in the plane of the plate. The fundamental equation taking into account the volume forces X, Y tangential to the plate surface takes the form

$$\Delta\Delta\Phi = -\frac{\partial^2}{\partial x^2}\int Y \mathrm{d}y + \nu\frac{\partial Y}{\partial y} - \frac{\partial^2}{\partial y^2}\int X \mathrm{d}x + \nu\frac{\partial X}{\partial x}. \tag{5.1}$$

The internal forces in the plate can be found from the formulae

$$N_{xx} = -\frac{\partial^2\Phi}{\partial y^2} - \int X \mathrm{d}x, \quad N_{yy} = -\frac{\partial^2\Phi}{\partial x^2} - \int Y \mathrm{d}y, \quad N_{xy} = \frac{\partial^2\Phi}{\partial x\,\partial y}. \tag{5.2}$$

The stresses are defined by $\sigma_{ij} = N/h_{ij}$, $i, j = x, y$. When the plate is not loaded over its surface, but only at its edges, the right hand-side of Eq. (5.1) vanishes. The solution takes the form of the biharmonic function which does not depend on material constants such as E, ν, G. Then the problem is reduced to the satisfaction of the boundary conditions. In the case where $X \neq 0$, $Y \neq 0$ the general solution of the problem can be obtained by the summation of the particular solution, resulting from Eq. (5.1) including the effect of the surface forces X, Y, and the general solution of the equation

$$\Delta\Delta\Phi = 0. \tag{5.3}$$

We obtain the displacements in the plane of the plate by solving the equations relating the strains and displacements. Eliminating u and v from Eq. (1.25) and expressing the strains ε_{xx}, ε_{yy}, ε_{xy} in terms of stresses, we obtain in the case of small displacements the following relations for the plane state of stress

$$Eu = \int (\sigma_{xx} - \nu\sigma_{yy})\,\mathrm{d}x + F_1(y),$$

$$Ev = \int (\sigma_{yy} - \nu\sigma_{xx})\,\mathrm{d}y + F_2(x),$$

where $F_1(y)$ and $F_2(x)$ are arbitrary functions of variables y and x, respectively. Expressing the stresses through the stress function, we have

$$\begin{aligned} Eu &= \int \frac{\partial^2\Phi}{\partial y^2}\,\mathrm{d}x - \nu\frac{\partial\Phi}{\partial x} + F_1(y), \\ Ev &= \int \frac{\partial^2\Phi}{\partial x^2}\,\mathrm{d}y - \nu\frac{\partial\Phi}{\partial y} + F_2(x). \end{aligned} \tag{5.4}$$

The equation

$$\gamma_{xy} = \frac{\partial u}{\partial y} + \frac{\partial v}{\partial x}$$

is an identity and determines the functions $F_1(y)$ and $F_2(x)$; on substitution we have

$$\int \frac{\partial^3\Phi}{\partial y^3}\,\mathrm{d}x + \int \frac{\partial^3\Phi}{\partial x^3}\,\mathrm{d}y + \frac{\mathrm{d}F_1(y)}{\mathrm{d}y} + \frac{\mathrm{d}F_2(x)}{\mathrm{d}x} \equiv -2\frac{\partial^2\Phi}{\partial x\,\partial y}. \tag{5.5}$$

Now we shall solve several fundamental problems concerning plates loaded in their planes by concentrated forces.

5.2 Infinite plate loaded at an interior point by a concentrated force

Let us begin with the case of an infinitely large plate carrying a concentrated load at the point with coordinates $x = 0$, $y = 0$ (Fig. 60). The force T is directed along the x-axis. Let us introduce the non-dimensional system of coordinates with a characteristic arbitrary length l. Expressing the force

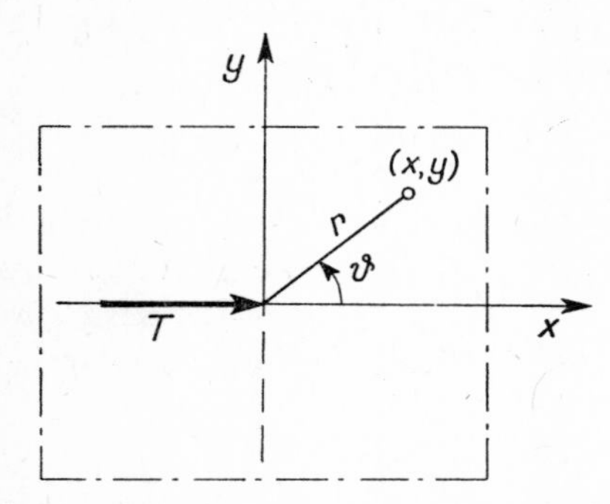

Fig. 60. Infinitely large sheet loaded at an interior point by a concentrated force T.

$X = T\delta(x, y)$ by means of the Fourier integral and introducing it into Eq.(5.1), we obtain

$$\Delta\Delta\Phi = \frac{T}{\pi^2 l^3} \int_0^\infty \int_0^\infty \alpha\left(\frac{\beta^2}{\alpha^2} - \nu\right) \sin \alpha x \cos \beta y \, \mathrm{d}\alpha \, \mathrm{d}\beta . \tag{5.6}$$

The stress function can also be found in the form of a Fourier integral. The substitution yields

$$\Phi = \frac{Tl}{\pi^2} \int_0^\infty \int_0^\infty \frac{\beta^2 - \nu\alpha^2}{\alpha(\alpha^2 + \beta^2)^2} \sin \alpha x \cos \beta y \, \mathrm{d}\alpha \, \mathrm{d}\beta . \tag{5.7}$$

The above integral can be evaluated with respect to α by means of calculus of residues.

$$\Phi = \frac{Tl}{4\pi} \int_0^\infty \left\{2 - [2 + (1+\nu)\beta x] \frac{\mathrm{e}^{-\beta x}}{\beta^2}\right\} \cos \beta y \, \mathrm{d}\beta .$$

The above integral is divergent but the stress function can be derived by integration between the limits B and ∞. Assuming that $B \to 0$ and neglecting constant terms, we obtain

$$\Phi = \frac{Tl}{4\pi}[(1-\nu)r\ln r\cos\vartheta - 2\vartheta r\sin\vartheta]. \tag{5.8}$$

It is interesting that the stress function contains the multivalued term. The integrals for stresses are convergent and unique, for example for σ_{xx} we have

$$\sigma_{xx} = -\frac{T}{4\pi h}\int_0^\infty [(1+\nu)\beta x + 2]\mathrm{e}^{-\beta x}\cos\beta y \,\mathrm{d}\beta. \tag{5.9}$$

On integration we infer

$$\left.\begin{matrix}\sigma_{xx}\\ \sigma_{xy}\end{matrix}\right\} = \mp\frac{T}{4\pi h}\left\{[2\pm(1+\nu)]\frac{x}{x^2+y^2} - (1+\nu)\frac{2xy^2}{(x^2+y^2)^2}\right\},$$

$$\sigma_{xy} = -\frac{T}{4\pi h}\left[(1-\nu)\frac{x}{x^2+y^2} + (1+\nu)\frac{2x^2y}{(x^2+y^2)^2}\right]. \tag{5.10}$$

The same stresses can be obtained directly from formula (5.8). Expressing the stresses in components σ_{rr}, $\sigma_{\vartheta\vartheta}$, $\sigma_{r\vartheta}$, we have

$$\sigma_{rr} = -(3+\nu)\frac{T}{4\pi hl}\frac{\cos\vartheta}{r},$$

$$\sigma_{\vartheta\vartheta} = (1-\nu)\frac{T}{4\pi hl}\frac{\cos\vartheta}{r}, \qquad \tau_{r\vartheta} = (1-\nu)\frac{T}{4\pi hl}\frac{\sin\vartheta}{r}. \tag{5.11}$$

The stresses have a singularity at the point $r = 0$. A very similar state of stress (however, not exactly the same) was obtained at the upper surface of the plate loaded by a concentrated bending moment (3.127). The displacements u and v can be obtained from Eq. (5.3). Upon introduction of (5.10) into (5.4), and the integration, we obtain

$$Eu = -\frac{T}{4\pi h}\left[(3+2\nu-\nu^2)\ln(x^2+y^2)^{1/2} + (1+\nu)^2\frac{y^2}{x^2+y^2}\right] + F_1(y),$$

$$Ev = \frac{T}{4\pi h}(1+\nu)^2\frac{xy}{x^2+y^2} + F_2(x). \tag{5.12}$$

Due to symmetry with respect to the x-axis, $F_2 = 0$. Condition (5.4) gives the equation

$$\frac{\mathrm{d}F_1}{\mathrm{d}y} + \frac{\mathrm{d}F_2}{\mathrm{d}x} = 0.$$

Then $\mathrm{d}F_1/\mathrm{d}y = 0$. Therefore, $F_1 = \text{const}$.

Assuming that the plate is supported at an infinite distance, we obtain infinitely large displacements in the plate. Let us calculate the displacements with respect to the straight line crossing two arbitrary points lying on the y-axis ($x = 0, y = \pm a$). Then we find

$$Eu = -\frac{T}{4\pi h}\left[(3-\nu)(1+\nu)\ln\frac{r}{a} - (1+\nu)^2\cos^2\vartheta\right]. \tag{5.13}$$

We see from the above formula that at the point of application of the load the displacement u is infinitely large. When the plate is loaded by the force uniformly distributed along a small length $2c$ (Fig. 61), the stresses can be ob-

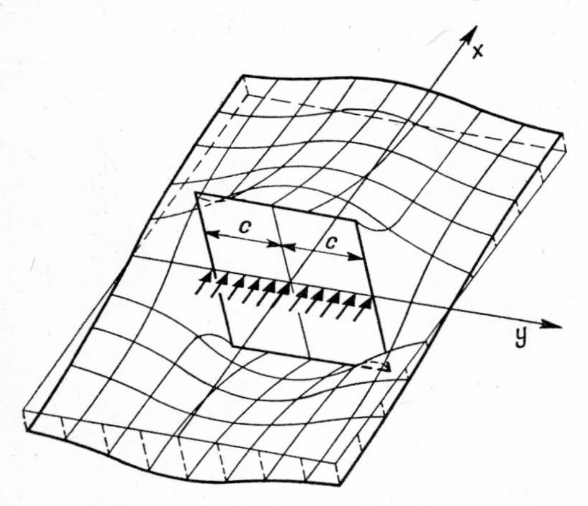

Fig. 61. Distribution of stresses σ_{xx} in a sheet loaded by a force uniformly distributed along the segment 2c.

tained directly from integrals (5.9), multiplying only the integrands by $(\sin\beta c)/\beta c$. Upon integration we obtain

$$\left.\begin{matrix}\sigma_{xx}\\ \sigma_{yy}\end{matrix}\right\} = -\frac{T}{8\pi hc}\left[[1+\nu\pm(1-\nu)](\vartheta_2-\vartheta_1)+\frac{1+\nu}{2}(\sin 2\vartheta_2-\sin 2\vartheta_1)\right],$$

$$\tau_{xy} = -\frac{T}{8\pi hc}\left[(1-\nu)\ln\frac{r_2}{r_1}-\frac{1+\nu}{2}(\cos 2\vartheta_2-\cos 2\vartheta_1)\right]. \tag{5.14}$$

We observe that in this case only the tangential stress τ_{xy} is singular at the points $y = \pm c$, $x = 0$. This follows from the rapid change of the load. We see that in order to obtain a finite value of τ_{xy} it is necessary not only to distribute the load along a certain length, but also to ensure the variations of these loads to be continuous and smooth. For $x = 0$, the stresses σ_{xx} and σ_{yy} attain the maximum value. For $-c \leqslant y \leqslant c$

$$\sigma_{xx} = \frac{T}{4ch}, \qquad \sigma_{yy} = \frac{T}{4ch},$$

for $y < -c$ and $y > c$

$$\sigma_{xx} = \sigma_{yy} = 0. \tag{5.15}$$

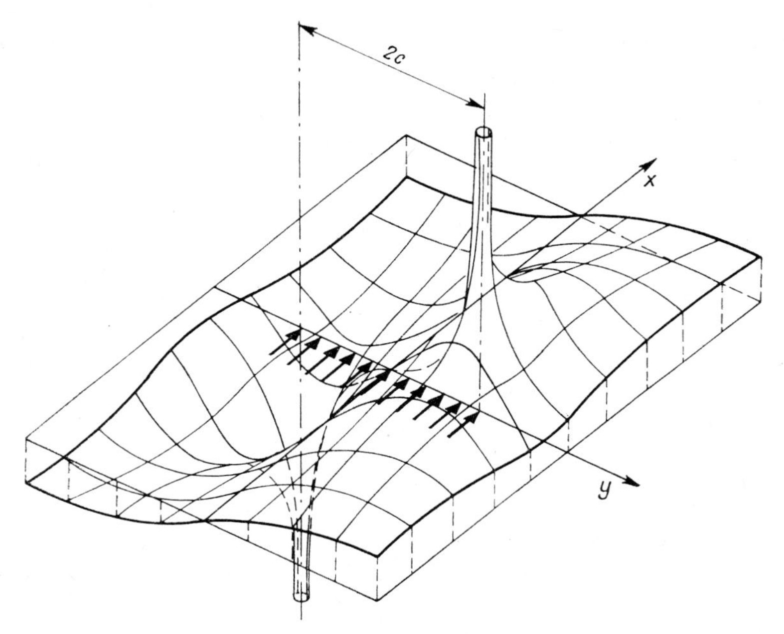

Fig. 62. Tangential stresses τ_{xy} in a sheet loaded by a force uniformly distributed along the segment 2c.

The stresses caused by the distributed force T are presented graphically in Fig. 61 and Fig. 62.

The solution for the case presented in Fig. 63 may also be interesting. The force is introduced in the plate by means of the tangential stresses acting along the distance $2c$. By a similar treatment we find the following results:

$$\left.\begin{matrix}\sigma_{xx}\\ \sigma_{yy}\end{matrix}\right\} = -\frac{T}{8\pi hc}\left[[2\pm(1+\nu)]\ln\frac{r_2}{r_1} \pm \frac{1+\nu}{2}(\cos 2\vartheta_2 - \cos 2\vartheta_1)\right],$$

$$\tau_{xy} = -\frac{T}{8\pi hc}\left[2(\vartheta_2 - \vartheta_1) - \frac{1+\nu}{2}(\sin 2\vartheta_2 - \sin 2\vartheta_1)\right]. \tag{5.16}$$

We observe that at the points $x = \pm c$ the stresses σ_{xx} and σ_{yy} are singular. (Fig. 64.) This results from the rapid change of the load. The stresses τ_{xy} have a maximum for $y = 0$.

For $-c \leqslant x \leqslant c, \quad \tau_{xy} = -\dfrac{T}{4hc}$.

For $y < -c$ and $c < y, \quad \tau_{xy} = 0$.

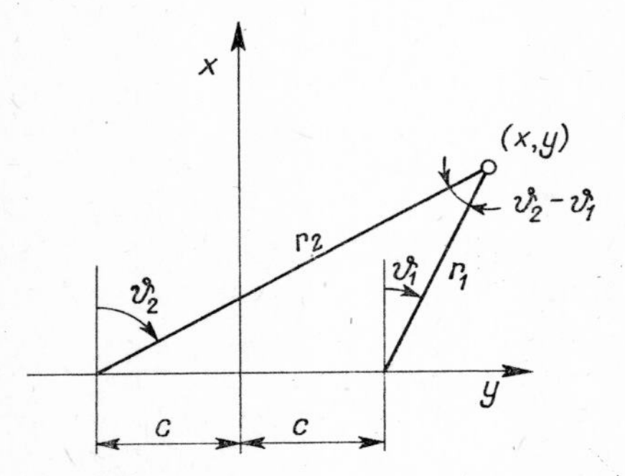

Fig. 63. Infinitely large sheet loaded by a force uniformly distributed along the segment 2c of x axis.

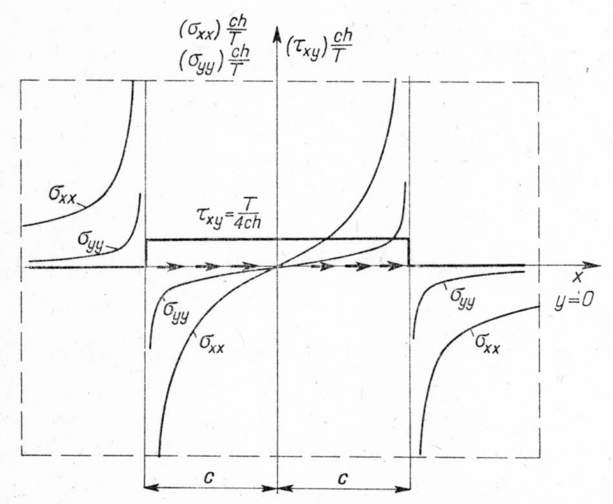

Fig. 64. Distribution of stresses in the sheet from Fig. 63.

The stresses σ_{xx}, σ_{yy}, τ_{xy} for $y = 0$ are presented in Fig. 64. The case where the force acts at the interior point of a semi-infinite strip plate was considered by E. Melan [5.12].

5.3 Circular plate loaded by a tangential load

Let us consider one example: a circular plate clamped at its outside radius R subject to a tangential load T acting at the centre and in the plane of the plate. We solve this case in a different way than previously. Namely, we satisfy certain boundary conditions at two radii of the plate, c and R. We assume that the resultant forces acting on a cylindrical surface of radius c must balance T. If $c \to 0$, we get the case of a concentrated load. We have said nothing here about the area $r < c$. But since c is small, the load may be assumed uniformly distributed over a circular area of radius c. Let us introduce a system of polar coordinates r, ϑ.

The stress function satisfying Eq. (5.3) may be assumed as follows (see (3.133)):

$$\begin{aligned} \Phi = {} & C_{10}+C_{20}r^2+C_{30}\ln r+C_{40}r^2\ln r+ \\ & +(C_{11}r+C_{21}r^3+C_{31}r^{-1}+C_{41}r\ln r)\cos\vartheta+ \\ & +\sum_{n=2}^{\infty}(C_{1n}r^n+C_{2n}r^{n+2}+C_{3n}r^{-n}+C_{4n}r^{-n+2})\cos n\vartheta+ \\ & +E_1\vartheta+E_2r^2\vartheta+E_3r\vartheta\sin\vartheta+E_4r\vartheta\cos\vartheta. \end{aligned} \tag{5.17}$$

In this case, due to the nature of the loading, we take the stress function corresponding to the action of the concentrated force T in the form:

$$\Phi = C_1 r\vartheta\sin\vartheta+(C_2r^3+C_3r^{-1}+C_4r\ln r)\cos\vartheta, \tag{5.18}$$

where C_i are the arbitrary constants. Then the stresses are

$$\begin{aligned} \sigma_{rr} &= \left(2\frac{C_1}{r}+2C_2r-\frac{2C_3}{r^3}+\frac{C_4}{r}\right)\cos\vartheta, \\ \sigma_{\vartheta\vartheta} &= \left(6C_2r+\frac{2C_3}{r^3}+\frac{C_4}{r}\right)\cos\vartheta, \\ \tau_{r\vartheta} &= \left(2C_2r-\frac{2C_3}{r^3}+\frac{C_4}{r}\right)\sin\vartheta. \end{aligned} \tag{5.19}$$

The displacement components are

$$\begin{aligned} u = {} & \frac{1}{E}\left[2C_1\ln r+C_2r^2(1-3\nu)+\frac{C_3}{r^2}(1+\nu)+C_4(1-\nu)\ln r\right]\cos\vartheta+ \\ & +F_1\sin\vartheta+F_2\cos\vartheta+[4C_4+2C_1(1-\nu)]\frac{\vartheta\sin\vartheta}{2E}, \end{aligned} \tag{5.20$_1$}$$

$$v = \frac{1}{E}\left[-2C_1(\nu+\ln r)+C_2 r^2(5+\nu)+\frac{C_3}{r^2}(1+\nu)+\right.$$

$$\left.+C_4(1-\nu)(1-\ln r)\right]\sin\vartheta+F_1\cos\vartheta-F_2\sin\vartheta-$$

$$-[4C_4+C_1(1-\nu)]\frac{\sin\vartheta-\vartheta\cos\vartheta}{2E}+F_3 r. \tag{5.20$_2$}$$

Equations (5.20) contain in all seven constants C_1, C_2, C_3, C_4 and E_1, E_2, E_3. These can be found by the following conditions:

$$T = -2\int_0^\pi (\sigma_{rr}\cos\vartheta-\tau_{r\vartheta}\sin\vartheta)r\,d\vartheta,$$

$$\text{hence}\quad 2C_1 = -\frac{T}{\pi h},$$

$$\vartheta = 0, \quad v = 0,$$

$$r = R, \quad u = v = 0, \tag{5.21}$$

$$r = c, \quad \varepsilon_{\vartheta\vartheta} = 0,$$

$$\frac{2(1+\nu)}{E}\tau_{r\vartheta} = \gamma_{r\vartheta} \equiv \frac{1}{r}\frac{\partial u}{\partial\vartheta}+\frac{\partial v}{\partial r}-\frac{v}{r}.$$

The above conditions yield the following expressions:

$$\sigma_{rr} = -\frac{T}{4\pi hr}\left\{(3+\nu)+\frac{(1+\nu)^2 r^2}{(3-\nu)(c^2+R^2)}-\frac{(1+\nu)c^2}{\left(1+\frac{c^2}{R^2}\right)r^2}\right\}\cos\vartheta, \tag{5.22$_1$}$$

$$\sigma_{\vartheta\vartheta} = -\frac{T}{4\pi hr}\left\{\frac{3r^2(1+\nu)^2}{(3-\nu)(c^2+R^2)}+\frac{c^2(1+\nu)}{\left(1+\frac{c^2}{R^2}\right)r^2}-(1+\nu)\right\}\cos\vartheta, \tag{5.22$_2$}$$

$$\tau_{r\vartheta} = \frac{T}{4\pi hr}\left\{(1-\nu)-\frac{r^2(1+\nu)}{(3-\nu)(c^2+R^2)}+\frac{(1+\nu)c^2}{\left(1+\frac{c^2}{R^2}\right)r^2}\right\}\sin\vartheta; \tag{5.22$_3$}$$

$$u = -\frac{T}{8\pi Gh}\left\{(3-\nu)\ln\frac{r}{R}-\left[\frac{\frac{r^2}{R^2}-1}{\frac{c^2}{R^2}+1}\right]\frac{1+\nu}{2}\left[\frac{c^2}{r^2}-\frac{1-3\nu}{3-\nu}\right]\right\}\cos\vartheta, \tag{5.22$_4$}$$

$$v = \frac{T}{8\pi Gh}\left\{(3-\nu)\ln\frac{r}{R} - \frac{1+\nu}{1+\dfrac{c^2}{R^2}}\,\frac{(5+\nu)\dfrac{r^2}{R^2}+(1-3\nu)}{2(3-\nu)} - \right.$$

$$\left. - \frac{1+\nu}{2}\,\frac{c^2}{r^2}\left[\frac{\dfrac{r^2}{R^2}+1}{\dfrac{c^2}{R^2}+1}\right] + (1+\nu)\right\}\sin\vartheta. \qquad (5.22)_5$$

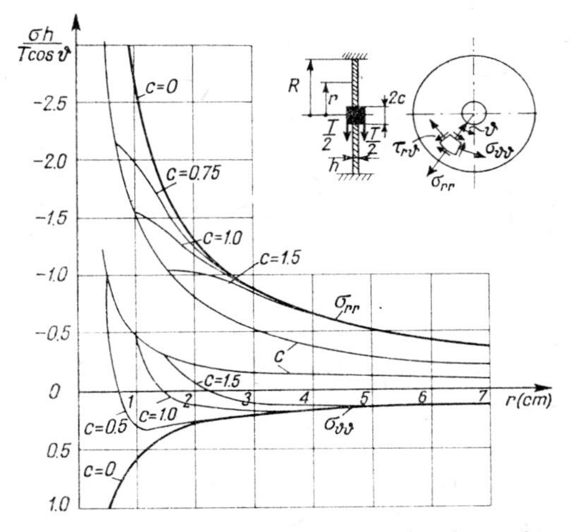

Fig. 65a. Normal stress variation in a circular sheet subjected to a tangential load.

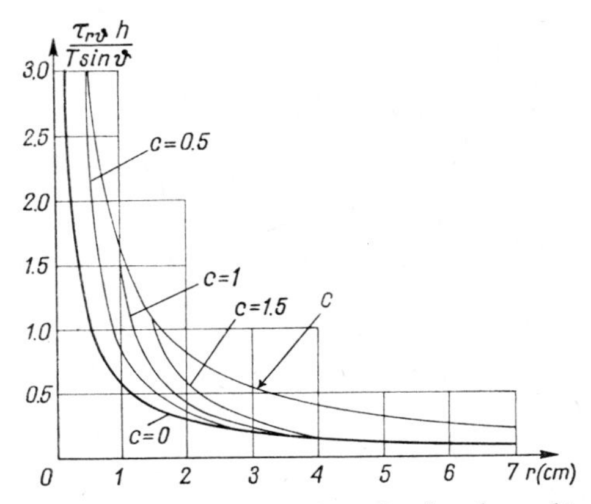

Fig. 65b. Shear stress variation in a circular plate subjected to a tangential load.

The expressions (5.22) are plotted in Figs. 65a and 65b for various ratio of c/R. For $c = 0$ and $R \to \infty$ we get the results obtained previously for a concentrated load.

5.4 Infinite orthotropic plate loaded at an interior point

Let us consider an orthotropic plate loaded by a concentrated force T lying in its plane. If we apply the classical theory, Eq. (1.76) becomes

$$\frac{1}{E_y}\frac{\partial^4\Phi}{\partial x^4}+\left(\frac{1}{G}-\frac{2\nu_x}{E_x}\right)\frac{\partial^4\Phi}{\partial x^2\partial y^2}+\frac{1}{E_x}\frac{\partial^4\Phi}{\partial y^4}$$

$$=-\frac{1}{E_y}\frac{\partial^2}{\partial x^2}\int Y\mathrm{d}y+\frac{\nu_y}{E_y}\frac{\partial Y}{\partial y}-\frac{1}{E_x}\frac{\partial^2}{\partial y^2}\int X\mathrm{d}x+\frac{\nu_x}{E_x}\frac{\partial X}{\partial x}. \tag{5.23}$$

The internal forces are defined by formulae (5.2) as previously. Representing the load $X = T$ and the stress function by means of Fourier integrals, we have

$$\Phi=\frac{Tl}{\pi^2}\int_0^\infty\int_0^\infty\frac{(\beta^2-\nu_x\alpha^2)\sin\alpha x\cos\beta y\,\mathrm{d}\alpha\,\mathrm{d}\beta}{\alpha\left[\dfrac{\alpha^4}{E_y}+\left(\dfrac{1}{G}-\dfrac{2\nu_x}{E_x}\right)\alpha^2\beta^2+\dfrac{\beta^4}{E_x}\right]}. \tag{5.24}$$

The above integral can be evaluated with respect to α by means of the calculus of residues. On integration we find

$$\Phi=\mathrm{re}\frac{TE_y l}{\pi E_x}\int_0^\infty\left[\frac{E_x}{E_y\beta^2}-\frac{(1-\nu_x\alpha_1^2)\mathrm{e}^{i\alpha_1\beta x}}{2\alpha_1^2(\alpha_1^2-\alpha_2^2)}+\right.$$

$$\left.+\frac{(1-\nu_x\alpha_2^2)\mathrm{e}^{i\alpha_2\beta x}}{2\alpha_2^2(\alpha_1^2-\alpha_2^2)}\right]\frac{\cos\beta y}{\beta^2}\,\mathrm{d}\beta, \tag{5.25}$$

where α_1, α_2 are the roots of the characteristic equation

$$\frac{\alpha^4}{E_y}+2\left(\frac{1}{2G}-\frac{\nu_x}{E_x}\right)\alpha^2+\frac{1}{E_x}=0. \tag{5.26}$$

Introducing the notations

$$\lambda=\sqrt[4]{\frac{E_y}{E_x}},\qquad \mu_2=\frac{\sqrt{E_xE_y}}{2G}-\nu_x\sqrt{\frac{E_y}{E_x}},$$

we obtain the roots in the form

$$\alpha_{1,2}^2 = -\lambda^2\left(\mu_2 \pm \sqrt{\mu_2^2-1}\right).$$

In order to obtain the stresses we differentiate the stress function according to (2.4). We obtain, on integration with respect to β,

$$\begin{aligned}
\sigma_{xx} &= -\text{re}\,\frac{TE_y}{2\pi hE_x l(\alpha_1^2-\alpha_2^2)}\left[\frac{(1-\nu_x\alpha_1^2)\,ix}{[(i\alpha_1 x)^2+y^2]\alpha_1} - \frac{(1+\nu_x\alpha_2^2)\,ix}{[(i\alpha_2 x)^2+y^2]\alpha_2}\right],\\
\sigma_{yy} &= -\text{re}\,\frac{TE_y}{2\pi hE_x l(\alpha_1^2-\alpha_2^2)}\left[\frac{(1-\nu_x\alpha_1^2)\,i\alpha_1 x}{(i\alpha_1 x)^2+y^2} - \frac{(1-\nu_x\alpha_2^2)\,i\alpha_2 x}{(i\alpha_2 x)^2+y^2}\right],\\
\tau_{xy} &= -\text{re}\,\frac{TE_y}{2\pi hE_x l(\alpha_1^2-\alpha_2^2)}\left[\frac{(1-\nu_x\alpha_1^2)\,iy}{[(i\alpha_1 x)^2+y^2]\alpha_1} - \frac{(1-\nu_x\alpha_2^2)\,iy}{[(i\alpha_2 x)^2+y^2]\alpha_2}\right],
\end{aligned} \tag{5.27}$$

where $i = \sqrt{-1}$. In the case where $\mu_2 > 1$ we have the roots in the form $\alpha_1 = ik_1$, $\alpha_2 = ik_2$, where $k_{1,2} = \lambda\sqrt{\mu_2 \pm \sqrt{\mu_2^2-1}}$ has a real value. Introducing it into Eq. (5.27), we find the formulae for the stresses.

In the second case where $\mu_2 = 1$, $k_1 = k_2 = k$, the stress function is represented by the integral

$$\Phi = \frac{TlE_y}{\pi^2 E_x}\int_0^\infty\int_0^\infty \frac{\beta^2-\nu_x\alpha^2}{\alpha(\alpha^2+k^2\beta^2)^2}\sin\alpha x\cos\beta y\,\mathrm{d}\alpha\,\mathrm{d}\beta. \tag{5.28}$$

Introducing the new variable $k\beta = \beta_1$, the above integral can be transformed to the integral (5.7) for the isotropic plate. Then we can write at once the following expression for the stresses:

$$\begin{aligned}
\sigma_{xx} &= -\frac{Tk}{4\pi h}\left\{(3+\nu_x k^2)\,\frac{x}{x^2k^2+y^2} - (1+\nu_x k^2)\,\frac{2xy^2}{x^2k^2+y^2}\right\},\\
\sigma_{xy} &= -\frac{Tk}{4\pi h}\left[(1-\nu_x k^2)\,\frac{y}{x^2k^2+y^2} + (1+\nu_x k^2)\,\frac{2x^2y}{x^2k^2+y^2}\right].
\end{aligned} \tag{5.29}$$

For the isotropic plate $k = 1$.

In the third case $\mu_2 < 1$, the roots of the characteristic equation take the form

$$\alpha_{1,2} = \lambda\left[\sqrt{\frac{1+\mu_2}{2}} \pm i\sqrt{\frac{1-\mu_2}{2}}\right],$$

$$\alpha_{1,2}^2 = -\lambda^2\left(\mu_2 \pm i\sqrt{1-\mu_2}\right).$$

The solution can be found by the introduction of the above in expressions (5.27).

Introducing the polar coordinates (r, ϑ), we find [3.87]

$$\begin{aligned}
\sigma_{rr} &= \frac{T}{2\pi h(\beta_1^2-\beta_2^2)l}\left[i\beta_1(1-\nu_y\beta_2^2)\frac{(1+\beta_1^2)\sin^2\vartheta+\beta_1^2}{\cos^2\vartheta-\beta_1^2\sin^2\vartheta}-\right.\\
&\left.-i\beta_2(1-\nu_y\beta_1^2)\frac{(1+\beta_2^2)\sin^2\vartheta+\beta_2^2}{\cos^2\vartheta-\beta_2^2\sin^2\vartheta}\right]\frac{\cos\vartheta}{r},\\
\sigma_{\vartheta\vartheta} &= -\frac{T}{2\pi hl}\left[\frac{1+\nu_y\beta_1\beta_2}{i(\beta_1+\beta_2)}\right]\frac{\cos\vartheta}{r},\\
\tau_{r\vartheta} &= -\frac{T}{2\pi hl}\,\frac{1+\nu_y\beta_1\beta_2}{i(\beta_1+\beta_2)}\,\frac{\sin\vartheta}{r},
\end{aligned} \tag{5.30}$$

where β_1, β_2 are the roots of the equation

$$\frac{\beta^4}{E}+\left(\frac{1}{G}-\frac{2\nu_x}{E_x}\right)\beta^2+\frac{1}{E_y}=0. \tag{5.31}$$

Despite the fact that expressions (5.27) or (5.30) include $i=\sqrt{-1}$ the stresses have real values as the product of the roots β_1, β_2 and the sum $i(\beta_1+\beta_2)$ have real values. The stresses $\sigma_{\vartheta\vartheta}$ and $\tau_{r\vartheta}$ depend in a simple way on the angle ϑ. Only the stresses σ_{rr} are expressed by a more complex function of ϑ. For the isotropic plate $\beta_1=\beta_2=i$ and we obtain Eq. (5.11).

5.5 Plates loaded at their edges

5.5.1 *Concentrated force normal to edge* [5.1]

Let us consider a plate loaded by a concentrated force N. The direction of the force is perpendicular to the free edge (Fig. 66). Since in this case the plate is loaded only at the edge, we need only the general solution of Eq. (5.3). This equation is identical with that for the laterally loaded plate and has the same solution (3.123) containing four arbitrary constants C_1 to C_4. The constants C_i are functions of the variable β. For the edge we have the following boundary conditions at $x=0$:

$$\sigma_{xx}=\delta(y)N/h; \qquad \tau_{xy}=0.$$

Because the plate is infinite we have for $x\to\infty$: $\sigma_{xx}\to 0,\ \sigma_{yy}\to 0,\ \tau_{xy}\to 0.$

It has been here assumed that the force N acts at the point $x = 0$, $y = 0$. Solving the above set of equations, we obtain the following expression for the stress function

$$\Phi = -\frac{Nl}{\pi h}\int_0^\infty (1+\beta x)\frac{e^{-\beta x}}{2}\cos\beta y \, d\beta . \tag{5.32}$$

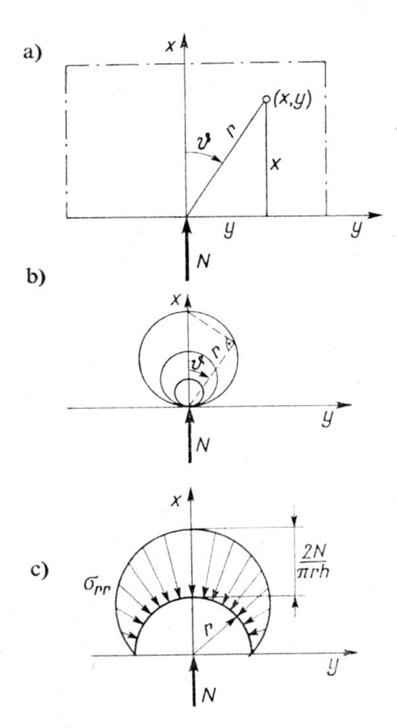

Fig. 66. Semi-infinite sheet loaded at its free edge: a) *coordinates*; b) *circle of constant shear stress*; c) *distribution of normal stress* σ_{rr}.

On integration we have

$$\Phi = -\frac{N}{\pi} r\vartheta \sin\vartheta ,$$

where

$$r^2 = x^2 + y^2 , \qquad \vartheta = \arctan\frac{y}{x} .$$

The stresses obtained from the above formulae have the form

$$\begin{aligned}
\sigma_{xx} &= -\frac{2N}{\pi hl}\frac{x^3}{(x^2+y^2)^2},\\
\sigma_{yy} &= -\frac{2N}{\pi hl}\frac{xy^2}{(x^2+y^2)^2},\\
\tau_{xy} &= -\frac{2N}{\pi hl}\frac{x^2y}{(x^2+y^2)^2}.
\end{aligned}\tag{5.33}$$

Expressing the stresses by means of components σ_{rr}, $\sigma_{\vartheta\vartheta}$, $\tau_{r\vartheta}$, we have

$$\sigma_{rr} = -\frac{2N}{\pi hl}\frac{\cos\vartheta}{r}, \qquad \sigma_{\vartheta\vartheta} = \tau_{r\vartheta} = 0, \qquad r^2 = x^2+y^2.$$

We see that in the cross-sections defined by the lines r = const, ϑ = const there appear only radial stresses creating the system whose focus is at the loading point. The stresses σ_{rr} are constant on the circle $\cos\vartheta/r$ = constant tangential to the plate's edge at the point $x = 0$, $y = 0$ (Fig. 66b). The other stresses are equal to zero. At the points on the circle $\cos\vartheta/r$ = constant, the tangential stresses $\tau_{\max}$ also have constant value. The maximum tangential stresses here are

$$\tau_{\max} = \frac{\sigma_{rr}}{2} = -\frac{N}{\pi hl}\frac{\cos\vartheta}{r}.$$

It is easy to prove that the above solution corresponds to the given boundary conditions while considering the equilibrium of that portion of the plate subjected to the action of the force N and stresses σ_{rr}. Integrating the projections of the stresses on the x-axis along the semi-circle r = constant, we obtain the resultant force equal to N (Fig. 66c). The displacements in the plate can be obtained by means of Eq. (5.4). Substituting the stresses, we obtain on integration

$$\begin{aligned}
Ev &= -\frac{N}{\pi h}\left[(1-\nu)\arctan\frac{y}{x} - (1+\nu)\frac{xy}{r^2}\right] + F_2(x),\\
Eu &= -\frac{N}{\pi h}\left[\ln r^2 + (1+\nu)\frac{y^2}{r^2}\right] + F_1(y).
\end{aligned}\tag{5.34}$$

On substitution in Eq. (5.5) we find

$$\frac{dF_2}{dx} + \frac{dF_1}{dy} = 0. \tag{5.35}$$

Due to symmetry, the displacement v should be equal to zero for $y = 0$. We have then $F_2 = 0$ and $F_1 =$ constant. Assuming the points of the plate to be immovable for $x = \infty$, we obtain the infinitely large displacement u. Let us calculate the displacement u with respect to two arbitrarily chosen points of the edge, $y = \pm a$, $x = 0$. The constant F_1 will be found from the condition that at these points the displacement is equal to zero. As a result we obtain

$$Eu = -\frac{N}{\pi h}\left[\ln\frac{r^2}{a^2} - (1+\nu)\frac{x^2}{r^2}\right]. \tag{5.36}$$

From the above solution we find the infinitely large displacement u at the loading point. In reality, all concentrated loads act over a certain finite surface and produce finite stresses and displacements. The stresses caused by such real loads can be obtained from the above results by means of the influence-line and surface methods or directly representing the load by a proper Fourier integral or series. When the load is uniformly distributed along the length $2c$ (Fig. 67), we directly obtain the following stresses:

$$\left.\begin{aligned}\left.\begin{matrix}\sigma_{xx}\\ \sigma_{yy}\end{matrix}\right\} &= -\frac{N}{2\pi hc}\left[\vartheta_2 - \vartheta_1 \mp \frac{1}{2}(\sin 2\vartheta_2 - \sin 2\vartheta_1)\right],\\ \tau_{xy} &= \frac{N}{4\pi hc}(\cos 2\vartheta_2 - \cos 2\vartheta_1).\end{aligned}\right. \tag{5.37}$$

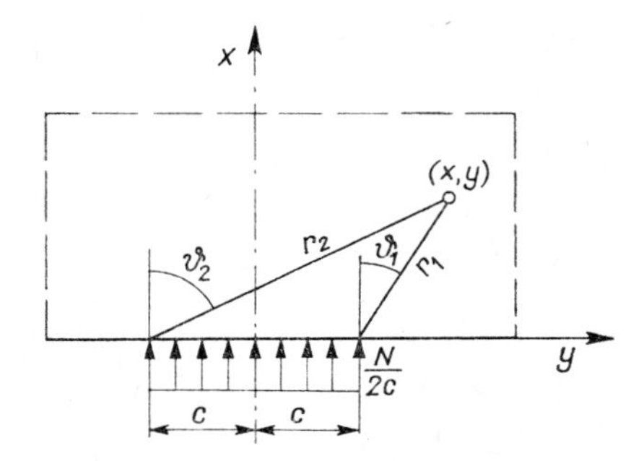

Fig. 67. Semi-infinite sheet loaded by a uniformly distributed force along the segment 2c.

For $x = 0$ we have

$$\vartheta_1 = -\frac{1}{2}\pi, \quad \vartheta_2 = \frac{1}{2}\pi,$$

and the stresses are

$$\sigma_{\max} = -\frac{N}{2hc}; \qquad \sigma_{yy} = \frac{N}{2hc}.$$

We see that these stresses in comparison to those appearing in the infinitely large plate loaded inside its area are here 2 and $2/\nu$ times larger. At points $y = \pm c$ the tangential stresses are not infinitely large, changing only rapidly from zero for $y < -c$ and $y > +c$ and for $-c < y < c$ to the value $\tau_{xy} = N/2hc$ for points $x = \pm c$. The maximum tangential stresses are

$$\tau_{\max} = -\frac{N}{2\pi hc}\sin(\vartheta_2 - \vartheta_1)$$

and are constant on the circles which cross the points of the edge $y = \pm c$ and whose centres are situated on the x-axis. These stresses are largest on the semi-circle whose centre is at the point $x = 0$, $y = 0$.

5.5.2 *Force tangential to edge*

In the similar way we can calculate the stresses in a plate loaded by a force tangential to the edge (Fig. 68). The difference consists only in the fulfilment of the other boundary conditions. As a result we have

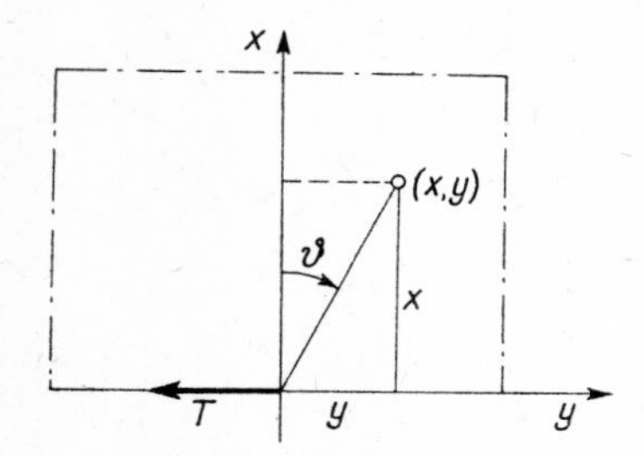

Fig. 68. Semi-infinite sheet loaded at the edge by the tangential force T.

$$\begin{aligned}
\sigma_x &= \frac{2T}{\pi hl}\,\frac{x^2 y}{(x^2+y^2)^2},\\
\sigma_{yy} &= \frac{2T}{\pi hl}\,\frac{y^3}{(x^2+y^2)^2},\\
\tau_{xy} &= \frac{2T}{\pi hl}\,\frac{xy^2}{(x^2+y^2)^2}.
\end{aligned} \tag{5.38}$$

In the system of coordinates r, ϑ we obtain

$$\sigma_{rr} = \frac{2T}{\pi hl}\frac{\sin\vartheta}{r}, \qquad \sigma_{\vartheta\vartheta} = \tau_{r\vartheta} = 0.$$

We obtain again the radial distribution of the stresses.

The maximum stresses σ_{rr} appear for $\vartheta = \pm\pi/2$, i.e. at the edge, and are equal to $\sigma_{rr} = \pm 2T/\pi rhl$.

When the force tangential to the edge of the plate is distributed along a length $2c$, the stresses are

$$\begin{aligned}
\sigma_{xx} &= \frac{T}{4\pi hl}(\cos 2\vartheta_2 - \cos 2\vartheta_1),\\
\sigma_{yy} &= \frac{T}{4\pi hc}\left(4\ln\frac{r_2}{r_1} + \cos 2\vartheta_2 - \cos 2\vartheta_1\right), \qquad (5.39)\\
\tau_{xy} &= \frac{T}{4\pi hc}[2(\vartheta_2 - \vartheta_1) - (\sin 2\vartheta_2 - \sin 2\vartheta_1)].
\end{aligned}$$

For $y = \pm c$ the stresses σ_{yy} are singular and increase to $\pm\infty$. The rapid variation of the load at these points accounts for this fact. The stresses σ_{xx} and τ_{xy} are finite in the whole area of the plate.

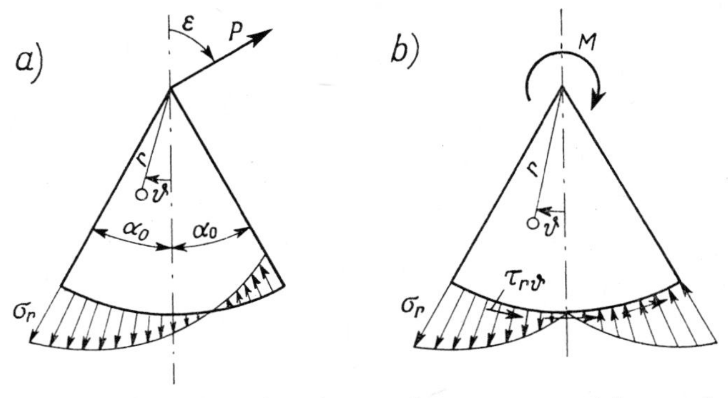

Fig. 69. Wedge sheet loaded at the apex by a concentrated force and concentrated bending moment.

5.5.3 *Wedge loaded by concentrated force at its apex*

The radial distribution of stresses is also obtained when the plate takes the form of a wedge and is loaded at its apex by a concentrated force (Fig. 69). The edges $\vartheta = \alpha_0$ and $\vartheta = -\alpha_0$ are free; hence $\sigma_{\vartheta\vartheta} = \tau_{r\vartheta} = 0$. Introducing the system of polar coordinates r, ϑ, we derive the following expressions for the concentrated force:

$$\sigma_{rr} = -\frac{P}{rhl}\left[\frac{\cos\vartheta\cos\varepsilon}{\alpha_0+\frac{1}{2}\sin 2\alpha_0} + \frac{\sin\vartheta\sin\varepsilon}{\alpha_0-\frac{1}{2}\sin 2\alpha_0}\right],$$

$$\sigma_{\vartheta\vartheta} = 0, \quad \tau_{r\vartheta} = 0.$$

The circles crossing the apex represent the curves of constant stress.

5.5.4 *Wedge loaded by concentrated moment at its apex*

The solution is derived by the introduction of the stress function $\Phi = \Phi(\vartheta)$ (not depending on r). On the satisfaction of boundary conditions we obtain the equations $\vartheta = -\alpha_0$, $\vartheta = \alpha_0$, $\sigma_{\vartheta\vartheta} = \tau_{r\vartheta} = 0$:

$$\sigma_{rr} = C\frac{\sin 2\vartheta}{hr^2l^2}, \quad \sigma_{\vartheta\vartheta} = 0, \quad \tau_{r\vartheta} = C\,\frac{\cos 2\vartheta - \cos 2\alpha_0}{2hr^2l^2}. \tag{5.41}$$

The constant C, defined from the equilibrium condition of the part of the wedge cut out along the circle, is

$$C = \frac{2M}{\int\limits_{-\alpha_0}^{\alpha_0}(\cos 2\vartheta - \cos 2\alpha_0)\,d\vartheta} = \frac{M}{\frac{1}{2}\sin 2\alpha_0 - \alpha_0\cos 2\alpha_0}.$$

In this case the distribution of the stresses is not radial. There also appear the tangential stresses $\tau_{r\vartheta}$ besides the normal stresses σ_{rr}. In the case of the wedge loaded by a concentrated force, the stresses decrease inversely proportionally to the radius r. In the case of the moment, the stresses decrease more rapidly — inversely proportionally to the second power of the radius. Corresponding solutions for the anisotropic plates were worked out by S. Lekhnitzky [5.3].

5.6 Orthotropic plate loaded at free edge

In the case where an orthotropic plate is loaded at the edge, the solution can also be found in the form of a Fourier integral. Assuming that the directions of orthotropy are parallel and perpendicular to the edge, the stresses caused by the concentrated force N normal to the edge are expressed by the following formulae:

$$\sigma_{ij} = -\frac{N\alpha_1\alpha_2}{\pi h}(\alpha_1+\alpha_2)\frac{x_1 x_i x_j}{(\alpha_1 x_1^2+x_2^2)(\alpha_2 x_1^2+x_2^2)}, \tag{5.42}$$

$$i \neq j, \quad i,j = 1,2, \quad x_1 = x, \quad x_2 = y,$$

where α_1 and α_2 are the roots of the characteristic equation

$$\frac{\alpha^4}{E_y} - \left(\frac{1}{G} - \frac{2\nu_x}{E_x}\right)\alpha^2 + \frac{1}{E_x} = 0. \tag{5.43}$$

Introducing the polar coordinates (r, ϑ), we obtain

$$\sigma_{rr} = -\frac{N(\alpha_1+\alpha_2)}{\pi h\sqrt{E_x E_y}}\frac{\cos\vartheta}{rD(\vartheta)}, \qquad \sigma_{\vartheta\vartheta} = \tau_{r\vartheta} = 0, \tag{5.44}$$

where

$$D(\vartheta) = \frac{\sin^4\vartheta}{E_y} + \left(\frac{1}{G} - \frac{2\nu_x}{E_x}\right)\sin^2\vartheta\cos^2\vartheta + \frac{\cos^4\vartheta}{E_x}.$$

The expression $D(\vartheta)$ has a physical meaning. It represents the reciprocal of Young modulus for tension and compression in the radial directions at the arbitrary angle ϑ.

As we see, also in this case only radial stresses appear. The points of equal stresses $\sigma_{rr} = \sigma_0$ are given by the equation

$$\frac{x^4}{E_x} + \left(\frac{1}{G} - \frac{2\nu_x}{E_x}\right)x^2y^2 + \frac{y^4}{E_y} + \frac{N(\alpha_1+\alpha_2)}{\pi h\sigma_0\sqrt{E_x E_y}}x(x^2+y^2) = 0 \tag{5.45}$$

and are presented graphically in Fig. 70. We notice three different types of diagrams: a), b) and c) depending on the ratios E_x/E_y, G/E_y.

In the case of an isotropic plate, $\alpha_1 = \alpha_2 = 1$, we obtain the solution (5.33). When the force P applied at the edge of the plate acts in the direction making an angle φ (Fig. 71) the stresses are

$$\sigma_{rr} = -\frac{P(\alpha_1+\alpha_2)}{\pi\sqrt{E_x E_y}}\frac{\cos\varphi\cos\vartheta + \sqrt{E_x/E_y}\sin\varphi\sin\vartheta}{rD(\vartheta)},$$

$$\sigma_{\vartheta\vartheta} = \tau_{r\vartheta} = 0.$$

The above problem was solved by M. Sokolovski [3.46] and S. Lekhnitzky [5.3]. H. D. Conway [3.47] solved the similar problem for the plate whose direction of orthotropy created a certain angle with the edge.

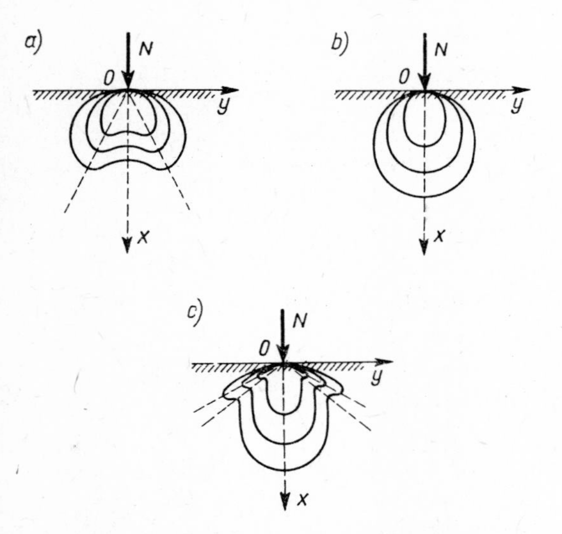

Fig. 70. Various types of stress distribution in an orthotropic sheet (S. Lekhnitzky [5.3]).

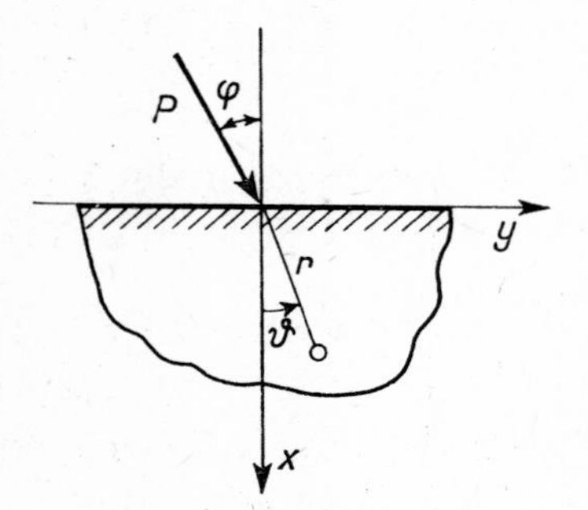

Fig. 71. Semi-infinite sheet subjected to an force of the slope φ.

5.7 Plates of various shapes under loads in their plane

Several problems of plates loaded at the edges by concentrated forces acting in their planes have been solved in the case of finite dimensions of the plates. For example, Filon [5.4] and Girkmann [5.2] considered an infinitely long plate of width $2b$ loaded by a pair of oppositely directed forces acting along one straight line and perpendicular to the edges. Expressing the stress function by means of formula (3.123) and fulfilling the boundary conditions, we obtain the stresses in the form of the complex Fourier integrals. Some of them could be evaluated analytically. The problem when these forces are acting towards each other was solved by Filon [5.4] and in another way by Seewald [5.5]. The distribution of the tangential stresses in the cross-sections of the plate was particularly examined. In the cross-section where the shear force is equal to zero the tangential stresses have no resultant. When the forces are situated one near another, the tangential stresses reach extremum values near the edge. With the increase of the distance of the forces between each other the distribution of the tangential stresses becomes very similar to the parabolic distribution, such as results from elementary theory.

For a plate loaded by a concentrated force at its edge (Fig. 72) and made of a material with the linear law of plastic hardening, the solution was found by Shevchenko [5.6]. It turns out that the distribution of the stresses in this case is the same as that in the exclusively elastic state, both in the plastic and elastic area. Inside the circle of diameter $d = 2P/\pi\sigma_{pl}$ a plasticized area appears, outside the circle the plate remains elastic (Fig. 72).

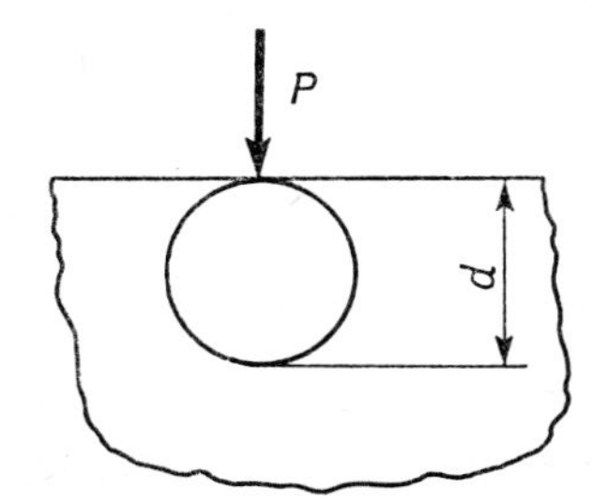

Fig. 72. Plastic region in semi-infinite sheet subjected to normal force at the edge.

The distiibution of the stresses in the circular plate (Fig. 73) loaded by two oppositely directed forces acting along the diameter can easily be obtained by a simple superposition of the radial distributions caused by each force separately. At any point A of the circle we obtain the constant compression in two mutual perpendicular directions r_1 and r_2 equal to

$$\frac{2P}{\pi}\frac{\cos\vartheta_1}{r_1} = \frac{2P}{\pi}\frac{\cos\vartheta_2}{r_2} = \frac{2P}{\pi d}.$$

In order to fulfil the condition that the edge of the plate is unloaded and has no stresses, it is sufficient to add the uniform tension in the plane of the plate of magnitude $2P/\pi d$ to the above two simple radial stress distributions. The stress in the horizontal diametral section is

$$\sigma_{yy} = -\frac{4P\cos^3\vartheta}{\pi rh} + \frac{2P}{\pi dh} = \frac{2P}{\pi dh}\left[1 - \frac{4d^4}{(d^2+4x^2)^2}\right].$$

The maximum compressive stress σ_{yy} is at the centre of the plate:

$$\sigma_{\max} = -\frac{6P}{\pi d}.$$

At the end of the diameter B–C the stresses σ_{yy} vanish.

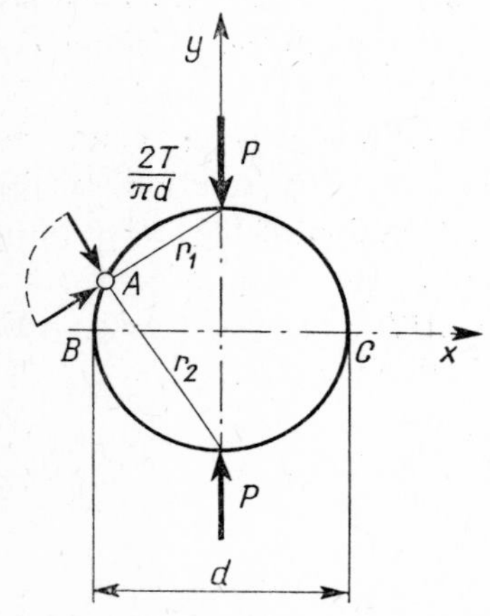

Fig. 73. Circular sheet subjected to two oppositely directed concentrated forces.

Other cases of circular plates loaded by concentrated forces can be treated in a similar way. The solution for the case of two equal and opposite forces acting along a chord and other cases can be found in the fundamental work by S. Timoshenko [5.7]. These problems were solved first by H. Hertz [5.8]. and J. H. Michell [5.9].

Certain new experimental results concerning the above-discussed problem are presented in the Appendix (p. 535).

References 5

[5.1] K. Girkmann: Angriff von Einzellasten in der vollen Ebene und in der Halbebene. *Ing.-Arch. 11* (1940) 415.

[5.2] K. Girkman: Angriff von Einzellasten in der streifenförmigen Scheibe, *Ing.-Arch. 13* (1943) 273.

[5.3] S. Lekhnitzky: Certain cases of plain problems of the theory of elastic anisotropic body, *Sb. eksperimentalnykh metod opredelenia napriazhenii i deformatsii v uprugoi i plasticheskoi zonakh 1* (1935) (in Russian).

[5.4] L. N. G. Filon: On an approximate solution of the bending of a beam of rectangular section, *Phil. Trans. Roy. Soc. Load.*, (*A*) *201* (1903) 63.

[5.5] F. Seewald: Die Spannungen und Formänderungen von Balken mit sechseckigen Querschnitt, *Abh. a. d. Aerodyn. Inst. a.d. Tech. Hochschule Aachen 7* (1927).

[5.6] K. N. Shevchenko: A concentrated force acting on a semifinite body, *Prikl. Mat. Mekh. 12*, *4* (1948) (in Russian).

[5.7] S. Timoshenko: *Theory of elasticity*, New York–London 1934.

[5.8] H. Hertz, *Z. Math. Phys. 28* (1883).

[5.9] J. H. Michell: *Proc. Lond. Math. Soc. 32* (1900), 44; *34* (1901) 134.

[5.10] R. E. Glover: Discussion of reference, 134, *J. Appl. Mech. Trans. ASME 72* (1950) 218.

[5.11] F. C. Monge, *Estado de Barra Sometida a Esfuezo Axial Soldata a una Placa Semi-indefinida*, Instituto Tecnico de la Construccion y Edificacion, Publicacion No 59.

[5.12] E. Melan, Der Spannungszustand der durch eine Einzelkraft in Innern beanspruchten Halbscheibe, *ZAMM 12* (1932) 343.

6

Plates under lateral loads and loads in the middle plane

6.1 Large elastic deflections of a plate

We shall consider now a plate subjected to a concentrated load and examine its behaviour for the case in which the deflections are no longer small in comparison with the thickness but are still small as compared to other dimensions. The plate under concentrated lateral load was already studied in Chapter 3. There, however, we worked with linear theory which was correct when the deflections of the plate were small in comparison to its thickness. The deflections being moderately large, the complete equations (2.1), (2.8) (taking into account the action of the forces in the plane of the plate) should be used. It can be noticed that a plate with large deflections works like a shell and the effects of membrane forces are then important. Equations (2.1) are the non-linear system of differential equations. Generally the exact solution of such a system is unknown and only possible in a few cases. Only approximate methods such as those due to Ritz or Galerkin, or a perturbation method, or the method of finite differences, etc. are successful. The solution consists then in the assumption of a deflection of the plate in the form of a series of functions in which a certain number of undefined parameters appears. These parameters can be found, for example, by means of the principle of virtual displacements. To solve the problem we shall use Galerkin's method where the main idea can be described in brief as follows. Let us imagine to have found approximate expressions for the deflection w in the form of a series

$$w = \sum_{i=1}^{\infty} f_i w_i(x, y) \tag{6.1}$$

satisfying the given boundary conditions. The series contains a certain number of unknown parameters f_i. These parameters can be determined in the following way. From the second plate equation

$$\frac{1}{Eh} \Delta^2 \Phi = \frac{1}{2} L(w, w),$$

we find the expression for Φ. Then we transform Eq. $(2.1)_1$ into

$$R = D\Delta^2 w - L(w, \Phi) - q = 0. \tag{6.2}$$

The effect of transverse shear deformation is neglected here. Now, let us substitute in Eq. (6.2) the assumed expressions for w and Φ. If these expressions were exact, Eq. (6.2) would be satisfied accurately at every point of the surface of the plate. As a result we would obtain the identity $0 = 0$. As the functions w and Φ do not satisfy Eq. (6.2) exactly, its left-hand side is not equal to zero. There remains a residue $R(x, y)$ which can be interpreted as a certain load distributed over the surface of the plate and being the result of an error of the assumed solution. Let us apply the principle of the virtual displacements and calculate the work of this load on the arbitrary virtual displacement δw over the surface of the plate. The plate is in equilibrium if

$$D \iint_S [\Delta^2 w - L(w, \Phi) - q]\, \delta w \, dx\, dy = \iint_S R\, \delta w \, dx\, dy = 0. \tag{6.3}$$

Let us express the variations of the deflection w as variations of parameters f_i. As variations of f_i are independent, Eq. (6.3) will be satisfied if the equations below are satisfied

$$\iint_A R(x, y) w_i \, dx\, dy = 0. \tag{6.4}$$

On evaluation of the integrals (6.4) for every f_i we obtain the set of algebraical equations, from which we find the parameters f_i.

6.2 Circular plate

Let us consider a plate supported at its edge and assume that the points lying on the circumference can move freely (Fig. 74). Let us neglect the effect of the displacements caused by the shear forces. According to classical linear theory, the solution has the form

$$w = f\left(1 - \frac{r^2}{a^2} + 2\frac{r^2}{a^2}\ln\frac{r}{a}\right), \tag{6.5}$$

where f is the deflection of the plate at its centre. We assume that the solution of the non-linear equations (6.2) has an analogous form. Then, only the parameter f is to be determined. Substituting the assumed solution into Eq. $(2.1)_2$ we can calculate the function Φ. In the symmetrical case, Eq. $(2.1)_2$ can easily be integrated once. We have

$$\frac{\mathrm{d}}{\mathrm{d}r}(\Delta\Phi) = -\frac{Eh}{2r}\left(\frac{\mathrm{d}w}{\mathrm{d}r}\right)^2. \tag{6.6}$$

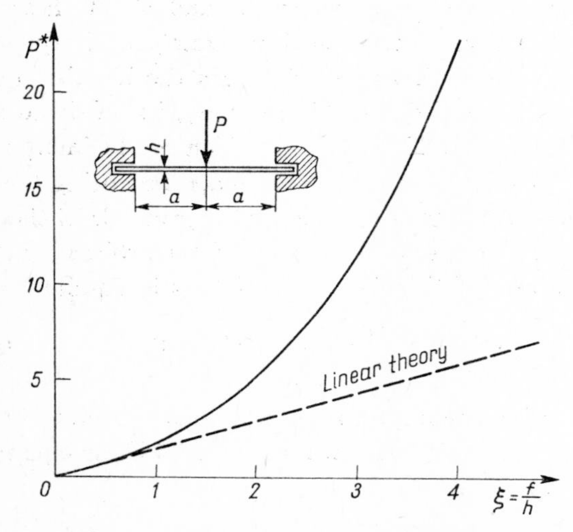

Fig. 74. Deflection at the centre of a circular plate loaded by a concentrated force resulting from linear and non-linear theory.

On substitution of (6.5) into (6.6) we have

$$\frac{\mathrm{d}}{\mathrm{d}r}(\Delta\Phi) = -\frac{8Ehf^2}{a^3}\frac{r}{a}\ln^2\frac{r}{a}.$$

Evaluating this equation we use the following recursion formula:

$$\int \varrho^m \ln^n \varrho \,\mathrm{d}\varrho = \frac{\varrho^{m+1}\ln^n \varrho}{m+1} - \frac{n}{m+1}\int \varrho^m \ln^{n-1}\varrho \,\mathrm{d}\varrho. \tag{6.7}$$

On evaluation we have

$$\frac{d\Phi}{dr} = \frac{4Ehf^2}{a}\left(\frac{r^3}{4a^3}\ln^2\frac{r}{a} - \frac{3}{8}\frac{r^3}{a^3}\ln\frac{r}{a} + \frac{7}{32}\frac{r^3}{a^3}\right) +$$

$$+ C_1\frac{r}{2} + C_2\frac{1}{r}. \tag{6.8}$$

At the edge $r = a$, the normal forces $N_{rr} = 0$; then

$$\frac{1}{r}\frac{d\Phi}{dr} = 0 \qquad (r = a).$$

For $r = 0$ the displacement $u = \frac{1}{Eh}\left(r\frac{d^2\Phi}{dr^2} - \nu\frac{d\Phi}{dr}\right) = 0$. From the above condition we have

$$C_1 = \frac{7}{4}E\frac{f^2}{a^2}; \qquad C_2 = 0. \tag{6.9}$$

In the symmetrical case, Eq. (6.2) can also be evaluated once. We have

$$D\frac{d}{dr}(\Delta^2 w) = \frac{1}{r}\int_0^r qr\,dr - \frac{1}{r}\frac{d\Phi}{dr}\frac{dw}{dr}. \tag{6.10}$$

Substituting yields

$$R = D\frac{8f}{a^2 r} - \frac{P}{2\pi r} - 4E\frac{hf^3}{a^3}\left(\frac{r^3}{a^3}\ln^3\frac{r}{a} - \frac{3}{2}\frac{r^3}{a^3}\ln^2\frac{r}{a} +\right.$$

$$\left. + \frac{7}{8}\frac{r^3}{a^3}\ln\frac{r}{a} - \frac{7}{8}\frac{r}{a}\ln\frac{r}{a}\right) r\ln\frac{r}{a}. \tag{6.11}$$

On substitution in (6.4) and integrating we have [6.1]

$$\frac{191}{648}\frac{Ehf^3}{D} + 16f = \frac{Pa^2}{\pi D}. \tag{6.12}$$

Let us introduce the non-dimensional magnitudes $\xi = f/h$; $P^* = Pa^2/Eh^4\pi$; then for $\nu = 0.3$ we have $0.294\xi^3 + 1.47\xi = P^*$. This relation is presented in Fig. 74. We observe that when the deflection is equal to the thickness of the plate, the rigidity is 1.21 times larger than that resulting from linear theory. The stresses in the plane plate can be found from (2.6).

At the centre we have the membrane stresses

$$\sigma_{rr}^{*} = \sigma_{\vartheta\vartheta}^{*} = \frac{7}{8}\,\xi^{2}.$$

At the edge ($r = a$)

$$\sigma_{rr}^{*} = 0; \qquad \sigma_{\vartheta\vartheta}^{*} = \frac{1}{4}\,\xi^{2}, \quad \text{where} \quad \sigma^{*} = \frac{\sigma a^{2}}{Eh^{2}}.$$

The bending stresses are

$$\begin{aligned} \sigma_{rr} &= -\frac{2Ehf}{(1-\nu^{2})a^{2}}\left(1+\ln\frac{r}{a}+\nu\ln\frac{r}{a}\right), \\ \sigma_{\vartheta\vartheta} &= -\frac{2Ehf}{(1-\nu^{2})a^{2}}\left(\nu+\ln\frac{r}{a}+\nu\ln\frac{r}{a}\right). \end{aligned} \tag{6.13}$$

The above formulae give infinitely large stresses at the plate centre, i.e. at the point of action of the concentrated force.

Assuming the load P to be distributed uniformly over a circular area with a small radius $r = c$, we can use a simple relation existing in plates with small deflections between the stresses at the centre of such an area, and the stresses caused at the point $r = 0$. Applying this relation to the plate with large deflections, we obtain, at the centre of the loaded area with a radius c, approximately

$$\sigma_{rr} = \sigma_{\vartheta\vartheta} = -\frac{2Ehf}{(1-\nu^{2})a^{2}}\left[(1+\nu)\ln\frac{c}{a}+1\right]+\frac{3}{2}\,\frac{P}{\pi h^{2}}. \tag{6.14}$$

The stresses at the edge are

$$\sigma_{rr}^{*} = -\frac{2}{1-\nu^{2}}\,\xi, \qquad \sigma_{\vartheta\vartheta}^{*} = -\frac{2\nu}{1-\nu}\,\xi.$$

The problems of plates with different support at the edges can be solved in the same way, i.e. assuming that a function for the deflection w satisfies the given boundary conditions. The results of the calculations for the circular plate with an immovable built-in edge, and a movable edge are shown in Table 3.

The following notations were used:

$$\begin{aligned} &A\xi^{3}+B\xi = P, \\ &\sigma_{rr_{m}}^{*} = k_{1}\,\xi^{2}, \qquad \sigma_{\vartheta\vartheta_{m}}^{*} = k_{2}\,\xi^{2}, \\ &\sigma_{rr_{b}}^{*} = k_{3}\,\xi, \qquad \sigma_{\vartheta\vartheta_{b}}^{*} = k_{4}\,\xi. \end{aligned} \tag{6.15}$$

Table 3.

	Boundary conditions	A	B	Center $k_1 = k_2$	k_1	Edge k_2	k_3	k_4
Plate clamped	Edge immovable	0.651	1.47	1.232	0.357	0.107	−2.198	−0.659
	Edge free to move	0.294	1.47	0.875	0	−0.250	−2.198	−0.659
Plate simply supported	Edge immovable	0.825	0.577	0.895	0.488	0.147	0	0.606
	Edge free	0.157	0.577	0.407	0	−0.341	0	0.606

The solution for the rectangular plates and strip plates with large deflections and various boundary conditions can be found in [6.3]. Also the book by A. S. Volmir [6.1] is devoted to solutions of these problems. All values contained in Table 3 are taken from this book.

6.3 Stability of plates

In the previous chapters we discussed the distribution of the stresses and displacements in plates loaded by concentrated forces. But these loads may also produce lateral buckling of thin-walled structures. Let us discuss briefly the general methods of calculation of the critical loads in plates.

First, we introduce certain notations. We assume that at the moment before the buckling the plate is in the state denoted by an index zero. Thus:

$$w = 0, \quad u = u_0, \quad v = v_0, \quad \Phi = \Phi_0, \quad \text{etc.} \tag{6.16}$$

After buckling we have

$$w = w, \quad u = u_0 + u_1, \quad v = v_0 + v_1, \quad \Phi = \Phi_0 + \Phi_1, \quad \text{etc.} \tag{6.17}$$

The values with the index 1 result from the deflection w which appears after buckling. The function Φ_1 satisfies the equation

$$\frac{1}{Eh}\Delta\Delta\Phi_1 = \frac{1}{2}L(w, w). \tag{6.18}$$

The stresses resulting from the stress function Φ_0 are called the *stresses of the basic state*. The stress function Φ_0 satisfies the equation

$$\Delta\Delta\Phi_0 = 0. \tag{6.19}$$

We usually apply two methods to solve the problem. One consists of the satisfaction of the differential equations (2.1). The second — the energy method — follows the variational principle of virtual displacements.

Let us discuss the first method.

6.3.1 *Method of differential equations*

The equations of equilibrium and compatibility make a set with two unknown functions w and Φ, which describe the behaviour of the plate subjected to the external loads. We have

$$\begin{aligned} D\Delta\Delta w &= N_{xx}\frac{\partial^2 w}{\partial x^2}+N_{yy}\frac{\partial^2 w}{\partial y^2}+2N_{xy}\frac{\partial^2 w}{\partial x\,\partial y}, \\ \frac{1}{Eh}\Delta\Delta\Phi &= \frac{1}{2}L(w,w). \end{aligned} \tag{6.20}$$

The solution to the problem consists in the determination of these two functions. But the above set of equations is non-linear with respect to the functions w and Φ. Usually, the exact solution is unknown. If our aim is to calculate the critical load only, we can assume that after buckling the plate deflects only to a small degree under the loads in its plane. We have after buckling

$$N_{xx} = N_{xx_0}+N_{xx_1},$$

where N_{xx_0} is found by solving (6.19).

Since N_{xx_1} can be an arbitrarily small value, we can neglect it. Then we obtain the linear equation

$$D\Delta\Delta w = N_{xx_0}\frac{\partial^2 w}{\partial x^2}+N_{yy_0}\frac{\partial^2 w}{\partial y^2}+2N_{xy_0}\frac{\partial^2 w}{\partial x\,\partial y}. \tag{6.21}$$

This simplified equation is independent of the second equation of (6.19) and is sufficient for the determination of the critical load.

6.3.2 *Energy method*

The variational principle of virtual displacements gives the following condition:

$$\delta(U-L) = 0, \tag{6.22}$$

where U is the total strain energy of the plate and L is the work done by the external loads applied at the edges and surfaces of the plate. The strain energy U is given by Eqs. (2.18), (2.19) and consists of the bending energy and the energy due to the strains produced in the middle plane:

$$U = U_1 + U_2.$$

But as we remember (see Eq. (2.20)), the last energy U_1 can be represented in the form

$$U_1 = L + \frac{1}{2}\iint \left[N_{xx}\left(\frac{\partial w}{\partial x}\right)^2 + N_{yy}\left(\frac{\partial w}{\partial y}\right)^2 + \right.$$

$$\left. + 2N_{xy}\frac{\partial w}{\partial x}\frac{\partial w}{\partial y}\right]\mathrm{d}x\mathrm{d}y, \text{ or} \tag{6.23}$$

$$U_1 = L + \frac{1}{2}\iint_S N_{\alpha\beta} w_{,\alpha} w_{,\beta} \mathrm{d}S, \qquad \alpha, \beta = x, y,$$

where L is the work done by the edge forces, and we have introduced the strains from Eq. $(17.1)_1$ and assumed that the plate is loaded only at the edges. Substituting (6.23) into Eq. (6.22), we obtain

$$\delta\left\{\frac{D}{2}\iint_S [(\Delta w)^2 - (1-\nu)L(w, w)]\mathrm{d}S + \frac{1}{2}\iint_S N_{0\alpha\beta} w_{,\alpha} w_{,\beta} \mathrm{d}S\right\} = 0. \tag{6.24}$$

This condition due to G. H. Bryan [6.4] contains the membrane forces of the basic state. Generally, these forces result from the solution of Eq. (6.19).

In the case of complicated edge loads, for example concentrated forces, this solution is usually very complex and not always possible. But the solution of the problem can be derived in the other way.

The second integral of (6.24) can be taken in the form (6.25) by introducing the relations

$$\frac{1}{2}\left(\frac{\partial w}{\partial x}\right)^2 = \varepsilon_{x_1} - \frac{\partial u_1}{\partial x}, \qquad \frac{1}{2}\left(\frac{\partial w}{\partial y}\right)^2 = \varepsilon_{y_1} - \frac{\partial v_1}{\partial y},$$

$$\frac{\partial w}{\partial x}\frac{\partial w}{\partial y} = \gamma_{xy_1} - \frac{\partial u_1}{\partial y} - \frac{\partial v_1}{\partial x}.$$

Let us substitute these values into Eq. (6.24), remembering that

$$\frac{1}{E}(\sigma_{xx_0}-\nu\sigma_{yy_0}) = \frac{\partial u_0}{\partial x}, \quad \frac{1}{E}(\sigma_{yy_0}-\nu\sigma_{xx_0}) = \frac{\partial v_0}{\partial y},$$

$$\frac{2(1+\nu)}{E}\tau_{xy_0} = \frac{\partial u_0}{\partial y}+\frac{\partial v_0}{\partial x}.$$

Writing

$$U_{11} = \frac{1}{2}\iint_S N_{0\alpha\beta}w_{,\alpha}w_{,\beta}\,\mathrm{d}S$$

we find that

$$U_{11} = \iint_S [u_{0\alpha,\beta}N_{1\alpha\beta}-u_{1\alpha,\beta}N_{0\alpha\beta}]\mathrm{d}S. \tag{6.25}$$

Integrating by parts, it can be shown that the expression (6.25) can be put to the following form:

$$U_{11} = \oint_C [u_{0\alpha}l_\beta N_{1\alpha\beta}-u_{1\alpha}l_\beta N_{0\alpha\beta}]\mathrm{d}s -$$

$$-\iint_S [u_{0\beta}N_{1\alpha\beta,\alpha}-u_{1\beta}N_{0\alpha\beta,\alpha}]\mathrm{d}S, \tag{6.26}$$

where $l_1 = l$, $l_2 = m$ (see (2.21)), $\alpha, \beta = 1, 2 = x, y$.

According to Eq. $(1.79)_1$ the double integrals on the right-hand side of Eq. (6.26) vanish if there are no body forces acting in the middle plane of the plate. The integrals along the boundary represent the work done by the forces $\overline{X}$, $\overline{Y}$ applied at the boundary. Then we can write

$$\delta\left\{U_2+\oint [u_0\overline{X}_1+v_0\overline{Y}_1]\mathrm{d}s-\oint [u_1\overline{X}_0+v_1\overline{Y}_0]\mathrm{d}s\right\} = 0, \tag{6.27}$$

where $\overline{X}_0$, $\overline{Y}_0$ and u_0, v_0 are the basic edge loads and displacements, respectively. $\overline{X}_1$, $\overline{Y}_1$ and u_1, v_1 are the additional edge loads and displacements appearing after buckling.

If the forces $\overline{X}_0$ and $\overline{Y}_0$ are given at the edge of the plate, the forces $\overline{X}_1$ and $\overline{Y}_1$ are equal to zero and condition (6.27) takes the form

$$\delta\left\{U_2-\oint [u_1\overline{X}_0+v_1\overline{Y}_0]\mathrm{d}s\right\} = 0. \tag{6.28}$$

If the displacements u_0, v_0 are given at the edge, the additional displacements u_1 and v_1 are equal to zero and we obtain

$$\delta\left\{U_2+\oint [u_0\overline{X}_1+v_0\overline{Y}_1]\mathrm{d}s\right\} = 0. \tag{6.29}$$

If we consider the first case, the plate loaded by the edge forces $\overline{X}_0$, $\overline{Y}_0$, then the solution requires the determination of the displacements u_1, v_1 at the edge of the plate. These displacements result from the expression $(1.71)_1$ which can be taken in the form

$$\tfrac{1}{2}(u_{1\alpha,\beta}+u_{1\beta,\alpha}) = -\frac{1+\nu}{Eh} d_\alpha^\lambda d_\beta^\mu \Phi_{1,\lambda\mu} - \frac{\nu}{Eh} a^{\lambda\mu}\Phi_{1,\lambda\mu}\delta_{\alpha\beta} - \tfrac{1}{2} w_{,\alpha} w_{,\beta}. \quad (6.30)$$

The displacements u_1, v_1 are determined by the additional stress function Φ_1 and the deflection w. It can be proved that the terms resulting from Φ_1 are of the same order as the first terms on the right-hand side of Eq. (6.30) containing the derivatives of w.

In numerous papers treating stability problems (see also [6.5]) the displacements at the edge u_1, v_1 are obtained from the assumption that the stretching of the middle plane of the plate after buckling is negligible:

$$\varepsilon_{xx_1} = \varepsilon_{yy_1} = \gamma_{xy_1} = 0.$$

We have then

$$u_{1\alpha,\beta}+u_{1\beta,\alpha} = -w_{,\alpha} w_{,\beta}. \quad (6.31)$$

It follows that the displacements u_1, v_1 at the edge can be represented by the deflection w only, and the solution of Eq. (6.19) is not necessary. In the case of a rectangular plate, we have, for example,

$$u_1 = -\frac{1}{2}\int_0^a \left(\frac{\partial w}{\partial x}\right)^2 \mathrm{d}x, \qquad v_1 = -\frac{1}{2}\int_0^b \left(\frac{\partial w}{\partial y}\right)^2 \mathrm{d}y, \quad (6.32)$$

where a and b are the dimensions of the plate in the x and y-axis directions, respectively.

Generally, this assumption is not correct, and is proper only if the condition $L(w, w) = 0$ is satisfied, i.e. if the right-hand side of Eq. (6.18) is equal to zero. This is possible only if the plate is bent to a cylindrical surface. In the case of concentrated loads this assumption is not proper and may involve certain errors.

In order to avoid solving the equation $\Delta\Delta\Phi_0 = 0$, N. A. Alfutov and L. I. Balabukh [6.6] proposed to introduce instead of real forces N_{xx_0}, N_{yy_0}, and N_{xy_0} a certain statically equivalent system of forces $N^*_{xx_0}$, $N^*_{yy_0}$ and $N^*_{xy_0}$ which satisfy the equations of equilibrium and boundary conditions, but do not satisfy the equation of compatibility. We have for an arbitrary statically equivalent system

$$\oint [u_1 \overline{X}_0 + v_1 \overline{Y}_0] \mathrm{d}s = \iint_S u_{1\alpha,\beta} N^*_{0\alpha\beta} \mathrm{d}S. \tag{6.33}$$

Using expressions (6.30), we obtain

$$\oint [u_1 \overline{X}_0 + v_1 \overline{Y}_0] \mathrm{d}s = \frac{1}{2} \iint_S N^*_{0\alpha\beta} w_{,\alpha} w_{,\beta} \mathrm{d}S -$$

$$- \frac{1}{Eh} \iint_S N^*_{0\alpha\alpha} N_{1\beta\beta} \mathrm{d}S - \frac{1+\nu}{Eh} \iint_S (N^*_{0\alpha\alpha} N_{1\beta\beta} - N^*_{0\alpha\beta} N_{1\alpha\beta}) \mathrm{d}S. \tag{6.34}$$

It can be shown by integrating by parts that the last integral on the right-hand side of Eq. (6.34) is equal to zero. Since $-N_{i\alpha\alpha} = \Delta\Phi_i$, we obtain the condition

$$\delta \left\{ U_2 + \frac{1}{2} \iint_S N^*_{0\alpha\beta} w_{,\alpha} w_{,\beta} \mathrm{d}S - \frac{1}{Eh} \iint_S \Delta\Phi^*_0 \Delta\Phi_1 \mathrm{d}S \right\} = 0, \tag{6.35}$$

where U_2 is the bending energy (2.18). If

$$N^*_{xx_0} = N_{xx_0}, \quad N^*_{yy_0} = N_{yy_0}, \quad N^*_{xy_0} = N_{xy_0},$$

this condition becomes identical with the previous one (6.24). Using it, is not necessary to find the solution of Eq. (6.19). We have to determine only the function $\Delta\Phi_1$ from Eq. (6.18) which in the case of concentrated loads is usually much simpler than solving Eq. (6.19).

6.4 Rectangular plate loaded by compressive forces

Let us consider the problem shown in Fig. 75 as the first example. The plate loaded by a pair of compressive forces can buckle when these forces exceed a certain crtical value. Let us assume that the dimension b is considerably smaller than the second dimension, the length of the plate. Thus we can consider the plate infinitely long in the x-axis direction. The edges $y=0$ and $y=b$ are jointly supported. Such a problem was solved approximately by A. Sommerfeld [6.7] using the linearized Eq. (6.21). We follow his treatment.

Due to the symmetry with respect to the y-axis we consider only the area for $x \geqslant 0$. Let us separate from this area the strip of ε width along the direction of the action of the forces P and assume that the forces N_{yy_0} remain unchanged during bending. For this area $0 < x < \varepsilon$, Eq. (6.21) is valid. Assuming

$$N_{yy_0} = -\frac{P}{2\varepsilon}, \qquad N_{xy_0} = N_{xx_0} = 0,$$

we find (by neglecting the effect of transverse shear deformation)

$$D\Delta\Delta w + \frac{P}{2\varepsilon}\frac{\partial^2 w}{\partial y^2} = 0.$$

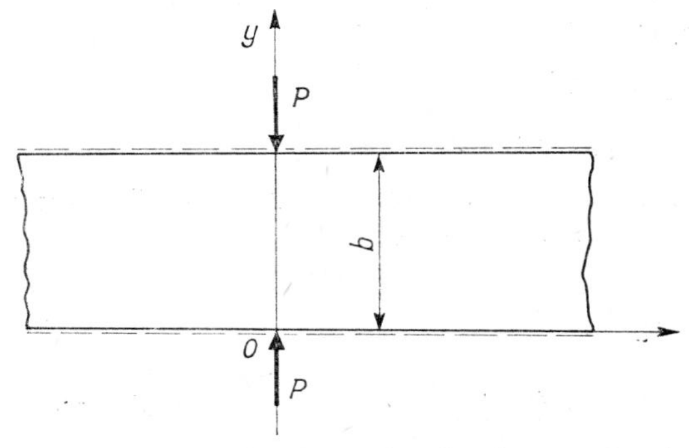

Fig. 75. Strip plate simply supported at the edges $y = 0$ and $y = b$, subjected to two compressed forces P.

Integrating with respect to x with $\partial w/\partial x = 0$ for $x = 0$ and taking $\varepsilon \to 0$, we find the equation

$$D\frac{\partial^3 w}{\partial x^3} + \frac{P}{2}\frac{\partial^2 w}{\partial y^2} = 0. \tag{6.36}$$

In the area $x > 0$, $0 \leqslant y \leqslant b$, Eq. (6.21) takes the form

$$\Delta\Delta w = 0. \tag{6.37}$$

The boundary conditions are: for $y = 0$ and $y = b$

$$w = 0, \qquad \frac{\partial^2 w}{\partial y^2} = 0,$$

for $x = 0$

$$\frac{\partial w}{\partial x} = 0 \quad \text{(due to symmetry)},$$

and for $x \to \infty$

$$w = 0, \qquad \frac{\partial w}{\partial x} = 0.$$

Let us assume that in the area for $x > 0$ the deflection is given by the equation

$$w = w(x)\sin\frac{\pi y}{b}. \tag{6.38}$$

Substituting this into Eq. (6.37) we have

$$\frac{d^4 w(x)}{dx^4} - 2\left(\frac{\pi}{b}\right)^2 \frac{d^2 w(x)}{dx^2} + \left(\frac{\pi}{b}\right)^4 w(x) = 0. \tag{6.39}$$

The solution of the above equation is

$$w(x) = (A+Bx)\,e^{-\pi x/b} + (C+Dx)\,e^{\pi x/b}.$$

From the condition $w \to 0$ for $x \to \infty$ we have $C = D = 0$. The condition of symmetry gives $B = \pi A/D$. Finally we have

$$w = A\left(1+\frac{\pi x}{b}\right)e^{-\pi x/b}\sin\frac{\pi y}{b}.$$

Substituting this expression into Eq. (6.36), we obtain the critical force

$$P_{cr} = \frac{4\pi D}{b}. \tag{6.40}$$

Let us consider now a plate of finite length a and use the energy method. An approximate solution can be obtained by taking the following series for the deflection of the buckled plate:

$$w = \sin\frac{\pi y}{b} \sum_{m=1,3,5,\dots}^{\infty} w_m \sin\frac{m\pi x}{a}. \tag{6.41}$$

The expression for the strain energy of bending becomes

$$U_b = \frac{abD}{8} \sum_{m=1,3,5,\dots}^{\infty} w_m^2 \left(\frac{m^2\pi^2}{a^2} + \frac{\pi^2}{b^2}\right)^2.$$

If we assume that the stretching of the middle plane of the plate is negligible, the work done by the compressive force P during buckling takes the form

$$L = \frac{P}{2}\int_0^b \left(\frac{\partial w}{\partial y}\right)^2 dy\bigg|_{x=a/2} = \frac{\pi^2 P}{4b}(w_1 - w_3 + w_5 - \dots)^2.$$

Applying the principle of virtual displacements, we give a variation δw to the deflection w. We can express the variation of the deflection w as the variations of the coefficients w_m. Then, equating the work to the strain energy, we obtain

$$P_{cr} = \frac{\pi^2 ab^2 D}{2} \frac{\sum\limits_{m=1,3,5,\dots}^{\infty} w_m^2 \left(\frac{m^2}{a^2} + \frac{1}{b^2}\right)^2}{(w_1 - w_3 + w_5 - \dots)^2}. \tag{6.42}$$

We are looking for the minimum value of the critical force P, so equating to zero the derivatives of this expression with respect to each coefficient w_m we obtain a system of linear algebraic equations. On solving it we find

$$P_{cr} = \frac{\pi^2 ab^2 D}{2} \frac{1}{\sum\limits_{n=1,3,5,\dots}^{\infty} \frac{1}{\left(\frac{m^2}{a^2} + \frac{1}{b^2}\right)^2}}. \tag{6.43}$$

The above series can easily be summed. We obtain the following numerical results:

$$P_{cr} = \alpha_c \frac{\pi D}{b}, \tag{6.44}$$

where the values of α_c for various a/b are given in Table 4.

Table 4.

$\alpha_c \quad \frac{a}{b} =$	0.5	0.75	1	1.5	2	3	∞
S. Timoshenko [6.5] Eq. (6.43)	18.80	9.00	6.00	4.46	4.11	4.02	4.0
N. Yamaki [6.8]	17.68	11.96	8.17	5.27	4.73	4.6	

We see that for the increasing ratio a/b the sum approaches the limiting value 4 for the infinitely long plate. If the sides $y = 0$ and $y = b$ of the plate are clamped, taking the deflection in the form of the similar series satisfying the boundary conditions, we obtain for the long plate

$$P_{cr} = \frac{8\pi D}{b}. \tag{6.45}$$

The above problems were solved with the simplified assumption that the middle plane of the plate was non-deformable after buckling. Some papers have recently been published based on condition (6.24) where the true stresses are determined. The authors, using computers, obtained different values of the critical load (usually larger than the results from the previous simplified calculations).

In order to compare the results obtained assuming the simplified and exact conditions (6.31), (6.30), let us consider the square plate loaded by four concentrated forces as shown in Fig. 76. We apply here the method due to

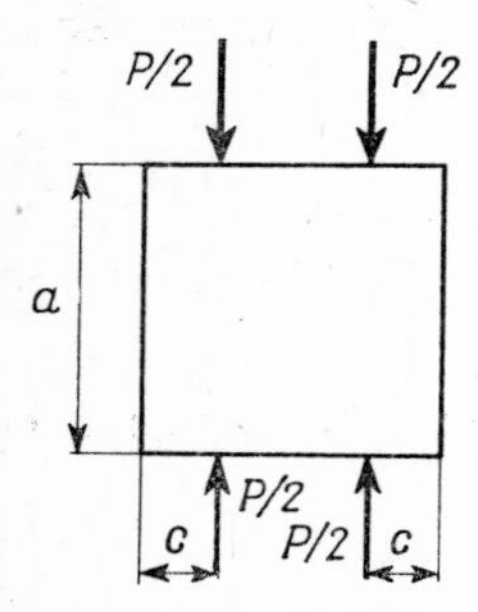

Fig. 76. Square plate simply supported at the edges loaded by four concentrated forces.

N. A. Alfutov and L. I. Balabukh [6.6]. The plate is simply supported, the boundary conditions are as follows:

$$w = 0, \quad \frac{\partial^2 w}{\partial x^2} = 0 \quad \text{for} \quad x = 0,\ x = a,$$

$$w = 0, \quad \frac{\partial^2 w}{\partial y^2} = 0 \quad \text{for} \quad y = 0,\ y = 0.$$

The following boundary conditions are for the stress function:

$$\frac{\partial \Phi_1}{\partial x} = \frac{\partial \Phi_1}{\partial y} = 0.$$

We take the expressions (6.46) for the deflection w and the stress function Φ_1.

$$w = f_1 \sin\frac{\pi x}{a} \sin\frac{\pi y}{b}, \quad \Phi_1 = C_1 \sin^2\frac{\pi x}{a} \sin^2\frac{\pi y}{b}. \tag{6.46}$$

Integrating Eq. (6.18) by means of Galerkin method, we find $C_1 = -f_1^2 E/4$.

According to Eq. (2.18) the strain energy of bending is $U_b = \pi^2 f_1 D/2a^2$. The statically equivalent state of stress can be taken as follows:

$$N^*_{yy_0} = -\frac{P}{2\varepsilon} \quad \text{for} \quad (c-\varepsilon/2) < x < (c+\varepsilon/2),$$

$$N^*_{yy} = 0 \quad \text{for} \quad 0 \leqslant x \leqslant (c-\varepsilon/2),\ (c+\varepsilon/2) \leqslant x \leqslant a/2,$$

$$N^*_{xx_0} = 0, \quad N^*_{xy_0} = 0.$$

Due to symmetry we need consider only half of the plate. On integration we find

$$\Delta\Phi_1 = C_1\left(\frac{\pi}{a}\right)^2\left(\cos\frac{2\pi x}{a} - 2\cos\frac{2\pi x}{a}\cos\frac{2\pi y}{a} + \cos\frac{2\pi y}{a}\right).$$

Equation (5.35) yields

$$\frac{\pi^4}{a^2}\frac{f_1 D}{2} - \frac{\pi^2 f_1^2 P_{cr}}{4a}\sin^2\frac{\pi c}{a} - \frac{\pi^2 f_1^2 P_{cr}}{16a}\cos\frac{2\pi c}{a} = 0.$$

Then

$$P_{cr} = \frac{4\pi^2 D}{a}\cdot\frac{1}{1-\frac{1}{2}\cos(2\pi c/a)}; \tag{6.47}$$

for $c = \frac{1}{2}a$ we find

$$P_{cr} = \frac{4\pi^2 D}{a}\cdot\frac{2}{3} = \frac{\pi D}{a}\cdot\frac{8\pi}{3} = 8.37\frac{\pi D}{a}. \tag{6.48}$$

The displacement at the edge resulting from Eq. (6.30) is

$$v_1 = \frac{\pi^2 f_1^2}{8a}\left(1 - \frac{1}{2}\cos\frac{2\pi x}{a}\right). \tag{6.49}$$

Using Eq. (6.49), we find the same result as (6.47).

Assuming the simplified condition (6.31) and calculating the displacement v_1 from Eq. (6.32), we obtain

$$v_1 = \frac{\pi^2 f_1^2}{8a}\left(1 - \cos\frac{2\pi x}{a}\right), \tag{6.50}$$

and the critical load

$$P_{cr} = \frac{4\pi^2 D}{a}\frac{1}{1-\cos(2\pi c/a)} \tag{6.51}$$

with $c = \frac{1}{2}a$, $P_{cr} = 2\pi^2 D/a = 6.28\pi D/a$.

If $c \to 0$, we obtain $P_{cr} \to \infty$ which is impossible. We observe that the results obtained by using the statically equivalent system differ considerably from the previous ones due to S. Timoshenko [6.5], A. R. Filipov [6.9], A. I. Lure [6.10], where the coefficients $\alpha_c = 6.00$ (see Table 4 on p. 267) was obtained instead of $\alpha_c = 8.37$. This difference results from the assumption of the simplified condition of a non-deformable middle plane in the earlier papers. The exact numerical computations performed in [6.8] and [6.11] gave the values $\alpha_c = 8.17$ and $\alpha_c = 8.5$.

6.5 Stability of circular plate

The problem of the stability of the circular plate subjected to two compressive forces acting along a diameter was solved by M. Rózša [6.13]. The plate was free at the edges. On the basis of the iteration method by H. A. Schwarz [6.14] the minimum magnitude of the critical load is determined by the formula

$$P_{cr} = \frac{w^{(k)}}{w^{(k+1)}},$$

where $w^{(k)}$ and $w^{(k+1)}$ are the successive approximate solutions-iterations, for the deflection w satisfying the differential equation of the plate

$$D\Delta\Delta w^{(k+1)} = \tfrac{1}{2}L(w^{(k)}, \Phi). \tag{6.52}$$

The stress function Φ represented by the Fourier series, in polar coordinates (r, ϑ) was defined in such a way as to satisfy the boundary conditions. At the edge the internal forces N_{rr} were equilibrated to the external load at two points $\vartheta = 0$ and $\vartheta = \pi/2$. On the assumption of the deflection $w^{(k)}$ in the form of a similar expression and solving Eq. (6.52), the deflection of the plate $w^{(k+1)}$ was obtained. In the numerical example treated, two iterations were made and it was proved that it was adequate to take four terms of the series for Φ in order to achieve the sufficient accuracy 1 per cent. For $\nu = 0.25$ the following result was obtained:

$$P_{cr} = 3.36\frac{D}{R}.$$

References 6

[6.1] A. S. Volmir: *Elastic plates and shells*, Gostekhizdat, Leningrad–Moskva 1956 (in Russian).

[6.2] A. S. Volmir: *Stability of deformable systems*, Izd. "Nauka", Moskva 1967 (in Russian).

[6.3] M. S. KORNISHIN, F. S. ISANBAEVA: *Elastic plates and shells*, Izd. "Nauka", Moskva 1958 (in Russian).
[6.4] G. H. BRYAN: *Proc. Lond. Math. Soc. 22* (1891) 54.
[6.5] S. TIMOSHENKO: *Theory of elastic stability*, NY, 1936; I. Gere, *Theory of elastic stability*, NY, 1961.
[6.6] N. A. ALFUTOV, L. I. BALABUKH: On the possibility to solve the problems of the stability of plates without determination of the initial state of stress, *Prikl. Mat. Mekh. 4* (1967) 716–722 (in Russian).
[6.7] A. SOMMERFELD: *Z. Math. Phys. 54* (1907) 113–121.
[6.8] N. YAMAKI: *Rep. Inst. High Speed. Mech. Tôhoku Univ. 4* (1954).
[6.9] A. R. FILIPOV: *Izv. AN SSSR, Otdel Mat. i Estestvennykh Nauk 7* (1933).
[6.10] A. J. LURE: *Trudy Leningradsk. Indust. In-ta 3* (1939).
[6.11] F. A. ROMANENKO: Stability of the plane form of equilibrium of nonuniform plates under action of concentrated loads, *Prikl. Mekh. Otd. Matem. i Kibernetiki AN SSSR 2, 1* (1966) (in Russian).
[6.12] A. V. ALEKSANDROV, B. I. LASHCHENIKOV: On the application of an energetic method to stability problems of elastic systems, *Stroit. Mekh. i Raschet Sooruzhenii 5* (1965) (in Russian).
[6.13] M. RÓZŠA: Stability analysis of a circular plate submitted to two compressive forces acting along a diameter, *Acta Techn. Acad. Scient. Hung. 55, No. 1–2* (1966) 153–172.
[6.14] H. A. SCHWARZ: *Gesammelte Werke*, 1, 231–265.

7

Membrane shells under concentrated loads

In many technical problems the shells are loaded in such a way that the bending stresses are small and can be neglected. Then only the membrane stresses and strains in the middle surface of the shell need be considered. These stresses can be considered as uniformly distributed across the thickness. The equations based on such assumptions are called the *equations of the membrane theory of shells*. The stresses in a thin spherical container, in which a gas or liquid is compressed, are uniform over the thickness and can serve as an example. In solving such a problem we neglect the bending rigidity. If this is admissible, depends mainly on the kind of loads and supports. Concentrated loads usually produce in the vinicity of the loading points the bending stresses and the calculations performed on the basis of the membrane theory give incorrect results. However, at a certain distance from the loading point, the bending stresses vanish and only the membrane stresses remain, which can easily be calculated by means of the membrane theory. If we neglect the bending, the problem of stress analysis is greatly simplified, since the resultant moments and resultant shearing forces in Eq. (3.9) vanished. Thus the only unknowns are the three quantities N_{11}, N_{22}, and $N_{12} = N_{21}$ which can be determined from the first three conditions of equilibrium of an element (1.79). Hence the problem becomes statically determinate.

7.1 Shells of revolution

Let us consider a shell of revolution obtained by the rotation of a plane curve $z_1 = f(r)$ around an axis lying in the plane of the curve. This generating curve is called a *meridian*. The position of the meridian is defined by an angle ϑ and the position of a parallel circle by angle φ between a normal to the shell and its axis of revolution. The main radii of curvature are here R_1 and R_2 which are measured on a normal to the meridian, and r is a radius of a parallel circle. We note from Fig. 77 that

$$r = R_2 \sin\varphi .$$

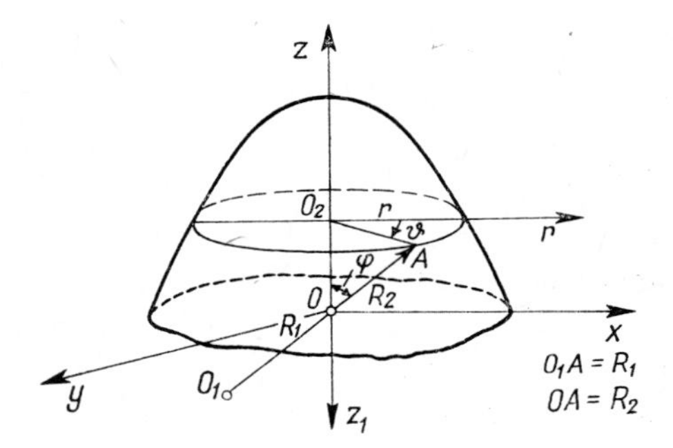

Fig. 77. Geometry of shell of revolution.

The principal coordinates α_1, α_2 which follow the lines of main curvatures of the shell are $\alpha_1 = \varphi$ and $\alpha_2 = \vartheta$. For the line element of the middle surface we have

$$\mathrm{d}s^2 = R_1^2\,\mathrm{d}\varphi^2 + r^2\,\mathrm{d}\vartheta^2, \tag{7.1}$$

and since $A_1 = R_1$, $A_2 = r$. Further we have the relations

$$\frac{\mathrm{d}r}{\mathrm{d}\varphi} = R_1 \cos\varphi, \quad \frac{\mathrm{d}z_1}{\mathrm{d}\varphi} = R_1 \sin\varphi, \quad \text{and} \quad \frac{1}{r}\frac{\mathrm{d}r}{\mathrm{d}\varphi} = \frac{R_1}{R_2}\cot\varphi .$$

Introducing the above expressions into the first three equilibrium conditions (1.79), we obtain the following equations:

$$\begin{aligned}
&\frac{\partial (rN_{\varphi\varphi})}{\partial\varphi} + R_1 \frac{\partial N_{\varphi\vartheta}}{\partial\vartheta} - R_1 N_{\vartheta\vartheta}\cos\vartheta + XrR_1 = 0,\\
&\frac{\partial (rN_{\varphi\vartheta})}{\partial\varphi} + R_1 \frac{\partial N_{\vartheta\vartheta}}{\partial\vartheta} + R_1 N_{\varphi\vartheta}\cos\varphi + YrR_1 = 0,\\
&\frac{N_{\varphi\varphi}}{R_1} + \frac{N_{\vartheta\vartheta}}{R_2} = Z.
\end{aligned} \tag{7.2}$$

It is to be noted that Eq. $(7.2)_3$ does not contain any derivatives of the unknowns. It may always be used to eliminate one of the normal forces and to reduce the problem to two differential equations.

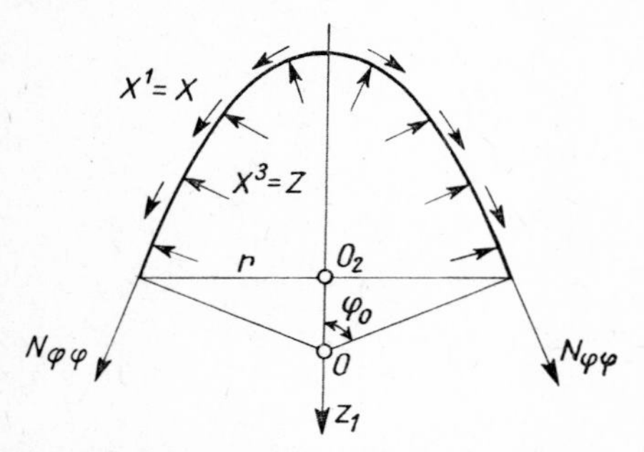

Fig. 78. Positive directions of loads in the shell of revolution.

Let us consider a symmetrical case. Then the stresses are independent of ϑ and all derivatives with respect to this coordinate vanish. We have then

$$\begin{aligned} &\frac{\mathrm{d}}{\mathrm{d}\varphi}(rN_{\varphi\varphi}) - R_1 N_{\vartheta\vartheta}\cos\varphi = -XrR_1, \\ &\frac{N_{\varphi\varphi}}{R_1} + \frac{N_{\vartheta\vartheta}}{R_2} = Z. \end{aligned} \tag{7.3}$$

Considering the equilibrium condition of a portion of a shell, we can obtain the equation which can replace Eq. $(7.3)_1$.

$$N_{\varphi\varphi} = \frac{1}{R_2 \sin^2\varphi}\int_0^{\varphi_0} R_1 R_2 (Z\cos\varphi - X\sin\varphi)\sin\varphi\,\mathrm{d}\varphi. \tag{7.4}$$

7.2 Spherical shell

In the case of a spherical shell loaded only at the edges ($X = Y = Z = 0$) the differential equations of the equilibrium (7.2) take the form

$$\begin{aligned} &N_{\varphi\varphi} = -N_{\vartheta\vartheta}, \\ &\sin\varphi\frac{\partial N_{\varphi\varphi}}{\partial\varphi} + \frac{\partial N_{\varphi\vartheta}}{\partial\vartheta} + 2N_{\varphi\varphi}\cos\varphi = 0, \\ &\sin\varphi\frac{\partial N_{\varphi\vartheta}}{\partial\varphi} - \frac{\partial N_{\varphi\varphi}}{\partial\vartheta} + 2N_{\varphi\vartheta}\cos\varphi = 0. \end{aligned} \tag{7.5}$$

The general solution of this set of equations may be written in the following form:

$$
\begin{aligned}
N_{\varphi\varphi} &= \frac{\cos n\vartheta}{\sin^2\varphi}\left[C_1\left(\tan\frac{\varphi}{2}\right)^n + C_2\left(\cot\frac{\varphi}{2}\right)^n\right] + \\
&\quad + \frac{\sin n\vartheta}{\sin^2\varphi}\left[C_3\left(\tan\frac{\varphi}{2}\right)^n + C_4\left(\cot\frac{\varphi}{2}\right)^n\right], \\
N_{\vartheta\varphi} &= \frac{\sin n\vartheta}{\sin^2\varphi}\left[-C_1\left(\tan\frac{\varphi}{2}\right)^n + C_2\left(\cot\frac{\varphi}{2}\right)^n\right] + \\
&\quad + \frac{\cos n\vartheta}{\sin^2\varphi}\left[C_3\left(\tan\frac{\varphi}{2}\right)^n - C_4\left(\cot\frac{\varphi}{2}\right)^n\right],
\end{aligned}
\tag{7.6}
$$

where n is an arbitrary integer.

Let us consider the case $n = 0$.

(a) Taking only the constants C_1 or C_2, we have

$$
N_{\varphi\varphi} = \frac{C_1}{\sin^2\varphi}, \qquad N_{\varphi\vartheta} = 0. \tag{7.7}
$$

(b) Taking only C_3 or C_4, we find

$$
N_{\varphi\varphi} = 0, \qquad N_{\varphi\vartheta} = \frac{C_3}{\sin^2\varphi}. \tag{7.8}
$$

At the poles of the sphere the factor $\sin\varphi$ in the denominator vanishes. At these points the stress resultants assume infinite values. We may imagine that these singularities correspond to the application of external forces or couples to the shell. To determine their magnitude we consider the equilibrium of the portion of the sphere in the vicinity of the pole. The vertical resultant of the forces transmitted through an arbitrary parallel circle is, in the first case,

$$
P = \int_{-\pi}^{\pi} N_{\varphi\varphi} \sin\varphi r \, \mathrm{d}\vartheta .
$$

Since $r = R\sin\varphi$,

$$
P = \int_{\pi}^{\pi} N_{\varphi\varphi} \sin^2\varphi R \, \mathrm{d}\vartheta .
$$

This equation determines the constants C_1 when P is given. We find that

$$C_1 = \frac{P}{2\pi R}$$

which gives

$$N_{\varphi\varphi} = \frac{P}{2\pi R \sin^2 \varphi}. \tag{7.9}$$

The solution obtained corresponds to the action of the normal concentrated force (Fig. 79a).

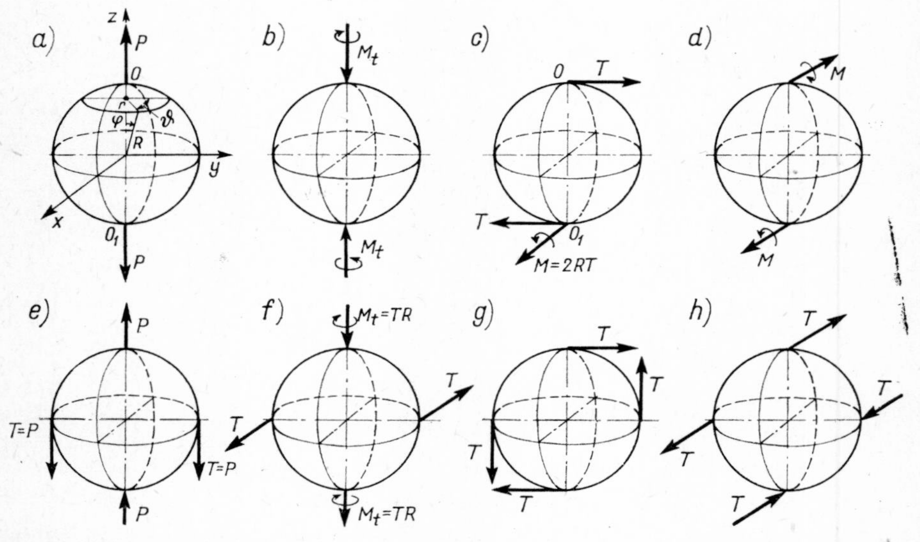

Fig. 79. Spherical shell subjected to various loads at the poles: a) *two radial forces*; b) *two twisting moments*; c) *two tangential forces and bending moment*; d) *two bending moments*; e) *two normal forces and two tangential forces at the equator*; f) *two tangential forces at the equator and two twisting moments at the poles*; g) *four tangential forces at the equator*; h) *two tangential forces at the equator and two at the poles.*

Calculating in a similar way the resultant of the stresses in the second case (b), we find

$$N_{\varphi\vartheta} = \frac{M_t}{2\pi R^2 \sin^2 \varphi}.$$

This solution corresponds to the action of the concentrated twisting moment M_t (Fig. 79b).

Now we take the case

(c) $n = 1$:

$$N_{\varphi\varphi} = C_1 \frac{\tan\frac{1}{2}\varphi \cos\vartheta}{\sin^2\varphi} = -N_{\vartheta\vartheta},$$
$$N_{\varphi\vartheta} = -C_1 \frac{\tan\frac{1}{2}\varphi \sin\vartheta}{\sin^2\varphi}. \tag{7.10}$$

Examining the equilibrium of the shell, we find the value of the constants and the magnitude of the external forces and couples which must be applied at the poles. From the antisymmetry of all stress resultants with respect to the meridian $\vartheta = 0$ it follows that the resultant force in the section $\varphi =$ constant must be perpendicular to the plane of this meridian,

$$T = \int_{-\pi}^{\pi} (N_{\varphi\varphi} \cos\varphi \sin\vartheta + N_{\varphi\vartheta} \cos\vartheta) R \sin\varphi \, d\vartheta .$$

Then we have

$$N_{\varphi\varphi} = \frac{T}{\pi R} \frac{\cos\vartheta}{\sin\varphi (1+\cos\varphi)},$$
$$N_{\vartheta\varphi} = -\frac{T}{\pi R} \frac{\sin\vartheta}{\sin\varphi (1+\cos\varphi)}. \tag{7.11}$$

This solution corresponds to the load presented in Fig. 79c. The singularities at both poles are different. In the vicinity of the pole O we have

$$\frac{N_{\varphi\varphi}}{\cos\vartheta} = -\frac{N_{\varphi\vartheta}}{\sin\vartheta} \cong \frac{T}{2\pi r},$$

and at the second pole O_1 we have

$$\frac{N_{\varphi\varphi}}{\cos\vartheta} = -\frac{N_{\varphi\vartheta}}{\sin\vartheta} \cong \frac{2TR^2}{\pi r^3}.$$

If we use the solution for C_2, we obtain the same result but the poles are displaced.

(d) By superposition we can find the solution for the case given in Fig. 79d.

$$N_{\varphi\varphi} = \frac{M}{\pi R^2} \frac{\cos\vartheta}{\sin^3\varphi},$$
$$N_{\varphi\vartheta} = \frac{M}{\pi R^2} \frac{\cos\varphi \sin\vartheta}{\sin^3\varphi}.$$

This corresponds to the action of two moments M at the poles of the sphere. Adding or dividing the above groups of loads, we can build solutions for various cases.

Case $n \geqslant 2$. The solution for $n \geqslant 2$ gives no resultants of the membrane force and likewise no singular loads appear at the apex. These terms can, however, be used to obtain the solution for other combinations of singular loads. Following the method due to G. Messmer [7.2], we may transform the solution (7.6) by using the following identities:

$$\sum_{n=0,2,4,\dots}^{\infty} A^n = \frac{1}{1-A^2}, \qquad \sum_{n=1,3,5,\dots}^{\infty} A^n = \frac{A}{1-A^2}, \qquad |A| < 1. \tag{7.12}$$

Introducing

$$A = \tan\frac{\varphi}{2}(\cos\vartheta + i\sin\vartheta),$$

we find the following expressions convergent for $|\varphi| < 90°$:

$$\frac{1}{\sin^2\varphi}\sum_{n=0,2,4,\dots}^{\infty}\left(\tan\frac{\varphi}{2}\right)^n e^{in\vartheta} = \frac{1}{2}\left[\frac{1}{\sin^2\varphi} + \frac{\cos\varphi + i\sin^2\varphi\sin\vartheta\cos\vartheta}{\sin^2\varphi\,(1-\sin^2\varphi\cos^2\vartheta)}\right],$$

$$\frac{1}{\sin^2\varphi}\sum_{n=1,3,5,\dots}^{\infty}\left(\tan\frac{\varphi}{2}\right)^n e^{in\vartheta} = \frac{1}{2}\,\frac{\cos\varphi\cos\vartheta + i\sin\vartheta}{\sin\varphi\,(1-\sin^2\varphi\cos^2\vartheta)}. \tag{7.13}$$

Separating real and imaginary parts, we find

(e) $$N_{\varphi\varphi} = \frac{C_5\cos\varphi}{\Delta\sin^2\varphi}, \qquad N_{\varphi\vartheta} = \frac{-C_5\sin\vartheta\cos\vartheta}{\Delta};$$

(f) $$N_{\varphi\varphi} = \frac{C_6\sin\vartheta\cos\vartheta}{\Delta}, \qquad N_{\varphi\vartheta} = \frac{C_6\cos\varphi}{\Delta\sin^2\varphi};$$

(g) $$N_{\varphi\varphi} = \frac{C_7\cos\varphi\cos\vartheta}{\Delta\sin\varphi}, \qquad N_{\varphi\vartheta} = \frac{-C_7\sin\vartheta}{\Delta\sin\varphi};$$

(h) $$N_{\varphi\varphi} = \frac{C_8\sin\vartheta}{\Delta\sin\varphi}, \qquad N_{\varphi\vartheta} = \frac{C_8\cos\varphi\cos\vartheta}{\Delta\sin\varphi},$$

where $\Delta = 1-\sin^2\varphi\cos^2\vartheta$.

The above solutions correspond to the action of the normal and tangential loads as presented in Figs. 79 e, f, g, h if we take the following values for the constants C_i:

$$C_5 = \frac{P}{2\pi R}, \qquad C_6 = \frac{M_t}{2\pi R^2}, \qquad C_7 = C_8 = \frac{T}{2\pi R}.$$

This can be proved by considering the equilibrium of the stress resultants in the vicinity of the loading points.

The above solutions contain singularities of the first, second, and third order and give infinite stresses at the points of application of the loads. They differ from the results obtained by means of the bending theory. But at some distance from the loading points the bending moments vanish and the membrane forces given by the above formulae represent the real state of stresses.

The other solutions for the concentrated loads based on the equations of membrane theory can be found in the works of F. Martin [7.1], and V. Z. Vlasov [1.12]. The case of a spherical shell subjected to the action of a system of concentrated loads has also been considered by W. Flügge [1.16]. He also considered a shell resting on a point supports and loaded by its own weight.

7.3 Conical shell

In the case of a conical shell we have $\alpha_1 = s$, $\alpha_2 = \vartheta$ and $R_1 = \infty$, $R_2 = s\cot\alpha$. If the shell is subjected to edge loads only ($X = Y = Z = 0$), the general solution of Eqs. (7.2) is

$$N_{ss} = \frac{n}{\cos\alpha}\frac{C_{1n}}{s^2} + \frac{C_{2n}}{s}, \qquad N_{s\vartheta_n} = \frac{C_{1n}}{s^2},$$

$$N_{\vartheta\vartheta_n} = 0.$$

If C_{1n}, $C_{2n} \neq 0$, there will always be infinite stresses at the vertex $s = 0$. For the first harmonic $n=1$ we obtain the case of a horizontal force T and the moment M with the horizontal axis. Determining the horizontal resultants of the forces transmitted through an arbitrary parallel circle

$$T = \int_{-\pi}^{\pi} (N_{ss_1}\cos\alpha\cos^2\vartheta - N_{s\vartheta_1}\sin^2\vartheta)\, s\cos\alpha\, d\vartheta = C_{21}\pi\cos^2\alpha,$$

we find that

$$N_{ss_1} = \frac{T\cos\vartheta}{s\pi\cos^2\alpha}, \qquad N_{\vartheta\vartheta} = N_{s\vartheta} = 0.$$

In the case of the action of the bending moment we calculate the resultant moment with respect to an axis through the vertex:

$$M = \int_{-\pi}^{\pi} N_{s\vartheta_1} \sin^2 \vartheta s \cos \alpha s \sin \alpha \, d\vartheta = C_{11} \pi \cos \alpha \sin \alpha .$$

Then we have

$$N_{ss_1} = \frac{M \cos \vartheta}{s^2 \pi \cos^2 \alpha \sin \alpha}, \qquad N_{s\vartheta_1} = \frac{M \sin \vartheta}{s^2 \pi \cos \alpha \sin \alpha}.$$

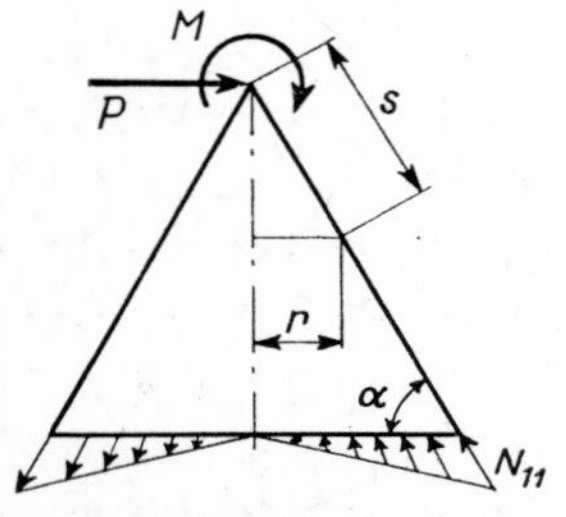

Fig. 80. Conical shell subjected to concentrated force and bending moment at the apex.

It is interesting to note that for the higher harmonics there exists a solution

$$N_{s_n} = \frac{C_n \cos n\vartheta}{s}$$

for any integer n having the same order of singularity $1/s$. But these singularities do not correspond to any external forces and moments.

References 7

[7.1] F. Martin: Die Membran-Kugelschale unter Einzellasten, *Ing. Arch. 17* (1949) 167.

[7.2] G. Messmer: Über eine Gruppe von Singularitäten im Membranspannungzustand der Kugelschale, *Ing. Arch. 28* (1959) 208.

[7.3] E. Cetmeli: Rotationssymmetrische Schale unter symmetrischer Belastung die in beiden Richtungen konstante Druckspannungen hervorruft, *Istanbul tekn. univ. Büll. Tech. Univ. Istanbul, 21.1* (1968).

[7.4] F. Lagomaggiore: Calcolo a membrana del giperboloide di rotazione ad una falda soggetto a carichi anulari, *G. genio civile 106, No. 7–8* (1968).

[7.5] O.I. Medvedkov: Solution of the problem of the membrane state of stress of paraboloid of revolution meeting the condition $p = e^x$ (p—analytical function) and loaded by concentrated load and moment, *Uch. Zap. Mosk. Obl. Ped. In-ta 227* (1970) (in Russian).

[7.6] A. A. Klunnik: On application of p-analytical functions for the solution of membrane shells of revolution subject to concentrated load, *Vychislitelnaia i Prikladnaia Matematika Mezhved. Nauchn. ab., 15* (1971) (in Russian).

8

Spherical shell
Solutions of the general theory

The action of concentrated loads on spherical shell has been widely studied. The solution for a concentrated normal force was first obtained by E. Reissner in 1946 [8.1] by using the classical theory of shallow shells. This problem was also considered by W. Flügge and D. A. Conrad [8.2], A. Kalnins and P. M. Naghdi [8.3], R. P. Nordgren [8.4], and W. T. Koiter [1.19]. Solutions for other concentrated loads, such as a tangential force and a bending moment, have been obtained first on the basis of the membrane approximation [7.1], [7.2] also, bending stresses and displacements in a spherical shell subjected to localized loads were determined by F. A. Leckie [8.6] on the basis of the simplified theory of spherical shells due to A. Havers [8.7]. Using Langer's technique [8.8] and working in spherical polar coordinates (ϑ, φ), Leckie obtained an asymptotic solution which is valid for all values of φ: In addition to integration constants, depending on the nature of the loading, this solution contains constants depending on the boundary conditions. It is known, however, that bending effects in a spherical shell are highly localized, except in the case of a shell with free edges. The stresses produced by concentrated forces can consequently be separated from those resulting from boundary effects and can be considered separately. This property is confirmed by numerical results given in [8.21].

If we consider only stresses and displacements due to concentrated loads, simple solutions corresponding to a normal force, tangential force, and a bending moment can be obtained. In deriving these solutions it is supposed that the stresses and displacements are very small at some distance from the loading point. The other boundary conditions can clearly be satisfied by the addition of a suitable solution of the homogeneous shell equations. But the influence of the boundary conditions is usually limited to a small area near the edge of the shell, and so does not disturb the state of stresses and displacements near the loading point if this point is remote from the edge. The solution within the scope of the improved theory given by Naghdi [1.8] in which the effect of transverse shear as well as that of rotary inertia are included was first obtained by J. P. Wilkinson and A. Kalnins [8.9] for the case of a normal concentrated force.

Let us solve our problem by means of linear shell equations, taking into account the effect of transverse shear deflections. We adopt the non-dimensional system of rectangular Cartesian coordinates x, y defined by $x=\bar{x}/l$, $y=\bar{y}/l$ where l is a certain characteristic length. The shell equations then take the form [8.13]

$$\left(\Delta+\frac{2}{R^2}\right)\left[D\left(\Delta+\frac{1+\nu}{R^2}\right)w-\frac{1}{R}\left(1-\frac{h^2}{5(1-\nu)}\Delta\right)\Phi\right].$$

$$=\left[1-\frac{(2-\nu)h^2}{10(1-\nu)}\Delta\right]Z, \tag{8.1}$$

$$\left(\Delta+\frac{2}{R^2}\right)\left[\frac{1}{Eh}\left(\Delta+\frac{1-\nu}{R^2}\right)\Phi+\frac{w}{R}\right]=-\frac{\nu}{2E}\Delta Z,$$

where $D=Eh^3/121-\nu^2)$ and

$$\Delta=\frac{1}{l^2}\left(\frac{\partial^2}{\partial x^2}+\frac{\partial^2}{\partial y^2}\right) \tag{8.2}$$

is the Laplacian operator. For an isotropic shell,

$$\bar{\varepsilon}=\frac{\nu h}{2R},\qquad \varepsilon=\frac{\nu h^2}{10(1-\nu)l^2},\qquad \eta=\frac{h^2}{5(1-\nu)l^2}.$$

As first of all we are interested in the state of stress near the loading point; we assume, to simplify the calculations, that the coefficients of the first quadratic form are constant. Then we can assume that $A_1=A_2=1$. This simplification will enable us to achieve the solution of the problem of the differential equations with constant coefficients. Let us further assume that the shell extends infinitely in all directions. This assumption is made to avoid the influence of the boundary effects on the stresses and displacements near the loading point and will enable us to derive the particular solution corresponding to the concentrated force. For some type of boundary conditions this solution gives already sufficiently accurate numerical results.

8.1 Spherical shell loaded by a concentrated normal force

Now we consider a shallow spherical shell loaded by a concentrated normal force acting at the point $x = 0$, $y = 0$ (Fig. 81) in the outwards direction. We represent this force by means of a Fourier integral. The displacement w and the stress function Φ are then obtained from Eqs. (8.1) in the form

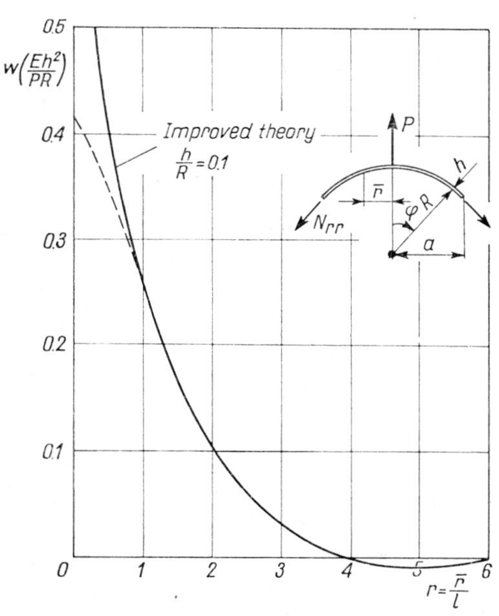

Fig. 81. Deflection of a spherical shell subjected to normal concentrated force. Continuous line results from equations of shells with the effect of transverse shear deformations. Dotted line represents classic theory with no transverse shear effects.

$$w = \frac{Pl^2}{\pi^2 D} \int_0^\infty \int_0^\infty [(1+(\eta-\varepsilon)(\alpha^2+\beta^2))(\alpha^2+\beta^2-(1-\nu)k_R)- \\ -\bar{\varepsilon}(1+\eta(\alpha^2+\beta^2))(\alpha^2+\beta^2)] \frac{\cos\alpha x \cos\beta y \, d\alpha \, d\beta}{(\alpha^2+\beta^2-2k_R)D(\alpha,\beta)},$$

$$\Phi = \frac{PR}{\pi^2} \int_0^\infty \int_0^\infty [1+(\eta-\varepsilon)(\alpha^2+\beta^2)+ \\ +\bar{\varepsilon}(\alpha^2+\beta^2)(\alpha^2+\beta^2-(1+\nu)k_R)] \frac{\cos\alpha x \cos\beta y \, d\alpha \, d\beta}{(\alpha^2+\beta^2-2k_R)D(\alpha,\beta)}, \tag{8.3}$$

where

$$D(\alpha, \beta) = (\alpha^2+\beta^2-k_R)^2+4\varkappa^4[1+\eta(\alpha^2+\beta^2)],$$

$$4\varkappa^4 = Ehl^4/R^2D = 12(1-\nu^2)l^4/R^2h^2.$$

The length l can be adjusted so that the coefficient $4\varkappa^4 = 1$. Then we have

$$l = \frac{\sqrt{Rh}}{\sqrt[4]{12(1-\nu^2)}}, \quad \text{then} \quad k_R = \frac{l^2}{R^2} = \frac{1}{\sqrt{12(1-\nu^2)}}\frac{h}{R}. \tag{8.4}$$

Introducing the polar coordinates (γ, φ) and $(r = \bar{r}/l, \vartheta)$, these integrals can easily be evaluated. We have

$$\alpha = \gamma\cos\varphi, \quad \beta = \gamma\sin\varphi,$$

$$x = r\cos\vartheta, \quad y = r\sin\vartheta.$$

Then on substitution and integration with respect to φ, we have

$$w = \frac{Pl^2}{2\pi D}\int_0^\infty \{[1+(\eta-\varepsilon)\gamma^2][\gamma^2-(1-\nu)k_R]-$$

$$-\bar{\varepsilon}[1+\eta\gamma^2]\gamma^2\}\frac{J_0(\gamma r)\gamma\,d\gamma}{(\gamma^2-2k_R)D(\gamma)}, \tag{8.5}$$

and a similar expression for the stress function Φ. Here $J_0(\gamma r)$ is the Bessel function of the first kind, and

$$D(\gamma) = (\gamma^2-k_R)^2+\eta\gamma^2+1. \tag{8.6}$$

Let us transform the denominator of (8.5) as follows:

$$\frac{1}{(\gamma^2-2k_R)D(\gamma)} = \left[\frac{1}{\gamma^2-2k_R}-\frac{\gamma^2+\eta}{D(\gamma)}\right]\frac{1}{1+k_R^2+2\eta k_R}.$$

The coefficient $(1+k_R^2+2\eta k_R)$ is only a small corection of the order $1+0(h/R)$ and in the case of thin shells it can be neglected. The notation $0(h/R)$ denotes here the value of the order (h/R). However the complicated form of the denominator makes the exact analytical evaluation of the above integrals difficult. Therefore we proceed as previously for the plate resting on an elastic foundation and evaluate it approximately. Let us transform the denominator $D(\gamma)$. Since $h^2/l^2 \ll 1$ and $k_R \ll 1$, we can write, as a good approximation,

$$\frac{1}{D(\gamma)} \approx \left[1-\frac{\gamma^2}{\gamma^4+1}(\eta-2k_R)\right]\frac{1}{\gamma^4+1} \tag{8.7}$$

Introducing the above into (8.5) and presenting the integrand in the form of simple fractions, we obtain on integration

$$w = \frac{Pl^2}{2\pi D}\left\{-\operatorname{kei} r - k_R\left[(1+\nu)\left(\frac{\pi}{2} Y_0(r\sqrt{2k_R}) + \ker r\right) + \frac{1}{2} r\ker' r\right] + \right.$$

$$\left. + (\eta - \varepsilon)\ker r + \frac{\eta}{4} r\ker' r\right\}, \tag{8.8}$$

where $Y_0(r\sqrt{2k_R})$ is the Bessel function of the second kind. Its numerical values can be found in the tables of special functions. This function for $r \ll 1$ takes the form

$$Y_0(r\sqrt{2k_R}) \simeq \frac{2}{\pi}\left[\left(1 - k_R\frac{r^2}{2}\right)\left(\ln\frac{r}{2} + \gamma_0 + \ln\sqrt{2k_R}\right) + k_R\frac{r^2}{2}\right].$$

The functions $\ker r$, $\operatorname{kei} r$ were described in the chapter devoted to the circular plates resting on an elastic foundation.

The first term in (8.8) corresponds to Reissner's solution obtained by means of the theory of shallow shells, the second term multiplied by the coefficient k_R appears as the result of use of the complete non-shallow shell equations (8.1). The third term in (8.8) presents the effect of transverse shear and normal deformations. These terms are significant only for small r. For small $r \ll 1$ we can replace the functions $\ker r$, $\operatorname{kei} r$ by the power series expansion. Taking only the first two terms of the series, we obtain an approximate formula for the displacement for $r \ll 1$.

$$w = \frac{\sqrt{3(1-\nu^2)}}{4\pi}\frac{PR}{Eh^2}\left[\pi - 4(1+\nu)k_R\ln\sqrt{2k_R} + 2k_R + \right.$$

$$\left. + \left(r^2 - 4(\eta - \varepsilon)\right)\left(\ln\frac{r}{2} + \gamma_0\right) - r^2 - \eta\right]. \tag{8.9}$$

We observe that near the loading point the deflection of the shell is defined by functions identical with those obtained previously for the plate resting on an elastic foundation. The terms containing the coefficient k_R which result from the effect of non-shallowness of the shell are of the order $(h/R)\ln(R/h)$. This means that they are only slightly larger than the previously discarded terms of the order h/R. Taking into account the deflection resulting from the effect of transverse shear deformation, we obtain the term containing $-\ln r$ which goes to infinity as $r \to 0$. This deflection is of a local character and for thin shells the effect of transverse shear deformations is limited to the area $0 < r < l$ (Fig. 81). However, the equation cannot be used to calculate the deflection at those points of the shell that are very close to the point of application of the load ($\bar{r} < h$). When $\bar{r}$ is of the same order of magnitude as the thickness of the shell, Eq. (8.9) is no longer applicable. We know from the theory of thick plates (see Sect. 2.1) that the deflection at the point $\bar{r} = 0$, $\xi = 0$ is finite. To calculate the maximum deflection we can proceed as previously, namely, calculate the deflection from the formula for $\bar{r} = h$ and add the deflection of the central portion of the shell ($\bar{r} \leqslant 2h$). This deflection can be estimated approximately as for thick plates resting on an elastic foundation.

Solving Eqs. (1.114) and taking into account only the small terms of order h^2/l^2, we obtain the shear forces

$$Q_r = -\frac{P}{2\pi l}\int_0^\infty \left[\gamma^2 - \frac{1}{2}\eta\right]\frac{\gamma^2 J_1(\gamma r)\,d\gamma}{D(\gamma)} \tag{8.10}$$

and on integration

$$Q_r = \frac{P}{2\pi l}\left[\operatorname{ker}'r + \frac{5}{4}\eta\left(\operatorname{kei}'r + \frac{1}{5}r\operatorname{ker}r\right) - \frac{1}{2}k_R(3\operatorname{kei}'r + r\operatorname{ker}r)\right].$$

For small $r \ll 1$ the above formula can be simplified to

$$Q_r = -\frac{P}{2\pi l}\left[\frac{1}{r} - \frac{\pi}{8}r + \left(\frac{7}{8}\eta - \frac{5}{4}k_R\right)r\cdot\ln\frac{r}{2}\right]. \tag{8.11}$$

The bending moments are calculated from,

$$\begin{aligned} M_{rr} = \frac{P}{2\pi}\Bigg\{&\operatorname{ker}r - (1-\nu)\frac{1}{r}\operatorname{kei}'r + \frac{1+\nu}{2}k_R\Bigg[\pi\left(Y''(r\sqrt{2k_R}\right) + \\ &+\frac{\nu}{r}\pi Y'(r\sqrt{2k_R}) - 2(1-\nu)\frac{1}{r}\operatorname{ker}'r - \operatorname{kei}r - \frac{r}{1+\nu}\operatorname{kei}'r\Bigg] + \\ &+\left[-(1-\nu)\,\varepsilon\frac{1}{r}\operatorname{ker}'r + \frac{\eta}{4}(1+5\nu)\operatorname{kei}r + \frac{r\eta}{4}\operatorname{kei}'r\right]\Bigg\}. \end{aligned}$$

$$\begin{aligned} M_{\vartheta\vartheta} = \frac{P}{2\pi}\Bigg\{&\nu\operatorname{ker}r + (1-\nu)\frac{1}{r}\operatorname{kei}'r + \\ &+\frac{1+\nu}{2}k_R\Bigg[\pi\left(\frac{1}{r}Y'(r\sqrt{2k_R}) + \nu\pi Y''(r\sqrt{2k_R})\right) + \\ &+2(1-\nu)\frac{1}{r}\operatorname{ker}'r - \nu\operatorname{kei}r - \frac{\nu r}{1+\nu}\operatorname{kei}'r\Bigg] + \\ &+\left[\frac{(1-\nu)\varepsilon}{r}\operatorname{ker}'r + \frac{\eta}{4}(5+\nu)\operatorname{kei}r + \nu\frac{r\eta}{4}\operatorname{kei}'r\right]\Bigg\} \end{aligned} \tag{8.12}$$

for small $r \ll 1$;

$$Y'(r\sqrt{2k_R}) = \frac{2}{\pi r}, \qquad Y''(r\sqrt{2k_R}) = -\frac{2}{\pi r^2}.$$

These terms cancel with the singular term of the function $2(1-\nu)\dfrac{1}{r}\operatorname{ker}' r$. Finally we obtain

$$\left.\begin{matrix} M_{rr} \\ M_{\vartheta\vartheta} \end{matrix}\right\} = \frac{P}{4\pi}\left[-(1+\nu)\left(\ln\frac{r}{2}+\gamma_0\right) \mp \frac{1-\nu}{2} \pm 2(1-\nu)\varepsilon\frac{1}{r^2}\right]. \tag{8.13}$$

The bending moments are singular at the point $r = 0$ and approach $+\infty$ and $-\infty$. Two singular terms appear in expressions (8.13). First, the classical $\ln r$, and second $1/r^2$ coming out of the improved theory which also takes

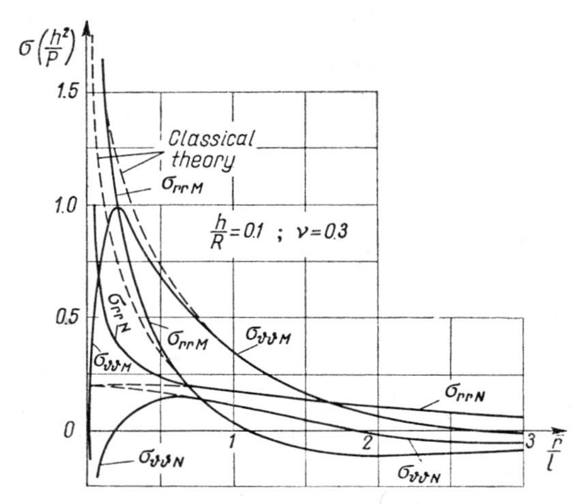

Fig. 82. Distribution of stresses in a spherical shell subjected to normal force. Continuous lines result from equations with the effect of transverse shear deformations.

into account the transverse shear and normal deformations. The terms containing $1/r^2$ have a stronger singularity of the second order. We see that the moments contain a singularity identical with that obtained for flat plates. When the load is in an outward direction, M_{rr} and $M_{\vartheta\vartheta}$ approach $+\infty$ and $-\infty$ respectively as $r \to 0$, according to improved theory. For the classical theory both approach $+\infty$ (Fig. 82).

We find the membrane forces from (8.3) and (1.112). The stress function can be obtained in a similar way as the deflection w. On integration of $(8.3)_2$, we find

$$\Phi = \frac{PR}{2\pi}\left\{-\frac{\pi}{2}Y_0(r\sqrt{2k_R}) - \operatorname{ker} r + \bar{\varepsilon}\operatorname{ker} r - \right.$$

$$\left. - \left[\left(\frac{\eta}{2} - \varepsilon\right)\operatorname{kei} r + \eta\frac{r}{4}\operatorname{kei}' r\right]\right\}. \tag{8.14}$$

For the shallow shell the coefficient $k_R \ll 1$ and we have

$$\frac{\pi}{2}Y_0(r\sqrt{2k_R}) = \ln\frac{r}{2} + \gamma_0.$$

The membrane forces obtained from (8.14) are

$$N_{rr} = -\frac{1}{r}\frac{d\Phi}{dr} - \frac{\Phi}{R^2} - \frac{M_{rr}}{R}$$

$$= \frac{PR}{2\pi l^2}\frac{1}{r}\left\{\frac{\pi}{2}Y_0'(r\sqrt{2k_R}) + \operatorname{ker}' r - \bar{\varepsilon}\operatorname{ker}' r + \right.$$

$$+ \left[\left(\frac{\eta}{2} - \varepsilon\right)\operatorname{kei}' r + \eta\frac{r}{4}\operatorname{ker} r\right] +$$

$$\left. + k_R\left[\frac{\pi}{2}Y_0(r\sqrt{2k_R}) + (1-\nu)\operatorname{kei}' r\right]\right\},$$

$$N_{\vartheta\vartheta} = -\frac{d^2\Phi}{dr^2} - \frac{\Phi}{R^2} - \frac{M_{\vartheta\vartheta}}{R}$$

$$= \frac{PR}{2\pi l^2}\left\{\frac{\pi}{2}Y''(r\sqrt{2k_R}) - \operatorname{kei} r - \frac{1}{r}\operatorname{ker}' r + \bar{\varepsilon}\left(\frac{1}{r}\operatorname{ker}' r + \operatorname{kei} r\right) + \right.$$

$$+ \left[\left(\frac{3}{4}\eta - \varepsilon\right)\operatorname{ker} r - \left(\frac{\eta}{2} - \varepsilon\right)\frac{1}{r}\operatorname{kei}' r + \frac{\eta}{4}r\operatorname{ker}' r\right] +$$

$$\left. + k_R\left[\frac{\pi}{2}Y_0(r\sqrt{k_R}) + (1-\nu)\left(\operatorname{ker} r - \frac{1}{r}\operatorname{kei}' r\right)\right]\right\}.$$

The formulae for the membrane force for $r \ll 1$ can be transformed to

$$\left.\begin{matrix} N_{rr} \\ N_{\vartheta\vartheta} \end{matrix}\right\} = \sqrt{3(1-\nu^2)}\frac{P}{8h}\left\{1 + \frac{4(1+\nu)}{\pi}k_R\left[\ln\frac{r}{2} + \gamma_0 + \frac{1-\nu}{2}\right] - \right.$$

$$\left. - 4(\eta - \varepsilon)\left[\ln\frac{r}{2} + \gamma_0 - \frac{1-\nu}{2(2-\nu)}\right] \pm \frac{8}{\pi}\bar{\varepsilon}\frac{1}{r^2}\right\}. \tag{8.15}$$

The first term in the brackets { } corresponds to the result from shallow shell equations. The remaining terms are the results of the effect of non-shallowness of the shell and the transverse normal stresses σ_{zz}. These terms contain the singular functions $\ln(r/2)$ and $1/r^2$. According to the theory of shallow shells the membrane forces for $r \to 0$ reach a constant value

$$N_{rr} = N_{\vartheta\vartheta} = \sqrt{3(1-\nu^2)}\,\frac{P}{8h},$$

which differs slightly from Eq. (8.9).

Analyzing all the solutions obtained above, we see that in a shell with bending rigidity the deflection under a normal load remains finite if we neglect the effect of transverse shear deformations. But there are still infinite bending moments and shear forces. The strongest singularities, those involving an infinite deflection, occur in structures lacking bending rigidity (see the case of membrane shell considered in Chapter 7). This indicates that the bending rigidity is of local importance.

The stresses on the upper surface of the shell caused by bending and membrane forces are presented graphically in Fig. 82. The case where the concentrated force acts on the surface of the circle of a radius c is interesting from the technical point of view. Based on the analogy with the plate on the elastic foundation we can write the following equations for the deflection and the stresses in the isotropic shell. At the point $r = 0$ we have

$$w_0 = \frac{\sqrt{12(1-\nu^2)}}{\pi}\,\frac{PR}{Eh^2}\left[\frac{1}{c^2}+\frac{1}{c}\operatorname{ker}' c-\frac{1}{2}(1+\nu)k_R\ln\sqrt{2k_R}-\frac{1}{4}k_R\right]+$$

$$+\frac{3}{5\pi}(1+\nu)(2-\nu)\frac{P}{Eh}\left[\operatorname{ker} c+\frac{c\operatorname{ker}' c}{2(2-\nu)}\right],$$

$$\sigma_{rrN} = \sigma_{\vartheta\vartheta N} = \frac{\sqrt{3(1-\nu^2)}}{\pi}\,\frac{P}{h^2}\left[\frac{1}{c^2}+\frac{1}{c}\operatorname{ker}' c\right]-$$

$$-\frac{P}{2\pi Rh}\left[(1-\nu)\operatorname{kei}' c+\frac{c}{2}\operatorname{ker}' c\right]+$$

$$+\frac{3(1+\nu)(2-\nu)}{10\pi}\,\frac{P}{Rh}\left[\operatorname{ker} c+\frac{c\operatorname{ker}' c}{2(2-\nu)}\right]+\nu\frac{\sqrt{3(1-\nu^2)}}{2\pi}\,\frac{P}{Rh}\,\frac{1}{c^2}, \quad (8.16)$$

$$\sigma_{rrM} = \sigma_{\vartheta\vartheta M} = \frac{3(1+\nu)}{\pi}\,\frac{P}{h^2}\,\frac{\operatorname{kei}' c}{c}-\frac{6\sqrt{3(1-\nu^2)}}{10\pi}\,\frac{P}{Rh}\,\nu\,\frac{\operatorname{ker}' c}{c},$$

where $c = \bar{c}/l = \sqrt[4]{12(1-\nu^2)}\bar{c}/\sqrt{Rh}$ is a non-dimensional value. The first terms in the above relations which correspond to the results of the shallow shell equations are the largest. The remaining terms appear as the result of taking into account the effects of transverse shear deformations and the complete equations of the spherical shell. These terms are negligible if $c > 1$.

It is interesting that in the case where the load is distributed over a certain small area, the membrane and bending stresses at the point $r = 0$ are equal $\sigma_{rrN} = \sigma_{\vartheta\vartheta N}$, $\sigma_{rrM} = \sigma_{\vartheta\vartheta M}$. In the case of the concentrated load we obtained, as previously, $\sigma_{rrN} \to \infty$, $\sigma_{\vartheta\vartheta N} \to -\infty$ and $\sigma_{rrM} \to \infty$, $\sigma_{\vartheta\vartheta M} \to -\infty$. This result can be explained in the following way. If the load acts over the surface πc^2, the membrane stresses and bending stresses change rapidly at $r = c$. Calculating the stresses in the circular plate, we obtained the difference between the moments $M_{\vartheta\vartheta}$ for $r > c$ and $r < c$ as $r \to c$ equal to $\nu Ph/10\pi c^2$. If $c \to 0$, this difference tends to infinity. The membrane force $N_{\vartheta\vartheta}$ behaves in a similar way, i.e. it changes rapidly from $\nu hP/4\pi c^2$ to $-\nu hP/4\pi c^2$ for $r = c$.

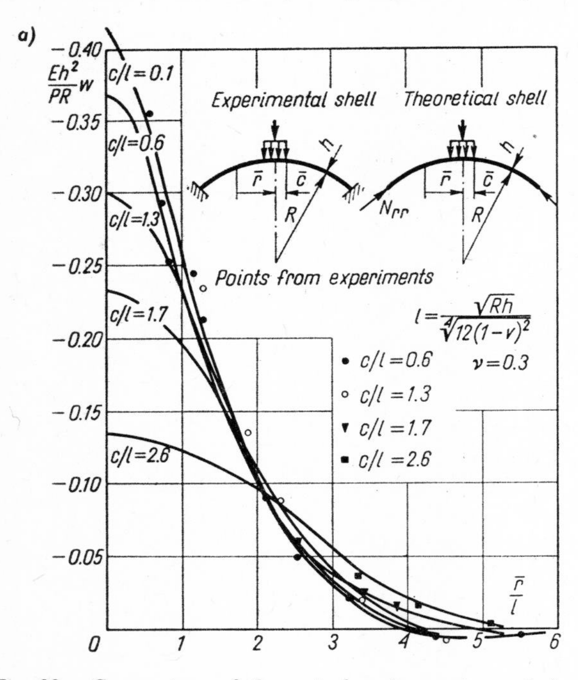

Fig. 83a. Comparison of theoretical and experimental data. Continuous lines present the deflection and stresses in the shell for various dimension of the loaded area (A. S. Tooth [8.16]).

If $c \to 0$, this difference tends also to infinity. This peculiar behaviour of the stress $\sigma_{\vartheta\vartheta}$ is caused by the imperfection of the theory of shells, which reduces the analysis of the three-dimensional problem of the theory of elasticity to a two-dimensional problem. If this problem were considered more accurately, we would obtain another solution also discontinuous, but with a similar distribution of stresses. (See the solution for the circular plate.) Some numerical

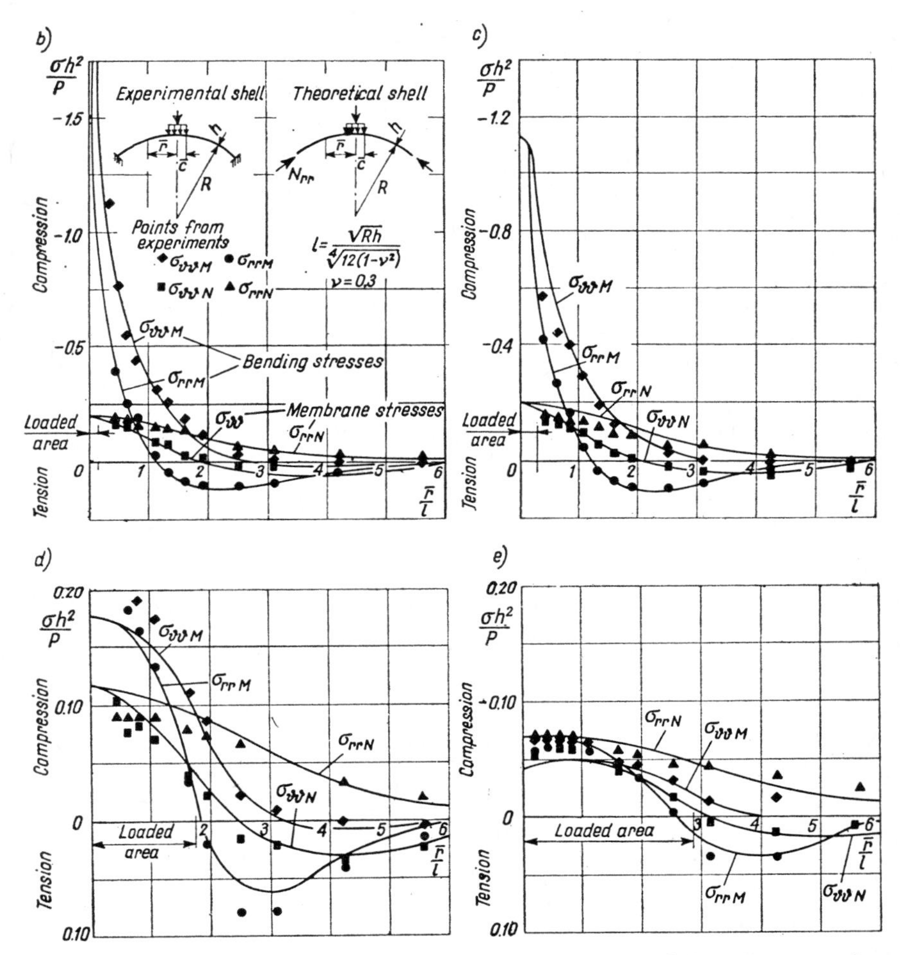

Fig. 83b, c, d, e. Comparison of theoretical results and experimental data. Continuous lines present the deflection and stresses in the shell for various dimension of the loaded area (A. S. Tooth [8.16]).

results for the deflections and stresses in the shell for different ratios $\bar{c}/l$ are shown in Fig. 83. Taking into account the effect of transverse normal stresses, we obtain, for $c \to 0$, $N_{rr} = \infty$, $N_{\vartheta\vartheta} = -\infty$. At a large distance from the point of application of the load the membrane forces approach the value which is obtained from the equations of the membrane theory (7.5). We see that when the load is distributed over a small area the logarithmic peak is replaced by a sharp finite cusp. The height of this cusp depends strongly on the area over which the load is distributed. The maximum moment can give only the precise information on the load distribution. Outside the direct vicinity of the loaded area the point load analysis usually gives an accurate answer if the loaded area is small enough.

The non-shallow spherical shell loaded at the vertex has been considered by W. T. Koiter [1.19] in spherical coordinates (this means not making the assumption $A_1 = A_2 = 1$), and based on the equations of the theory of non-shallow spherical shell. Analyzing the singular solutions of Eq. (1.122) in terms of Legendre functions, he obtained the following solution of the problem:

$$\psi = -\frac{12(1-\nu^2)PR^2}{2\pi(2i\lambda^2-1)Eh^3}\left[Q_1(\cos\varphi)+\frac{\pi}{2}\cot\left(\frac{1}{2}n\pi\right)P_n(\cos n\varphi)-Q_n(\cos\varphi)\right],$$

where n is the root of the equation

$$n(n+1) = 2i\lambda^2+1, \qquad \lambda^2 = \sqrt{3(1-\nu^2)}\frac{R}{h} = \frac{1}{2k_R};$$

$Q_k(\cos\varphi)$ and $P_k(\cos\varphi)$ are the Legendre functions of order k. The deflection and the stress function are related to the function ψ by means of the equation

$$\psi = w+\frac{2i\lambda^2-\nu}{EhR}\Phi \quad \text{then} \quad w = \mathrm{Re}(\psi)+\frac{\nu}{2\lambda^2}\mathrm{Im}(\psi).$$

The results obtained in this way are almost equal to the results obtained by means of the Fourier integral. For example, W. T. Koiter obtained the following expression for the normal deflection under the point of application of the load:

$$w(0) = \frac{\sqrt{3(1-\nu^2)}}{4}\frac{PR}{Eh^2}\left[1+\frac{2(1+\nu)}{\pi\lambda^2}\left(\ln\lambda+\gamma_0-1+\frac{1}{2}\ln 2\right)+\right.$$
$$\left.+\frac{4}{3\pi\lambda^2}+O(\lambda^{-3})\right].$$

When the ratio $\bar{c}/2h$ is very small the portion of the shell near the point of application of the load should be treated as a body all three dimensions of which are of the some order of magnitude. The stresses in this area will be different than given by Eqs. (8.16). The tensile stresses can be obtained approximately from Westergard's formula (Sect. 3.7) for the plate resting on an elastic foundation. For the spherical shell the coefficient $k = Eh/R^2$.

Then, adding the effect of the membrane forces we have the tensile stresses

$$\sigma_{rr} = \frac{P}{h^2}\left[[0.275(1+\nu)\ln\frac{h^2R^2}{\bar{b}^4} - \frac{\sqrt{3(1-\nu^2)}}{8}\right]$$

where

$$\bar{b} = \begin{cases} \sqrt{1.6\bar{c}^2+h^2}-0.675h & \text{when} \quad \bar{c} < 1.724h, \\ \bar{c} & \text{when} \quad \bar{c} > 1.724h. \end{cases}$$

The general conclusions concerning the application of shallow shell theory resulting from the above calculations are as follows. In the case where the load is distributed over the surface πc^2, where c is of the order of the characteristic length l or when the deflection or stresses are calculated at points lying several distances l apart from the point of application of the load, the differences between the results of shallow shell theory and the improved theory are negligible.

In the case where the load is distributed over the surface πc^2, where $h < \bar{c} < l$ and the stresses are calculated at points lying at a distant of the same order, the complete equations improved by the influence of transverse shear and normal deformations should be applied. If $c < 0.5h$, the exact stress distribution in the vicinity of the loaded area can only be obtained from the three-dimensional equations of the theory of elasticity.

The last two remarks refer to thicker shells where $0.02 < h/R < \sim 0.1$. In the case of thin shells the non-linear effect of large deflections quickly modifies the results obtained by means of linear theory.

The results (§ 8.1) were checked by an experiment. A. S. Tooth [8.16] carried out the experiments with a shallow spherical shell loaded at its apex by a concentrated normal force and obtained the data which confirmed the theoretical calculations. The points ● ■ ▲ ♦ in Fig. 83 present the results obtained using strain gages. There is seen a good conformity between theory and experiment both for stresses and displacements. The aim of Tooth's experiments was also to define the influence of the boundary conditions and finite sizes of the shell on the stresses and displacements near the loading point. It was found that the influence of the edge of the shell is very small when the loading point is at a distance larger than $5l$ to $6l$ from the edge. Even if the distance is smaller, only the part of the stressed area nearest the edge is deformed.

The displacement of the shell shows greater deviation from the theoretical curves than the stresses. It was also found that the stress and displacement fields near the point of application of the load did not undergo changes with the change of the position of the loaded point on the surface of the shell. Those experiments were performed for loads acting on the circular surface with the radius $\bar{c}$ for different ratios $\bar{c}/l$.

The results described make it possible to form a solution for a more complicated load. It is possible to find the solution for the case of an arbitrary load by using the singular solution as the Green function of the problem. For example, we can calculate the case of the load acting along a certain length and distributed in a definite way, by the superposition of the solution for the simple concentrated force. This can be done by an analytical or graphical method. This method was proposed for the spherical shell by R. M. Kenedi [8.17] who calculated in this way a few cases of shells under complex loads.

8.2 Spherical shell loaded by a concentrated moment

If the concentrated moment acting on a shell is represented as the resultant of two oppositely directed forces of infinite magnitude acting at points, a zero distance apart all effects produced by the moment can be obtained from the solutions for the concentrated normal and tangential forces.

Let us consider first a more general case: an arbitrary shell loaded by two oppositely directed normal forces P (Fig. 84). However, the load presented in Fig. 84 is not equivalent to the concentrated bending moment. The moment should be represented by the couple of two oppositely directed, parallel forces, but the forces P in Fig. 84 create an angle $2\mathrm{d}\varphi$. Summing the projections of both forces on the directions normal and parallel to the middle surface of the shell, we find

$$Z(x, y) = \frac{P}{\pi^2 l^2} \int_0^\infty \int_0^\infty [\cos\alpha(x+\varepsilon) - \cos\alpha(x-\varepsilon)] \cos\beta y \,\mathrm{d}\alpha\,\mathrm{d}\beta$$

$$= -\frac{2P\varepsilon}{\pi^2 l^2} \int_0^\infty \int_0^\infty \alpha \sin\alpha x \cos\beta y \,\mathrm{d}\alpha\,\mathrm{d}\beta = \frac{2P\varepsilon}{\pi^2 l^3} \frac{\partial}{\partial x} \delta(x, y),$$

$$X(x, y) = \frac{P}{\pi^2 l^2} \int_0^\infty \int_0^\infty [\cos \alpha(x+\varepsilon) + \cos \alpha(x-\varepsilon)] 2 \sin(\mathrm{d}\varphi) \cos \beta y \, \mathrm{d}\alpha \, \mathrm{d}\beta$$

$$= \frac{2P\varepsilon}{\pi^2 l^2 R_1} \int_0^\infty \int_0^\infty \cos \alpha x \cos \beta y \, \mathrm{d}\alpha \, \mathrm{d}\beta = \frac{2P\varepsilon}{\pi l^2 R_1} \delta(x, y),$$

where $\sin(\mathrm{d}\varphi) = \sin(\varepsilon/R_1) \cong \varepsilon/R_1$.

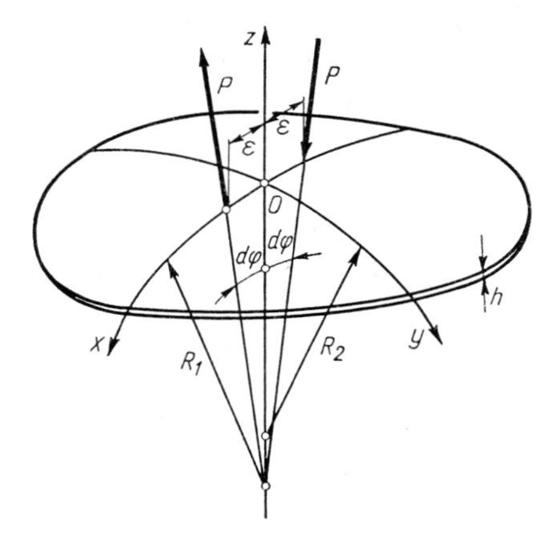

Fig. 84. Concentrated moment as the resultant of two oppositely directed normal forces.

Since, for $\varepsilon \to 0$, $2P\varepsilon\cos(\mathrm{d}\varphi) = 2P\varepsilon = M =$ constant, we have

$$Z(x, y) = \frac{M}{l} \frac{\partial}{\partial x} [\delta(x, y)],$$

$$X(x, y) = \frac{M}{R_1} \delta(x, y).$$

Then the action of two normal forces P produces in the shell the effects corresponding to the loads $Z(x, y)$, $X(x, y)$. Let us denote by $w(M_1)$ the deflection of the shell produced by the concentrated bending moment M_1 being the couple of parallel forces whose vector representation is parallel to the y-axis, $w(P_x)$ is the deflection produced by the force P_x tangential to the surface of the shell and acting in the direction of the x-axis, and $w(P_z)$ is the deflection produced by the concentrated normal force P. Keeping in mind the principle of superposition, we find the deflection $w(M_1)$ from the formula

$$w(M_1) = \frac{M_1}{P}\left\{\frac{1}{l}\frac{\partial}{\partial x} w(P_z) - \frac{1}{R_1} w(P_x)\right\}.$$

A similar relation is valid for the concentrated bending moment M_2 whose vector representation is tangential to the y-axis.

Applying the set of curvilinear coordinates α_1, α_2, we have

$$w(M_1) = \frac{M_1}{P}\left\{\frac{1}{A_1}\frac{\partial}{\partial \alpha_1}[w(P_z)] - \frac{1}{R_1} w(P_x)\right\}.$$

The internal forces and moments in the shell can be obtained in a similar way. We see that the solution of the problem of the shell loaded by a concentrated bending moment can be obtained from the solutions for the concentrated normal and tangential forces. The second term in the above brackets, presenting the deflection produced by the tangential force, is usually much smaller than the first term and is often neglected. Let us write

$$w_1(M_1) = \frac{M_1}{P}\frac{1}{A_1}\frac{\partial}{\partial \alpha_1}[w(P_z)],$$

$$w_2(M_1) = -\frac{M_1}{PR_1} w(P_x);$$

then

$$w(M_1) = w_1(M_1) + w_2(M_1).$$

It can be proved that in the case of a spherical shell the ratio $w_2(M_1)/w_1(M_1)$ is of the order $\sqrt{h/R}$. The solution for the spherical shell resulting from the first term $w_1(M)$ only will be presented in the following section. On differentiation of the results for the concentrated normal force we find

$$w_1 = -\frac{Ml}{2\pi D}\left\{\text{kei}' r - (\eta - \varepsilon)\,\text{ker}' r + \frac{\eta}{4} r\,\text{kei}\, r\right\}\cos\vartheta,$$

$$\Phi_1 = -\frac{MR}{2\pi l}\left\{\frac{1}{r} + \text{ker}' r + \frac{\eta}{2}\left[\left(\frac{3}{2} - \nu\right)\text{kei}' r + \frac{r}{2}\,\text{ker}\, r\right]\right\}\cos\vartheta - \frac{MR\bar{\varepsilon}}{2\pi l}\frac{1}{r}\cos\vartheta. \tag{8.17}$$

Equations (8.17) yield the following expressions for the normal forces, the shear forces, and the bending moments:

$$
\begin{aligned}
N_{rr1} = & -\frac{MR}{2\pi l^3}\left\{\frac{2}{r^2}\left(\frac{1}{r}+\mathrm{ker}' r\right)+\frac{1}{r}\,\mathrm{kei}\, r-\right.\\
& \left.-\frac{\eta}{2}\left[\left(\frac{3}{2}-\nu\right)\frac{1}{r}\left(\mathrm{ker}\, r-\frac{2}{r}\,\mathrm{kei}' r\right)+\frac{1}{2}\,\mathrm{ker}' r\right]\right\}\cos\vartheta-\\
& -\frac{MR\bar{\varepsilon}}{\pi l^3}\frac{\cos\vartheta}{r^3},
\end{aligned}
\tag{8.18}
$$

$$
\begin{aligned}
N_{\vartheta\vartheta 1} = & \frac{MR}{2\pi l^3}\left\{\frac{2}{r^2}\left(\frac{1}{r}+\mathrm{ker}' r\right)+\frac{1}{r}\,\mathrm{kei}\, r-\mathrm{kei}' r-\right.\\
& -\frac{\eta}{2}\left[\left(\frac{3}{2}-\nu\right)\frac{1}{r}\left(\mathrm{ker}\, r-\frac{2}{r}\,\mathrm{kei}' r\right)-\right.\\
& \left.\left.-(2-\nu)\,\mathrm{ker}' r+\frac{1}{2}\,r\,\mathrm{kei}\, r\right]\right\}\cos\vartheta+\frac{MR\bar{\varepsilon}}{\pi l^3}\frac{\cos\vartheta}{r^3},
\end{aligned}
$$

$$
\begin{aligned}
N_{r\vartheta 1} = & -\frac{MR}{2\pi l^3}\left\{\frac{2}{r^2}\left(\frac{1}{r}+\mathrm{ker}' r\right)+\frac{1}{r}\,\mathrm{kei}\, r-\right.\\
& \left.-\frac{\eta}{2}\left[\left(\frac{3}{2}-\nu\right)\frac{1}{r}\left(\mathrm{ker}\, r-\frac{2}{r}\,\mathrm{kei}' r\right)\frac{1}{2}\,\mathrm{ker}' r\right]\right\}\sin\vartheta-\\
& -\frac{MR\bar{\varepsilon}}{\pi l^3}\frac{\sin\vartheta}{r^3}.
\end{aligned}
$$

Solving the set of Eqs. (1.114), we obtain the shear forces in the form of the following integrals:

$$
Q_{r1} = \frac{M}{2\pi l^2}\int_0^\infty\left(\gamma^4-\frac{1}{2}\eta\gamma^2\right)\frac{J_1'(\gamma r)\cos\vartheta\,\mathrm{d}\gamma}{D(\gamma)},
$$

where $J_1' = \dfrac{\mathrm{d}}{\mathrm{d}r}\,[J_1(\gamma r)]$. For the shear force Q_ϑ a similar expression can be found. On integration we obtain

$$
Q_{r1} = \frac{M}{2\pi l^2}\left[\frac{1}{r}\,\mathrm{ker}' r+\mathrm{kei}\, r+\frac{5\eta}{4}\left(\frac{1}{r}\,\mathrm{kei}' r-\mathrm{ker}\, r-\frac{1}{5}\,\mathrm{ker}' r\right)\right]\cos\vartheta,
$$

$$
Q_{\vartheta 1} = \frac{M}{2\pi l^2}\left[\frac{1}{r}\,\mathrm{ker}' r+\frac{5\eta}{4}\left(\frac{1}{r}\,\mathrm{kei}' r+\frac{1}{5}\,\mathrm{ker}\, r\right)\right]\sin\vartheta.
$$

By differentiation according to formulae (1.113) we obtain the following expressions for the bending moments:

$$M_{rr1} = \frac{M}{2\pi l}\left\{\ker' r-(1-\nu)\left(\frac{1}{r}\ker r-\frac{2}{r^2}\,\mathrm{kei}' r\right)+\right.$$

$$\left.+(1-\nu)\eta\left[\frac{2\varepsilon}{r^2\eta}\ker' r-(1-\nu)\frac{1}{r}\,\mathrm{kei}\, r+\frac{(1-\nu)\,\mathrm{kei}' r+r\ker r}{4(1-\nu)}\right]\right\}\cos\vartheta,$$

$$M_{\vartheta\vartheta 1} = \frac{M}{2\pi l}\left\{\nu\ker' r+(1-\nu)\left(\frac{1}{r}\ker r-\frac{2}{r^2}\,\mathrm{kei}' r\right)-\right.$$

$$-(1-\nu)\eta\left[\frac{2\varepsilon}{r^2\eta}\ker' r-(1-\nu)\frac{1}{r}\,\mathrm{kei}\, r-\right. \tag{8.19}$$

$$\left.\left.-\frac{(1+\nu-2\nu^2)\,\mathrm{kei}' r+\nu r\ker r}{4(1-\nu)}\right]\right\}\cos\vartheta,$$

$$M_{r\vartheta 1} = -\frac{M}{2\pi l}\left\{(1-\nu)\left(\frac{1}{r}\ker r-\frac{2}{r^2}\,\mathrm{kei}' r\right)-\right.$$

$$\left.-(1-\nu)\eta\left[\frac{2\varepsilon}{r^2\eta}\ker' r+\frac{\nu}{2}\frac{1}{r}\,\mathrm{kei}\, r-\frac{1}{4}\,r\ker r\right]\right\}\sin\vartheta.$$

When r is small simple approximate forms of Eqs. (8.17)-(8.19) can be given. For the displacement we have

$$w_1 = \frac{Ml}{4\pi D}\left[r\left(\ln r-\frac{1}{2}+\ln 2+\gamma_0-\frac{1}{32}\pi r^2\right)-2(\eta-\varepsilon)\cdot\left(\frac{1}{r}+\frac{\pi}{8}r\right)\right]\cos\vartheta \tag{8.20}$$

where $\gamma_0 = 0.5772$ is Euler's constant. The transverse shear deformation produces singularities of higher order. For $r \to 0$ the deflection increases to infinity. When we neglect this effect the deflection remains finite and for $r = 0$ we have $w = 0$. The approximate forms for small r of Eqs. (8.17), (8.19) are

$$N_{rr1} = -\frac{MR}{16\pi l^3}\left[r\left(-\ln\frac{r}{2}-\gamma_0+\frac{3}{4}\right)+(5\eta-4\varepsilon)\frac{1}{r}+16\bar{\varepsilon}\frac{1}{r^3}\right]\cos\vartheta,$$

$$N_{\vartheta\vartheta 1} = -\frac{MR}{16\pi l^3}\left[3r\left(-\ln\frac{r}{2}-\gamma_0+\frac{5}{12}\right)+(5\eta-4\varepsilon)\frac{1}{r}-16\bar{\varepsilon}\frac{1}{r^3}\right]\cos\vartheta,$$

$$N_{r\vartheta 1} = -\frac{MR}{16\pi l^3}\left[r\left(-\ln\frac{r}{2}-\gamma_0+\frac{3}{4}\right)+(5\eta-4\varepsilon)\frac{1}{r}+16\bar{\varepsilon}\frac{1}{r^3}\right]\sin\vartheta.$$

$$\tag{8.21}$$

The second terms in (8.21) result from the effect of shear deformation and the third from the effect of stresses σ_{zz}. Both are of order h/R and have influence only near the point of application of the load. For small r the approximate forms of expressions for shear forces and bending moment are

$$Q_{r1} = \frac{M}{2\pi l^2}\left[\frac{1}{r}\left(\frac{1}{r}+\frac{\pi}{8}r\right)+\frac{5}{8}\eta\left(\ln\frac{r}{2}+\gamma_0+\frac{9}{10}\right)\right]\cos\vartheta,$$

$$Q_{\vartheta 1} = \frac{M}{2\pi l^2}\left[\frac{1}{r}\left(\frac{1}{r}-\frac{\pi}{8}r\right)+\frac{7}{8}\eta\left(\ln\frac{r}{2}+\gamma_0-\frac{5}{14}\right)\right]\sin\vartheta, \qquad (8.22)$$

$$M_{rr1} = -\frac{M}{4\pi l}\left(\frac{1+\nu}{r}-\frac{3+\nu}{16}\pi r+4(1-\nu)\frac{\varepsilon}{r^3}\right)\cos\vartheta,$$

$$M_{\vartheta\vartheta 1} = -\frac{M}{4\pi l}\left(\frac{1+\nu}{r}-\frac{1+3\nu}{16}\pi r-4(1-\nu)\frac{\varepsilon}{r^3}\right)\cos\vartheta, \qquad (8.23)$$

$$M_{r\vartheta 1} = \frac{1-\nu}{4\pi l}M\left(\frac{1}{r}-\frac{\pi r}{16}+4(1-\nu)\frac{\varepsilon}{r^3}\right)\sin\vartheta.$$

Comparing these results with the corresponding expressions for a flat plate subjected to a concentrated bending moment, we see that they differ only in the second terms in brackets in Eqs. (8.21) and (8.22). This means that at sufficiently small distances from the loading point the shell behaves as a flat plate.

The numerical value of displacement w_1 and stresses in the shell are presented graphically in Fig. 85. The above results have been obtained by means of the equations of the shallow shell theory by the assumption that the shell is infinitely large. It is interesting to compare the above results with those obtained on the basis of more exact equations for a shell of finite dimensions. Such a calculation was performed by Leckie [8.6] in the system of spherical coordinates, taking as a starting point Haver's equations [8.7]. His calculations prove that for thin shells the stresses resulting from a single force have significant values only in the range $0 < \varphi < 20°$, i.e. near the loading point. In the areas of the shell near the edge there appear stresses produced by boundary effects.

Comparing the present solution (by neglecting the effect of the shear deflections) with Leckie's results we have Table 5 (p. 300).

As the second example we take the numerical results obtained by Leckie [8.6] for the case presented in Fig. 86. A hemispherical shell which rests on a rigid

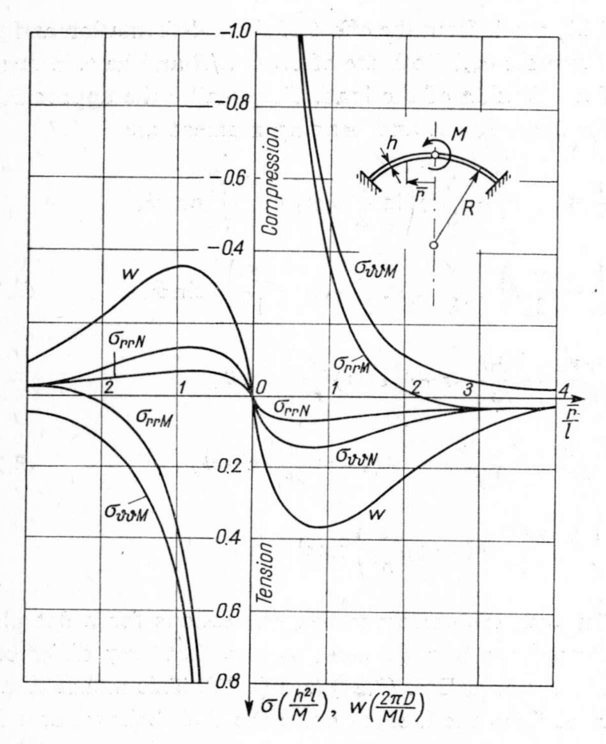

Fig. 85. Membrane and bending stress variation in a spherical shell subjected to concentrated bending moment.

Table 5.

r/l	Present solution	Leckie's solution
1	−0.384	−0.375
2	−0.0143	−0.0137
3	0.0491	0.049

$$\sigma_{rrM} = \frac{h^2 l}{M}, \quad \nu = 0.3, \quad \varkappa = 10.$$

foundation is intersected by an elastic pipe carrying a moment M. The intersection angle is $\varphi_0 = 0.1$ rad, $\varkappa = 10$, for both sphere and pipe and $\nu = 0.3$.

The following boundary conditions were taken for the built-in edge of the sphere:

$$u_s = v_s = w_s = \partial w_s / \partial \varphi = 0.$$

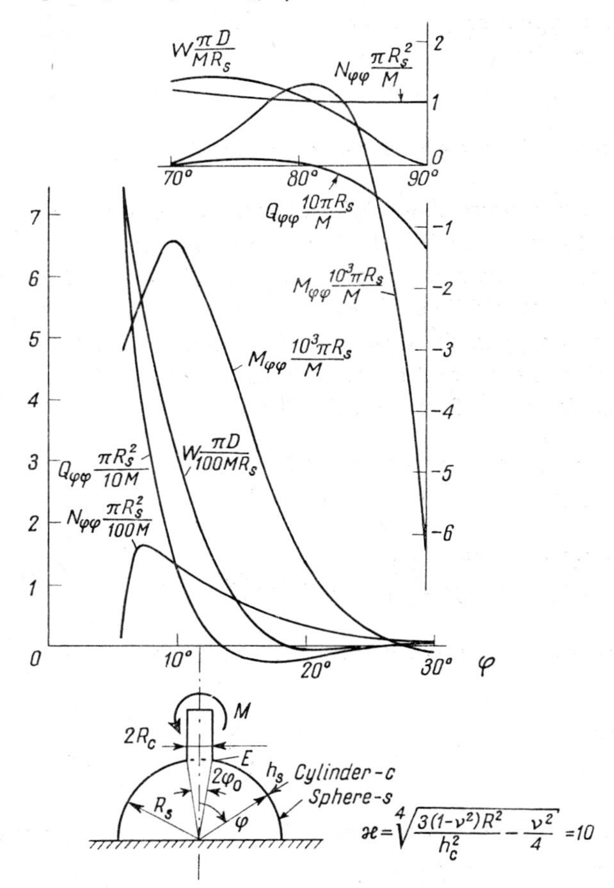

Fig. 86. Displacement and internal forces in a spherical shell loaded through an elastic pipe carrying a moment M (F. A. Leckie [8.6]).

The conditions to be applied at the cylinder-sphere junction were

$$v_c = v_s, \quad u_c = u_s \sin\varphi_0 - w_s \cos\varphi_0,$$

$$w_c = w_s \sin\varphi_0 + u_s \cos\varphi_0.$$

Also, tangent rotations $\beta_c = \beta_s$, shear forces $Q_{xc} = Q_{\varphi s} \sin\varphi_0 - N_{\varphi\varphi} \cos\varphi_0$, and moments $M_{xxc} = M_{\varphi\varphi}$. Here the index c denotes the cylinder, and s the sphere.

The variation of radial deflection and stress resultants shows (Fig. 86) that the bending stresses appear only in the area near the point of application of the load and near the edge.

8.3 Spherical shell loaded by a concentrated tangential force

In this case the effect of transverse shear deformation on the bending of the shell is insignificant and can be neglected. Then we solve the problem by means of classical theory. We assume that the shell is shallow and of infinite extent. The force T acts in its middle surface at the origin of the non-dimensional coordinate system (x, y) in the direction of the x-axis. This force can be expressed as a Fourier integral and substituted into Eq. (8.1).

We find the solution also in the form of a Fourier integral:

$$w = (1+\nu)\frac{Tl^3}{\pi^2 RD}\int_0^\infty\int_0^\infty \frac{\alpha \sin \alpha x \cos \beta y \, d\alpha \, d\beta}{(\alpha^2+\beta^2)[(\alpha^2+\beta^2)^2+1]}, \tag{8.24}$$

$$\Phi = \frac{Tl}{\pi^2}\int_0^\infty\int_0^\infty \frac{1+(\alpha^2+\beta^2)(\beta^2-\nu\alpha^2)}{\alpha(\alpha^2+\beta^2)[(\alpha^2+\beta^2)^2+1]} \sin \alpha x \cos \beta y \, d\alpha \, d\beta. \tag{8.25}$$

On introducing the new variables γ, φ the first of these integrals can be evaluated in the form

$$w = (1+\nu)\frac{Tl^3 \cos\vartheta}{\pi^2 RD}\int_0^\infty \frac{J_1(\gamma r)\, d\gamma}{\gamma^4+1} = (1+\nu)\frac{Tl^3}{2\pi RD}\left(\frac{1}{r}+\mathrm{ker}' r\right)\cos\vartheta. \tag{8.26}$$

Then, we obtain the following expressions for the bending moments:

$$M_{rr} = -(1+\nu)\frac{Tl}{2\pi R}\left\{(1-\nu)\frac{2}{r^2}\left(\frac{1}{r}+\mathrm{ker}' r\right)+(1-\nu)\frac{1}{r}\,\mathrm{kei}\, r-\mathrm{kei}' r\right\}\cos\vartheta,$$

$$M_{\vartheta\vartheta} = -(1+\nu)\frac{Tl}{2\pi R}\left[-(1-\nu)\frac{2}{r^2}\left(\frac{1}{r}+\mathrm{ker}' r\right)-\right. \tag{8.27}$$

$$\left.-(1-\nu)\frac{1}{r}\,\mathrm{kei}\, r-\nu\,\mathrm{kei}' r\right]\cos\vartheta,$$

$$M_{r\vartheta} = -(1-\nu^2)\frac{Tl}{2\pi R}\left[\frac{2}{r^2}\left(\frac{1}{r}+\mathrm{ker}' r\right)+\frac{1}{r}\,\mathrm{kei}\, r\right]\sin\vartheta.$$

Replacing $\text{ker}' r$, $\text{kei}\, r$ by the first two terms of the power series expansion, we obtain from (3.110) an approximate formula for the displacement of the shell near the loading point ($r \ll 1$):

$$w = \frac{1+\nu}{16} \frac{Tl^3}{RD} r\cos\vartheta. \tag{8.28}$$

The corresponding approximate forms of Eqs. (8.27) are

$$\begin{aligned} M_{rr} &= -(1+\nu)\frac{Tl}{16\pi R} r\left[(3+\nu)\left(\ln\frac{r}{2}+\gamma_0\right)-\frac{5+3\nu}{4}\right]\cos\vartheta, \\ M_{\vartheta\vartheta} &= -(1+\nu)\frac{Tl}{16\pi R} r\left[(1+3\nu)\left(\ln\frac{r}{2}+\gamma_0\right)-\frac{3+5\nu}{4}\right]\cos\vartheta, \\ M_{r\vartheta} &= -(1-\nu^2)\frac{Tl}{16\pi R} r\left(\ln\frac{r}{2}+\gamma_0-\frac{3}{4}\right)\sin\vartheta. \end{aligned} \tag{8.29}$$

Equations (8.29) show that all bending moments vanish at $r = 0$.

The membrane forces are calculated from expression (8.25) for the stress function. We can write $\Phi = \Phi_1 + \Phi_2$ where

$$\begin{aligned} \Phi_1 &= \frac{Tl}{\pi^2}\int_0^\infty\int_0^\infty \frac{\sin\alpha x\cos\beta y\,\mathrm{d}\alpha\,\mathrm{d}\beta}{\alpha(\alpha^2+\beta^2)}, \\ \Phi_2 &= -(1+\nu)\frac{Tl}{\pi^2}\int_0^\infty\int_0^\infty \frac{\alpha\sin\alpha x\cos\beta y\,\mathrm{d}\alpha\,\mathrm{d}\beta}{(\alpha^2+\beta^2)^2+1}. \end{aligned} \tag{8.30}$$

The integral for Φ_1 is not convergent and so cannot be evaluated. However, the integration with respect to α only is possible. We find

$$\Phi_1 = \frac{Tl}{2\pi}\int_0^\infty (1-\mathrm{e}^{-\beta x})\frac{\cos\beta y}{\beta^2}\,\mathrm{d}\beta. \tag{8.31}$$

Integral (8.31) is still divergent, but by integrating from some parameter B to infinity, we obtain

$$\begin{aligned} \Phi_1 = -\frac{Tl}{2\pi}\Bigg[& x(\ln r+\ln B+\gamma_0)-y\vartheta+\frac{\mathrm{e}^{-Bx}-1}{B}-\mathrm{Si}(By)+ \\ &+\sum_{k=1}^{\infty}(-B)^k\frac{r^{k+1}\cos[(k+1)\vartheta]}{kk!}\Bigg], \end{aligned} \tag{8.32}$$

where Si denotes the sine integral. As $B \to 0$, $\ln B \to -\infty$ and the other terms in B approach constant values independent of x and y. On calculating the stresses in the shell, the stress function Φ is differentiated twice and the terms which have just been mentioned vanish. This argument shows that the terms containing B can be omitted.

We then arrive at the following expression for Φ_1:

$$\Phi_1 = -\frac{Tl}{2\pi}(r\ln r\cos\vartheta - \vartheta r\sin\vartheta) + \text{const.} \tag{8.33}$$

The integral (8.30) for Φ_2 is convergent and can be evaluated by transforming to polar coordinates, viz.:

$$\Phi_2 = -(1+\nu)\frac{Tl}{2\pi}\int_0^\infty \frac{J_1(\gamma r)\gamma^2}{\gamma^4+1}\,\mathrm{d}\gamma\cdot\cos\vartheta = -(1+\nu)\frac{Tl}{2\pi}\cos\vartheta\,\mathrm{kei}'\,r. \tag{8.34}$$

Summing Eqs. (8.33) and (8.34) gives

$$\Phi = -\frac{Tl}{2\pi}[r\ln r\cos\vartheta - \vartheta r\sin\vartheta + (1+\nu)\mathrm{kei}'r]\cos\vartheta + \text{const.} \tag{8.35}$$

By differentiation we obtain from Eq. (8.35) the following expressions for the membrane forces:

$$\begin{aligned}
N_{rr} &= -\frac{T}{2\pi l}\left[1-(1+\nu)\left(\mathrm{ker}\,r - \frac{2}{r}\,\mathrm{kei}'r\right)\right]\frac{\cos\vartheta}{r},\\
N_{\vartheta\vartheta} &= \frac{T}{2\pi l}\left[[1-(1+\nu)\left(\mathrm{ker}\,r - \frac{2}{r}\,\mathrm{kei}'r - r\,\mathrm{ker}'r\right)\right]\frac{\cos\vartheta}{r},\\
N_{r\vartheta} &= \frac{T}{2\pi l}\left[1+(1+\nu)\left(\mathrm{ker}\,r - \frac{2}{r}\,\mathrm{kei}'r\right)\right]\frac{\sin\vartheta}{r}.
\end{aligned} \tag{8.36}$$

The approximate forms of these results for $r \ll 1$ are

$$\begin{aligned}
N_{rr} &= -(3+\nu)\frac{T}{4\pi l}\frac{\cos\vartheta}{r},\\
N_{\vartheta\vartheta} &= (1-\nu)\frac{T}{4\pi l}\frac{\cos\vartheta}{r},\\
N_{r\vartheta} &= (1-\nu)\frac{T}{4\pi l}\frac{\sin\vartheta}{r}.
\end{aligned} \tag{8.37}$$

These expressions correspond to those derived for an infinite flat plate subject to a concentrated force acting tangentially to its surface (5.11). In particular, the singularities at the loading point are the same.

Some numerical results for the stresses and displacements in a shell with Poisson's ratio $\nu = 0.3$ are presented graphically in Fig. 87.

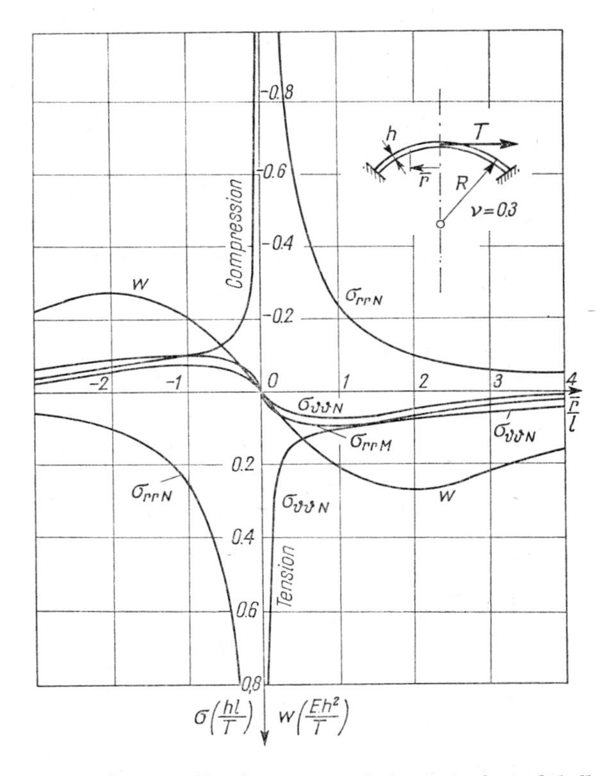

Fig. 87. Membrane and bending stress variation in a spherical shell subjected to tangential load.

8.4 Spherical shell subjected to a twisting moment

Let us consider a spherical shell subjected to a twisting moment M_t about its axis of revolution. The shell is fixed at the boundary $\bar{r} = a$. This case can be solved very simply by considering the equilibrium of the cup cut out by the plane perpendicular to the shell axis. We find that there exists the shear stresses only.

$$\tau_{r\vartheta} = \frac{M_t}{2\pi\bar{r}^2 h}. \tag{8.38}$$

Since the equilibrium equations (1.79) and the equations of compatibility are identically satisfied, we conclude that the membrane and bending stresses are equal to zero. Then nothing but the shear strain $\gamma_{r\vartheta}$ appears in the shell and the displacements $u = w = 0$. The displacement v can be found from Eq. $(1.72)_3$. In spherical coordinates $\alpha_1 = \varphi R$, $\alpha_2 = \bar{r}\vartheta$, and the coefficients of the first quadratic form are $A_1 = R$, $A_2 = \bar{r}$ and we have $u = 0$ and

$$\gamma_{r\vartheta} = \frac{r}{R}\frac{\mathrm{d}}{\mathrm{d}\varphi}\left(\frac{v}{r}\right) = \frac{\tau_{r\vartheta}}{G}. \tag{8.39}$$

On substitution $\tau_{r\vartheta}$ from (8.38) into (8.39) and integrating we find

$$\frac{v}{r} = \frac{M_t}{2\pi R^2 hG}[-\tfrac{1}{2}\operatorname{cosec}\varphi\cot\varphi + \tfrac{1}{2}\ln\tan\tfrac{1}{2}\varphi] + C. \tag{8.40}$$

Since at $r = a$, $v = 0$, we have

$$C = \frac{M_t}{4\pi RhG}\left[\frac{R}{a^2}\sqrt{R^2-a^2} - \ln\left(\frac{a}{R+\sqrt{R^2-a^2}}\right)\right]$$

giving

$$v = \frac{M_t r}{4\pi hG}\left[\frac{\sqrt{R^2-a^2}}{Ra^2} - \frac{\sqrt{R^2-r^2}}{Rr^2} + \right.$$
$$\left. + \frac{1}{R^2}\ln\left(\frac{r}{a}\right)\frac{R+\sqrt{R^2-a^2}}{R+\sqrt{R^2-r^2}}\right]. \tag{8.41}$$

If $R = \infty$, i.e. for the case of the flat plate,

$$v = \frac{M_t r}{4\pi hG}\left[\frac{1}{a^2} - \frac{1}{r^2}\right]. \tag{8.42}$$

In all the above expressions, $G = E/2(1+\nu)$.

If the twisting moment is introduced in the shell by a rigid circular insert, the distribution of stresses does not change.

The shear stresses and displacement variation are presented graphically in Fig. 88.

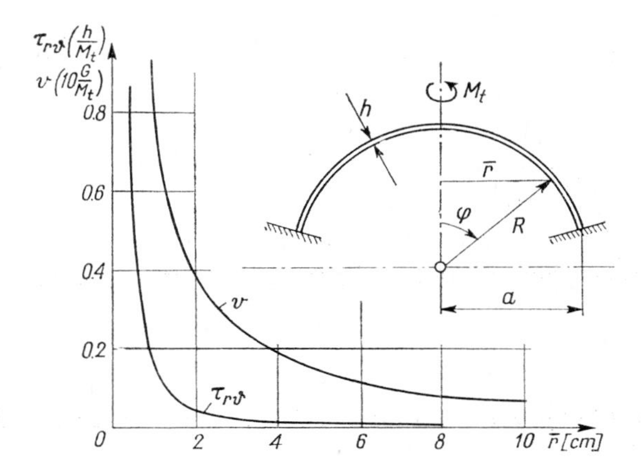

Fig. 88. Shear stress and the displacement v variations in a spherical shell subjected to the twisting moment M_t.

8.5 The effect of boundary conditions. General solution

8.5.1 *Shallow shell*

If we want to satisfy other boundary conditions than those resulting from the assumed particular solution, we have to add to this solution the solution of the homogeneous shell equations. This solution includes a certain number of arbitrary constants. For a shallow spherical shell, Eqs. (8.1) take the following form in non-dimensional coordinates (x, y):

$$\frac{D}{l^4}\Delta\Delta w - \frac{1}{Rl^2}\Delta\Phi = 0,$$
$$\frac{1}{Ehl^4}\Delta\Delta\Phi + \frac{1}{Rl^2}\Delta w = 0. \tag{8.43}$$

In order to solve the above set of equations we introduce a new function $F(x, y)$ and express the stress function Φ and the deflection w in the following way:

$$w = \Delta\Delta F, \qquad \Phi = -\frac{Ehl^2}{R}\Delta F. \tag{8.44}$$

Introducing w and Φ into the second equation of the group (8.43), we find that it is satisfied automatically. Introducing w and Φ in the first equation of the set (8.43), we have the equation

$$\Delta\Delta\left(\Delta\Delta+\frac{Ehl^4}{DR^2}\right)F=0. \tag{8.45}$$

We assume as previously that

$$\frac{Ehl^4}{DR^2}=1, \qquad l=\frac{\sqrt{Rh}}{\sqrt[4]{12(1-\nu^2)}}.$$

We shall look for the general solution of the above equation in the form

$$F=\sum_n F(r)_n \cos n\vartheta. \tag{8.46}$$

Introducing this into Eq. (8.45), we find that the function $F(r)$ must satisfy the equations

$$\begin{aligned}
&\text{(a)}\quad \left(\frac{d^2}{dr^2}+\frac{1}{r}\frac{d}{dr}-\frac{n^2}{r^2}\right)\left(\frac{d^2F(r)}{dr^2}+\frac{1}{r}\frac{dF(r)}{dr}-\frac{n^2}{r^2}F(r)\right)=0,\\
&\text{(b)}\quad \left(\frac{d^2}{dr^2}+\frac{1}{r}\frac{d}{dr}-\frac{n^2}{r^2}\right)\left(\frac{d^2F(r)}{dr^2}+\frac{1}{r}\frac{dF(r)}{dr}-\frac{n^2}{r^2}F(r)\right)+F(r)=0.
\end{aligned} \tag{8.47}$$

The first equation of (8.47) is satisfied when the function $F(r)$ is given by

$$F^{(1)}(r)_n=A_1r^n+A_2r^{-n}+A_3r^{n+2}+A_4r^{-n+2}. \tag{8.48}$$

The second equation can be factored and we obtain two independent equations

$$\begin{aligned}
&\frac{d^2F(r)}{dr^2}+\frac{1}{r}\frac{dF(r)}{dr}+\left(i-\frac{n^2}{r^2}\right)F(r)=0,\\
&\frac{d^2F(r)}{dr^2}+\frac{1}{r}\frac{dF(r)}{dr}-\left(i+\frac{n^2}{r^2}\right)F(r)=0,
\end{aligned} \tag{8.49}$$

where $i=\sqrt{-1}$. If we introduce $x=\sqrt{i}\,r$ into the first equation of (8.49) and $x=\sqrt{-i}\,r$ in the second one, we obtain *Bessel's equation*:

$$\frac{d^2F^{(2)}}{dx^2}+\frac{1}{x}\frac{dF^{(2)}}{dx}+\left(1-\frac{n^2}{x^2}\right)F^{(2)}=0.$$

The solutions of this equation are the Bessel functions of the first and second type and nth order. The Bessel functions of nth order are considered the functions of the first type and the Hankel functions of the nth order — the functions of the second type.

If we introduce again the real variable r and separate the imaginary part from the real one, we obtain four independent functions representing the solution of Eqs. (8.47). These functions are the Thomson functions of nth order. The general solution of Eq. (8.47) takes the form

$$F(r)_n = C_1 \operatorname{ber}_n r + C_2 \operatorname{bei}_n r + C_3 \operatorname{ker}_n r + C_4 \operatorname{kei}_n r + \\ + C_5 r^{n+2} + C_6 r^{-n+2} + C_7 r^n + C_8 r^{-n}. \tag{8.50}$$

The deflection w is given by $w = \Delta\Delta F$. But from Eq. (8.45) we see that $\Delta\Delta F^{(2)} = - F^{(2)}$ and $\Delta\Delta F^{(1)} = 0$. Then

$$w_n = -(C_1 \operatorname{ber}_n r + C_2 \operatorname{bei}_n r + C_3 \operatorname{ker}_n r + C_4 \operatorname{kei}_n r) \cos n\vartheta \tag{8.51}$$

and

$$\Phi_n = -\frac{Ehl^2}{R} \Delta F = \frac{Eh^2}{\sqrt{12(1-\nu^2)}} [-C_1 \operatorname{bei}_n r + C_2 \operatorname{ber}_n r - \\ - C_3 \operatorname{kei}_n r + C_4 \operatorname{ker}_n r + C_5 r^n + C_6 r^{-n}] \cos n\vartheta. \tag{8.52}$$

If $n = 0$, i.e. in the case of symmetrical loading, the solution takes the form

$$\begin{aligned} w &= -(C_1 \operatorname{ber} r + C_2 \operatorname{bei} r + C_3 \operatorname{ker} r + C_4 \operatorname{kei} r), \\ \Phi &= \frac{Eh^2}{\sqrt{12(1-\nu^2)}} [-C_1 \operatorname{bei} r + C_2 \operatorname{ber} r - C_3 \operatorname{kei} r + C_4 \operatorname{ker} r + C_5 + C_6 \ln r], \end{aligned} \tag{8.53}$$

where $\operatorname{ker}_0 r = \operatorname{ker} r$, $\operatorname{kei}_0 r = \operatorname{kei} r$, etc.

The functions $\operatorname{ber}_n r$, $\operatorname{bei}_n r$, $\operatorname{ker}_n r$, $\operatorname{kei}_n r$ are the Thomson functions of nth order and are related to the Bessel functions by the formulae

$$\begin{aligned} \operatorname{ber}_n r + i \operatorname{bei}_n r &= J_n(re^{3\pi i/4}) = i^n I_n(r\sqrt{i}), \\ \operatorname{ker}_n r + i \operatorname{kei}_n r &= \tfrac{1}{2}\pi H_n(re^{-3\pi i/4}) = i^n K_n(r\sqrt{i}). \end{aligned} \tag{8.54}$$

The second derivatives of these functions are given by the relations

$$\begin{aligned} \operatorname{ber}_n'' r &= -\operatorname{bei}_n r - \frac{1}{r} \operatorname{ber}_n' r + \frac{n^2}{r^2} \operatorname{ber}_n r, \\ \operatorname{bei}_n'' r &= \operatorname{ber}_n r - \frac{1}{r} \operatorname{bei}_n' r + \frac{n^2}{r^2} \operatorname{bei}_n r, \\ \operatorname{ker}_n'' r &= -\operatorname{kei}_n r - \frac{1}{r} \operatorname{ker}_n' r + \frac{n^2}{r^2} \operatorname{ker}_n r, \\ \operatorname{kei}_n'' r &= \operatorname{ker}_n r - \frac{1}{r} \operatorname{kei}_n' r + \frac{n^2}{r^2} \operatorname{kei}_n r. \end{aligned} \tag{8.55}$$

Unfortunately, the use of these functions depends on having the tables of Thomson's functions of higher orders, which are not well tabulated in the literature. (See Tables 13, 14, pp. 547, 549.)

The above solutions of Eqs. (8.43) enable one to solve the problem of the shell arbitrarily loaded and with arbitrary boundary conditions. For example, in the case of the concentrated bending moment applied to the shell at the apex the satisfaction of the boundary conditions requires the solution of Eqs. (8.47) for $n = 1$.

8.5.2 *Non-shallow spherical shell*

Let us consider a non-shallow spherical shell in geographic coordinates (φ, ϑ). The differential equations (1.222) can be replaced by the three differential equations (8.56) for three unknown functions ψ_i, viz.

$$\begin{aligned} \Delta\psi_1+\mu_1\psi_1 &= 0, \\ \Delta\psi_2+\mu_2\psi_2 &= 0, \\ (\Delta+2)^2\psi_3 &= 0, \end{aligned} \tag{8.56}$$

where

$$\mu_{1,2} = 1 \pm \frac{2R}{h} i\sqrt{3(1-\nu^2)}.$$

Since $\alpha_1 = \varphi$, $\alpha_2 = \vartheta$ (Fig. 2), the coefficients of the first quadratic form are $A_1 = R$, $A_2 = R\sin\varphi$. The Laplace operator takes the form

$$\Delta = \frac{1}{\sin\varphi R^4}\left[\frac{\mathrm{d}}{\mathrm{d}\varphi}\left(\sin\varphi\frac{\mathrm{d}}{\mathrm{d}\varphi}\right)+\frac{1}{\sin\varphi}\left(\frac{\mathrm{d}^2}{\mathrm{d}\vartheta^2}\right)\right]. \tag{8.57}$$

The solution of Eqs. (8.56) can be obtained in the form of the series

$$\psi_k = \sum_{n=0}^{\infty} \psi_{kn} \begin{matrix}\cos n\vartheta \\ \sin n\vartheta\end{matrix}. \tag{8.58}$$

Substituting the above relation into Eqs. (8.56), we find that the functions ψ_{kn} are the solutions of the following ordinary differential equation:

$$\frac{1}{\sin\varphi}\frac{\mathrm{d}}{\mathrm{d}\varphi}\left[\sin\varphi\frac{\mathrm{d}\psi_{kn}}{\mathrm{d}\varphi}\right]+\left(\mu_k-\frac{n^2}{\sin^2\varphi}\right)\psi_{kn} = 0, \tag{8.59}$$

where μ_k is the parameter which takes the values μ_1, μ_2 and 2.

Introducing the new variable $x = \cos\varphi$, the above equation can be written in the form

$$\frac{\mathrm{d}}{\mathrm{d}x}\left[(1-x^2)\frac{\mathrm{d}\psi_{k,n}(x)}{\mathrm{d}x}\right]+\left[m_k(m_k+1)-\frac{n^2}{1-x^2}\right]\psi_{kn}(x) = 0, \tag{8.60}$$

where

$$\left.\begin{matrix} m_1 \\ m_2 \end{matrix}\right\} = -\frac{1}{2}+\sqrt{\frac{R}{h}\sqrt{3(1-\nu^2)}+\frac{5}{8}} \pm \mathrm{i}\sqrt{\frac{R}{h}\sqrt{3(1-\nu^2)}-\frac{5}{8}},$$

$$m_3 = 1.$$

Equation (8.60) for $k = 3$ has a solution in the form of Legendre polynomials. The general solution can be presented in the form

$$\psi_{kn}(x) = A_k P^n_{mk}(x)+B_k Q^n_{mk}(x) \tag{8.61}$$

where A_k, B_k are arbitrary complex constants, and P^n_{mk}, Q^n_{mk} are the Legendre functions of first and second kind with complex argument. This means that the general solution of the fundamental equation for the spherical shell takes the form

$$\psi = \psi_0+\sum_{k=1}^{3}\sum_{n=1}^{\infty}[A_k P^n_{mk}(x)+B_k Q^n_{mk}(x)]^{\cos n\vartheta}_{\sin n\vartheta}, \tag{8.62}$$

The functions $P^n_m(x)$, $Q^n_m(x)$ for arbitrary m and integral values n can be presented in the form of the hypergeometrical function:

$$\begin{aligned} P^n_m(x) &= \frac{1}{\Gamma(1-n)}\left(\frac{x+1}{x-1}\right)^{n/2} F\left(-m, m+1; 1-n; \frac{1-x}{2}\right), \\ Q^m_n(x) &= \frac{\mathrm{e}^{n\pi i}}{2^{m+1}}\frac{\Gamma(m+n+1)}{\Gamma(m+\frac{3}{2})}\frac{\Gamma(\frac{1}{2})(x^2-1)^{n/2}}{x^{m+n+1}}\times \\ &\times F\left(\frac{m+n}{2}+1; \frac{m+n+1}{2}; m+\frac{3}{2}; \frac{1}{x^2}\right), \end{aligned} \tag{8.63}$$

where

$$\begin{aligned} F(\alpha, \beta, \gamma, x) = 1&+\frac{\alpha\beta}{\gamma\cdot 1}x+\frac{\alpha(\alpha+1)\beta(\beta+1)}{\gamma(\gamma+1)1\cdot 2}x^2+ \\ &+\frac{\alpha(\alpha+1)(\alpha+2)\beta(\beta+1)(\beta+2)}{\gamma(\gamma+2)1\cdot 2\cdot 3}x^3+\ldots \end{aligned}$$

is the hypergeometrical function.

The derivatives of the functions P_m^n and Q_m^n can be obtained from the recurrent relations:

$$(x^2-1)\frac{dP_m^n}{dx} = \sqrt{x^2-1}\,P_m^{n+1}+xnP_m^n,$$

$$\frac{dP_m^n}{d\varphi} = -(m+1)\cot\varphi P_m^n+\frac{m+1-n}{\sin\varphi}P_{m+1}^n.$$

The same relations are valid for the functions $Q_m^n(x)$. For thin shells $|m| \gg 1$ the functions P_m^n and Q_m^n can be presented in the form of the asymptotic solution:

$$P_m^n(\cos\varphi) = m^n\sqrt{\frac{2}{m\pi\sin\varphi}}\left\{1+\frac{(2n-1)(2n+3)}{8m}\cos\left[\left(m+\frac{1}{2}\right)\varphi+\right.\right.$$

$$\left.\left.+\left(n-\frac{1}{2}\right)\frac{\pi}{2}\right]-\frac{n^2-\frac{1}{4}}{2m\sin\varphi}\cos\left[\left(m+\frac{3}{2}\right)\varphi+\left(n-\frac{3}{2}\right)\frac{\pi}{2}\right]+\ldots\right\},$$

$$Q_m^n(\cos\varphi) = m^n\sqrt{\frac{\pi}{2m\sin\varphi}}\left\{1+\frac{(2n-1)(2n+3)}{8m}\cos\left[\left(m+\frac{1}{2}\right)\varphi+\right.\right.$$

$$\left.\left.+\left(n+\frac{1}{2}\right)\frac{\pi}{2}\right]+\frac{n^2-\frac{1}{4}}{2m\sin\varphi}\cos\left[\left(m+\frac{3}{2}\right)\varphi+\left(n+\frac{3}{2}\right)\frac{\pi}{2}\right]+\ldots\right\}.$$

In the axisymmetrical case, in relations (8.63) one should assume that $n = 0$.

Solution by means of the auxiliary complex function

The solution of Eq. (1.222) can be presented also in another form, namely by means of an auxiliary complex function. If we multiply Eq. $(1.120)_2$ by $(\pm 2i\varkappa^2-\nu)/EhR$ where $\varkappa^2 = \sqrt{3(1-\nu^2)}\,R/h$ and add both Eqs. (1.120), we find:

$$\left(\Delta+\frac{2}{R^2}\right)\left[\left(\Delta+\frac{1\pm 2i\varkappa^2}{R^2}\right)\left(w-\frac{\nu\mp 2i\varkappa^2}{EhR}\Phi\right)+\frac{\nu^2}{EhR^3}\right]=\frac{Z}{D}.$$

Since for thin shells h^2/R^2 is small in comparison with unity, we can neglect the last term in the brackets. Introducing the auxiliary function ψ,

$$\psi = w-\frac{\nu\mp 2i\varkappa^2}{EhR}\Phi, \tag{8.64}$$

we find the equation

$$\left(\Delta+\frac{2}{R^2}\right)\left(\Delta+\frac{\pm 2i\varkappa^2+1}{R^2}\right)\psi = \frac{Z}{D}. \tag{8.65}$$

Let us consider the homogeneous equation. Its solution is given by the sum

$$\psi = \psi_1+\psi_2+\psi_3,$$

where ψ_1, ψ_2 and ψ_3 are the solutions of the equations

$$\left(\Delta+\frac{\pm 2\,i\,\varkappa^2+1}{R^2}\right)\psi_{1,2} = 0, \qquad \left(\Delta+\frac{2}{R^2}\right)\psi_3 = 0.$$

The asymptotic solution of the first of the above equations can be obtained as the combination of the trigonometric functions and the Kelvin functions $\operatorname{ker} r$, $\operatorname{kei} r$.

If we represent the solution in the form of the series (8.58) we find the following solutions for the second equation, $k = 3$:
for $n = 0$:

$$\psi_{3,0} = A_0\left[\cos\varphi \ln\left(\frac{1-\cos\varphi}{1+\cos\varphi}\right)+2\right]+B_0\cos\varphi,$$

for $n = 1$:

$$\psi_{3,1} = A_1\left[\sin\varphi \ln\left(\frac{1-\cos\varphi}{1+\cos\varphi}\right)-2\cot\varphi\right]+B_1\sin\varphi,$$

for $n = 2$:

$$\psi_{3,n\geqslant 2} = \frac{A_n}{n(n^2-1)}\left(\frac{1-\cos\varphi}{1+\cos\varphi}\right)^{n/2}(n+\cos\varphi)+$$

$$+\frac{B_n}{n(n^2-1)}\left(\frac{1+\cos\varphi}{1-\cos\varphi}\right)^{n/2}(n-\cos\varphi), \tag{8.66}$$

where A_n and B_n are constants of integration which can be complex. Let us analyze the above solutions. For $n = 0$ and $n = 1$ they are singular at the point $\varphi = 0$. This can mean that it corresponds to certain concentrated loads. For example, let us take into consideration the solution for $n = 0$

$$\psi = A_0\left[\cos\varphi \ln\left(\frac{1-\cos\varphi}{1+\cos\varphi}\right)+2\right]+B_0\cos\varphi,$$

which gives the following value of the membrane forces:

$$N_{\varphi\varphi} = -\frac{2A_0}{\sin^2\varphi} = -N_{\vartheta\vartheta}, \qquad N_{\varphi\vartheta} = 0.$$

Assuming $2A_0 = -P/2\pi R$ we obtain results identical with those obtained in § 7.2 by means of the membrane theory. The normal deflection corresponding to this state of stress is

$$w = \frac{(1+\nu)P}{2\pi Eh}\left[1+\frac{1}{2}\cos\varphi\ln\left(\frac{1-\cos\varphi}{1+\cos\varphi}\right)\right]-C\cos\varphi.$$

The deflections u and v can be found by integrating the equations

$$\varepsilon_{\varphi\varphi} = \frac{1}{Eh}(N_{\varphi\varphi}-\nu N_{\vartheta\vartheta}) = \frac{1}{R}\left(\frac{du}{d\varphi}+w\right),$$

$$\varepsilon_{\vartheta\vartheta} = \frac{1}{Eh}(N_{\vartheta\vartheta}-\nu N_{\varphi\varphi}) = \frac{1}{R}(u\cot\varphi+w),$$

which gives

$$u = \frac{(1+\nu)P}{2\pi Eh}\left\{\left(\cot\varphi-\frac{1}{2}\sin\varphi\ln\left(\frac{1-\cos\varphi}{1+\cos\varphi}\right)\right\}+C\sin\varphi,\right.$$

$$v = 0.$$

The constant C corresponds to the vertical displacement of the shell as a rigid body and can be disregarded. It can be shown that the bending moments and shear forces resulting from this solution are equal to zero, viz.:

$$M_{\varphi\varphi} = M_{\vartheta\vartheta} = M_{\varphi\vartheta} = Q_\varphi = Q_\vartheta = 0.$$

If $n = 1$, the singular solution corresponding to the action of a concentrated bending moment M and tangential force T can be obtained. We find

$$N_{\varphi\varphi} = -N_{\vartheta\vartheta} = \frac{M\cos\vartheta}{\pi R^2\sin^3\varphi}+\frac{T(1-\cos\varphi)\cos\vartheta}{\pi R\sin^3\varphi},$$

$$M_{\varphi\vartheta} = \frac{M\cos\varphi}{\pi R^2\sin^3\varphi}\sin\vartheta-\frac{T(1-\cos\varphi)\sin\vartheta}{R\sin^3\varphi},$$

$$w = \frac{(1+\nu)T}{4\pi Eh}\left[\sin\varphi\ln\left(\frac{1-\cos\varphi}{1+\cos\varphi}\right)-2\cot\varphi\right]\cos\vartheta, \tag{8.67}$$

$$u = \frac{1+\nu}{4\pi Eh}\left(T+\frac{M}{R}\right)\left[\ln\left(\frac{1-\cos\varphi}{1+\cos\varphi}\right)-\frac{2\cot\varphi}{\sin\varphi}\right]\cos\vartheta,$$

$$v = -\frac{1+\nu}{4\pi Eh}\left(T+\frac{M}{R}\right)\left[2+\frac{2}{\sin^2\varphi}+\cos\varphi\ln\left(\frac{1-\cos\varphi}{1+\cos\varphi}\right)\right]\sin\vartheta.$$

The shear forces Q_r, Q_ϑ resulting from the above solution are equal to zero, and the bending moments are nearly zero (of the order h^2/R^2)

$$M_{\varphi\varphi} = -M_{\vartheta\vartheta} = O\left(\frac{h^2}{R^2}\right)\left(T+\frac{M}{R}\right).$$

J. G. Simmonds [8.24] also obtained the Green function for closed spherical shell. The Green function was built of two parts, Γ and G, each of them being obtained as the fundamental solution of the differential equation of second order with complex coefficients. Γ denotes a rapidly decreasing function expressed by Legendre functions, and G a slowly variable function given by elementary functions. Γ and G differ slightly from the solutions Γ^s and G^s obtained by means of the shallow shell theory. Simmonds proved that $|\Gamma-\Gamma^s| = O(h/R)$ and $|G-G^s| = O[h/R\ln(h/R)]$. The first estimation concerns itself with the whole area of the shell, the second — only the region of its shallow part surrounding the singular point.

L. I. Liotko [8.25] solved the problem of a closed spherical shell loaded by two concentrated normal forces at the poles by using the equations of three-dimensional theory of elasticity. He found that at the points distant from the point of the application of the load, the state of stress and displacement agrees with that resulting from the equations of the shell theory. However, at the region surrounding the singular point, the problem cannot be considered as two-dimensional.

Iu. A. Ustinov [8.26] considered the problem of the spherical shell loaded at the poles by the concentrated normal forces based also on the three-dimensional theory of elasticity. He obtained the solution in the form of the series of Legendre functions. The papers by N. A. Poliakov and Iu. A. Ustinov [8.27, 8.28] are devoted to the asymptotic solutions of the theory of elasticity for the problem of the closed spherical shell under concentrated loads. The term $\varepsilon = \ln(R_2/R_1)$ was asumed to be a small parameter, where R_1 and R_2 were the external and internal radii of the shell.

References 8

[8.1] E. Reissner: Stresses and small displacements of shallow spherical shells, II, *J. Math. Phys. 25* (1946) 279.

[8.2] W. Flügge and D. A. Conrad: Singular solutions in the theory of shallow shells, *Stanford University Div. Eng. Mech. Tech. Report No. 101* (1956) 10.

[8.3] A. Kalnins and P. M. Nahdi: Propagation of axisymmetric waves in an unlimited elastic shell, *J. Appl. Mech.* 27, *Trans. ASME 82, ser. E* (1960) 690–695.

[8.4] R. P. NORDGREN: On the method of Green's function of the thermoelastic theory of shallow shells, *Int. J. Eng. Sci.* 1 (1963) 279–308.

[8.5] A. KALNINS: On fundamental solutions and Green's function in the theory of elastic plates, *J. Appl. Mech.* 33, *Trans. ASME, 88, serie E* (1966) 31–38.

[8.6] F. A. LECKIE: Localized loads applied to spherical shells, *J. Mech. Eng. Sci.* 3, 2 (1961) 111–118.

[8.7] A. HAVER: Asymptotische Biegetheorie der unbelasteten Kugelschale, *Ing.-Arch. 6* (1935) 282–308.

[8.8] R. E. LANGER: On the asymptotic solutions of ordinary differential equations with reference to the Stokes' phenomenon about a singular point, *Trans. Amer. Math. Soc.* 37 (1938).

[8.9] J. P. WILKINSON and A. KALNINS: On non-symmetric dynamics problems of elastic spherical shells, *J. Appl. Mech. 32, Trans. ASME* (1965) 525–529.

[8.10] P. P. BIJLAARD: On the stresses from local loads in spherical pressure vessels, *Bull. U. S. Weld. Res. Comm. 34* (1956).

[8.11] J. CHINN: *Influence charts for deflection and stresses in spherical shells*, Ph. D. Thesis, Cornell University, 1958.

[8.12] J. P. WILKINSON and A. KALNINS: Deformation of open spherical shells under arbitrary located concentrated loads, *J. Appl. Mech. Trans. ASME 6* (1966) 305–312.

[8.13] S. ŁUKASIEWICZ: Concentrated loads on shallow spherical shells, *Quart. J. Mech. Appl. Math. 20, 3* (1967) 293–305.

[8.14] V. F. CHIZHOV: Deformation of non-shallow spherical shell under local load, *Izv. Vuzov. Mashinostroenie 2* (1973) (in Russian).

[8.15] F. A. LECKIE: Solutions for the spherical shell subjected to axially symmetric loading, *Symp. Nuclear Reactor Containtment Buildings and Pressure Vessels*, Glasgow 1960.

[8.16] A. S. TOOTH: An experimental investigation of shallow spherical domes subjected to a variety of load actions, *Symp. Nuclear Reactor Containment Buildings and Pressure Vessels*, Glasgow 1960.

[8.17] R. M. KENEDI: Influence line methods of shell analysis, *Symp. Nuclear Reactor Containment Buildings and Pressure Vessels*, Glasgow 1960.

[8.18] A. S. TOOTH and R. M. KENEDI: The influence line technique of shell analysis, *Proc. Coll. Simpl. Calcul. Meth.*, Brussels, 1961.

[8.19] M. A. MEDICK: On the initial response of a spherical shell to a concentrated force, *J. Appl. Mech. 29, 4* (1962) 689.

[8.20] B. M. GATKOV: Shallow spherical shell under action of concentrated forces, *Sb. Rab. Aspir. Mekh. Mat. Fiz. Fak. Lvov. Un-ta*, 1963 (in Russian).

[8.21] H. AINSO and M. A. GOLDBERG: Analysis of a shallow spherical shell under eccentrically applied concentrated load. Pressure vessel technology, Part I: Design and analysis, *Proceedings of the First International Conference on Pressure Vessel Technology*, Delft, 1967.

[8.22] S. ŁUKASIFWICZ: The effect of transverse shear deformations in the locally loaded spherical shell, *Arch. Bud. Masz. 16, 4* (1969).

[8.24] J. G. SIMMONDS: Green's functions for closed, elastic spherical shells, exact and accurate aproximate solutions, *Proc. Koninkl. Ned. Akad. Wet. 71.3* (1968).

[8.25] L. I. LIOTKO: The action of two normal concentrated forces on the closed spherical shell applied at the poles, *Izv. AN SSSR, MTT 2* (1970) (in Russian).

[8.26] Iu. A. USTINOV: Transition from the three-dimensional problem of the theory of elasticity to a two-dimensional one for the closed spherical shell under singular external loads, in: *Tr. VI Vses. Konf. po Teor. Obolochek i Plastinok*, „Nauka", Moscow (in Russian).

[8.27] N. A. POLIAKOV, Iu. A. USTINOV: Asymptotic solution of the problem of the theory of elasticity for a spherical shell under singular loads, *P.M.M. 32.1* (1968) (in Russian).

[8.28] N. A. POLIAKOV, Iu. A. USTINOV: Investigation of the asymptotic solution of the problem of the theory of elasticity near the concentrated force for a closed shell, in: *Tr. VII Vses. Konf. po Teor. Obolochek i Plastinok*, „Nauka", Moscow, 1970 (in Russian).

[8.29] N. G. GURIANOV: Spherical shell under loading uniformly distributed over the plane, in: *Issl. po teor. plast.* i *obolochek, vyp. 5*, Kazan 1967 (in Russian).

[8.30] A. V. SACHENKOV, Iu. G. KONOPLEV: Investigation of strength and stability of shallow spherical shells locally loaded, in. *Issl. po teor. plast. i obolochek*, vyp. 5, Kazan 1967 (in Russian).

[8.31] D. N. MITRA; Bending of simply-supported shallow spherical shell under a uniform line load on a part of a parallel circle, *Proc. Cambridge Philos. Soc. 63.1* (1967).

[8.32] J. N. ROSSETOS: Deformation of shallow spherical shell sandwich under local loading, *NASA Techn. Note D-3855*, 1967.

[8.33] A. A. VESELYI: On determination of internal forces and moments in a spherical pneumatic shell produced by the forces normal to the middle surface, *Stroit. konstruktsii, Mezhved. Resp. Nauchn. sb. 7* (1968) (in Russian).

[8.34] V. N. GNATYKIV, E. I. TRUSH: Shallow sandwich spherical shells under the action of concentrated forces, *Dinamika i prochn. Mashin. Resp. mezhved. Nauch. tekhn. sb. 8* (1968) (in Russian).

[8.35] N. G. GURIANOV: Shallow spherical shell loaded by a force distributed over circular surface whose center is at an arbitrary point, *Sb. aspir. rabot. Kazansk. un-ta, Tochnye Nauki*, Kazan 1968 (in Russian).

[8.36] A. N. DOBROMYSLOV: Investigation of spherical shells of revolution under concentrated circular loads, *Beton i zhelezobeton 7* (1968) (in Russian).

[8.37] A. F. ULITKO: State of stress of a sphere with cavity loaded by concentrated forces, *Prikladnaia Mekhanika 4.5* (1968) (in Russian).

[8.38] P. J. HARRIS: Stresses in a sperical shell under narrow ring loads, *Engng. J. (Canada) 51.11* (1968).

[8.39] V. N. ZALESOV, G. A. KOLOSOV: Axisymmetrical displacement caused by concentrated forces in thin spherical shell under internal pressure, *Sb. mater. 6 Nauch. tekhn. konf. Mogilevsk. Mashinostroit. In-ta*, Mogilev 1969 (in Russian).

[8.40] E. REISSNER, F. WAN: Rotationally symmetric stress and strain in shells of revolution, *Stud. Appl. Math. 48.1* (1969).

[8.41] L. E. VASILEVA, A. S. KHRISTENKO: Spherical shell under the action of local load, *Sudost. i morsk. sooruzh., Resp. Mezhved. Nauch. Tekhn. sb. 14* (1970) (in Russian).

[8.42] N. G. GURIANOV: Action of local loads on non-shallow spherical shell, in: *Issledovania po teorii plastin i obolochek, Kazan, 6–7* (1970) (in Russian).

[8.43] A. R. YACOUD, M. HANNA: Shallow spherical shells under hydrostatic pressure or couple nucleus, *Bull. Calcutta Math. Soc. 62.4* (1970).

[8.44] A. K. GALINSH, N. G. GURIANOV: Action of local loads on shallow transverse-isotropic spherical shell, *Sb. aspir. rabot Kazansk. Un-ta, Teoria Plastin i Obolochek 1* (1971) (in Russian).

[8.45] A. S. KHRISTENKO, A. N. KHOMCHENKO: On the action of a concentrated load on an orthotropic shallow sphere, *Tr. Nikolaevsk. Korablestroi. In-ta 60* (1971) (in Russian).

[8.46] A. K. GALINSH, N. G. GURIANOV: Displacement of transverse isotropic spherical plate and shallow spherical segment caused by local load, *Soprotivl. Materialov i Teor. Sooruzj. Resp. Mezhved. Nauch.-Tekhn. sb. 16* (1972) (in Russian).

[8.47] A. K. GALINSH, N. G. GURIANOV: On mathematical analogies in theory of shallow spherical shells and plates, taking into account shear loads, *Mekh. polimerov 2* (1972) (in Russian).

[8.48] N. A. POLIAKOV: On compression of a spherical shell, in: *Matematika, nekotorye ee prilozhenia i metodika prepodavania*, Rostov n/Donu 1972 (in Russian).

[8.49] R. C. GWALTNEY: Localized loads applied to torispherical shells, *Nucl. Eng. and Des. 23.1* (1972).

[8.50] A. K. GALINSH, N. G. GURIANOV: Deflection of transverse isotropic spherical plates and shallow spheres by local load, in *Teoria obolochek i plastin, M. "Nauka"*, 1973 (in Russian).

[8.51] E. I. TRUSH: State of stresses of components made of nonhomogeneous in thickness material having form of shallow spherical shells, in *Lesn. Khoz-vo, lesnaia bumazh. i derevoobr. promyshlennost, Kiev, 2* (1973) (in Russian).

9

An arbitrary shell loaded by a normal concentrated force

Let us consider now an arbitrary shell (Fig. 89) loaded by a concentrated orce P normal to the middle surface of the shell. Let us introduce the local. fnon-dimensional system of coordinates (x, y) with a characteristic length l The origin of the coordinates is at the point of application of the load.

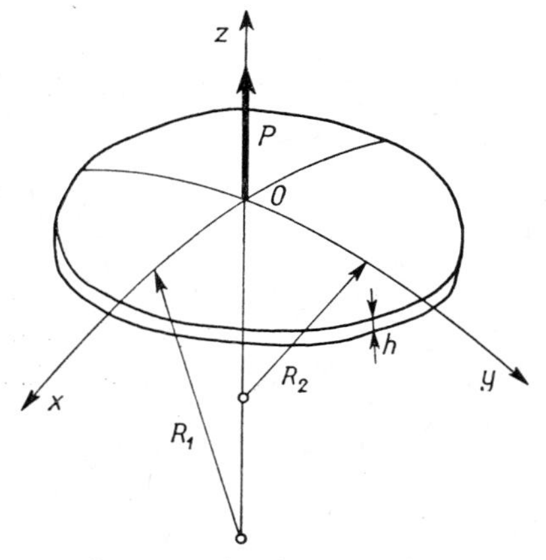

Fig. 89. Arbitrary shell subjected to a normal concentrated force.

The directions of the axes follow the directions of the main curvature of the middle surface. The main radii of curvature are denoted by R_1, R_2. We known from the previous chapter that the effects produced by the concentrated load on the spherical shell are of local character.

It can be expected that also in the case of an arbitrary shell the deflections and stresses decrease rapidly with increasing distance from the point of application of the load. As first of all we are interested in the state of stress at this point, we assume, to simplify the calculations, that in the area close to this point the radii R_1 and R_2 and the coefficients of the first quadratic form are constant. Then we can write $A_1 = A_2 = 1$. This simplification will enable us to achieve the solution of the problem by means of differential equations with constant coefficients [9.1]. We assume further that the shell extends infinitely in all directions. This assumption is made on order to avoid the influence of boundary effects on the state of stress and deflection near the point of application of the load. In order to simplify the numerical calculations we present the basic equations (1.99), (1.101) in a more simple form. Namely, we can write

$$D\left(\Delta+\frac{1}{R_1^2}-\frac{1-\nu}{2R_1R_2}+\frac{1}{R_2^2}\right)^2 w-(1-\eta\Delta)\Delta_k\Phi = [1-(\eta-\varepsilon)\Delta]Z,$$

$$\frac{1}{Eh}\left(\Delta+\frac{1}{R_1R_2}\right)^2\Phi+\Delta_k w = -\frac{\bar{\varepsilon}R_2}{Eh}\Delta Z, \quad \text{where} \quad \bar{\varepsilon} = \frac{\nu h}{2R_2}. \tag{9.1}$$

It is easy to prove that the error appearing while replacing Eq. (1.103) by the new Eqs. (9.1) is smaller than the ratio L^4/R^4 which is usually very small in comparison with unity. Here L is the characteristic wave length (see § 1.11.1).

9.1 Deflection and stress function

The force P acting on the shell can be expressed as a double Fourier integral (3.65). We assume that the solution of the shell equations can also be presented in the form of Fourier integrals.

On substitution in Eq. (9.1) we obtain the deflection of the shell w and the stress function Φ in the form of integrals:

$$w = \frac{Pl^2}{\pi^2 D}\int_0^\infty\int_0^\infty \frac{(\alpha^2+\beta^2-\mu_R)^2}{D(\alpha,\beta)}\cos\alpha x\cos\beta y\,d\alpha\,d\beta,$$

$$\Phi = \frac{PR_2}{\pi^2}\int_0^\infty\int_0^\infty \frac{(\alpha^2+\lambda_R\beta^2-\varphi_R)}{D(\alpha,\beta)}\cos\alpha x\cos\beta y\,d\alpha\,d\beta, \tag{9.2}$$

where

$$D(\alpha,\beta) = (\alpha^2+\beta^2-\chi_R)^2(\alpha^2+\beta^2-\mu_R)^2+4\varkappa^4(\alpha^2+\lambda_R\beta^2-\varphi_R)^2,$$

$$4\varkappa^4 = \frac{Ehl^4}{R_2^2 D} = 12(1-\nu^2)\frac{l^4}{R_2^2 h^2}.$$

The effect of transverse shear and normal stresses is neglected here ($\eta = \varepsilon = \bar{\varepsilon} = 0$),

$$\varkappa = \sqrt[4]{3(1-\nu^2)}\,\frac{l}{\sqrt{R_2 h}}, \qquad \lambda_R = \frac{R_2}{R_1},$$

$$\varphi_R = (1+\lambda_R)\,\lambda_R\,\frac{l^2}{R_2^2}, \qquad \chi_R = \left(1-\frac{1-\nu}{2}\lambda_R+\lambda_R^2\right)\frac{l^2}{R_2^2},$$

$$\mu_R = \frac{l^2}{R_2^2}\lambda_R.$$

The length l can be adjusted in the same way as for a spherical shell, so that the coefficient $4\varkappa^4 = 1$. Then we obtain

$$l = \frac{\sqrt{R_2 h}}{\sqrt[4]{3(1-\nu^2)}}, \qquad k_R = \frac{l^2}{R_2^2} = \frac{1}{\sqrt{3(1-\nu^2)}}\,\frac{h}{R_2}, \tag{9.3}$$

and

$$\varphi_R = (1+\lambda_R)\,\lambda_R k_R, \qquad \chi_R = \left(1-\frac{1-\nu}{2}\,\lambda_R+\lambda_R^2\right)k_R, \qquad \mu_R = \lambda_R k_R.$$

It can be observed from the above that the coefficients φ_R, χ_R, and μ_R have a value of the order of h/R which means that they are very small in comparison with unity. However, as they appear together with the variables α and β which can also be very small, their influence may be significant. At the origin of integration, i.e. for $\alpha = 0$ and $\beta = 0$, the numerator and denominator of the integrands (9.2) are equal to zero when $\varphi_R = \chi_R = \mu_R = 0$. Then the whole integrand is indeterminate. Even the addition of very small coefficients may considerably change the value of integrand. In conclusion, the coefficients φ_R, χ_R, and μ_R should not be neglected during integration with small values of α and β. For larger values of α and β their influence is insignificant and can be neglected. The results obtained by neglecting φ_R, χ_R, and μ_R are identical with those obtained by means of shallow shell theory. The expressions (9.2) can easily be integrated by means of the calculus of residues. Then we obtain

$$w = -\mathrm{im}\frac{Pl^2}{4\pi D}\sum_{k=1}^{2}\int_0^\infty\left[\frac{(\alpha_k^2+\beta^2-\mu_R)e^{i\alpha_k x}\cos\beta y\,d\beta}{\alpha_k(\alpha_k^2+\beta^2-\chi_R)\left(\alpha_k^2+\beta^2-\frac{1}{2}(\chi_R+\mu_R)-(-1)^k i/2\right)}\right],$$

$$\Phi = -\mathrm{im}\frac{PR_2}{4\pi}\sum_{k=1}^{2}\int_0^\infty\left[\frac{(-1)^k e^{i\alpha_k x}\cos\beta y\,d\beta}{\alpha_k i\left(\alpha_k^2+\beta^2-\frac{1}{2}(\chi_R+\mu_R)-(-1)^k i/2\right)}\right], \tag{9.4}$$

where α_1, α_2 are the roots of the characteristic equation

$$(\alpha^2+\beta^2-\mu_R)^2(\alpha^2+\beta^2-\chi_R)^2+(\alpha^2+\lambda_R\beta^2-\varphi_R)^2 = 0 \tag{9.5}$$

and have the following form:

$$\left.\begin{matrix}\alpha_1^2\\ \alpha_2^2\end{matrix}\right\} = -\beta^2\pm\eta_1+\frac{\chi_R+\mu_R}{2}+i\left(\eta_2\mp\frac{1}{2}\right), \tag{9.6}$$

and

$$\alpha_1 = a+bi, \qquad \alpha_2 = c+di,$$

where

$$\left.\begin{matrix}a\\ b\end{matrix}\right\} = \frac{1}{\sqrt{2}}\left\{\left[\left(\beta^2-\frac{\chi_R+\mu_R}{2}-\eta_1\right)^2+\left(\eta_2-\frac{1}{2}\right)^2\right]^{1/2}\pm\right.$$
$$\left.\pm\left(-\beta^2+\frac{\chi_R+\mu_R}{2}+\eta_1\right)\right\}^{1/2},$$

$$\left.\begin{matrix}c\\ d\end{matrix}\right\} = \frac{1}{\sqrt{2}}\left\{\left[\left(\beta^2-\frac{\chi_R+\mu_R}{2}+\eta_1\right)^2+\left(\eta_2+\frac{1}{2}\right)^2\right]^{1/2}\pm\right.$$
$$\left.\pm\left(-\beta^2+\frac{\chi_R+\mu_R}{2}-\eta_1\right)\right\}^{1/2},$$

$$\left.\begin{matrix}\eta_1\\ \eta_2\end{matrix}\right\} = \frac{1}{2\sqrt{2}}[\xi_\beta\mp1\pm(\chi_R-\mu_R)^2]^{1/2},$$

$$\xi_\beta = \left\{1-(\chi_R-\mu_R)^2+16\left[(1-\lambda_R)\beta^2-\frac{\chi_R+\mu_R}{2}+\varphi_R\right]^2\right\}^{1/2}.$$

The roots of the characteristic equation calculated for different values of β are collected in Table 10, p. 539. By numerical integration of expressions (9.4), the deflection of the shell for the different ratios $\lambda_R = R_1/R_2$ has been obtained. The results are presented graphically in Fig. 90. The curves presenting equal deflections of the shell are shown in these figures. For the spherical shell we have a symmetrical deflection and the lines presenting the equal deflection make a system of concentric circles (Fig. 90a). With the decrease of λ_R the deflections become less symmetrical, so that for the cylindrical shell the lines of equal deflection form a set of curves similar in shape to an elongated ellipse.

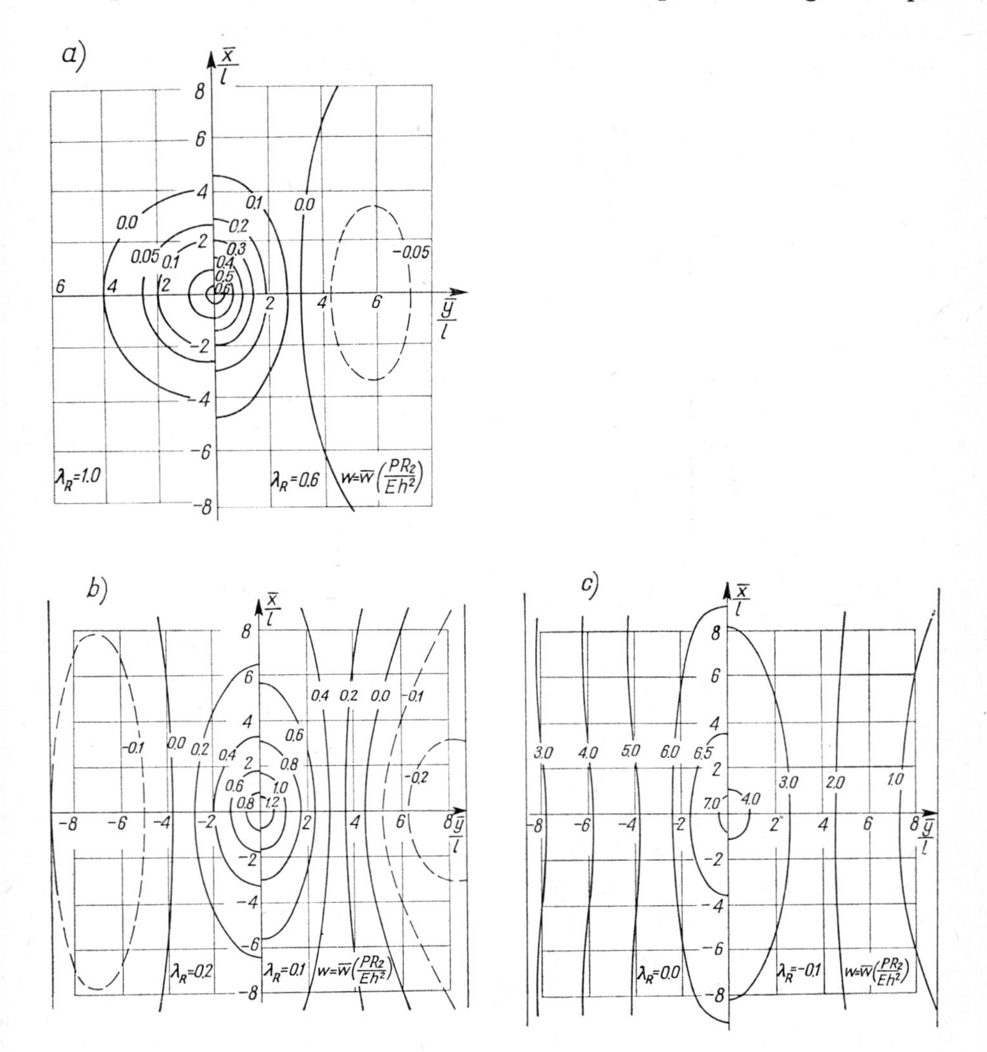

Fig. 90. Deflection variation in a shell of positive Gauss curvature for various parameters λ_R.

Simultaneously, with the expansion of the displaced area of the shell, the value of the maximum displacement increases, reaching for the cylindrical shell a value about 17 times larger than for the spherical shell. The figures show that the displacements spread much further in the direction of the small curvature of the shell than in the direction of the large one. The ratio w/w_c given in Fig. 91 is the ratio of the maximum displacement w of the double curvature shell to displacement of the cylindrical shell w_c calculated by means

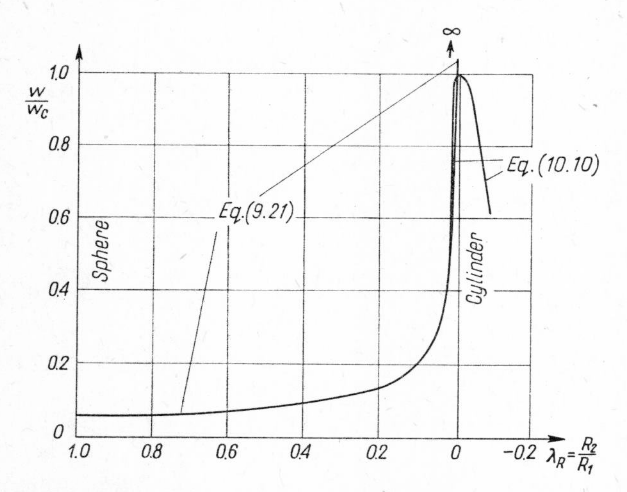

Fig. 91. Deflection of a double curvature shell referred to deflection of a cylindrical shell for various parameters λ_R.

of the numerical integration of (9.4). The thickness of the cylindrical shell and the radius of curvature are equal to the thickness and smaller radius of curvature of the shell measured at the loading point. As in this solution the boundary conditions have not been considered, it may seem that the computed data do not concern any concrete problem and can be treated only as approximate values determining the behaviour of the shell in the vicinity of the loading point. However, we see from Figs. 90 a–d that the displacements of the shell of double curvature appear practically in a very limited area only. These displacements increase slowly at first. After, when the ratio of the radii of curvatures $\lambda_R = R_1/R_2$ decreases from 1 to 0 they increase more rapidly. The area in which the displacements appear changes in a similar way. For the shells with $1 \geqslant \lambda_R \geqslant 0.2$, the character of the displacement does not depend on the curvature of the shell and the area where the displacements appear does not exceed 20 l. These results enable us to make the following statement: The displacements of the doubly curved shells, when $1 \geqslant \lambda_R > 0$, loaded by concentrated forces have a local character and appear in the areas resembling in shape the ellipse, whose longer semiaxis is directed towards the smaller curvature of the shell. The smaller is λ_R, the larger is the displaced area and the larger are the displacements of the shell.

However, for $\lambda_R = 0.2$ already at the distance $x = 10\,l$ the deflection amounts only to about 10 per cent of the maximum displacement ($x = 0$). For $\lambda_R = 0.4$ they are about 4 per cent. Thus, if the load does not act at the edge of the shell and when the distance of the loading point from the edge is larger than several multiples of l, the effect of the boundary conditions can be neglected.

The experiments proved that the above statement is not valid for the case of free edge. In this case the influence of the boundary conditions on the deflection is considerable. It means that the solution expressed by integrals can be considered accurate for the doubly curved shell when $0.2 \leqslant \lambda_R \leqslant 1$ and when the shell has no free or movable edge. For the cylindrical shell ($\lambda_R = 0$) we have obtained the maximum deflection and the maximum deflected area. The deflections of such a shell disappear (practically) only at a distance $x = 500\,l$. The solution expressed by integrals (9.2) is more exact when a cylindrical shell is long, i.e. when $L = 20\,R$. Should the dimensions of the shell be smaller, its finite length must be taken into consideration by the introduction of functions satisfying the boundary conditions. Shells of negative Gaussian curvature $1/R_1 R_2 \leqslant 0$ behave similarly. Their deflections also spread very far. Therefore, the determination of the boundary conditions is necessary.

9.2 Stresses in the shell

To calculate the stress in a shell we must first of all determine bending moments and normal forces. We calculate them from formulae (1.108) and (1.109) by means of the deflection and stress function. In the assumed coordinate system (x, y) we have the following equations:

$$M_{xx} = -\frac{D}{l^2}\left[\frac{\partial^2 w}{\partial x^2} + \nu\frac{\partial^2 w}{\partial y^2} + k_R(\lambda_R^2 + \nu)w\right] + (1-\nu)\eta\frac{\partial Q_x}{\partial x} - \varepsilon(Z + 2\Delta_k \Phi), \tag{9.7}$$

$$M_{yy} = -\frac{D}{l^2}\left[\frac{\partial^2 w}{\partial y^2}+\nu\frac{\partial^2 w}{\partial x^2}+k_R(\nu\lambda_R^2+1)\,w\right]+$$
$$+(1-\nu)\eta\frac{\partial Q_y}{\partial y}-\varepsilon(Z+2\Delta_k\Phi),$$
$$M_{xy} = -\frac{D}{l^2}(1-\nu)\frac{\partial^2 w}{\partial x\,\partial y}+\frac{(1-\nu)\eta}{2}\left(\frac{\partial Q_y}{\partial x}+\frac{\partial Q_x}{\partial y}\right),$$
$$Q_x = -\frac{D}{l^3}\frac{\partial}{\partial x}\{\Delta w+k_R[1+\lambda_R^2]\,w\}-\frac{\partial}{\partial x}\left[(\eta-\varepsilon)Z+\eta\Delta_k\Phi\right]+$$
$$+\frac{1-\nu}{2}\eta\frac{\partial\psi}{\partial y},\tag{9.8}$$
$$Q_y = -\frac{D}{l^3}\frac{\partial}{\partial y}\{\Delta w+k_R[1+\lambda_R^2]\,w\}-\frac{\partial}{\partial y}[(\eta-\varepsilon)Z+\eta\Delta_k\Phi]-$$
$$-\frac{1-\nu}{2}\eta\frac{\partial\psi}{\partial x},$$
$$N_{xx} = -\frac{1}{l^2}\left\{\frac{\partial^2\Phi}{\partial y^2}+\lambda_R k_R\Phi-\frac{D}{R_1}\left[\left(\Delta w+\underline{(1+\lambda_R^2)k_R w}\right)\lambda_R-(1-\nu)\frac{\partial^2 w}{\partial y^2}\right]\right\},$$
$$N_{yy} = -\frac{1}{l^2}\left\{\frac{\partial^2\Phi}{\partial x^2}+\lambda_R k_R\Phi-\frac{D}{R_2}\left[\Delta w+\underline{(1+\lambda_R^2)k_R w}-\lambda_R(1-\nu)\frac{\partial^2 w}{\partial x^2}\right]\right\},$$
$$S = \frac{1}{l^2}\frac{\partial^2\Phi}{\partial x\,\partial y},\qquad N_{xy} = S-\frac{M_{xy}}{R_1},\qquad N_{yx} = S-\frac{M_{xy}}{R_2},$$

where $\lambda_R = R_2/R_1$, $R_2 \leqslant R_1$, $k_R = l^2/R_2^2$. The underlined terms are one order smaller and can be neglected. If we neglect the effect of transverse shear deformations, the moments M_{xx}, M_{yy} are related by the equation

$$M_{xx}+M_{yy} = (1+\nu)\,[M_{xx}]_{\nu=1} = (1+\nu)\,[M_{yy}]_{\nu=1}\tag{9.9}$$

which permits the direct evaluation of the expression for one of the moments when the other is known. For example, $M_{xx} = (1+\nu)\,[M_{yy}]_{\nu=1}-M_{yy}$. After substitution of the previously calculated displacement w and the stress function into Eqs. (9.7) we get the bending moments

$$M_{xx} =$$
$$-\operatorname{im}\frac{P}{4\pi}\int\limits_0^\infty\sum_{k=1}^{2}\frac{(\alpha_k^2+\beta^2-\mu_R)[\alpha_k^2+\nu\beta^2-k_R(\lambda_R^2+\nu)]\,e^{i\alpha_k x}\cos\beta y\,d\beta}{\alpha_k(\alpha_k^2+\beta^2-\chi_R)\left(\alpha_k^2+\beta^2-\frac{\chi_R+\mu_R}{2}-(-1)^k\frac{i}{2}\right)},\tag{9.10}$$

$$M_{xy} =$$

$$-\operatorname{im}(1-\nu)\frac{P}{4\pi}\int_0^\infty \sum_{k=1}^{2} \frac{(\alpha_k^2+\beta^2-\mu_R)i\beta\, e^{i\alpha_k}\sin\beta y\, d\beta}{(\alpha_k^2+\beta^2-\chi_R)\left(\alpha_k^2+\beta^2-\dfrac{\chi_R+\mu_R}{2}-(-1)^k\dfrac{i}{2}\right)}.$$

Membrane forces

$$N_{xx} = -\operatorname{im}\frac{PR_2}{4\pi l^2}\times$$

$$\times\int_0^\infty \sum_{k=1}^{2} \frac{(-1)^k[\beta^2-\lambda_R k_R - i\lambda_R k_R(\alpha_k^2+\beta^2)-ik_R(1-\nu)\beta^2]\, e^{i\alpha_k x}\cos\beta y\, d\beta}{i\alpha_k\left(\alpha_k^2+\beta^2-\frac{1}{2}(\chi_R+\mu_R)-(-1)^k i/2\right)},$$

$$N_{yy} = -\operatorname{im}\frac{PR_2}{4\pi l^2}\times$$

$$\times\int_0^\infty \sum_{k=1}^{2} \frac{(-1)^k[\alpha_k^2-\lambda_R k_R - ik_R(\alpha_k^2+\beta^2)-ik_R\lambda_R(1-\nu)\alpha_k^2]\, e^{i\alpha_k x}\cos\beta y\, d\beta}{i\alpha_k\left(\alpha_k^2+\beta^2-\frac{1}{2}(\chi_R+\mu_R)-(-1)^k i/2\right)},$$

$$S = \operatorname{im}\frac{PR_2}{4\pi l^2}\int_0^\infty \sum_{k=1}^{2} \frac{(-1)^k e^{i\alpha_k x}\beta\sin\beta y\, d\beta}{\alpha_k^2+\beta^2-\dfrac{(\chi_R+\mu_R)}{2}-(-1)^k\dfrac{i}{2}}. \tag{9.11}$$

The effects of transverse shear and normal stresses are neglected in Eqs. (9.10) and (9.11) ($\eta = \varepsilon = \bar{\varepsilon} = 0$).

As has been observed previously, the coefficients φ_R, μ_R, χ_R, k_R are very small (of order h/R) and are important only when α and β are also very small, i.e. at the origin of integration. When considering the shell subjected to concentrated forces the integrals determining the internal forces, especially the integrals for the bending moments, converge very slowly. Then the magnitude of the integrand at the origin of integration has a relatively small importance for the evaluation of the integral between definite limits. In conclusion, the omission of the coefficients φ_R, μ_R, χ_R in the integrals for internal forces does not cause large inaccuracies.

This conclusion is confirmed by J. Holand [10.21] who considered the cylindrical shell loaded by concentrated normal forces. Comparing the results of the simplified Donnell equations and complete Flügge equations, he found that the differences for bending moments do not exceed 1 per cent. The results of the numerical calculation of integrals (9.10), (9.11) are given in Fig. 92a. The curves express the lines of constant values of the σ_{xx} and σ_{yy} near the loading point ($x = 0, y = 0$). We see that with the change of parameter λ_R the membrane forces do not change as much as the displacements. For the doubly curved shells, when $0 < \lambda_R \leqslant 1$ these stresses are almost

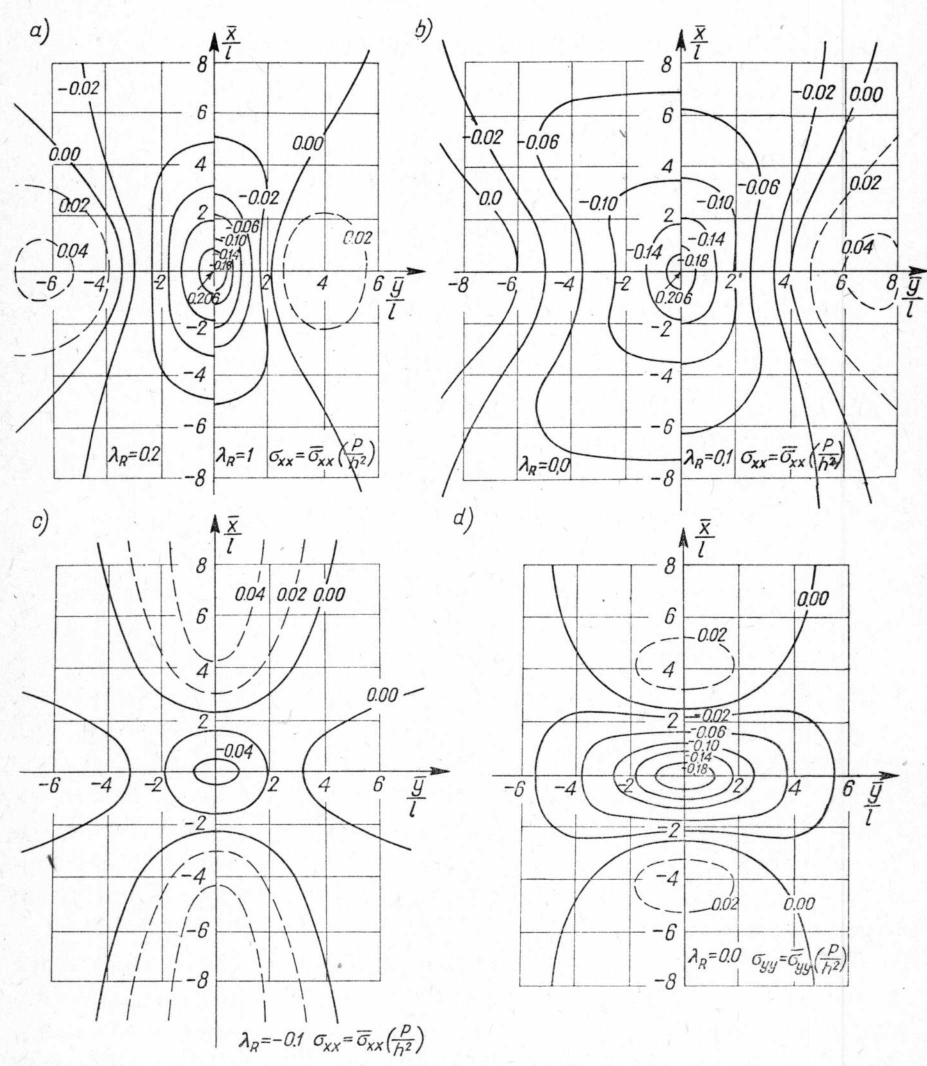

Fig. 92. Membrane stress variation in a double curvature shell referred to deflection of a cylindrical shell for various parameters λ_R.

independent of λ and disappear quickly with the distance from the loading point. At the distance of 10 l these forces constitute only 10 per cent of the maximum value. It is interesting to note that the maximum value of the membrane stresses appearing at the point $x = 0$, $y = 0$ amounting to $\sigma_{xx} = \sigma_{yy} = \sqrt{3(1-\nu^2)}\,P/8h^2$ is independent of the parameter λ_R when $\sim 0.1 \leqslant \lambda_R \leqslant 1$. These stresses are shown in Fig. 93. We see that only for $\lambda_R > 0$ the maximum

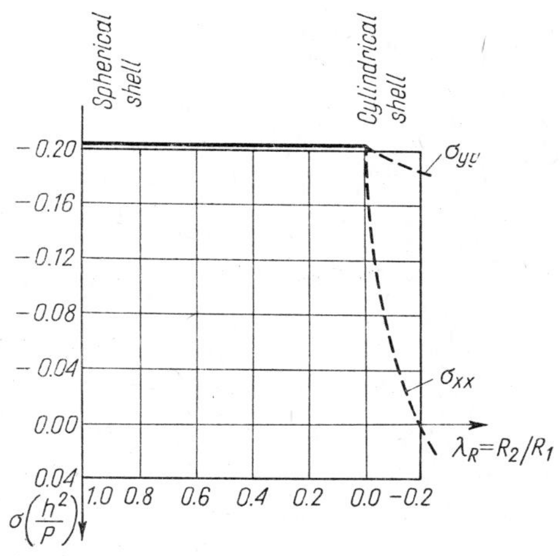

Fig. 93. Membrane stresses at the point of application of the load in the shell for various parameters λ_R.

stresses σ_{xx} and σ_{yy} are unequal. The stresses σ_{xx} decrease more rapidly than σ_{yy} and for $\lambda_R < -0.2$ they change their sign and become compressive stresses. We see that the shells of negative curvature carry the loads differently than those of positive curvature.

The membrane stresses of the shell of positive curvature are of a hydrostatic compression type near the loading point. In the negative curvature shells, the stresses σ_{xx} acting towards the center of curvature corresponding to the radius R_1 (which changes its sign) also change their sign. The stresses σ_{yy} do not undergo any significant change. The behaviour of the normal stresses can be seen in Fig. 94. That figure shows the element cut from the shell of positive and negative curvature loaded by a concentrated force and equilibrated by membrane forces. The element of the shell of the first type is supported uniformly on all sides by normal compressive forces.

The element cut from the shell of the second type is equilibrated by simultaneous action of tension towards the center of negative curvature and compression towards the center of positive curvature.

The evaluation of the integrals (9.10) has been performed numerically. As the exact analytical evaluation of the integral for the bending moment is not possible either, only methods of numerical integration can be applied. But we meet here an important difficulty. For small values of x and y the integrals (9.9) converge very slowly and require the extension of the interval

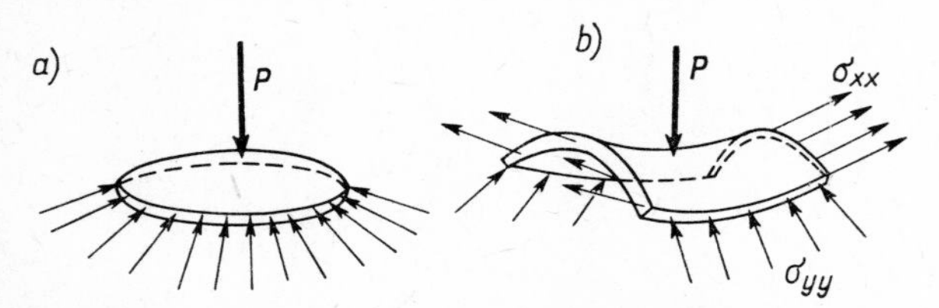

Fig. 94. Equilibrium of an element of a shell of positive and negative Gauss curvatures.

of integration for a large area. We know that in order to obtain sufficient accuracy the value of the integrands must be calculated in short intervals taking the small value of $\Delta\beta$. If the area of integration is large, the number of the intervals $\Delta\beta$ must also be numerous. The problem is complicated by the fact that, for increasing β, $\cos\beta$ becomes a quickly varying function. Integrating numerically, the interval $\Delta\beta$ should be taken to be very small. All these effects may give rise to great difficulties in numerical calculations of the integrals (9.10). Even the use of computers is not very successful here. These difficulties made it necessary to find an effective method of evaluation of slowly convergent integrals. One of such method is described in § 9.6. Its main idea consists of the division of the integral in two and subsequent separate integration between the limits $0-B$ and $B-\infty$. Here B is an arbitrary value. This method has already been used in Chapter 2 for the evaluation of the integrals for the bending moments in the cantilever plate.

9.3 Solution in polar coordinates

9.3.1 *Deflection and stress function*

We obtained in the previous chapter the deflection and the stresses in the shell in the form of single integrals whose numerical evaluation was difficult because of their slow convergence. Now we shall evaluate them approximately, introducing the system of polar coordinates (r, ϑ) related to the rectangular coordinates (x, y) by means of the equations $x = r\cos\vartheta$, $y = r\sin\vartheta$. The solution will be found by the use of the shallow shell equations. Then we neglect the parameters; $\chi_R = \mu_R = \varphi_R = 0$. The integral (9.2) takes the form [9.4]

$$w = \frac{Pl^2}{\pi^2 D}\int_0^{\infty}\int_0^{\pi/2}\{1+(\eta-\varepsilon)\gamma^2-\bar{\varepsilon}[1+\eta\gamma^2](\cos^2\varphi+ + \lambda_R\sin^2\varphi)\}\cos(\gamma x\cos\varphi)\cos(\gamma y\sin\varphi)\frac{\gamma\,d\gamma\,d\varphi}{D(\gamma,\varphi)}, \tag{9.12}$$

where

$$D(\gamma, \varphi) = \gamma^4+[1+\eta\gamma^2](\cos^2\varphi+\lambda_R\sin^2\varphi)^2.$$

For an isotropic shell:

$$\bar{\varepsilon} = \nu h/2R, \qquad \varepsilon = \nu h^2/10(1-\nu)l^2, \qquad \eta = h^2/5(1-\nu)l^2.$$

The above integral is very complex and its exact solution is unknown. We evaluate it approximately by the transformation of the integrand to the form permitting analytical integration. Let us transform the denominator $D(\gamma, \varphi)$ of the integrand to the form

$$D(\gamma, \varphi) = (\gamma^4+1)(1-R), \tag{9.13}$$

where

$$R = 2(1-\lambda_R)\frac{\sin^2\varphi}{\gamma^4+1}\left(1-\frac{1-\lambda_R}{2}\sin^2\varphi\right)(1+\eta\gamma^2)-\eta\frac{\gamma^2}{\gamma^4+1}.$$

As R is always smaller than unity if $0 < \lambda_R \leqslant 1$, we can develop the fraction $1/(1-R)$ in the power series

$$\frac{1}{1-R} = 1+R+R^2+\ldots \tag{9.14}$$

We obtain the integral (9.12) in the form

$$w = \frac{Pl^2}{\pi^2 D}\int_0^{\pi/2}\int_0^{\infty}[1+(\eta-\varepsilon)\gamma^2-\bar{\varepsilon}(1-\eta\gamma^2)(1-(1-\lambda_R)\sin^2\varphi)]\times \times\frac{1+R+R^2+\ldots}{\gamma^4+1}\cos(\gamma x\cos\varphi)\cos(\gamma y\sin\varphi)\gamma\,d\gamma\,d\varphi. \tag{9.15}$$

We see now that the evaluation of the displacement consists in the summation of the series whose terms are defined by the integrals of the form

$$\int_0^{\pi/2}\int_0^{\infty} \frac{\gamma^k}{(\gamma^4+1)^m} \cos(\gamma x \cos\varphi)\cos(\gamma y \sin\varphi)\sin^{2n}\varphi \,\mathrm{d}\varphi \cdot \gamma \,\mathrm{d}\gamma . \tag{9.16}$$

The integrals can be evaluated exactly first with respect to φ and then with respect to γ. Their values for various parameters m and n are given in Tables 11, 12.

The effect of the stresses σ_{33} normal to the middle surface of the shell is presented in (9.15) by the term $\bar{\varepsilon}$ which is usually much smaller than unity. In the case where this term gives only the numerical correction and does not introduce singularities of higher orders and a significant change in the behaviour of the shell, it can be neglected. Integrating with respect to φ, we obtain

$$H_{2n} = \int_0^{\pi/2} \cos(\gamma x \cos\varphi)\cos(\gamma y \sin\varphi)\sin^{2n}\varphi \,\mathrm{d}\varphi$$

$$= \frac{\pi}{2}\left[\varrho^{2n,\,n}\frac{J_n(\gamma r)}{(\gamma r)^n} + \varrho^{2n,\,n-1}\frac{J_{n-1}(\gamma r)}{(\gamma r)^{n-1}} + \ldots + \varrho^{2n,\,0} J_0(\gamma r)\right] \tag{9.17}$$

where $\varrho^{2n,\,m}$ are coefficients independent of γ, but including the functions of coordinates x and y (see Table 14). Integrating with respect to γ, we have to evaluate the following integrals:

$$I^k_{m,\,n} = \int_0^{\infty} \frac{\gamma^{1+k}}{(\gamma^4+1)^{1+n}} \frac{J_m(\gamma r)}{\gamma^m} \mathrm{d}\gamma . \tag{9.18}$$

These integrals can also be evaluated without using numerical calculations. Their solutions are given in Table 15. By substitution of the calculated coefficients into (9.15), we obtain the displacement w in the form of the sum of the following series [9.6]:

$$w = \frac{Pl^2}{2\pi D}\sum_{m=0}^{2n}\sum_{n=0}^{\infty} [2(1-\lambda_R)]^n \times$$

$$\times a_{m,n}\left[\frac{I^0_{m,\,n}}{r^m} + (\eta-\varepsilon)\frac{I^2_{m,\,n}}{r^m} - (1+n)\eta\frac{I^2_{m,\,n+1}}{r^m}\right], \tag{9.19}$$

where the coefficients $a_{m,n}$ are defined by the recurrent formulae

$$a_{0,0} = 1, \qquad a_{m,1} = \varrho^{2,m} - \frac{1-\lambda_R}{2}\varrho^{4,m},$$

$$a_{m,n} = a_{m,n-1} \,\overline{\times}\, \left((\varrho^{2,0} - \frac{1-\lambda_R}{2}\varrho^{4,0}\right)$$

for every

$$a_{1,2} = a_{1,1} \,\overline{\times}\left(\varrho^{2,0} - \frac{1-\lambda_R}{2}\varrho^{4,0}\right) = \varrho^{4,1} - (1-\lambda_R)\varrho^{6,1} + \frac{(1-\lambda_R)^2}{4}\varrho^{8,1}.$$

The symbol $\overline{\times}$ indicates the multiplication which for the coefficient $\varrho^{m,n}$ is $\varrho^{m,n} \,\overline{\times}\, \varrho^{k,1} = \varrho^{m+k,n+1}$. We should remember that in the expression (9.17) n can have only the value $n \geqslant m$ (see (9.7)) and therefore we must put $\varrho^{2n,m} = 0$ if $m > n$. Hence, $a_{m,n} = 0$ if $m > 2n$.

Let us calculate now the shell displacement w at the point where the force acts. Then $x = 0$, $y = 0$ and $\cos(\gamma x \cos\varphi) = \cos(\gamma y \sin\varphi) = 0$.

The coefficients H_{2n} can be defined using the relations

$$\int_0^{\pi/2} \sin^{2n}\varphi \, d\varphi = \frac{(2n-1)!!}{2^n n!}\frac{\pi}{2}$$

where $(2n-1)!! = 1 \cdot 3 \cdot 5 \cdot \ldots \cdot (2n-1)$ and

$$I_{0,n} = \int_0^{\infty} \frac{\gamma \, d\gamma}{(\gamma^4+1)^{1+n}} = \frac{(2n-1)!!}{2^{n+1} n!}\frac{\pi}{2}. \tag{9.20}$$

By substitution of the above coefficients in expression (9.19) we obtain the displacement in the form of the series

$$w_{0,0} = \frac{Pl^2}{8D}\left[1 + \sum_{n=1}^{\infty} (1-\lambda_R)^n \frac{(2n-1)!!}{2n!!}\right] = \frac{Pl^2}{8D}\frac{1}{\sqrt{\lambda_R}},$$

where $2n!! = 2 \cdot 4 \cdot 6 \cdot \ldots \cdot 2n$. The sum of the series is known when $(1-\lambda_R) < 1$. Then the displacement of the shell at the loaded point becomes [9.6]

$$w_{0,0} = \frac{\sqrt{3(1-\nu^2)}}{4} \cdot \frac{P\sqrt{R_1 R_2}}{Eh^2}. \tag{9.21}$$

This formula is valid only when $\lambda_R > 0$, i.e. for the shell of positive double curvature. For the cylinder, when $\lambda_R = 0$ we obtain the displacement $w \to \infty$ which is incorrect. For $R_1 = R_2$ we get the same formula as that developed by E. Reissner [8.1] for the spherical shell. The shallow shell loaded by a normal force was solved also in [9.5] wherein the identical formula for the displacement $w_{0,0}$ was obtained by means of numerical summation of the series.

9.3.2 *Internal forces and moments*

The stress function Φ can be obtained by a similar treatment. We find that:

$$\Phi = \frac{PR_2}{\pi^2} \int_0^{\infty} \int_0^{\pi/2} \left\{ [(1+(\eta-\varepsilon)\gamma^2)(\cos^2\varphi + \lambda_R \sin^2\varphi)] \frac{1}{D(\gamma,\varphi)} + \right.$$

$$\left. + \bar{\varepsilon} \frac{1}{\gamma^2} \right\} \cos(\gamma x \cos\varphi) \cos(\gamma y \sin\varphi) \gamma \, d\gamma \, d\varphi. \tag{9.22}$$

The effect of the normal stresses σ_{33} is represented by the term $\nu h/2R_2$ which s very small in comparison with unity, and usually can be neglected. In the above interval it introduces singularities of higher order and therefore it has been taken into consideration.

The stresses in the shell are defined by the moments and normal forces: The membrane forces can be obtained by using the stress function Φ. We have, for example,

$$N_{xx} = \frac{PR_2}{\pi^2 l^2} \int_0^{\infty} \int_0^{\pi/2} \left\{ [1+(\eta-\varepsilon)\gamma^2](\cos^2\varphi + \lambda_R \sin^2\varphi) \frac{1}{D(\gamma,\varphi)} + \bar{\varepsilon} \frac{1}{\gamma^2} \right\} \times$$

$$\times \sin^2\varphi \cos(\gamma x \cos\varphi) \cos(\gamma y \sin\varphi) \gamma^3 \, d\gamma \, d\varphi.$$

The normal forces N_{yy} and N_{xy} can be obtained in a similar way as the force N_{xx}. But a new integral appears in the formula for the tangential force N_{xy}. The integral D_{2n} can also be evaluated in a closed form. The result is given in Table 14. After integration we obtain the following series for the membrane forces: For example, for N_{xx} we have

$$\left.\begin{matrix} N_{xx} \\ N_{yy} \end{matrix}\right\} = \frac{PR_2}{2\pi l^2} \sum_{m=0}^{2n+2} \sum_{n=0}^{\infty} [2(1-\lambda_R)]^n b_{m,n} \left[\frac{I^0_{m,n}}{r^m} + (\eta-\varepsilon) \frac{I^2_{m,n}}{r^m} - \right.$$

$$-(1+n)\eta\frac{I^2_{m,n+1}}{r^m}\Bigg]\pm\frac{\nu Ph}{4\pi l^2}\frac{1}{r^2}, \tag{9.23}$$

where the coefficients $b_{m,n}$ are defined by the recurrent formulae: For N_{xx}

$$b_{m,0} = \varrho^{2,m}-(1-\lambda_R)\varrho^{4,m},$$

$$b_{m,n} = b_{m,n-1}\times\left(\varrho^{2,0}-\frac{1-\lambda_R}{2}\varrho^{4,0}\right),$$

$$b_{m,n} = 0 \quad \text{if} \quad m > 2n+2.$$

We must remember here that $\varrho^{2n,m} = 0$ for $m > n$. The formula for the internal force N_{yy} is the same as that for N_{xx}, only the coefficient $b_{m,n}$ is different. For N_{yy} we have $b_{0,0} = \varrho^{0,0}-(2-\lambda_R)\varrho^{2,0}-(1-\lambda_R)\varrho^{4,0}$. A similar expression can be written with the help of (9.23) for the tangential force N_{xy}. The value of the membrane forces at the loading point is of particular interest. In order to calculate it, let us carry out a treatment similar to that used for calculating the displacement $w_{(0,0)}$. Integrating the terms of series (9.23) for $x = 0$, $y = 0$, we get a series whose sum is known and equal to π. In conclusion if we neglect the effect of transverse shear deformations, the membrane forces can be calculated from the simple formula

$$N_{xx\,(0,0)} = N_{yy\,(0,0)} = \frac{\sqrt{3(1-\nu^2)}}{8}\frac{P}{h}; \qquad N_{xy} = 0.$$

We see that these forces are independent of the coefficient λ_R. This result confirms the numerical calculations performed in [9.3] where the membrane forces were found to be constant for $x = 0$, $y = 0$ and $0 < \lambda_R \leqslant 1$. The bending moments take the form

$$M_{xx} = \frac{P}{2\pi}\sum_{m=0}^{2n+1}\sum_{n=0}^{\infty}[2(1-\lambda_R)]^n\Bigg[\left(\frac{I^2_{m,n}}{r^m}-(1+n)\eta\frac{I^2_{m,n+1}}{r^m}\right)c_{m,n}+$$

$$+(1-\nu)\varepsilon\frac{I^4_{m,n}}{r^m}\bar{c}_{m,n}\Bigg], \tag{9.24}$$

where

$$c_{m,0} = \varrho^{0,m}-(1-\nu)\varrho^{2,m},$$

$$\bar{c}_{m,0} = \varrho^{2,m},$$

$$c_{m,n} = c_{m,n-1}\times\left(\varrho^{2,0}-\frac{1-\lambda_R}{2}\varrho^{4,0}\right).$$

For the bending moment M_{yy} we have the same formulae as previously, only the coefficient $c_{0,0}$ is different and equal to

$$c_{0,0} = \nu\varrho^{0,0} - (1-\nu)\varrho^{2,0}, \qquad \text{etc.}$$

The expressions for the twisting moment M_{xy} can be written in a similar way. The functions $I^k_{m,n}$ in the above formulae are given in Table 12.

9.3.3 *Nearly spherical shell*

When the shell does not differ very much from the sphere at the loading point, we can take into account only the first two terms of the series (9.19). Then the deflection of the shell takes the form

$$w = \frac{Pl^2}{2\pi D}\Bigg\{-\operatorname{kei} r + 2(1-\lambda_R)\Bigg[\left(-\frac{1}{2}\operatorname{kei} r + \frac{r}{4}\operatorname{kei}' r\right)\sin^2\vartheta + \\ + \frac{1}{r}\left(\frac{1}{r} + \operatorname{ker}' r + \frac{1}{4} r \operatorname{kei} r\right)\cos 2\vartheta + \\ + \left[(\eta-\varepsilon)\operatorname{ker} r + \frac{r\eta}{4}\operatorname{ker}' r\right]\Bigg]\Bigg\}. \tag{9.25}$$

We see that in the vicinity of the point of application of the load the deflection of the shell is described by the same functions as in the case of the spherical shell. The effect of transverse shear deformation gives the singular term $\ln r$ which tends to infinity as $r \to 0$. Taking into account the effect of the non-shallowness of the shell, we would obtain the term $k_R \ln\left(1/\sqrt{2k_R}\right)$ giving a correction of the order $h/R \ln(R/h)$.

The bending moment M_{rr} takes the form

$$M_{rr} = \frac{P}{2\pi}\Bigg\{\operatorname{ker} r - \frac{1-\nu}{r}\operatorname{ker}' r + \\ + \frac{1-\lambda_R}{4}\left[(1-\nu)\left(\operatorname{ker} r - \frac{2}{r}\operatorname{kei}' r\right) - r\operatorname{ker}' r\right] - \\ - 2(1-\lambda_R)\Bigg[\frac{6(1-\nu)}{r^2}\left(\frac{1}{r^2} + \frac{1}{r}\operatorname{ker}' r + \frac{1}{2}\operatorname{kei} r\right) - \\ - \frac{3-2\nu}{2r}\operatorname{kei}' r + \frac{3-\nu}{8}\operatorname{ker} r - \frac{r}{8}\operatorname{ker}' r\Bigg]\cos^2\vartheta -$$

$$-\frac{\varepsilon(1-\nu)}{r}\ker' r+\frac{\eta(1+5\nu)}{4}\operatorname{kei} r+\frac{r\eta}{4}\operatorname{kei}' r\bigg\},$$

(9.26)

$$M_{\vartheta\vartheta}=\frac{P}{2\pi}\bigg\{\nu\ker r+\frac{1-\nu}{r}\operatorname{kei}' r+$$

$$+\frac{1-\lambda_R}{4}\left[-(1-\nu)\left(\ker r-\frac{2}{r}\operatorname{kei}' r\right)-\nu r\ker' r\right]-$$

$$-2(1-\lambda_R)\left[-\frac{6(1-\nu)}{r^2}\left(\frac{1}{r^2}+\frac{1}{r}\ker' r+\frac{1}{2}\operatorname{kei} r\right)-\right.$$

$$\left.-\left(1-\frac{3}{2}\nu\right)\frac{1}{r}\operatorname{kei}' r-\frac{1-3\nu}{8}\ker r-\frac{\nu r}{8}\ker' r\right]\cos 2\vartheta+$$

$$+\frac{\varepsilon(1-\nu)}{r}\ker' r+\frac{\eta(5+\nu)}{4}\operatorname{kei} r+\frac{\eta r}{4}\operatorname{kei}' r\bigg\}.$$

For the membrane forces we have

$$N_{xx}=\frac{PR}{2\pi l^2}\bigg\{-\operatorname{kei} r\sin^2\vartheta+\frac{1}{r}\left(\frac{1}{r}+\operatorname{kei}' r\right)\cos 2\vartheta+$$

$$+\left[(\eta-\varepsilon)\ker r+\frac{\eta r}{4}\ker' r\right]\sin^2\vartheta+$$

$$+\left[\left(\frac{\eta}{2}-\varepsilon\right)\frac{1}{r}\operatorname{kei}' r+\frac{\eta}{4}\ker r\right]\cos 2\vartheta+$$

$$+(1-\lambda_R)\left[\frac{r}{2}\operatorname{kei}' r-\frac{2}{r}\left(\frac{1}{r}+\ker' r\right)-\operatorname{kei} r\right]\sin^4\vartheta+$$

$$+(1-\lambda_R)\left[\frac{1}{r}\left(\frac{1}{r}+\ker' r\right)+\frac{1}{2}\operatorname{kei} r\right]6\sin^2\vartheta\cos^2\vartheta+$$

$$+(1-\lambda_R)\frac{3}{r^2}\left[\frac{1}{2}-\frac{4}{r}\operatorname{kei}' r+2\ker r-\frac{r}{2}\ker' r\right]\cos 4\vartheta\bigg\}+$$

$$+\frac{\nu Ph}{4\pi l^2}\frac{1}{r^2}\cos 2\vartheta,$$

$$N_{yy}=-\frac{PR}{2\pi l^2}\bigg\{\operatorname{kei} r\sin^2\vartheta+\frac{1}{r}\left(\frac{1}{r}+\operatorname{kei}' r\right)\cos 2\vartheta-$$

$$-\left[(\eta-\varepsilon)\ker r+\frac{\eta r}{4}\ker' r\right]\cos^2\vartheta+$$

$$+\left[\left(\frac{\eta}{2}-\varepsilon\right)\frac{1}{r}\mathrm{kei}' r+\frac{\eta}{4}\ker r\right]\cos 2\vartheta+$$

$$+(1-\lambda_R)\left[-\frac{r}{2}\mathrm{kei}' r\sin^2\vartheta\cos^2\vartheta+\right.$$

$$+\left(\frac{1}{r^2}+\frac{1}{r}\ker' r+\frac{1}{2}\mathrm{kei}\, r\right)(8\sin^2\vartheta\cos^2\vartheta-1)+$$

$$\left.\left.+\frac{3}{r^2}\left(\frac{1}{2}-\frac{4}{r}\mathrm{kei}' r+2\ker r-\frac{r}{2}\ker' r\right)\cos 4\vartheta\right]\right\}-$$

$$-\frac{\nu Ph}{4\pi l^2}\frac{1}{r^2}\cos 2\vartheta.$$

We can easily write similar expressions for the bending moment $M_{r\vartheta}$ and for the membrane force N_{xy}. For small $r \ll 1$, the bending moments M_{rr}, $M_{\vartheta\vartheta}$ are

$$\left.\begin{matrix} M_{rr} \\ M_{\vartheta\vartheta} \end{matrix}\right\} = \frac{P}{4\pi}\left[-(1+\nu)\left(\ln\frac{r}{2}+\gamma_0\right)\mp\frac{1-\nu}{2}+\right.$$

$$\left.+\frac{1-\lambda_R}{4}\left(1+\nu-\frac{1-\nu}{2}\cos 2\vartheta\right)\pm 2(1-\nu)\frac{\varepsilon}{r^2}\right].$$

From the results obtained the bending moments are singular at the point of application of the load. The formulae contain two singular terms, the classical $\ln\frac{1}{2}r$ and $1/r^2$ resulting from the effect of transverse normal stresses. When the load acts in an outwards direction the moments $M_{rr}\to+\infty$, $M_{\vartheta\vartheta}\to-\infty$ as $r\to 0$. According to the classical theory, both the moments $M_{rr}\to+\infty$, $M_{\vartheta\vartheta}\to+\infty$. We obtain this result based on the equations of shallow shell theory. Applying the complete equations, we would obtain some additional terms of smaller importance. The membrane forces $N_{rr}\to+\infty$ and $N_{\vartheta\vartheta}\to-\infty$ as $r\to 0$. According to the classical theory of shallow shells the membrane forces are equal at the point of application of the load

$$N_{rr}=N_{\vartheta\vartheta}=\sqrt{3(1-\nu^2)}\,\frac{P}{8h}.$$

Applying the complete equations, we obtain an additional singular term $\ln\frac{1}{2}r$, but of smaller importance than the singular term resulting from the effect of transverse normal and shear stresses.

9.4 Load distributed over small surface

From the practical point of view the case of the force distributed over a small surface is very interesting. If we want to calculate the deflection and maximum stresses, we should take into account the way in which the force is distributed over the surface. This calculation can be done in a similar way as previously in the case of the concentrated force, i.e. representing the load by means of a Fourier integral. Then the problem reduces to the solution of the double integrals. In the case where the load is uniformly distributed over a small circular surface of the radius c, integrals (3.45) can be used. We obtain the solution by the simple multiplication of all integrals from the previous section by $2J_1(\gamma c)/\gamma c$. Then on integration with respect to φ we obtain the following relation for the deflection of the shell.

$$w = \frac{Pl^2}{\pi D} \sum_{m=0}^{2n} \sum_{n=0}^{\infty} [2(1-\lambda_R)]^n a_{mn} \int_0^{\infty} \frac{\gamma}{(\gamma^4+1)^{1+n}} \frac{J_m(\gamma r)}{\gamma^m r^m} \frac{J_1(\gamma c)}{\gamma c} \mathrm{d}\gamma .$$

The analytical evaluation, although complex, is possible. If the area on which the load is applied is not too large, we can easily write the following relation for the deflection at the point $r = 0$ which differs only slightly from that for the spherical shell.

$$w_{(0,0)} = \frac{\sqrt{12(1-\nu^2)}}{\pi} \frac{P\sqrt{R_1 R_2}}{Eh^2} \left[\frac{1}{c^2} + \frac{1}{c} \operatorname{ker}' c\right] +$$

$$+ \frac{3}{5}(1+\nu)(2-\nu) \frac{P}{\pi Eh} \left[\operatorname{ker} c + \frac{c \operatorname{ker}' c}{2(2-\nu)}\right].$$

The effects of transverse normal and shear stresses are here taken into account. We obtain the following series for the bending moments:

$$M_{xx} = \frac{P}{2\pi} \sum_{m=0}^{2n+1} \sum_{n=1}^{\infty} [2(1-\lambda_R)]^n c_{m,n} \int_0^{\infty} \frac{\gamma^3 J_m(\gamma r) J_1(\gamma c)}{(\gamma^4+1)^{1+n} \gamma^m r^m \gamma c} \mathrm{d}\gamma ,$$

where $c_{m,n}$ are the coefficients (9.24). If we need only the maximal stress at the point $r = 0$, these relations can be considerably simplified. For $r = 0$ we have the relation

$$\frac{J_m(\gamma r)}{\gamma^m r^m} \to \frac{1}{2^m \Gamma(m+1)} = \frac{1}{2^m m!}, \qquad r \to 0 .$$

Introducing this into the above series, we obtain, on integration with respect to γ,

$$M_{xx} = \frac{P}{\pi} \sum_{m=0}^{2n+1} \sum_{n=0}^{\infty} [2(1-\lambda_R)]^n \frac{I^2_{1,n}(c)}{c} \frac{c_{m,n}}{2^m m!}$$

$$= \frac{P}{2\pi} \sum_{n=0}^{\infty} \left\{ (1-\lambda_R)^n \left[1-(1-\nu)\frac{2n+1}{2(n+1)}\right] \frac{I^2_{1,n}(c)}{c} \frac{(2n-1)!!}{n!} - \right.$$

$$\left. -(1-\lambda_R)^{n+2} \left[1-(1-\nu)\frac{2n+5}{2n+6}\right] \frac{I^2_{1,n+1}(c)}{c} \frac{(2n-1)!!}{2^n n!} \right\}.$$

Since

$$I^2_{1,0}(c) = \mathrm{kei}'\, c,$$

$$I^2_{1,1}(c) = \tfrac{1}{2}(\mathrm{kei}'\, c - \tfrac{1}{2} c \,\mathrm{ker}\, c),$$

$$I^2_{1,2}(c) = \tfrac{3}{8}(\mathrm{kei}'\, c - \tfrac{1}{2} c \,\mathrm{ker}\, c + \tfrac{1}{12} c^2 \,\mathrm{ker}'\, c),$$

$$I^2_{1,n}(c) = \frac{(2n-1)!!}{2^n n!}(\mathrm{kei}'\, c - \tfrac{1}{2} c \,\mathrm{ker}\, c + \tfrac{1}{12} c^2 \,\mathrm{ker}'\, c + \ldots),$$

on introduction and subtraction of the series we find

$$M_{xx} = \frac{P}{2\pi c} \left\{ (1+\nu)\mathrm{kei}'\, c + \sum_{n=0}^{\infty} (1-\lambda_R)^n \left[1-(1-\nu)\frac{2n+1}{2(n+1)}\right] + \right.$$

$$+ \frac{(2n-1)!!}{2n!!}(\mathrm{kei}'\, c - \tfrac{1}{2} c \,\mathrm{ker}\, c) + \sum_{n=2}^{\infty} (1-\lambda_R)^n \times$$

$$\left. \times \left[1-(1-\nu)\frac{2n+1}{2(n+1)}\right] \frac{(2n-1)!!}{(2n)!!} \frac{1}{12} c^2 \,\mathrm{ker}'\, c + \ldots \right\}.$$

This series can be summed using the relation

$$1 - \frac{2n+1}{2(n+1)} = \frac{1}{2(n+1)}$$

and the series

$$\sum_{n=1}^{\infty} (1-\lambda_R)^n \frac{(2n-1)!!}{(2n)!!(n+1)} = \frac{1-\sqrt{\lambda_R}}{1+\sqrt{\lambda_R}}, \qquad 0 < \lambda_R < 1.$$

As a result we obtain the following relation for the bending moments M_{xx} and M_{yy} at the point $r = 0$; $\lambda_R \neq 0$:

$$\left.\begin{matrix} M_{xx} \\ M_{yy} \end{matrix}\right\} = \frac{P}{2\pi}\left\{(1+\nu)\left[\frac{\text{kei}' c}{c} - \frac{1-\lambda_R}{48} c \,\text{ker}' c\right] + \right.$$

$$+ \frac{1-\sqrt{\lambda_R}}{2\sqrt{\lambda_R}}\left(1+\nu \mp \frac{1-\nu}{1+\sqrt{\lambda_R}}\right) \times$$

$$\left. \times \left(\frac{\text{kei}' c}{c} - \frac{1}{2}\,\text{ker}\, c + \frac{1}{12} c^2\, \text{ker}' c\right)\right\}. \tag{9.27}$$

We have taken here only three terms of the series for the functions $I^2_{1,n}$ because of their rapid convergence.

If c is small, $c \ll 1$, the functions $\text{ker}\, c$, $\text{kei}\, c$, ... can be replaced by the series expansions. Then the bending stresses are defined in the following way:

$$\left.\begin{matrix} \sigma_{xx} \\ \sigma_{yy} \end{matrix}\right\} = \frac{3P}{2\pi h^2}\left\{-(1+\nu)\left[\left(\ln\frac{c}{2} + \gamma_0\right) - \frac{1}{2} + \frac{1-\lambda_R}{48}\right] + \right.$$

$$\left. + \frac{1-\sqrt{\lambda_R}}{4\sqrt{\lambda_R}}\left(1+\nu \mp \frac{1-\nu}{1+\sqrt{\lambda_R}}\right)\right\},$$

$$\lambda_R \neq 0,$$

where $c = \bar{c}/l$ is the non-dimensional value referred to the characteristic length l. The twisting moment $M_{xy} = 0$ at the point $r = 0$.

From Eq. (9.23) for the case of the concentrated force the maximum membrane stresses do not depend on the coefficient λ_R if $1 \geqslant \lambda_R \geqslant \sim 0$. In the case of the load distributed over a small surface, these stresses depend to a small degree on λ_R and can be obtained in good approximation from the same formulae as for the spherical shell.

9.5 Experimental investigation

In order to check the results obtained from the theory, experiments with the nearly spherical shell were performed. Their aim was the determination of the stresses in the shell in the vicinity of the point of application of the load. With regard to the fact that the stresses have a local character, i.e. they vanish rapidly with increasing distance from the point of application of the load, the method of measurement of the stresses by means of strain gauges was ignored and a modified photostress method was applied [9.4]. A model of a double curvature shell was made of plexiglass. The principal radii of curvature were $R_1 = 86$ cm, $R_2 = 69$ cm, the ratio $\lambda_R = R_2/R_1 = 0.8$. The shell was rigidly clamped at the edges in a steel frame made of channel bars. In order to permit the observation of the stresses in the external layers of the shell, two cavities 0.25 cm deep were made in the central portion of the shell. Two shells made of epoxy resin "Epidian 5" 0.23 cm thick were cemented in the cavities by means of a reflective cement. Owing to that, observing one side of the shell, the retardation of the polarized light caused by the twofold traversing of the optically sensitive layers is measured. The shell was loaded by the forces distributed over a small circular surface. The nearly uniformly distributed pressure between the stamp and the shell was ensured by a rubber pad. The experiment was performed for diffeıent diameters of stamp ranging from $D/h = 0.55$ to $D/h = 1.8$. The photoelastic effect observed in the external layers of the shell consists of the sum of the effects produced by the membrane and bending stresses. Without additional information these two effects cannot be separated.

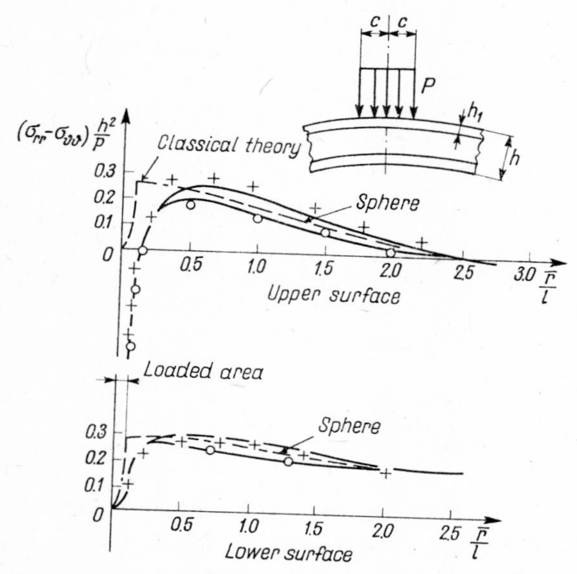

Fig. 95. Variation of isochromatics along principal directions of curvatures on upper and lower surface of shell, $\lambda_R = 0.8$, loaded by a force distributed over a small circular area.
+ experiment along the larger curvature; o experiment along the smaller curvature.

As the aim was to check the method of calculations, the isochromatic fringes calculated and those obtained by experiment were compared. The isochromatics were calculated for the distance corresponding to the middle surface of the optically sensitive layer.

A nearly elliptical distribution of isochromatics was found on the upper side of the shell. The isochromatics on the bottom side had a nearly circular shape. Both for the upper and lower surfaces the maximum fringe order appeared at the same distance from the point of application of the load. The distribution of the isochromatics along the principal directions of the curvatures is presented in Fig. 95. For comparison the distribution of the isochromatics obtained from the classical theory is given in the same diagram. It can be noticed that the classical theory gives large errors in the vicinity of the point of application of the load. On the contrary, the diagram obtained from the improved theory agrees better with the experimental results.

9.6 Simplified solution. Division of loads

From previous chapters it follows that the exact evaluation of the integrals defining the internal forces in the shell is possible only for the spherical shell. In the case of the shell of double curvature when $R_1 \neq R_2$ the solution can be obtained in the form of double series whose terms consist of Thomson's functions. These series converge rapidly in the case of the shells of nearly spherical shape. This convergence becomes poorer with the increase of one of the radii of curvature and when the shape of the shell becomes nearly cylindrical. For the cylindrical shell the series become divergent. In order to solve such problems we must integrate numerically. This integration is, however, very difficult because of the slow convergence of the integrals, so that sometimes the normal treatment does not give any result. For example, for a shell loaded by a concentrated force at its edge we obtain at this edge the stresses described by divergent integrals and the normal method of calculation fails. To obtain the solution satisfactory for engineering purposes, we shall apply the approximate method consisting of calculation in two stages. The main idea of the method is a division of the concentrated load in two line loads. First, local, a self-equilibrating load and second, general, a total distributed line load. The general load produces distortions in the entire shell; the local load, on the contrary, affects the area near the loading point only. We shall obtain the stresses and displacements caused by local loads using simplifications permitting the derivation of the closed formulae useful for further numerical computations. Since the action of the local load is practically limited to a small area, it does not affect the boundary conditions and, therefore, in this stage we can neglect them.

However, as the general load produces stresses in the entire shell, it affects the boundary conditions and in this stage they must be taken into account. It follows from the above that the shell loaded by an arbitrary system of concentrated forces can be investigated in the following way: First, we find the displacements and stresses produced by the distributed general load. In this case we investigate the shell simultaneously for all distributed loads, taking into account all necessary boundary conditions. Next, we add to these results the local effects computed from simple closed formulae.

This method facilitates considerably the calculation of the stresses in the shell. The solution for the shell subjected to a general load can be derived by the usual method of the theory of thin shells; for example, by the use of series, numerical integration, or more modern computational matrix devices. The functions describing the state of stress and displacement change here slowly, have no singularities and therefore the above-mentioned method can be applied. We integrate only between finite limits or sum the series only for a finite number of terms.

Now, we shall discuss the division of the concentrated loads. Let us express the concentrated force P as two integrals [9.1]

$$Z(x, y) = \frac{P}{\pi^2 l^2} \int_0^\infty \left[\int_0^B \cos\beta y \, d\beta + \int_B^\infty \cos\beta y \, d\beta\right] \cos\alpha x \, d\alpha = Z^\circ + Z^B$$

where B has a certain arbitrary value to be determined later.

The first integral can easily be evaluated and we have the following expression for the load $Z(x, y)$:

$$Z(x, y) = \frac{P}{\pi^2 l^2} \int_0^\infty \left[\frac{\sin By}{y} + \int_B^\infty \cos\beta y \, d\beta\right] \cos\alpha x \, d\alpha .$$

The first and second integrals are presented graphically in Figs. 96 a–c. The first one (Fig. 96b) expresses the general load which brings about stresses in the whole shell and the second one the local self-equilibrating load. Both of them act along the y-axis. It can be stated on the basis of numerical computations that only the first type of load affects the whole shell causing first of all the membrane stresses. On the contrary, all effects of the second load

(Fig. 96c) decrease very rapidly with x and y, and are important only near the loading point. It is easy to find that the local load is really self-equilibrating from calculating its resultant.

This method of solving the problem, consisting of the division of the concentrated load and separating the singular terms does not concern shells

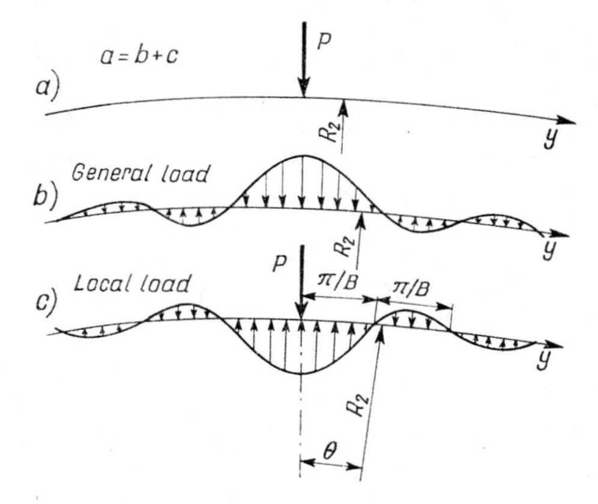

Fig. 96. Division of concentrated load into a local self-equilibrating load and total load.

only. This method can be used also for plates. We have already adopted it in § 4.1. However, it is not always worth using. If we find the solution directly, the division of the calculation in two parts is aimless. However, when the solution consists of the numerical computation of the singular functions this method enables us to reduce the amount of work. The stresses resulting from the local loads have the same character in shells and plates and depend only on the kind of load and the shape of structure in the vicinity of the point of application of the load.

9.7 Solutions for local load. Shell loaded by a normal concentrated force

Now we investigate the case of a shell loaded by a local load. In order to obtain the expressions for this case it is sufficient to change the lower limit of integration from 0 to B in the integrals given in § 9.2. We can also simplify the integrals by omitting the coefficients μ_R, x_R, and φ_R. These coefficients have influence at the beginning of the segment of integration when α and β are small. We have therefore the following expression for the displacement w:

$$w^B = -\text{im}\frac{Pl^2}{4\pi D}\int\limits_B^\infty \sum_{k=1}^2 \frac{e^{i\alpha_k x}\cos\beta y\,d\beta}{\alpha_k(\alpha_k^2+\beta^2-(-1)^k i/2)}, \tag{9.29}$$

and similar to (9.9) for the bending moments:

$$M_{xx}^B = -\text{im}\frac{P}{4\pi}\int\limits_B^\infty \sum_{k=1}^2 \frac{(\alpha_k^2+\nu\beta^2)e^{i\alpha_k x}\cos\beta y\,d\beta}{\alpha_k(\alpha_k^2+\beta^2-(-1)^k i/2)},$$

$$M_{xy}^B = -\text{im}(1-\nu)\frac{P}{4\pi}\int\limits_B^\infty \sum_{k=1}^2 \frac{i\,e^{i\alpha_k x}\sin\beta y\,d\beta}{\alpha_k^2+\beta^2-(-1)^k i/2}. \tag{9.30}$$

The formula for M_{yy} can easily be written with the help of (9.9).

The shear forces have the following form after simplifications:

$$Q_x^B = -\text{im}\frac{P}{4\pi l}\int\limits_B^\infty \sum_{k=1}^2 \frac{i(\alpha_k^2+\beta^2)e^{i\alpha_k x}\cos\beta y\,d\beta}{\alpha_k^2+\beta^2-(-1)^k i/2},$$

$$Q_y^B = -\text{im}\frac{P}{4\pi l}\int\limits_B^\infty \sum_{k=1}^2 \frac{(\alpha_k^2+\beta^2)e^{i\alpha_k x}\beta\sin\beta y\,d\beta}{\alpha_k(\alpha_k^2+\beta^2-(-1)^k i/2)}. \tag{9.31}$$

The membrane forces are:

$$N_{xx}^B = -\text{im}\frac{PR_2}{4\pi l^2}\int\limits_B^\infty \sum_{k=1}^2 \frac{(-1)^k e^{i\alpha_k x}\beta^2\cos\beta y\,d\beta}{i\alpha_k(\alpha_k^2+\beta^2-(-1)^k i/2)},$$

$$N_{yy}^B = -\text{im}\frac{PR_2}{4\pi l^2}\int\limits_B^\infty \sum_{k=1}^2 \frac{(-1)^k e^{i\alpha_k x}\alpha_k\cos\beta y\,d\beta}{i(\alpha_k^2+\beta^2-(-1)^k i/2)}, \tag{9.32}$$

$$N_{xy}^B = \text{im}\frac{PR_2}{4\pi l^2}\sum_{k=1}^2 \int\limits_B^\infty \frac{(-1)^k e^{i\alpha_k x}\beta\sin\beta y\,d\beta}{\alpha_k^2+\beta^2-(-1)^k i/2}.$$

In order to obtain the solution of the above results in a closed form, the integrals can be simplified by substituting the simplified expressions for the roots of the characteristic equation.

Namely, these roots can be developed in power series with respect to $1/\beta$ and only the two first terms of the series can be taken into account. Such a treatment is correct and gives a good approximation when β is large enough, i.e. if one can be neglected with comparison to $16\beta^4(1-\lambda_R)^2$ in expression

$\xi_\beta = \sqrt{1+16\beta^4(1-\lambda_R)^2} \approx 4\beta^2(1-\lambda_R)$. From that condition we find the lower limit of integration B which should be larger than, for example, $B > 1\Big/\sqrt{\sqrt{2}\,(1-\lambda_R)}$. This is possible only when $\lambda_R \neq 1$, i.e. when the shell is not a sphere. The spherical shell was, however, solved separately without this simplification.

Using similar simplifications for the development of the relations (9.6) in series, we obtain the following simplified expressions for the roots of the characteristic equations:

$$a = k_1 - \frac{k_2}{\beta}, \qquad c = k_1 + \frac{k_2}{\beta}, \qquad \eta_1 = \eta_2 = 2k_1\beta,$$

$$b = \beta - k_1 + \frac{k_1}{\beta}, \quad d = \beta + k_1 + \frac{k_1}{\beta}, \quad k_1 = \sqrt{\frac{1}{8}(1-\lambda_R)}, \tag{9.33}$$

$$k_2 = \frac{1+\lambda_R}{8}.$$

Substituting the above into (9.30) and neglecting certain small terms, we obtain simple and easy to integrate integrands.

The bending moments are expressed then as the following integrals:

$$\left.\begin{matrix} M^B_{xx} \\ M^B_{yy} \end{matrix}\right\} = \frac{P}{4\pi}\int_B^\infty [(1+\nu)C_1(k_1x) \mp (1-\nu)S_1(k_1x)\beta]\frac{e^{-\beta x}}{\beta}\cos\beta y\,d\beta, \tag{9.34}$$

where

$$S_1(k_1x) = [\sinh k_1x\cos(k_1x) + \sin(k_1x)\cosh(k_1x)]\frac{1}{2k_1},$$

$$C_1(k_1x) = \cosh(k_1x)\cos(k_1x).$$

For small $x \ll 1$, we have $S_1(k_1x) \cong x$ and $C_1(k_1x) \simeq 1$. When $x \to \infty$ both $S_1(k_1x) \to \infty$, $C_1(k_1x) \to \infty$.

The integrals (9.34) can easily be evaluated (see Table 18 of integrals):

$$M^B_{xx} = -(1+\nu)\frac{P}{4\pi}\left[\ln r + \ln B + \gamma_0 + \sum_{k=1}^{\infty}(-B)^k\frac{r^k\cos k\vartheta}{kk!}\right]C_1(k_1x) -$$

$$-(1-\nu)\frac{P}{4\pi}S_1(k_1x)\frac{e^{-Bx}}{r}\cos(By+\vartheta). \tag{9.35}$$

We have a similar expression for M_{yy}. The moment M_{xy}^B has the form

$$M_{xy}^B = (1-\nu)\frac{P}{4\pi} S_1(k_1 x)\frac{e^{-Bx}}{r}\sin(By+\vartheta), \qquad \text{where} \quad \vartheta = \arctan\frac{y}{x}.$$

For small $r \ll 1$ we can neglect the infinite sum in expressions (9.35).

When r is small in comparison with 1, the second term of Eq. (9.35) behaves as $\frac{1-\nu}{4\pi} P \frac{x^2}{r^2}$ and for $r = 0$ it has a finite value. The other expressions include the singular terms $(1+\nu)\frac{P}{4\pi} \ln r$ identical with those obtained previously for the plate or spherical shell.

For $x = 0$ and arbitrary y these expressions can be written in the form

$$M_{xx}^B = M_{yy}^B = -(1+\nu)\frac{P}{4\pi}\mathrm{Ci}(By)$$

where $\mathrm{Ci}(y)$ is a cosine integral. When $r > 2l$ the solution can be simplified to the form

$$\left.\begin{matrix} M_{xx}^B \\ M_{yy}^B \end{matrix}\right\} = \frac{P}{4\pi}\left[\frac{1+\nu}{B} C_1(k_1 x) \mp (1-\nu) S_i(k_1 x)\right]\frac{e^{-Bx}}{r}\cos(By+\vartheta). \tag{9.36}$$

The integrals determining the membrane forces converge more rapidly than the integrals for moments, and it is not necessary to divide them into two parts by numerical integration. However, this work can be accelerated by evaluating the integrals between the limits B and ∞ by means of closed formulae. With this aim the integrals between the limits B and ∞ for normal forces are given. Performing a treatment similar to that used previously, we obtain them in the form

$$\begin{aligned} \left.\begin{matrix} N_{xx}^B \\ N_{yy}^B \end{matrix}\right\} &= \frac{PR_2}{4\pi l^2}\int_B^\infty \left[\mp S_2(k_1 x) + S_3(k_1 x)\frac{1}{\beta}\right] e^{-\beta x}\cos\beta y \, d\beta, \\ N_{xy}^B &= \frac{PR_2}{4\pi l^2}\int_B^\infty S_2(k_1 x)\, e^{-\beta x}\sin\beta y \, d\beta, \end{aligned} \tag{9.37}$$

where

$$S_2(k_1 x) = [\sinh(k_1 x)\cos(k_1 x) - \cosh(k_1 x)\sin(k_1 x)]\frac{1}{2k_1},$$

$$S_3(k_1 x) = \sinh(k_1 x)\sin(k_1 x).$$

The above integrals can be evaluated in a similar way as previously. For small and large value of r we can write similar simplified formulae.

It is worth noting that membrane forces do not tend to infinity as $r \to 0$. They have no singularities other than the bending moments which are at this point infinitely large. The results obtained in the previous chapter confirm this. Near the point $r = 0$ for small values of x, the membrane forces behave like the function $(x^2 \ln r + \text{constant})$. The membrane stresses calculated for various parameters λ_R are given in Figs. 92. The above calculations prove that the shell near the loaded point behaves as a plate and all singularities produced by a concentrated force are the same as in the plates. It may be interesting to compare expressions (9.35) with (3.70) obtained in § 3.51. When the shell becomes more flat, $R_1 = R_2 \to \infty$, the parameter $k \to 0$ and the functions $S_1(k_1 x) \to x$, $C_1(k_1 x) \to 1$. We see that expressions (3.70) and (9.35) differ from one another only by constant values and the functions of the coordinates (x, y) are equal. When the curvature does not equal zero, the local solution for the shell can be obtained by the multiplication of the results obtained for plates by the functions $S_1(k_1 x)$ and $C_1(k_1 x)$ depending on curvature. Bending moments for several values of $\lambda_R = R_2/R_1$ were found by the use of the described method, namely, by the numerical evaluations of integrals between limits 0 and B, and next by adding to these results the values calculated from B to ∞. It turns out that in a large range of the parameter $0.2 < \lambda_R \leqslant 1$, the state of bending stress differs little from the symmetrical one, characteristic for the spherical shell. With the decrease of λ_R the difference

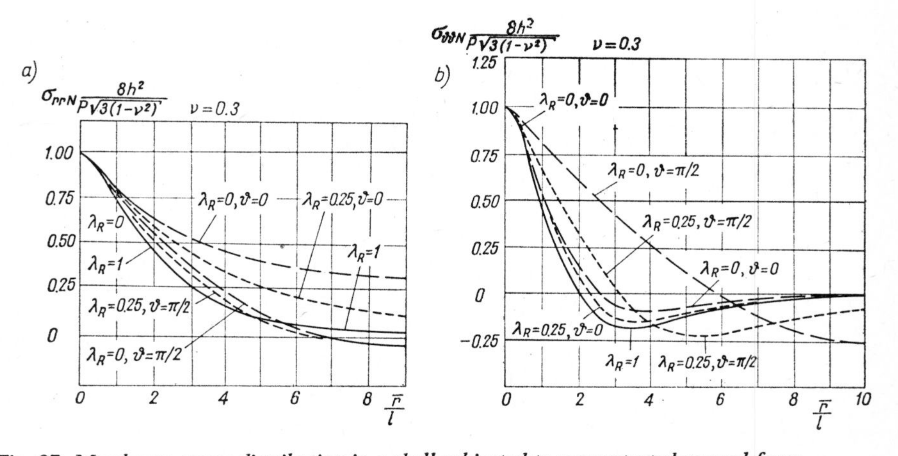

Fig. 97. Membrane stress distribution in a shell subjected to concentrated normal force.

between the state of stress in the shell treated and the spherical shell increases slightly at the bending and only for $\lambda_R = 0.2$ becomes larger. The bending stresses σ_{rr} and $\sigma_{\vartheta\vartheta}$ for the spherical shell, $\lambda_R = 1$ and infinitely long cylindrical shell, $\lambda_R = 0$ are presented graphically in Fig. 97a, b. In the case of the cylindrical shell the stresses spread further and disappear only at the distance of about $30l$ (in the direction of zero curvature) whereas in the spherical shell, they are very small already at the distance of $6l$. We see besides that the bending stresses (similarly as membrane stresses) spread further in the direction of the smaller curvature of the shell. The state of bending stress in the shell of negative curvature deviates much from the symmetrical one, spreading over the whole shell.

9.8 The effect of variability of the shell curvature

We have previously assumed the radii of curvature to be constant in the vicinity of the loading point. In the case of a double curvature shell this condition is satisfied for the sphere only. Should we consider other double curvature shells, the constancy of one of the radii of curvature requires the variability of the other one. This means that only the internal forces vanishing rapidly enough with increasing distance from the loading point can be calculated

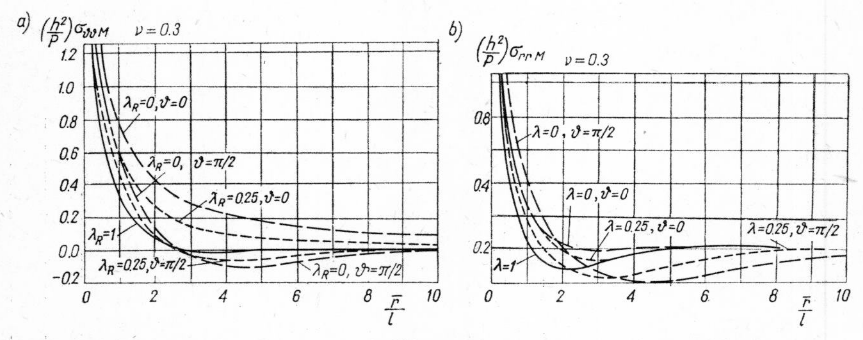

Fig. 98. Bending stress distribution in a shell loaded by a concentrated normal force.

from the integrals (9.2) with sufficient accuracy. If, however, the stresses extend far over the shell, the assumption of the constant radii of curvature may lead to too large errors. This effect is most important in the shells of negative Gaussian curvature.

If the radii of curvature of the shell are variable, the solutions previously obtained can be improved by the adoption of the perturbation method. In the considered area of the shell the radii of curvatures satisfy the condition

$$\frac{1}{R_{j1}} \leqslant \frac{1}{R_j(\alpha_1, \alpha_2)} \leqslant \frac{1}{R_{j2}}, \qquad j = 1, 2.$$

$R_{j1} = R_j(a, b)$ and $R_{j2} = R_j(c, d)$ denote here the maximum and minimum

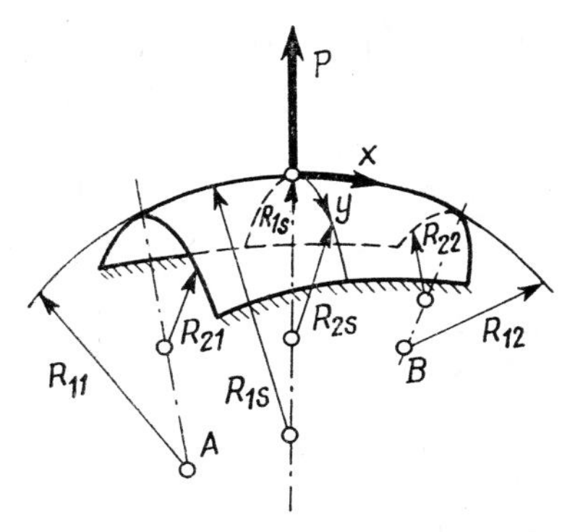

Fig. 99. Geometry of a shell of variable curvatures.

values of the radii $R_j(\alpha_1, \alpha_2)$. Therefore they can be expressed by the following power series:

$$\frac{1}{R_j(\alpha_1, \alpha_2)} = \frac{1}{R_{js}} \sum_{m=0}^{\infty} [1 - \varepsilon_j \delta_j(\alpha_1, \alpha_2)]^m = \frac{1}{R_{js}} \cdot \frac{1}{1 + \varepsilon_j \delta_j(\alpha_1, \alpha_2)} \tag{9.38}$$

where $\delta_j(\alpha_1, \alpha_2)$ are the functions of α_1, α_2 satisfying the conditions

$$\begin{aligned} &\delta_j(a, b) = 1, \qquad \delta_j(c, d) = -1; \\ &R_{1s} = \tfrac{1}{2}(R_{11} + R_{12}), \qquad R_{2s} = \tfrac{1}{2}(R_{21} + R_{22}) \end{aligned} \tag{9.39}$$

are the middle radii of curvature in the considered area of the shell (Fig. 99). The parameters

$$\varepsilon_1 = \frac{R_{12} - R_{11}}{R_{12} + R_{11}}, \qquad \varepsilon_2 = \frac{R_{22} - R_{21}}{R_{22} + R_{21}}$$

are always smaller than unity.

Let us assume that the deflection w and the stress function Φ can also be represented by the series and $\varepsilon_1 > \varepsilon_2 = \varepsilon$, $\varepsilon_1 = k\varepsilon$

$$w = w_0+\varepsilon w_1+\varepsilon^2 w_2+\dots,$$
$$\Phi = \Phi_0+\varepsilon\Phi_1+\varepsilon^2\Phi_2+\dots. \tag{9.40}$$

Substituting these series (9.40), (9.38) into Eqs. (9.1) and equating the coefficients by the same power of the parameter ε, we obtain the following set of differential equations of constant coefficients:

$$D\left(\Delta+\frac{1}{R_{1s}^2}-\frac{1-\nu}{2R_{1s}R_{2s}}+\frac{1}{R_{2s}^2}\right)^2 w_0-\Delta_{ks}\Phi_0 = Z,$$

$$\frac{1}{Eh}\left(\Delta+\frac{1}{R_{1s}R_{2s}}\right)^2\Phi_0+\Delta_{ks}w_0 = 0, \tag{9.41}$$

$$D\left(\Delta+\frac{1}{R_{1s}^2}-\frac{1-\nu}{2R_{1s}R_{2s}}+\frac{1}{R_{2s}^2}\right)^2 w_1-\Delta_{ks}\Phi_1 =$$
$$= -\frac{1}{A_1A_2}\left[\frac{1}{R_{2s}}\frac{\partial}{\partial\alpha_1}\left(\delta_2\frac{\partial\Phi_0}{\partial\alpha_1}\right)+\frac{1}{R_{1s}}\frac{\partial}{\partial\alpha_2}\left(k\delta_1\frac{\partial\Phi_0}{\partial\alpha_2}\right)\right]-$$
$$-\frac{1}{R_{1s}R_{2s}}\left[\frac{2\delta_1k+\delta_2}{R_{1s}}+\frac{\delta_1k+2\delta_2}{R_{2s}}\right]\Phi_0+4D\left(\frac{\delta_1k}{R_{1s}^2}-\right.$$
$$\left.-\frac{(1-\nu)(k\delta_1+\delta_2)}{4R_{1s}R_{2s}}\frac{\delta_2}{R_{2s}^2}\right)\left(\Delta+\frac{1}{R_{1s}^2}-\frac{1-\nu}{2R_{1s}R_{2s}}+\frac{1}{R_{2s}^2}\right)w_0,$$

$$\frac{1}{Eh}\left(\Delta+\frac{1}{R_{1s}R_{2s}}\right)^2\Phi_1+\Delta_{ks}w_1$$
$$= -\frac{1}{A_1A_2}\left[\frac{1}{R_{2s}}\frac{\partial}{\partial\alpha_1}\left(\delta_2\frac{\partial w_0}{\partial\alpha_1}\right)+\frac{1}{R_{1s}}\frac{\partial}{\partial\alpha_2}\left(k\delta_1\frac{\partial w_0}{\partial\alpha_2}\right)\right]-$$
$$-\frac{1}{R_{1s}R_{2s}}\left[\frac{2\delta_1k+\delta_2}{R_{1s}}+\frac{\delta_1k+2\delta_2}{R_{2s}}\right]w_0+$$
$$+\frac{2(\delta_1k+\delta_2)}{R_{1s}R_{2s}}\left(\Delta+\frac{1}{R_{1s}R_{2s}}\right)\frac{\Phi_0}{Eh},$$

where

$$\Delta_{ks} = \frac{1}{A_1A_2}\left[\frac{\partial}{\partial\alpha_1}\left(\frac{1}{R_{2s}}\frac{\partial}{\partial\alpha_1}\right)+\frac{\partial}{\partial\alpha_2}\left(\frac{1}{R_{1s}}\frac{\partial}{\partial\alpha_2}\right)\right]+$$
$$+\frac{1}{R_{1s}R_{2s}}\left(\frac{1}{R_{1s}}+\frac{1}{R_{2s}}\right). \tag{9.42}$$

It is assumed here that the system of coordinates α_1, α_2 can be applied in such a way that the coefficients of the first quadratic form of the middle surface of the shell A_1 and A_2 can be taken as constant.

Each equation of the set concerns the shell of constant curvature. We observe that the equations of the first approximation are identical with the equations used previously. To solve the set (9.41) we start by solving first two equations for w_0 and Φ_0. Having w_0 and Φ_0, we can solve next two equations for w_1 and Φ_1. The treatment should be repeated until sufficient accuracy is achieved. If $\varepsilon < \frac{1}{2}$, only two approximations are usually sufficient for engineering purposes.

If we neglect the terms containing the radii of curvature as parameters in Eq. (9.41) with the exception of the operator Δ_k, the set (9.42) simplifies and we find

$$
\begin{aligned}
&D\Delta\Delta w_0 - \Delta_{ks}\Phi_0 = Z,\\
&\frac{1}{Eh}\Delta\Delta\Phi_0 + \Delta_{ks}w_0 = 0,
\end{aligned}
\tag{9.43}
$$

$$
\begin{aligned}
&D\Delta\Delta w_1 - \Delta_{ks}\Phi_1\\
&\quad = -\frac{1}{A_1A_2}\left[\frac{1}{R_{2s}}\frac{\partial}{\partial\alpha_1}\left(\delta_2\frac{\partial\Phi_0}{\partial\alpha_1}\right) + \frac{1}{R_{1s}}\frac{\partial}{\partial\alpha_2}\left(\delta_1 k\frac{\partial\Phi_0}{\partial\alpha_2}\right)\right],\\
&\frac{1}{Eh}\Delta\Delta\Phi_1 + \Delta_{ks}w_1\\
&\quad = -\frac{1}{A_1A_2}\left[\frac{1}{R_{2s}}\frac{\partial}{\partial\alpha_1}\left(\delta_2\frac{\partial w_0}{\partial\alpha_1}\right) + \frac{1}{R_{1s}}\frac{\partial}{\partial\alpha_2}\left(\delta_1 k\frac{\partial w_0}{\partial\alpha_2}\right)\right],
\end{aligned}
$$

. . . .

$$
\begin{aligned}
D\Delta\Delta w_n - \Delta_{ks}\Phi_n &= \sum_{m=0}^{n-1}\left\{\frac{1}{R_{2s}}\frac{\partial}{\partial\alpha_1}\left[(-\delta_2)^{n-m}\frac{\partial\Phi_m}{\partial\alpha_1}\right] + \right.\\
&\quad \left. + \frac{1}{R_{1s}}\frac{\partial}{\partial\alpha_2}\left[(-\delta_1 k)^{n-m}\frac{\partial\Phi_m}{\partial\alpha_2}\right]\right\},\\
\frac{1}{Eh}\Delta\Delta\Phi_n + \Delta_{ks}w_n &= \sum_{m=0}^{n-1}\left\{\frac{1}{R_{2s}}\frac{\partial}{\partial\alpha_1}\left[(-\delta_2)^{n-m}\frac{\partial w_m}{\partial\alpha_1}\right] + \right.\\
&\quad \left. + \frac{1}{R_{1s}}\frac{\partial}{\partial\alpha_2}\left[(-\delta_1 k)^{n-m}\frac{\partial w_m}{\partial\alpha_1}\right]\right\}, \qquad n = 1, 2, \ldots.
\end{aligned}
$$

9 Arbitrary shell loaded by a normal concentrated force

Let us consider a shell loaded by a concentrated force P presented in Fig. 99. We introduce the system of non-dimensional orthogonal coordinates (x, y) whose origin is at the point of application of the load. We assume that the principal radius of curvature $R_1 = \text{constant}$ and the coefficients $A_1 = A_2 = 1$, The second radius R_2 can be approximately presented by the series

$$\frac{1}{R_2(x)} \cong \frac{1}{R_{2s}} \cdot \frac{1}{1+\varepsilon \sin \delta x} = \frac{1}{R_{2s}} \sum_{m=0}^{\infty} (-\varepsilon \sin \delta x)^m. \tag{9.44}$$

The parameter δ is assumed in such a way that the function $\sin \delta x$ changes from -1 to $+1$ in the considered area. The functions of the first approximation w_0 and Φ_0 were obtained in Section 9.1. Now we introduce them into Eqs. (9.43); we have

$$\begin{aligned} D\Delta\Delta w_1 - \Delta_{ks}\Phi_1 &= \frac{P}{2\pi^2} \int_0^\infty \int_0^\infty \frac{(\alpha^2+\lambda_s\beta^2)\alpha\,[\gamma_1 \sin\gamma_1 x - \gamma_2 \sin\gamma_2 x]\cos\beta y\, d\alpha\, d\beta}{(\alpha^2+\beta^2)^4 + (\alpha^2+\lambda_s\beta^2)^2}, \\ \frac{1}{Eh}\Delta\Delta\Phi_1 + \Delta_{ks} w_1 &= \frac{Pl^2}{2\pi^2 D R_{2s}} \int_0^\infty \int_0^\infty \frac{(\alpha^2+\beta^2)^2\alpha\,[\gamma_1 \sin\gamma_1 x - \gamma_2 \sin\gamma_2 x]\cos\beta y\, d\alpha\, d\beta}{(\alpha^2+\beta^2)^4 + (\alpha^2+\lambda_s\beta^2)^2}, \end{aligned} \tag{9.45}$$

where

$$\gamma_1 = \delta + \alpha, \quad \gamma_2 = \delta - \alpha, \quad \lambda_s = \frac{R_{2s}}{R_{1s}}.$$

Solving this set, we find the following approximate relation:

$$\begin{aligned} w_1 = \frac{Pl^4}{2\pi^2 D} \sum_{k=1}^{2} \int_0^\infty \int_0^\infty [-(-1)^k] \times \\ \times \frac{[(\alpha^2+\lambda_s\beta^2)(\gamma_k^2+\beta^2) - (\alpha^2+\beta^2)^2(\gamma_k^2+\lambda_s\beta^2)]\,\alpha\gamma_k \sin\gamma_k x \cos\beta y\, d\alpha\, d\beta}{[(\alpha^2+\beta^2)^4 + (\alpha^2+\lambda_s\beta^2)^2][(\gamma_k^2+\beta^2)^4 + (\gamma_k^2+\lambda_s\beta^2)^2]}, \end{aligned} \tag{9.46}$$

where l is the characteristic length. The boundary effects are here neglected.

For the stress function Φ_1 a similar integral can easily be derived. We observe that the above integral converges more rapidly than the integral $(9.2)_1$ for w_0. It can be evaluated by numerical methods. It can also be evaluated once by means of calculus of residues. For $x=0$ the above integral is equal to zero. We see then that the maximum deflection $w_{(0,0)}$ depends only on the average radii R_{1s}, R_{2s}.

9.9 Method of singular solutions

The calculation of the stresses in the shell of positive Gaussian curvature loaded by a concentrated normal force was performed by K. Forsberg and W. Flügge [9.9] in a different way. The method which Forsberg and Flügge used to investigate the effect of the concentrated loads in the shell consisted in studying the singular solutions of the homogeneous differential shell equations. These solutions can be interpreted as the effect of concentrated loads.

The environment of the point of application of the load was approximated by an elliptic paraboloid which has principal radii of curvature at its vertex equal to those of the shell at the loading point. The shallow shell equations were applied to the solution which was assumed in the form of the infinite series:

$$\eta = \sum_{n=0}^{\infty} f_n(r) \cos n\vartheta,$$

where $\eta(r, \vartheta)$ is the function solving the shell equations. Then, applying the perturbation method, the function $f_n(r)$ was assumed in the form of power series:

$$f_n = \sum_{m}^{\infty} f_{nm}(r)\gamma^m$$

where $\gamma = \dfrac{1}{2}\left(\dfrac{1-\lambda_R}{1+\lambda_R}\right)$; $\lambda_R = \dfrac{R_2}{R_1}$ is the parameter depending on the curvature of the shell. As a result the infinite set of recurrent ordinary differential equations was obtained. The conditions for a concentrated load were evaluated by the use of the following conditions:

(1) Displacement normal to the shell surface w to remain finite as $r \to 0$.
(2) Slope at origin $\lim(\partial w/\partial r) = 0$.
(3) Hoop strain ε_ϑ to remain finite as $r \to 0$.

(4) The shell element cut out at the vertex is in equilibrium under the normal load and the stress resultants.

On the numerical evaluations the results, as for example presented in Fig. 100, were obtained. These do not differ in a drastic way from those obtained by means of Fourier integrals.

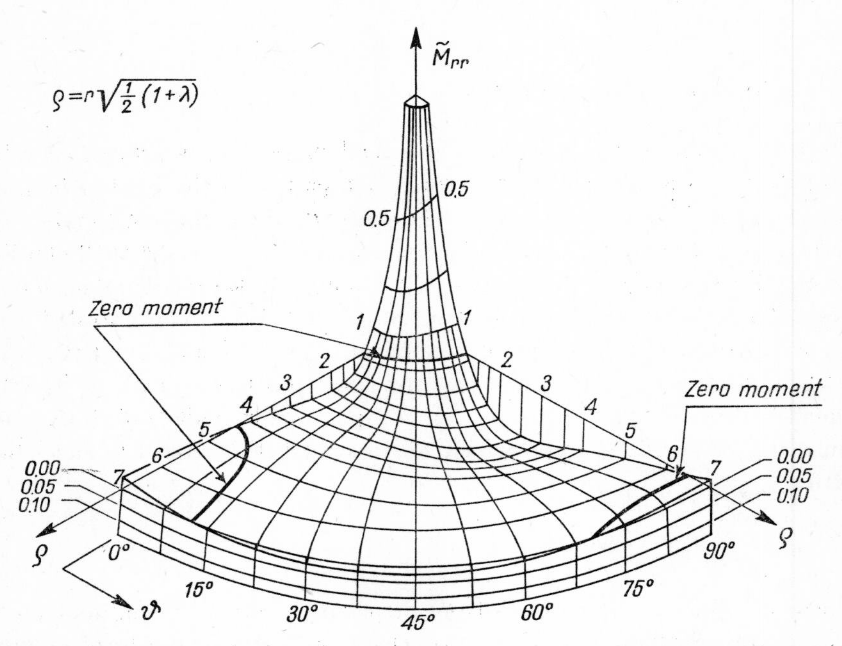

Fig. 100. Variation of the bending moment M_{rr} *in a shell of positive Gaussian curvature* $\lambda = 0.25$. *Results of Forsberg and Flügge* [9.9].

9.10 Load distributed along a small length

The solution for the shell loaded by a force distributed over a small segment of the length $2c$ can be derived directly from the previous results. The solution will be found by multiplying the integrands by the expression $\sin\beta c/\beta c$. For small magnitudes of β and c this expression is approximately equal to 1 and the integrals evaluated between 0 and B do not undergo a change.

Integrating from B to ∞, we obtain the following expression for the bending moments produced by the local loads. For small $r \ll 1$ we have

$$\left.\begin{matrix} M^B_{xx} \\ M^B_{yy} \end{matrix}\right\} = \frac{P}{8\pi c}\left\{(1+\nu)\,C_1(k_1 x)\left[\frac{e^{-Bx}}{B}[\sin B(y+c)-\sin B(y-c)]+\right.\right.$$
$$+x(\vartheta_2-\vartheta_1)-(y+c)\ln r_2+(y-c)\ln r_1-2c(\ln B+\gamma_0)\pm$$
$$\left.\left.\pm(1-\nu)\,S_1(k_1 x)\,(\vartheta_2-\vartheta_1)\right]\right\}. \qquad (9.47)$$

We observe that for $r = 0$ the moments do not increase to infinity but have a finite value.

$$M^B_{xx} = M^B_{yy} = -(1+\nu)\frac{P}{4\pi}\left[\ln c+\ln B+\gamma_0-\frac{\sin Bc}{Bc}\right].$$

The moments at the point $r = 0$ do not depend on the curvature of the shell. For large r the moments should be calculated from exact formulae. For $x = 0$ and arbitrary y we have

$$M^B_{xx} = M^B_{yy} = -(1+\nu)\frac{P}{4\pi}\left\{\frac{1}{2c}[(y+c)\,\mathrm{Ci}\,(y+c)\,B-\right.$$
$$\left.-(y-c)\,\mathrm{Ci}\,(y-c)\,B]-\frac{\sin Bc}{Bc}\cos By\right\}.$$

Let us see now how the load in this case is divided into a local and general part. The integrals evaluated between the limits 0 and B refer to a general load determined by

$$Z^0 = \frac{P}{\pi l}\int_0^B \frac{\sin \beta c}{\beta c}\cos \beta y\, d\beta.$$

This integral can be evaluated

$$Z^0 = \frac{P}{2\pi l c}[\mathrm{Si}\,(y+c)\,B-\mathrm{Si}\,(y-c)\,B],$$

where Si( ) is a sine integral.

The division of the load into local and general parts is presented graphically in Fig. 101. Figure 101a presents the whole load determined by the integral between the limits 0 and ∞. The general load is presented in Fig. 101b and the local load in Fig. 101c. It was assumed here that $c = 0.5$, $B = 2$. In Fig. 101b also a general load referred to a concentrated force for $B = 2$ was given (thin line). We observe that these two general loads differ slightly which let us draw the conclusion that the stresses produced by them will be identical. The fact that the integrals between the limits 0 and B do not change when one local load is replaced by another permits the use of the following approximate method. Let us imagine we have the solution for the shell loaded by a certain concentrated load, all effects by this load being known. We know then all values of the integrals between 0 and B. Now we want to calculate the stresses in the same shell subjected to another concentrated load, for example

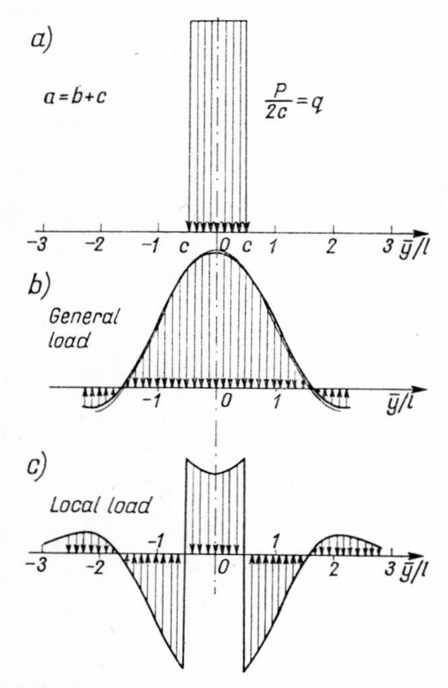

Fig. 101. Division of a distributed load into local and total parts.

a force distributed over a small area and acting at the same place of the shell. As the stresses caused by the second load will differ only in the local part (integrals between B and ∞), the solution can be obtained directly by subtracting from the whole state of stress the local part referring to the first load and adding the local part referring to the second load.

9.11 Experimental investigation

To prove the correctness of the theoretical results concerning shells of double positive curvatures ($\lambda_R \geqslant 0.2$) obtained in previous chapters, the following experiment was carried out with a shell of revolution, built-in at one edge and free at the second one.

The shells whose shape is given in Fig. 102a was made of aluminium having Young's modulus $E = 7.3 \cdot 10^5$ kG/cm^2 and the thickness $h = 0.11$ cm $\pm$

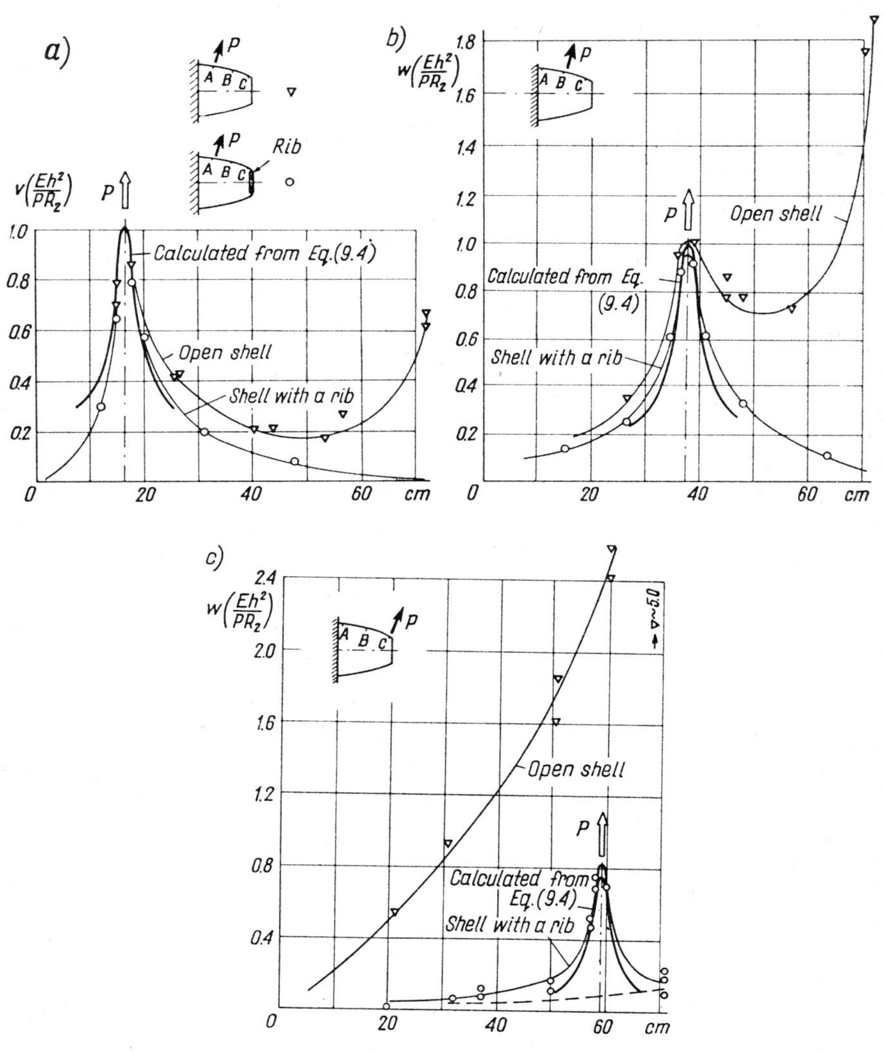

Fig. 102. Comparison of theoretical results and experimental data; thick line presents deflection of a shell calculated from Eq. (9.4) for various positions of the point of application of the load.

± 0.01 cm. The main radii of curvature changed in the area of the shell from $R_1 = 28$ cm and $R_2 = 17$ cm on the free edge to $R_1 = 200$ cm and $R_2 = 25$ cm on the built-in edge. The geometric coefficient λ_R changed in connection with the above, from $\lambda_R = 0.553$ to $\lambda_R = 0.150$. The shell was loaded by concentrated forces at three points laying on one generator and its displacement was measured twice for two different boundary conditions on the right-hand edge of the shell. In the first case this edge was free and the other time the edge was strengthened in its plane with a very rigid rib. The rib was jointly connected with the shell. The edge was freely supported.

The results of the measurements are given in Figs. 102a–c; Fig. 102a represents the displacemement of the shell when the concentrated force acts at a relatively small distance from the built-in edge. We see that the maximum displacement is smaller than that calculated directly from formula (9.21). The difference for the shell strengthened by a rib is about 15 per cent and for the shell with free edge 10 per cent. The difference is probably caused by the fact that the built-in edge is near the loading point. If the coefficient λ_R is rather small ($\lambda_R = 0.15$), the influence of the built-in edge spreads far. However, the character of the displacement in the case of the shell reinforced with a rib is very similar to the displacement calculated theoretically. The shell with free edges behaves differently and large displacements appear over a large area. However, the maximum displacement differs little from the displacement calculated from formula (9.21). The displacements of the same shell loaded at a point B situated at a great distance from both edges are presented graphically in Fig. 102b. The maximum displacement differs from that calculated only by a few per cent both when the right-hand side edge of the shell is clamped with a rigid rib and when it is free. In the case of a clamped edge, the calculated curve is similar to that obtained by experiment. When the right-hand edge is free, the shell changes its shape quite differently. Large displacements covering the whole cross-section of the shell appear on this edge. We see a similar phenomenon in Fig. 102c.

The concentrated force acts near the edge. It turns out that only in the case where the edge is clamped with a rib, the displacements calculated from (9.4) are in agreement with the experiment. When the edge is free, large displacements resulting from the change of the cross-section of the shell add to the local displacement. Then, the displacements calculated from formula (9.4) are many times smaller than the real displacement. In order to obtain numerical results for the last case in agreement with experiments, real boundary conditions should be taken into consideration. This would involve adding a particular solution with a number of constants of integration to that obtained.

In addition the stresses in the shell were measured by means of strain gauges. The state of stresses is most interesting near the three loaded points A, B, and C. The results of measurement are given in Fig. 103. The lines present the stress calculated from formulae (9.11) and (9.12). This calculation

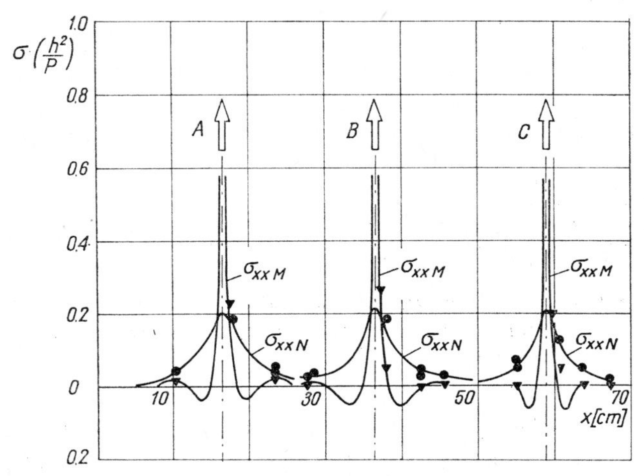

Fig. 103. Comparison of stresses calculated and stresses obtained from experiments for various positions of the points of application of the load.

was done in a similar way as the calculation of the displacement taking every time the local system of non-dimensional coordinates whose origin was at the point of loading. In the case of the shell where one edge is free or movable, the results of experiments confirmed the local character of the displacements and stresses in the shell of double positive curvature only if the concentrated force acts at a great distance from this edge. It can be observed from experiments that the free edge strongly influences the displacements and a little less the stresses. The state of membrane stress which is shown in Fig. 94a cannot appear near the free edge. The equilibrium of the internal forces requires additional stresses which produce a strong distortion of the shell. Therefore such shells should always be considered with real boundary conditions. On the contrary, the stresses in the shells of the positive double curvature, clamped or jointly supported at the edges, can be found with relatively good accuracy from the equations of § 9.2 without taking into consideration their real boundary conditions if the load acts at the distance of the order of some scores of l.

9.12 Conclusions

Based on the theoretical and experimental results given in this chapter, we can draw the following conclusions. If we consider a shell of positive Gaussian curvature under the action of the concentrated load applied to the shell at a satisfactory distance from the edge, we can divide its surface in four regions: I, II, III, and IV (Fig. 104).

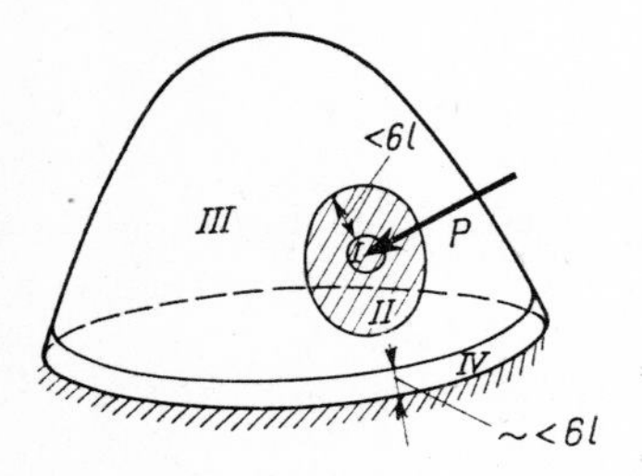

Fig. 104. Regions in a shell of positive Gaussian curvature subjected to a concentrated force: I thick plate theory; II shallow shell equations; III equations of membrane theory; IV edge effect.

In region I, which is in the vicinity of the point of application of the load, the shell behaves like a flat plate. Here the bending stresses reach the largest values.

In region II surrounding region I the membrane and bending stresses appear as well. The calculations in this region can be based on the shallow shell equations.

In region III, external to regions I and II, appear only the membrane stresses which can be calculated by means of the membrane theory of shells.

The last region IV is in the neighbourhood of the edge. The stresses in this area consist of membrane stresses and stresses produced by the edge effects.

The calculation of the stresses in the shell can be performed in several stages:

First we calculate the membrane stresses, then we take into account the edge stresses which can be determined by means of the equations of the edge effect. Finally, we consider the local effects in the vicinity of the loaded places. On the boundary between regions II and III the membrane stresses should coincide with the results of the equations of the membrane theory. The dimensions of region II in the case of doubly curved shells are from about $6l$ to $30l$.

References 9

[9.1] S. ŁUKASIEWICZ: Concentrated loadings on shells, *Proc. XI Int. Congr. Appl. Mech.*, Springer Verlag, 1965.

[9.2] W. FLÜGGE: Concentrated forces on shells, *Proc. XI Int. Congr. Appl. Mech.*, Springer Verlag, 1965.

[9.3] S. ŁUKASIEWICZ: Concentrated loads in shells, *Zesz. Nauk. Polit. Warsz., Mechanika 21* (1967) (in Polish).

[9.4] S. ŁUKASIEWICZ and J. STUPNICKI: Effect of transverse shear deformations in the locally loaded shell, *Arch. Bud. Masz. 17. 2* (1971).

[9.5] HO KWANG-CHIEN and CHEN FU: A simplified method for calculating double curvature shallow shells under the action of concentrated loads, *Acta Mech. Sinica 6.1* (1963) 19–37.

[9.6] S. ŁUKASIEWICZ: The solution for concentrated loads on shells by means of Thomson functions, *ZAMM 48* (1968) 247–254.

[9.7] P. M. VELICHKO, V. P. SHEVCHENKO: On the action of concentrated forces and moments on the shell of positive Gaussian curvature, *AN SSSR 12* (1969) (in Russian).

[9.8] A. L. GOLDENVEIZER: On the problem of shells under concentrated loads, *Prikl. Mat. Mekh. 18.2* (1954) (in Russian).

[9.9] K. FORSBERG and W. FLÜGGE: Point load on a shallow elliptic paraboloid, *J. Appl. Mech. 3* (1966) 575–584.

[9.10] G. N. CHERNYSHEV: On the action of concentrated forces and moments of an elastic thin shell of arbitrary shape, *Prikl. Mat. Mech. 27* (1963) 126–134.

[9.11] V. I. FEODOSEV, S. N. CHERNAKOV: On the transfer of concentrated forces on thin shell, *Inj. Sb. Mekh. Tverdego Tela 6* (1966) 57–63 (in Russian).

[9.12] V. M. DAREVSKII: Contact problems in the theory of shells (Action of local loads on shells), *Trudy VI Vses. Konf. po Teorii Obolochek i Plastinok*, Izd. Nauka, Moscow 1966, pp. 927–934 (in Russian).

[9.13] N. A. KILII: On the action of local loads on thin shells, *Izv. buzob. Stroit. i architektura 9* (1968) (in Russian).

[9.14] L. J. ERNST: De draagwerking van dubbelgekromde schalen, belast met een puntlast, *Rapport 8-73-4, Stevin- Laboratorium, Technische Hogeschool Delft.*

[9.15] J. L. SANDERS: Singular solutions to the shallow shell equations, *Trans. ASME, vol. E 37.2* (1970).

[9.16] J. DUNDUSS, M. G. SAMUCHIN: Transmission of concentrated forces into prismatic shells, *Int. J. Solids and Struct. 7. 12* (1971).

[9.17] Strength, stability, vibrations, *Spravochnik v 3 tomakh*, M. "Mashinostroenie" 1968, tom 2, gl. 2 (in Russian).

[9.18] A. S. KHRISTENKO: On the action of concentrated forces on orthotropic shell of arbitrary shape, *Trudy Nikolaevskogo korablestroit. in-ta 25* (1968) (in Russian).

[9.19] A. S. KHRISTENKO, M. A. CHEREMUSHEVA: The action of local loads on orthotropic shell of arbitrary shell, *Tr. Nikolaevskogo korablestroitel. in-ta 25* (1968) (in Russian).

[9.20] G. N. CHERNYSHEV: Presentation of Green's type solutions for shell equations using small parameter method, *PMM 32.6* (1968) (in Russian).

[9.21] G. N. CHERNYSHEV: On contact problems in the theory of shells, *in*: *Tr. VII Vses. konf. po teorii obolochek i plastinok*, M. "Nauka", 1970 (in Russian).

[9.22] G .N. CHERNYSHEV: Character of the solutions of shell equations of zero curvature under concentrated loads, *V sb. Tr. VII Vses. konf. po teorii obolochek i plastinok*, M. "Nauka", 1970 (in Russian).

[9.23] P. M. VELICHKO, V. P. SHEVCHENKO: On the solution of the problem of shell displacement under concentrated loads, *Teor. i Prikl. Mekh. Resp. Mizhvid. Temat. Nauk. Tekhn., zb., 4* (1973) (in Ukrainian).

[9.24] M. G. SAMUCHIN, J. DUNDURS: Transmission of concentrated forces into prismatic shells, *I. Int. J. Solids and Struct. 9.2* (1973).

10

Shells of cylindrical and nearly cylindrical shape subjected to normal force

In previous chapters we considered the shell of double positive Gaussian curvature and the load, deflection and stress function being expressed as double Fourier integrals. In the case of a sphere and shells of nearly spherical shape, the integrals determining deflection and stresses converge rapidly, and the deflection and stresses decrease rapidly with increasing distance from the loading point. We have found that if the loading point is not very close to the edge, the boundary conditions can be neglected.

On the contrary, when considering shells of zero and also small Gaussian curvature, the distortions decrease slowly, spreading throughout the whole shell. In this situation the real boundary conditions must be taken into consideration. When calculating stresses in the nearly cylindrical shell the deflection of the circumference should be described by periodic functions. Only then will the displacements and stresses have the same value when we arrive at the same point by traversing the circumference. In solving such shell problems it is convenient to express the deflections in cylindrical coordinates $(\bar{x}, \varphi)$, $\bar{x}$ being the distance measured along the generator, and φ the angle measured in the plane of the cross-section of the shell. Then the solution is usually found in Fourier series developing the functions with respect to the variable φ.

10.1 The nearly cylindrical shell loaded by two oppositely directed radial forces

Let us find a solution for a shell of nearly cylindrical shape subjected to two oppositely directed concentrated forces (Fig. 105). The local shape of the shell can be determined by the ratio of the radii of curvature $\lambda_R = R_2/R_1$ at the loading point. Here, $R_2 < R_1$.

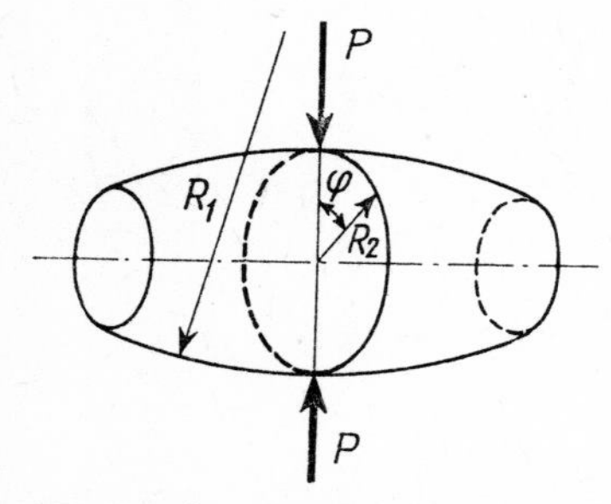

Fig. 105. Nearly cylindrical shell subjected to two oppositely directed radial forces.

For the circular cylinder, $\lambda_R = 0$.

We assume that both radii of curvature are constant near the loading point, and introduce a non-dimensional system of coordinates related to the smaller radius of curvature R_2; $x = \bar{x}/R_2$; $y = \varphi$. Then $A_1 = R_2$; $A_2 = R_2$ are constant.

Now the characteristic length is $l = R_2$. The origin of coordinates can be assumed to be at the point of application of the load. Let us represent the radial concentrated load by means of a distributed load q expressed as a function of the coordinates x, φ

$$P = q(x, \varphi) = \frac{P}{(\pi R_2)^2} \int_0^\infty \left[1 + 2 \sum_{n=2,4,\ldots,}^{\infty} \cos n\varphi\right] \cos \alpha x \, d\alpha. \tag{10.1}$$

We assume that the shell is infinitely long in the x-axis direction. This assumption can be satisfied exactly only in the case of a cylindrical shell. It is possible that for shells of double curvature it may have not any meaning. However, in shells of positive Gaussian curvature all distortions are damped more rapidly and are restricted to small areas. So, when the length of the shell L is larger than certain multiples of the wave length of distortion, this assumption can be correct. The limits may be defined from numerical computation.

Substituting Eq. (10.1) in (9.2) and neglecting the effect of transverse shear deformation yields [10.1]

$$w_{(x,\varphi)} = \frac{2PR_2^2}{\pi^2 D}\left\{\frac{1}{2}\int_0^\infty \frac{\cos \alpha x \, d\alpha}{(\alpha^2-\chi_R)^2+4\varkappa^4} + \right.$$

$$\left. + \sum_{n=2,4,\dots}^{\infty}\int_0^\infty \frac{(\alpha^2+n^2-\mu_R)^2 \cos\alpha x\, d\alpha \cos n\varphi}{(\alpha^2+n^2-\mu_R)^2(\alpha^2+n^2-\chi_R)^2+4\varkappa^4(\alpha^2+\lambda_R n^2-\varphi_R)^2}\right\} \tag{10.2}$$

and similar series for the stress function Φ where $4\varkappa^4 = 12(1-\nu^2)R_2^2/h^2$. The above integral can easily be evaluated with the aid of the calculus of residues in a similar way as in § 9.1.

We observe that the solutions (10.2) have a form similar to those obtained previously. If in formulae (9.2) the sign of the integral is replaced by the sign of the sum and βy by the variable $n\varphi$, we obtain a new solution. Now the coefficient $4\varkappa^4 \neq 1$ which yields the different relations for the roots of characteristic equation is

$$(\alpha^2+n^2-\mu_R)^2(\alpha^2+n^2-\chi_R)^2+4\varkappa^4(\alpha^2+\lambda_R n^2-\varphi_R)^2 = 0. \tag{10.3}$$

Solving this equation, we have

$$\left.\begin{matrix}\alpha_1^2\\ \alpha_2^2\end{matrix}\right\} = -n^2+\frac{\chi_R+\mu_R}{2} \mp i\varkappa^2\left[1\mp\left\{1+\frac{[2n^2(1-\lambda_R)-\chi_R-\mu_R+2\varphi_R]i}{\varkappa^2}\right\}^{1/2}\right]$$

or

$$\left.\begin{matrix}\alpha_1^2\\ \alpha_2^2\end{matrix}\right\} = -n^2\pm\eta_1+\frac{\chi_R+\mu_R}{2}+i(\eta_2\mp\varkappa^2). \tag{10.4}$$

We have

$$\alpha_1 = a+bi,$$

$$\alpha_2 = c+di,$$

where

$$\left.\begin{matrix}a\\ b\end{matrix}\right\} = \frac{1}{\sqrt{2}}\left\{\left[\left(n^2-\frac{\chi_R+\mu_R}{2}-\eta_1\right)^2+(\eta_2-\varkappa^2)^2\right]^{1/2}\pm\right.$$

$$\left.\pm\left(-n^2+\frac{\chi_R+\mu_R}{2}+\eta_1\right)\right\}^{1/2},$$

$$\left.\begin{matrix} c \\ d \end{matrix}\right\} = \frac{1}{\sqrt{2}} \left\{ \left[\left(n^2 - \frac{\chi_R + \mu_R}{2} + \eta_1 \right)^2 + (\eta_2 + \varkappa^2)^2 \right]^{1/2} \pm \right.$$
$$\left. \pm \left(-n^2 + \frac{\chi_R + \mu_R}{2} - \eta_1 \right) \right\}^{1/2},$$

$$4\varkappa^4 = 12(1-\nu^2) R_2^2/h^2,$$

$$\left.\begin{matrix} \eta_1 \\ \eta_2 \end{matrix}\right\} = \frac{\varkappa^2}{\sqrt{2}} \left[\xi_n \mp 1 \pm \frac{(\chi_R - \mu_R)^2}{4\varkappa^4} \right]^{1/2},$$

$$\xi_n = \left\{ \left[1 - \frac{(\chi_R - \mu_R)^2}{4\varkappa^4} \right]^2 + \frac{4}{\varkappa^4} \left[(1-\lambda_R) n^2 - \frac{\chi_R + \mu_R}{2} + \varphi_R \right]^2 \right\}^{1/2},$$

where

$$\varphi_R = (1+\lambda_R)\lambda_R, \quad \chi_R = 1 + \frac{1-\nu}{2}\lambda_R + \lambda_R^2, \quad \mu_R = \lambda_R, \quad \lambda_R = \frac{R_2}{R_1}.$$

It is worth noting here that in the case where $4\varkappa^4 \neq 1$ the roots α_1, α_2 can be obtained from the roots of Eq. (9.5) where it has been assumed that $4\varkappa^4 = 1$. Namely, for every integer $n = \sqrt{2}\varkappa\beta$ the roots are

$$\alpha_1 = \sqrt{2}\varkappa(a+bi), \quad \alpha_2 = \sqrt{2}\varkappa(c+di),$$

where a, b, c, d are the coefficients computed previously for the case $4\varkappa^4 = 1$.

The calculation of the shell deflection is then reduced to the numerical summation of the Fourier series whose terms contain the functions of the coordinate x. This series converges rapidly and no additional methods are necessary for its summation. It is, however, troublesome because of the complicated construction of the formulae defining the roots of the characteristic equation. We can facilitate these calculations by using the approximate method proposed by L. S. D. Morley [10.2]. For thin shells the non-dimensional structural parameter $\varkappa$ must have a numerical value $5 < \varkappa < 50$. Bearing these values in mind and making the substitution

$$\delta_n = \frac{2n^2(1-\lambda_R) - (\chi_R + \mu_R - 2\varphi_R)}{4\varkappa^2}$$

for $\delta_n \ll 1$, and $\lambda_R \ll 1$, the roots of Eq. (10.3) can be rewritten

$$\begin{aligned} \alpha_1^2 &= -n^2\lambda_R + 2i\varkappa^2\delta_n(1+2\delta_n i), \\ \alpha_2^2 &= -n^2(2-\lambda_R) + 2\varkappa^2 i + 1. \end{aligned} \tag{10.5}$$

The roots α_1 and α_2 can be approximated by

$$\alpha_1 = (1+i)\varkappa\delta_n f(\gamma_n), \qquad \alpha_2 = (1+i)\varkappa\left[1+\frac{(1-\frac{1}{2}\lambda_R)n^2 i}{2\varkappa^2}\right], \tag{10.6}$$

provided that it is agreed to neglect $5\delta_n^2$ in comparison with unity. In the above equation

$$f(\gamma_n) = \frac{1}{\sqrt{2}}[(1+\gamma_n^2)^{1/2}+1]^{1/2} + \operatorname{sign}(\lambda_R)\cdot i\frac{1}{\sqrt{2}}[(1+\gamma_n^2)^{1/2}-1]^{1/2},$$

$$\gamma_n = \frac{8n^2\varkappa^2\lambda_R}{[2n^2(1-\lambda_R)-(\lambda_R+\mu_R-2\varphi_R)]^2}. \tag{10.7}$$

The accuracy of Eqs. (10.6) and (10.7) is best for small values of n and these terms are usually the most important constituents of the final result for w.

The above approximation is not sufficiently accurate for larger values of the integer n. But for larger n it is permissible to neglect χ_R, μ_R, φ_R in comparison with n^2. Then the roots can be expressed in the form

$$\left.\begin{matrix}\alpha_1^2\\ \alpha_2^2\end{matrix}\right\} = -n^2 \pm \eta_1 + i(\eta_2 \mp \varkappa^2), \tag{10.8}$$

where

$$\left.\begin{matrix}\eta_1\\ \eta_2\end{matrix}\right\} = \varkappa n\sqrt{\frac{1-\lambda_R}{2}}[(k_n+1)^{1/2} \mp (k_n-1)^{1/2}],$$

$$k_n = \left[1+\frac{\varkappa^4}{4(1-\lambda_R)^2 n^4}\right]^{1/2}. \tag{10.9}$$

After some manipulation it is possible, for large n and small λ_R, to obtain the expression similar to that derived by S. W. Yuan [10.3]. The final result for the radial displacement w $(x = 0)$ is simplified to

$$\frac{1}{\varkappa^3}\left(\frac{Eh}{P}\right)w_{(0,\varphi)} = \frac{1}{2\pi\varkappa^2} + \frac{2}{\pi}\sum_{n=2,4,\ldots}^{m-2}\left\{\frac{1}{2}\frac{1+\varphi_1(\gamma_n)}{\varkappa^2} + \right.$$

$$\left. +\varphi_2(\gamma_n)\frac{2n^2(1-\lambda_R)}{\varkappa[n^2(1-\lambda_R)-1][2n^2(1-\lambda_R)-1]}\right\}\cos n\varphi +$$

$$+\frac{\varkappa\sqrt{2}}{\pi}\sum_{n=m,m+2,\ldots}^{\infty}\frac{(k_n+1)^{1/2}}{n^3 k_n}\cos n\varphi, \tag{10.10}$$

where m is the first even integer satisfying

$$5m^2[2m^2(1-\lambda_R)-1] > 16\varkappa^4. \tag{10.11}$$

The function $\varphi_1(\gamma_n)$, $\varphi_2(\gamma_n)$ has the following form:

$$\left.\begin{matrix}\varphi_1(\gamma_n)\\ \varphi_2(\gamma_n)\end{matrix}\right\} = \frac{1}{\sqrt{2}(1+\gamma_n^2)^{1/2}}\{[(1+\gamma_n^2)^{1/2}+1]^{1/2} \pm [(1+\gamma_n)^{1/2}-1]^{1/2}\}. \tag{10.12}$$

The deflections of the shell for various magnitudes of the parameter λ_R calculated by the summation of series (10.10) are presented graphically in Fig. 91. These deflections are compared with those calculated previously by means of numerical evaluation of the integral (9.1) and from formulae (9.21). The large increase of the rigidity of the shell appears with the increase of the double curvature in the direction of a generator. For $x = 0$ and $y = 0$ and $R/h = 100$, $\varkappa = 12.85$ we have the following numerical values:

Table 6. $w_{(0,0)} \cdot \left(\frac{Eh^2}{PR_2}\right)\sqrt{\frac{h}{R}}$.

λ_R	+0.10	0.00	−0.10
w_Σ Eq. (10.10)	0.127	0.740	0.424
$w_\int$ Eq. (9.4)	0.129	0.709	0.423
Difference per cent	1.6	4.2	0.2

We see that the differences between the deflections calculated by means of both the above methods are small, the largest differences being for the cylindrical shell. The deflection w_Σ was calculated on the assumption that the shell is an infinitely long tube whose cross-section is a circle. But in the calculations performed by means of the double integrals $w_\int$, the conditions ensuring the continuity of the displacements on the circumference of the cross-section of the tube are not taken into account. This second deflection given in Table 6 is calculated by the other assumptions. Namely, it was assumed that the shell extended infinitely in all directions, no conditions concerning the cross-section being imposed. In spite of it the deflections obtained do not differ much one from another. In conclusion we see that for cylindrical shells the deflection resulting from the concentrated force has a local character and depends mainly on the shape and rigidity of the shell in the vicinity of the loading point. However, the condition that the displacements of the whole shell decrease with increase of the distance from the loading point must be satisfied. This condition is approximately fulfilled for the long shell of built-in or jointly supported edges. It is not fulfilled when the shell has free edges. In this case the local deflection and the deflections of the shell as a whole add.

In the cylindrical shell a larger area undergoes a deflection than in the double curvature shells and this is the reason for the larger differences between the calculated magnitudes. The closer λ_R is to unity the smaller is the deflected area. Also the deflection near the loaded point depends to a lesser degree on the deflections of remote points. For the shell of double curvature we can, therefore, sum the deflection caused by the forces acting at various points of the shell, using expressions (9.1) or (9.21) and the results obtained will have sufficient accuracy for technical purposes.

When h/R decreases and the wall of the shell becomes thinner and thinner, the difference between the deflections w_Σ and w_ζ increases. Therefore, for thin cylindrical shells the calculations should be performed rather on the basis of expression (10.10) taking into account the real boundary conditions. When the cylindrical shell or a shell of nearly cylindrical shape is not long the effect of the boundary conditions cannot be neglected. The displacement of such shells extends in the direction of the smaller curvature and the boundary conditions effect the deflection even at points lying at a relatively large distance.

The various numerical values of deflection obtained on the basis of the various shell theories are collected in Table 7. These values were calculated for $R/h = 100$ and Poisson's coefficient $\nu = 0.3$. The parameter is $\varkappa = 12.85$. From (10.11) m is found equal to 6. As shown in this table the differences between the results obtained from the exact Flügge's equations and those from Eqs. (9.2) are very small and the maximum error does not exceed 0.5 per cent.

Table 7. $w_{(0,0)} \cdot \left(\frac{Eh^2}{PR_2}\right)\sqrt{\frac{h}{R}}$.

	n	0	2	4	6	P
$\lambda_R = 0$	Equations (9.2)	0.02	0.522	0.100	0.044	0.740
	Flügge		0.521	0.099	0.044	0.737
	Donnell–Vlasov		0.346	0.091	0.043	0.555
$\lambda_R = 0.1$	Equations (9.2)	0.02	0.016	0.024	0.024	
	Donnell–Vlasov		0.016	0.024	0.024	
$\lambda_R = -0.1$	Equations (9.2)	0.02	0.218	0.086	0.046	
	Donnell–Vlasov		0.169	0.080	0.044	

10.2 Stresses in nearly cylindrical shells

The deflection w and function Φ being known, the stresses in the shell can be found by substituting w and Φ in (9.7) and differentiating. In these expressions the length l should be replaced by R_2 due to the assumed system of coordinates. Since we are considering shells for which λ_R is small, we can neglect expressions including λ_R^2. As a result of the calculation we obtain formulae for internal forces similar to the previous (see (9.9), (9.10)). Summing the Fourier series, we find the numerical values of stresses. However, contrary to the series for the deflection which converge rapidly, those for bending moments converge very slowly. In order to facilitate numerical calculations we use the method described in Section 9.4 which consists of the division of the load into local and general parts. In this case this method will be reduced to the summation of the series between certain finite limits and addition of local effects calculated from Section 9.5.

We start from the general load. The calculations consist here in summing up the series between zero and a certain value $n_{\lim} = m$. It is easy to establish the approximate relation between $n_{\lim}$ and $\beta_{\lim} = B$ by comparing expressions (3.65) and (10.1). For a local load we have

$$P^B = \frac{P}{\pi l}\int_B^\infty \cos\beta y \, d\beta = P\delta(y) - P^\circ = P\delta(y) - \frac{P}{\pi l}\frac{\sin By}{y} \tag{10.13}$$

or

$$P^B = \frac{P}{\pi R_2}\sum_m^\infty \cos n\varphi = P\delta(y) - \frac{P}{\pi R_2}\sum_{n=0}^{m-1}\cos n\varphi$$

$$= P\delta(y) - \frac{P}{\pi R_2}\frac{1}{2}(\sin\varphi + \sin(m-1)\varphi + \sin m\varphi)\sin^{-1}\varphi.$$

As the non-dimensional coordinate is $y = R_2\varphi/l$, we have

$\beta = nl/R_2$ then $B \simeq ml/R_2$.

Introducing this into expression for P^B, $(10.13)_1$, we find that for small φ both expressions are equivalent.

Using a treatment similar to that for the calculation of the deflection, we find the following approximate formulae for the bending moments caused by the general loading.

$$M_{\varphi\varphi}^{0} = -\frac{P}{2\pi}\left\{\frac{\nu e^{-\varkappa x}}{2\varkappa}(\sin\varkappa x - \cos\varkappa x) - \right.$$

$$-\operatorname{im}\sum_{n=2,4,\ldots}^{m-2}\left[\frac{2(1+i)}{f(\gamma_n)}\left(\frac{n^2(1-\lambda_R)}{[2n^2(1-\lambda_R)-1]\varkappa}\left(1+\frac{(1-\nu)n^2\lambda_R}{n^2(1-\lambda_R)-1}\right)+\right.\right.$$

$$\left.\left.\left.+\frac{\nu n^2(1-\lambda_R)}{\varkappa^3}\right)e^{i\alpha_1 x} - \frac{1-i}{2\varkappa}\left(2\nu - \frac{n^2}{\varkappa^2}i\right)e^{i\alpha_2 x}\right]\cos n\varphi\right\}. \qquad (10.14)$$

For $x = 0$, this expression simplifies to

$$M_{\varphi\varphi(0,\varphi)}^{0} = \frac{P}{2\pi}\left\{\frac{\nu}{2\varkappa} + \sum_{n=2,4,\ldots}^{m-2}\left(\frac{2n^2(1-\lambda_R)}{2n^2(1-\lambda_R)-1}\times\right.\right.$$

$$\times\left[1+\frac{(1-\nu)n^2\lambda_R}{n^2(1-\lambda_R)-1}\right]\frac{\varphi_2(\gamma_n)}{\varkappa} + \frac{\nu}{\varkappa} +$$

$$\left.\left.+\frac{n^2}{\varkappa^3}\left[\varphi_1(\gamma_n)(1-\lambda_R)\nu + \frac{1}{2}\right]\right)\cos n\varphi\right\}. \qquad (10.15)$$

The bending moment M_{xx} can be obtained by substituting the moment $M_{\varphi\varphi}$ into (9.8).

We find the membrane forces by the differentiation of the stress function from (9.7). Substituting the simplified expressions (10.6), we have, for example,

$$N_{xx(x,\varphi)}^{0} = -\frac{P}{\pi R_2}\sum_{n=2,4,\ldots}^{m-2}\operatorname{im}\left\{[\varphi_1(\gamma_n) - i\varphi_2(\gamma_n)]\left(1-\frac{1-\lambda_R}{\varkappa^2}i\right)\times\right.$$

$$\left.\times\frac{2n^2\varkappa e^{ix\alpha_1}}{2n^2(1-\lambda_R)-1} + (1-i)\frac{n^2}{2\varkappa}e^{ix\alpha_2}\right\}\cos n\varphi. \qquad (10.16)$$

The complete formulae for the calculation of the stresses in the shell will be obtained by the addition to the above formulae of the expressions (9.36) calculated previously for the local load (see (9.27)).

Calculating the stresses which result from the local load by means of Fourier series (for example, from the deflection (10.10)), we find the following results:

The moments are

$$M_{xx(0,\varphi)}^{B} = M_{\varphi\varphi(0,\varphi)}^{B} = \frac{P}{2\pi}\frac{(1+\nu)}{\sqrt{2}}\sum_{n=m,m+2,\ldots}^{\infty}\frac{(k_n+1)^{1/2}}{nk_n}\cos n\varphi. \qquad (10.17)$$

The membrane forces are

$$N^B_{xx\,(0,\varphi)} = N^B_{\varphi\varphi\,(0,\varphi)} = -\frac{P}{2\pi R_2}\varkappa^2\sqrt{2}\sum_{n=m,m+2,\dots}^{\infty}\frac{(k_n-1)^{1/2}}{nk_n}\cos n\varphi .$$

As for large n, $k_n \to 1$, the series (10.17), for the moments, converges slowly as the series $\sum \frac{\cos n\varphi}{n}$. The convergence of the series (10.17) can be accelerated by making use of the sum of the series

$$\sum_{n=2,4,\dots}^{\infty}\frac{\cos n\varphi}{n} = -\frac{1}{2}\ln(2\sin\varphi)$$

and separating a singular part from the solution. As a result we have the moment in the form of a singular expression and the rest expressed by the rapidly convergent series. Let us take into account that

$$\sum_{n=2,4,\dots}^{\infty}\frac{(k_n+1)^{1/2}}{nk_n}\cos n\varphi = -\frac{\sqrt{2}}{2}\ln(2\sin\varphi)-$$

$$-\frac{\varkappa^4}{4(1-\lambda_R)^2}\sum_{n=2,4,\dots}^{\infty}\frac{(2k_n+1)\cos n\varphi}{n^5k_n(k_n+1)[(k_n+1)^{1/2}+\sqrt{2k_n}]}\,. \tag{10.18}$$

For small magnitudes of the angle φ, $\sin\varphi \approx \varphi \simeq y/R_2$, the singular part of the bending moment $-\frac{1}{4}(1+\nu)\ln(y/R_2)$ for $x \ll 1$ differs from (9.35) only by a constant value.

We observe that the stresses resulting from the local load can be calculated by the summation of the series between the limits $n = m$ and $n \to \infty$ or by means of integrals expressed as approximate closed formulae.

The local load effects to a large degree the shape of the bending moment diagram. In order to obtain the same diagram for the moment, summing up directly the series (10.14), we would have to take into account more than several hundred terms, which would imply a prohibitive amount of work.

10.3 Cylindrical shells

All necessary relations for the calculation of the stresses and deflections in the cylindrical shell can be obtained by substitution of $\lambda_R = 0$ in the results of the previous chapters. However, cylindrical shells are very often met in engineering practice, so it is useful to present the fundamental equations for them separately.

10.3.1 *General equations of the cylindrical shell*

Solving the problem of the circular cylindrical shell we can work from the fundamental equations given in § 1.11.2 where are presented the relations in the rectangular Cartesian coordinates (x, y). It is more convenient in this case to introduce the cylindrical non-dimensional coordinates $x = \bar{x}/R$, φ, where $\bar{x}$ is the coordinate along the generator of the cylinder and φ is the angle determining the position on the circumference. We have then the relations $y = R\varphi$ and $A_1 = A_2 = R$. The set of fundamental equations of he shell can be presented in the form

$$D(\Delta+1)^2 w-(1-\eta\Delta)R\frac{\partial^2\Phi}{\partial x^2} = \Big\{Z-\frac{1}{R^4}L(w,\Phi)- -w\frac{1}{R^2}\Delta_R\Phi-\Delta\Big[(\eta-\varepsilon)Z-\frac{\eta}{R^4}L(w,\Phi)\Big]\Big\}R^4, \tag{10.19}$$

$$\frac{1}{Eh}\Delta^2\Phi+R\frac{\partial^2 w}{\partial x^2} = \frac{1}{2}L(w,w)-\bar{\varepsilon}\frac{R^3}{Eh}\Delta Z+R^2 w\Delta_R w,$$

where

$$\eta = h^2/5(1-\nu)R^2, \qquad \varepsilon = \nu h^2/10(1-\nu)R^2, \qquad \bar{\varepsilon} = \nu h/2R,$$

$$L(w,\Phi) = \frac{\partial^2 w}{\partial x^2}\frac{\partial^2\Phi}{\partial\varphi^2}-2\frac{\partial^2 w}{\partial x\,\partial\varphi}\frac{\partial^2\Phi}{\partial x\,\partial\varphi}+\frac{\partial^2 w}{\partial\varphi^2}\frac{\partial^2\Phi}{\partial x^2},$$

$$\Delta = \frac{\partial^2}{\partial x^2}+\frac{\partial^2}{\partial\varphi^2}, \qquad \Delta_R = \frac{1}{R^2}\frac{\partial^2}{\partial x^2}.$$

The membrane forces are given by the relations

$$N_{xx} = -\frac{1}{R^2}\frac{\partial^2}{\partial\varphi^2}\Big[\Phi+(1-\nu)D\frac{w}{R}\Big],$$

$$N_{\varphi\varphi} = -\frac{1}{R^2}\frac{\partial^2\Phi}{\partial x^2}+\frac{D}{R^3}(\Delta+1)w, \tag{10.20}$$

$$N_{x\varphi} = \frac{1}{R^2}\frac{\partial^2\Phi}{\partial x\partial\varphi}; \qquad N_{\varphi x} = \frac{1}{R^2}\frac{\partial^2}{\partial x\,\partial\varphi}\Big(\Phi+\frac{1-\nu}{R}Dw\Big).$$

The bending moments are

$$M_{xx} = -\frac{D}{R^2}\left[\frac{\partial^2 w}{\partial x^2}+\nu\left(\frac{\partial^2 w}{\partial \varphi^2}+w\right)\right]+$$
$$+(1-\nu)\eta R\frac{\partial Q_x}{\partial x}-\varepsilon R^2\left[Z+2\left(-L(w,\Phi)\frac{1}{R^4}+\frac{1}{R^3}\frac{\partial^2 \Phi}{\partial x^2}\right)\right], \tag{10.21}$$
$$M_{\varphi\varphi} = -\frac{D}{R^2}\left[\frac{\partial^2 w}{\partial \varphi^2}+w+\nu\frac{\partial^2 w}{\partial x^2}\right]+(1-\nu)\eta R\frac{\partial Q_\varphi}{\partial \varphi}-$$
$$-\varepsilon R^2\left[Z+2\left(-L(w,\Phi)\frac{1}{R^4}+\frac{1}{R^3}\frac{\partial^2 \Phi}{\partial x^2}\right)\right],$$
$$M_{x\varphi} = M_{\varphi x} = -(1-\nu)\frac{D}{R^2}\frac{\partial^2 w}{\partial x\,\partial \varphi}+(1-\nu)\frac{\eta}{2}R\left(\frac{\partial Q_x}{\partial \varphi}+\frac{\partial Q_\varphi}{\partial x}\right).$$

A different method, not discussed previously, is often used in the case of the cylindrical shell. This method is particularly convenient in the case where the boundary conditions are determined by the displacements. We do not introduce the stress function, but we eliminate the internal forces from the equilibrium equations, expressing them in terms of the displacements u, v, w by means of the formulae (1.76), (1.75), (1.71). Then we obtain a set of three differential equations for the three unknown functions u, v, w. This set is presented below. The effect of the variable temperature is taken into account and the influence of the transverse normal and shear stresses has been neglected here. According to [10.10] we have

$$\Delta^2(\Delta+1)^2 w+4\varkappa^4\frac{\partial^4 w}{\partial x^4} = P_w,$$
$$\Delta^2 u = \frac{\partial^3 w}{\partial x\,\partial \varphi^2}-\nu\frac{\partial^3 w}{\partial x^3}+P_u, \tag{10.22}$$
$$\Delta^2 v = -(2+\nu)\frac{\partial^3 w}{\partial x^2\,\partial \varphi}-\frac{\partial^3 w}{\partial \varphi^3}+P_v,$$

where P_w, P_u, P_v are functions depending on the loading and the temperature

$$P_w = \Delta^2 P_z+\nu\frac{\partial P_u}{\partial x}+12\frac{R^2}{h^2}\frac{\partial P_v}{\partial \varphi}-\frac{3-\nu}{2}\frac{\partial^3 P_v}{\partial x^2\,\partial \varphi}-\frac{\partial^3 P_u}{\partial x^3}+$$
$$+\frac{1}{2}(1-\nu)\frac{\partial^3 P_u}{\partial x\,\partial \varphi^2}, \tag*{(10.23)$_1$}$$

$$P_u = \frac{\partial^2 P_x}{\partial x^2} + \frac{2}{1-\nu}\frac{\partial^2 P_x}{\partial \varphi^2} - \frac{1+\nu}{1-\nu}\frac{\partial^2 P_\varphi}{\partial x\,\partial \varphi} + \frac{3}{12}\left(\frac{h^2}{R^2}\right)\frac{\partial^2 P_x}{\partial x^2},$$

$$P_v = \frac{2}{1-\nu}\frac{\partial^2 P_\varphi}{\partial x^2} + \frac{\partial^2 P_\varphi}{\partial \varphi^2} - \frac{1+\nu}{1-\nu}\frac{\partial^2 P_x}{\partial x\,\partial \varphi} + \frac{1}{12}\left(\frac{h^2}{R^2}\right)\frac{\partial^2 P_\varphi}{\partial \varphi^2},$$

$$P_z = \frac{R^4}{D}Z - (1+\nu)\alpha_t\frac{R^2}{h}\Delta T_2 - 12\left(\frac{R^2}{h^2}\right)(1+\nu)\alpha_t RT_1, \qquad (10.23)_{2-6}$$

$$P_x = (1+\nu)R\alpha_t\frac{\partial T_1}{\partial x} - \frac{1-\nu^2}{Eh}R^2 X,$$

$$P_\varphi = (1+\nu)R\alpha_t\frac{\partial T_1}{\partial \varphi} - \frac{1+\nu}{12}\alpha_t h\frac{\partial T_2}{\partial \varphi} - \frac{(1-\nu^2)R^2}{Eh}Y,$$

where u, v, w are the displacements of the middle surface of the shell in the directions of $x = \bar{x}/R$, $y = \varphi$ and normal direction. The stress resultants and moments can be expressed in terms of the displacements.

$$N_{xx} = \frac{Eh}{(1-\nu^2)R}\left[\frac{\partial u}{\partial x} + \nu\left(\frac{\partial v}{\partial \varphi} + w\right) - \frac{1}{12}\left(\frac{h}{R}\right)^2\frac{\partial^2 w}{\partial x^2} - (1+\nu)\alpha_t RT_1\right],$$

$$N_{\varphi\varphi} = \frac{Eh}{(1-\nu^2)R}\left[\frac{\partial v}{\partial \varphi} + w + \nu\frac{\partial u}{\partial x} + \frac{1}{12}\left(\frac{h}{R}\right)^2\left(\frac{\partial^2 w}{\partial \varphi^2} + w\right) - (1+\nu)\alpha_t RT_1\right],$$

$$N_{x\varphi} = \frac{Eh}{2(1+\nu)R}\left[\frac{\partial u}{\partial \varphi} + \frac{\partial v}{\partial x} + \frac{1}{12}\left(\frac{h}{R}\right)^2\left(-\frac{\partial^2 w}{\partial x\,\partial \varphi} + \frac{\partial v}{\partial x}\right)\right],$$

$$N_{\varphi x} = \frac{Eh}{2(1+\nu)R}\left[\frac{\partial u}{\partial \varphi} + \frac{\partial v}{\partial x} + \frac{1}{12}\left(\frac{h}{R}\right)^2\left(\frac{\partial^2 w}{\partial x\,\partial \varphi} + \frac{\partial u}{\partial \varphi}\right)\right],$$

$$M_{xx} = -\frac{D}{R^2}\left[\frac{\partial^2 w}{\partial x^2} + \nu\left(\frac{\partial^2 w}{\partial \varphi^2} - \frac{\partial v}{\partial \varphi}\right) - \frac{\partial u}{\partial x} + (1+\nu)\alpha_t RT_2\right],$$

$$M_{\varphi\varphi} = -\frac{D}{R^2}\left[\frac{\partial^2 w}{\partial \varphi^2} + w + \nu\frac{\partial^2 w}{\partial x^2} + (1+\nu)\alpha_t RT_2\right],$$

$$M_{x\varphi} = -(1-\nu)\frac{D}{R}\left[\frac{\partial^2 w}{\partial x\,\partial \varphi} - \frac{\partial v}{\partial x}\right],$$

$$M_{\varphi x} = -(1-\nu)\frac{D}{R}\left[\frac{\partial^2 w}{\partial x\,\partial \varphi} - \frac{1}{2}\left(\frac{\partial v}{\partial x} - \frac{\partial u}{\partial \varphi}\right)\right],$$

$$Q_x = -\frac{D}{R^2}\left[\frac{\partial^3 w}{\partial x^3} + \frac{\partial^3 w}{\partial x\,\partial \varphi^2} - \frac{\partial^2 u}{\partial x^2} + \frac{1}{2}(1-\nu)\frac{\partial^2 u}{\partial \varphi^2} - \right.$$

$$-\frac{1}{2}(1+\nu)\frac{\partial^2\vartheta}{\partial x\,\partial\varphi}+(1+\nu)\alpha_t R\frac{\partial T_2}{\partial x}\Bigg],$$

$$Q_\varphi=-\frac{D}{R^2}\left[\frac{\partial^3 w}{\partial\varphi^3}+\frac{\partial w}{\partial\varphi}+\frac{\partial^3 w}{\partial x^2\partial\varphi}-(1-\nu)\frac{\partial^2 v}{\partial x^2}+(1+\nu)\alpha_t R\frac{\partial T_2}{\partial\varphi}\right],$$

where

$$T_1=\frac{1}{h}\int_{-h/2}^{h/2}T\mathrm{d}\xi,\qquad T_2=\frac{12}{h^2}\int_{-h/2}^{h/2}T(\xi)\,\mathrm{d}\xi,$$

$T(\xi)$ represents the temperature field in the shell.

If the shell is loaded only by the normal force Z and the temperature of the shell is constant, the first equation $(10.22)_1$ is identical with equations (1.119) derived previously. The method presented here was used by many authors in the presentation of the equations for the cylindrical shell. They obtained many variants of the set presented here depending on the constitutive equations assumed and simplifications employed. W. Flügge derived a very exact but complex set of equations for the displacements. Eliminating the displacements u and v from the second and third equation of the set given by him, we find the following equation for the displacement w:

$$\Delta^2(\Delta+1)^2 w+4\varkappa^4\frac{\partial^4 w}{\partial x^4}+2(1-\nu)\left[\frac{\partial^6 w}{\partial x^2\partial\varphi^4}+\frac{\partial^4 w}{\partial x^2\,\partial\varphi^2}-\frac{\partial^4 w}{\partial\varphi^4}\right]$$

$$=\frac{R^4}{D}\Delta^2 Z. \tag{10.24}$$

Morley [1.18] discarded the terms in brackets multiplying the coefficient $2(1-\nu)$ in (10.24) and obtained an equation identical with (1.119). He found that this equation is sufficiently accurate and proposed to use it instead of the Donnell equations (1.123). The Donnell equations give results which differ of about 10–30 per cent from the results of the exact equations of the cylindrical shell. We obtain this equation neglecting unity in the operator $(\Delta+1)^2$ in $(10.22)_1$. Basing the computations on the Donnell equation is equivalent to the assumption that the shell considered is a shallow shell. This equation can also be applied for the case where the functions w and Φ are varying rapidly.

Later Donnell proposed replacing Eq. (1.123) by the so-called *complete Donnell equation* which differs slightly from the Morley equation. According to L. Ting and S. W. Yuan [10.8] (1958) this equation is as exact as the Flügge equation.

Assuming the solution of Eq. (10.22) in the form $w = e^{i\alpha x}\cos n\varphi$, we find the characteristic equation

$$(\alpha^2+n^2-1)^2(\alpha^2+n^2)^2+4\varkappa^4\alpha^4 = 0. \tag{10.25}$$

The calculation of the roots of this characteristic equation can be considerably simplified. Holand [10.12] found that in the case of the cylindrical shell the following simple closed formulae exist for the roots of the characteristic equation (10.22):

$$\begin{aligned}\alpha_1 &= \pm(a\pm bi),\\ \alpha_2 &= \pm(c\pm di).\end{aligned} \tag{10.26}$$

where

$$a = \varkappa e^{-\delta}\cosh\delta, \qquad b = \varkappa e^{\delta}\cosh\delta,$$

$$c = \varkappa e^{-\delta}\sinh\delta, \qquad d = \varkappa e^{\delta}\sinh\delta,$$

$$a>0, \quad b>0, \quad c>0, \quad d>0.$$

The parameter δ is defined by

$$\sinh(4\delta) = \frac{2}{\varkappa^2}n\sqrt{n^2-1}.$$

The solution (10.26) is obtained in the following manner. The characteristic equation (10.25) may be rewritten in the form

$$(\alpha^2+n\sqrt{n^2-1})^4+4\varkappa^4\alpha^4-\alpha^2[2\alpha^4-\alpha^2(6n^2-1)-6n^4-2n^2] = 0.$$

Holand found that the terms in the square brackets are of little importance. When these terms are neglected, we find the approximate values for the roots in the form (10.26). The following identity was used during derivations.

$$\pm\sqrt{\pm i+\sinh 4\delta} = \pm\frac{1}{\sqrt{2}}(e^{2\delta}\pm ie^{-2\delta}).$$

Inspection of these roots will show that they satisfy accurately, within the limitation $h/R \ll 1$, the so-called *complete Donnell equation* which corresponds to the characteristic equation

$$\alpha^8+4n^2\alpha^6+(4\varkappa^4+6n^4)\alpha^4+4n^6\alpha^2+n^4(n^2-1)^2 = 0.$$

Holand stated [10.12] that with the same degree of accuracy the roots (10.26) are also the roots of Eq. (10.25). The characteristic equation of the Donnell equations corresponding to the shallow shell is obtained when unity is neglected in comparison with n^2. A corresponding simplification of the roots leads to a change of the parameter δ to δ_0 defined by

$$\sinh 4\delta_0 = \frac{2n^2}{\varkappa^2}\,.$$

Similar expressions to Eq. (10.26) can be written for the roots of Flügge's characteristic equation [1.16].

10.3.2 *Infinitely long shell loaded by two oppositely directed radial forces*

The deflections in the plane $\varphi = 0$ (Fig. 106) of an infinitely long cylindrical shell loaded by two oppositely directed radial forces obtained from formula (10.2) by substituting $\lambda_R = 0$ are presented graphically in Fig. 107. Two curves are given there. One resulting from the improved Eqs. (10.19) (the effect of transverse shear deformation is neglected) and the second from the

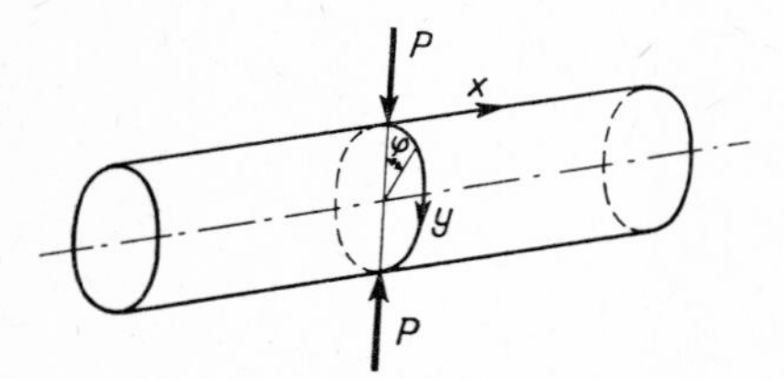

Fig. 106. Cylindrical shell subjected to two oppositely directed radial forces.

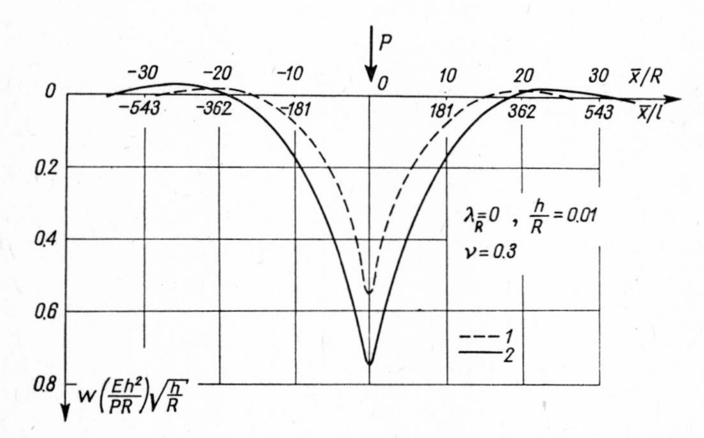

Fig. 107. Deflection variation along generator. The line ---------- presents the result of Donnell–Vlasov equations; the continuous line ———— presents the Flügge and Morley equations.

Donnell–Vlasov equations. We observe a difference between both results of about 25 per cent.

Displacements, bending moments, and membrane forces are presented graphically in Fig. 108. These diagrams were calculated for two values of $h/R = \sqrt{12} \cdot 10^{-3}$ and $h/R = \sqrt{12} \cdot 10^{-2}$ by K.W. Bieger [10.16] in 1959. The numerical results obtained by Bieger differ little from these obtained by the previously described methods. Bieger obtained his solution by means of series of Fourier integrals of the type (10.2). The exact Flügge's equations for the cylindrical shell are the starting point of this work. The determination of the exact distribution of the moments in the vicinity of the loading point has been possible owing to the separation of the singular terms similarly as it was done in the previous section. Bieger proved that the distribution of the displacements and stresses in the cylindrical shell depends on the ratio R/h.

The shell deflection for two values of the ratio h/R is given in Fig. 108. Comparing the maximum deflections appearing at the point of application of the load with the results collected in the table from the previous chapter and calculated for $h/R = 10^{-2}$, $\nu = 0.3$, we have

$$w^* = w(0,0)\frac{Eh^2}{PR}\sqrt{\frac{h}{R}} \approx 0.737.$$

We see that the numerical value of w^* is almost constant. However, the deflection w at other points of the shell changes considerably and depends on the parameter h/R. In a thinner shell the deflection vanishes more slowly with distance x/R than in a thicker one.

The diagrams given in Figs. 109, 110, 111 present the influence surfaces for internal forces. Comparing them with the influence surfaces for the bending moments and shear forces for plates, we can be convinced that in the vicinity of the loading point the contour lines have a similar shape. This results also from the previous considerations that the singularities in the plate and shell are the same. The membrane forces are presented graphically in Figs. 111a, b. The shear forces Q_x and Q_y are given in Figs. 110a, b and displacements u and v in Figs. 108c, d.

The infinitely long cylindrical shell loaded by two opposite radial forces was considered by many authors. S. W. Yuan [10.3] in 1949 obtained the series (10.17) for $\lambda_R = 0$ taking Donnell's equations as a starting point.

The same problem was considered for a second time by S.W. Yuan and L. Ting [10.7] who replaced Donnell's equations by the very accurate but complex Flügge's equation. The results obtained differed from previous ones

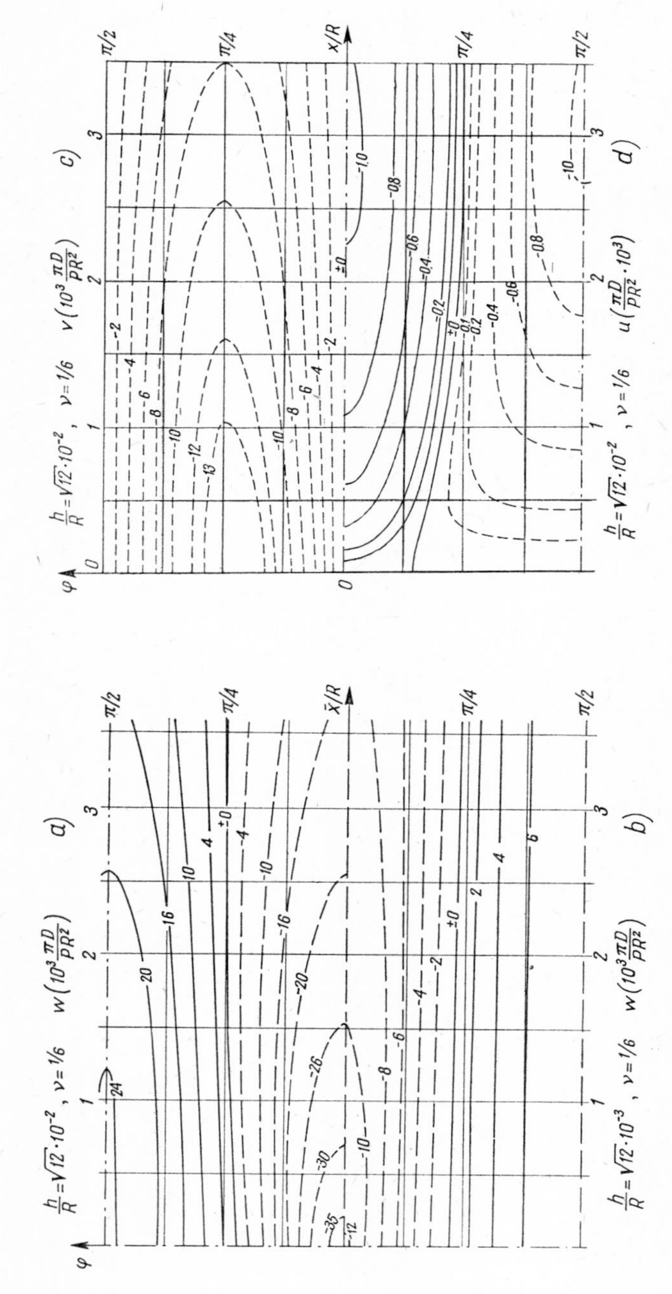

Fig. 108. Influence surfaces for the displacements: a), b) deflection for two values of the coefficient h/R; c), d) displacements u and v (from K. Bieger [10.16]).

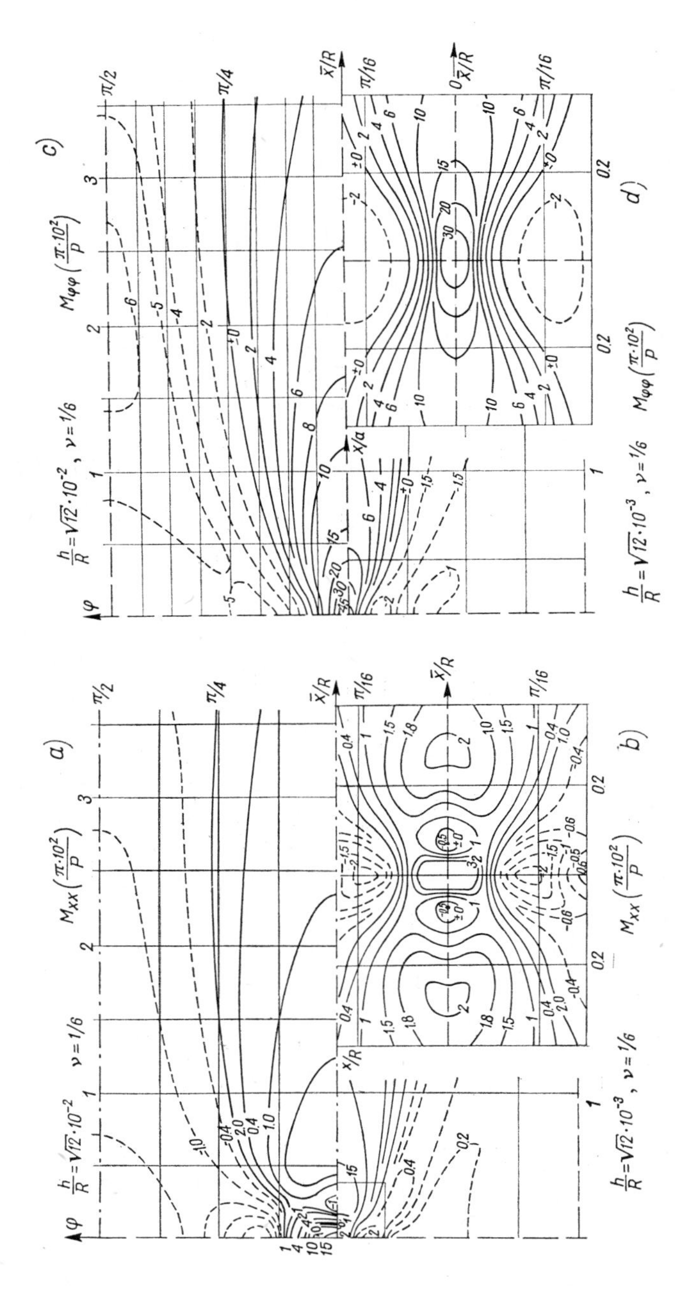

Fig. 109. Influence surfaces for bending moments in a cylindrical shell subjected to concentrated normal force; a), b) Bending moment M_{xx} for two values of the coefficient h/R; c), d) Bending moment $M_{\varphi\varphi}$ for two values of the coefficient h/R (from [10.16]).

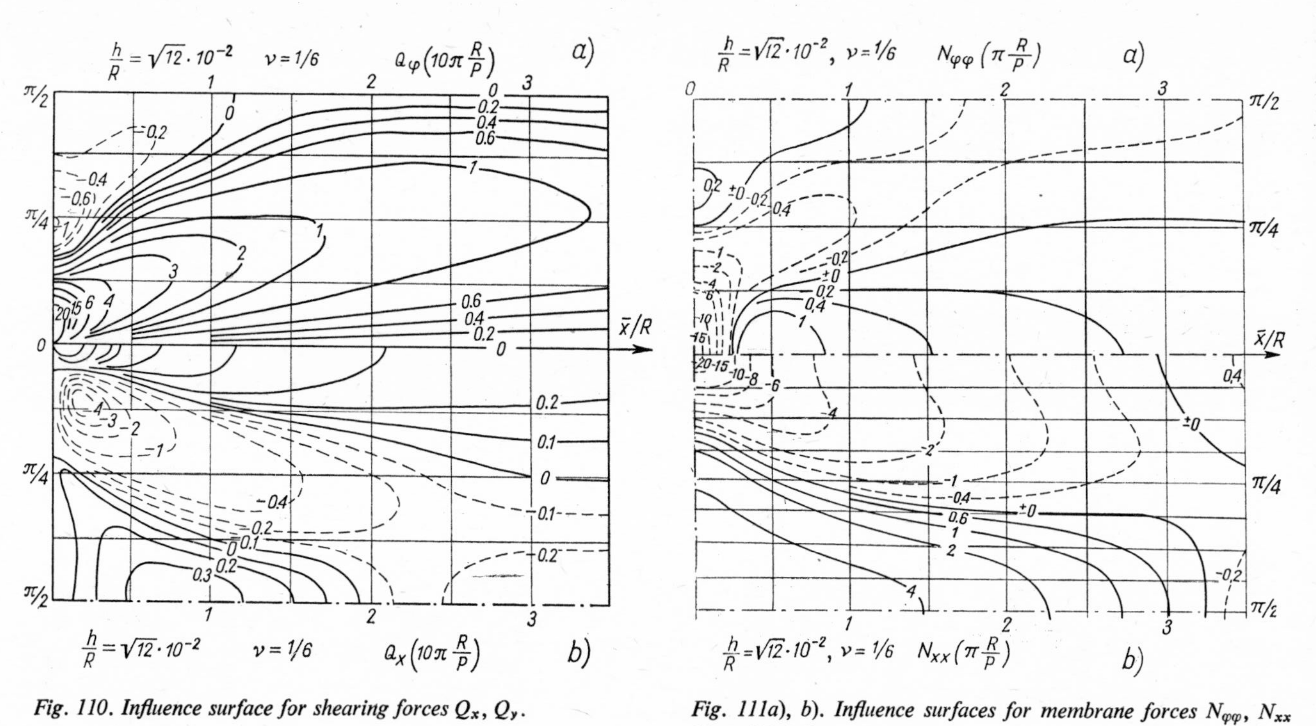

Fig. 110. Influence surface for shearing forces Q_x, Q_y.

Fig. 111a), b). Influence surfaces for membrane forces $N_{\varphi\varphi}$, N_{xx} (from [10.16]).

by about 25 per cent. Morley [10.2], based on his equations proposed in [1.18], considered the shell loaded in the same way. The results obtained in § 10.1 are for $\lambda = 0$ identical with his results. In paper [10.4] by Bijlaard a number of tables and curves were presented for determining bending moments, membrane forces, and radial displacements in a cylinder due to radial loads distributed over small areas. The circular cylinder of finite length was assumed to be simply supported at the ends.

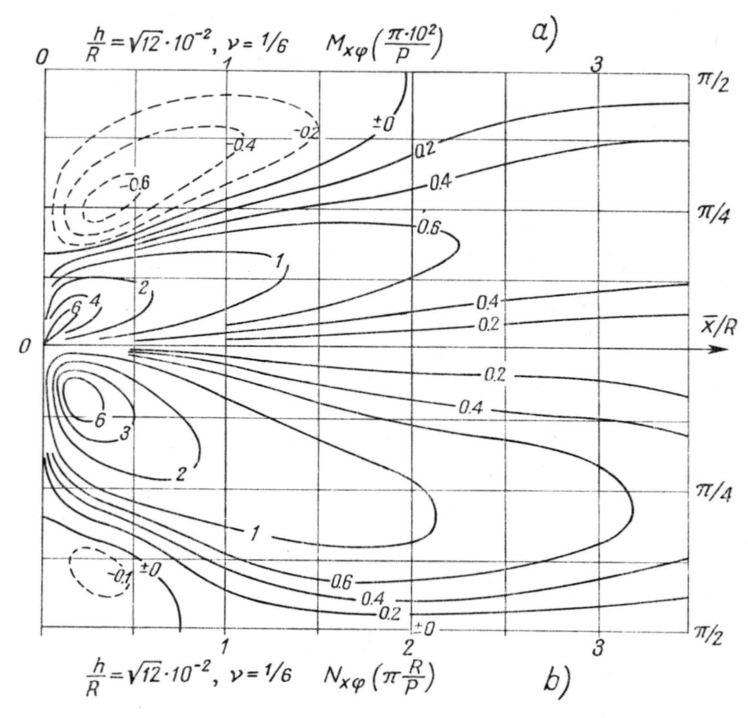

Fig. 112. Influence surfaces for twisting moment and tangential force N_{xy} (from [10.16]).

Kempner, Sheng and Pohle [10.6] obtained displacements, rotations, membrane and bending stresses corresponding to line loadings on a simply supported thin-walled circular cylinder for the cases:

(1) A radial load uniformly distributed along a line segment.

(2) Uniformly distributed bending moment whose vector representation acts in the direction of the axis of the cylinder. Klein [10.20] presented numerous diagrams to determine the deflections and stresses in the region of

localized loaded areas of an infinitely long circular cylinder subjected to internal pressure. The results were derived by assuming double Fourier series for the radial deflection.

10.3.3 *The cylindrical segment of infinite length*

Holand [10.21] also gave influence diagrams for moments, forces, and displacements of a circular cylindrical segment of finite width. The analysis was based on Donnell's theory. In [10.21] Holand investigated the same problem and gave influence diagrams and curves for bending moments well suited for practical purposes. The shell considered (Fig. 113) was simply supported at its

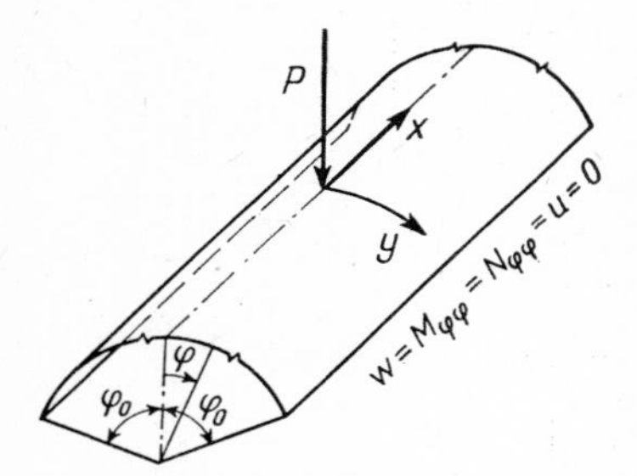

Fig. 113. Cylindrical segment of infinite length simply supported at edges.

edges and loaded by a concentrated radial force. The solution was obtained by dividing the infinitely long shell into two parts in the loaded cross-section perpendicular to the shell axis and subdividing the load into two equal parts acting on the two shell halves respectively. Stresses and displacements were then obtained as edge disturbances originating from this section. In this case similar expressions in the form of trigonometric function of x result. The bending moments M_{xx} and M_{yy} may be written:

$$M_{11} = D(\varkappa_{11}+\nu\varkappa_{22}), \quad M_{22} = D(\varkappa_{22}+\nu\varkappa_{11}). \tag{10.27}$$

$D\varkappa_{11}$ and $D\varkappa_{22}$ depend very mildly on ν and on other parameters. These quantities have been evaluated for $\nu = 0.3$ and are given in Figs. 114a and b. $D\varkappa_{11}$ is damped out very rapidly and is practically equal to zero outside the region $-5l < \bar{x} < 5l$ and $-7l < \bar{y} < 7l$, and $D\varkappa_{22}$ is very slowly damped in the x-direction. For $\bar{x} > 45$, l varies as a cosine function in the y-direction. The variation with $\bar{x}$ for $\bar{y} = 0$ is given in Fig. 115.

The above diagrams were calculated for the angle $\varphi_0 = 34.4°$ and give

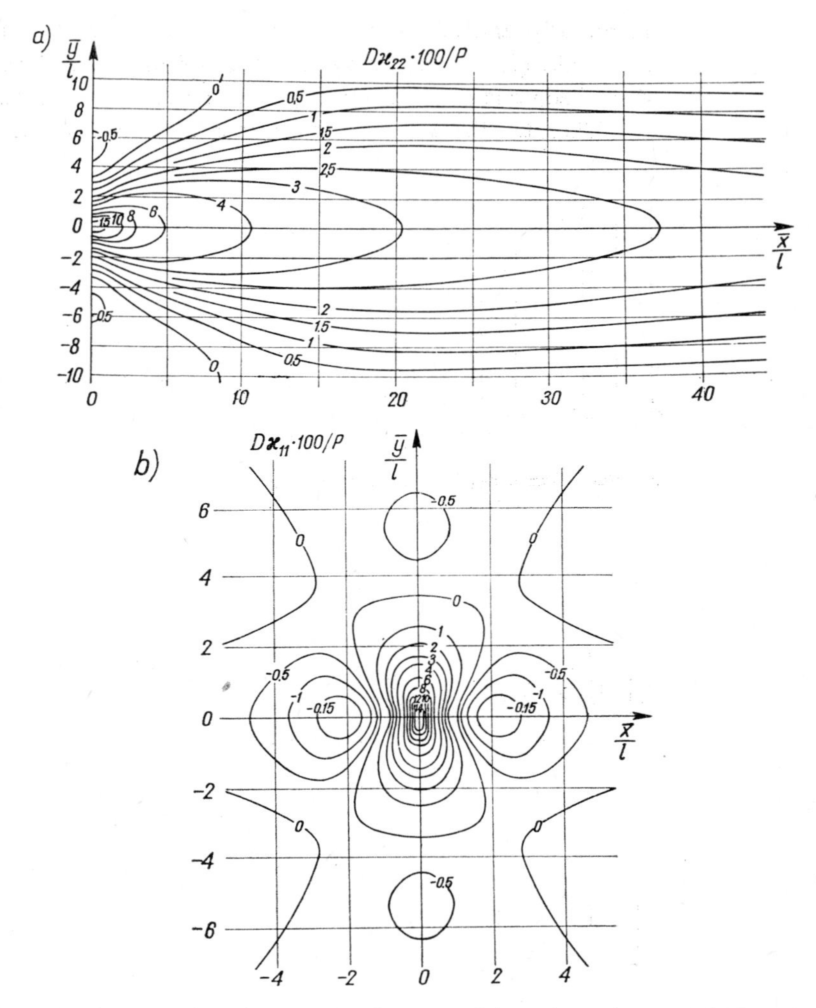

Fig. 114. Influence surfaces for curvature changes $\varkappa_{11}D/P$ *and* $\varkappa_{22}D/P$; $\nu = 0.3$ *(from J. Holand* [10.21]).

for $h/R = 0.01$ the value of $\bar{y} \cong 6\,l$. But we see from Figs. 114 that the changes of curvatures resulting from a concentrated load are damped very rapidly in the circumferential direction. As a consequence the width of the shell and the location of the straight edges is of minor importance unless the load is very close to the edge.

The bending moments M_{11} and M_{22} can be found by the use of the diagrams together with formula (10.27). The slow damping of $\varkappa_{11}$ means that the influence of the curved edges is more important for this quantity and must be taken into account. When the force is distributed over the area $2c_1 \times 2c_2$,

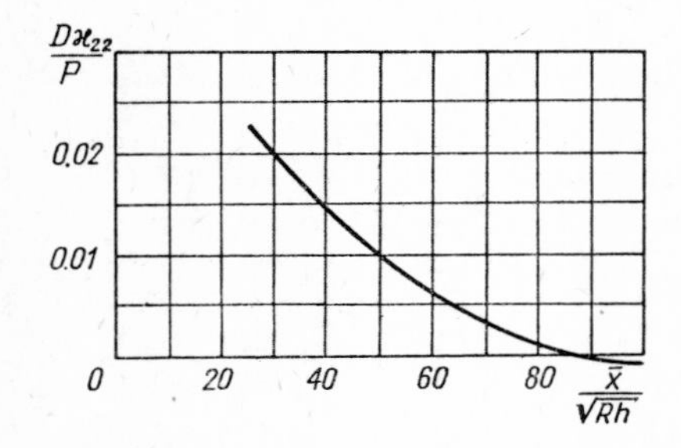

Fig. 115. Variation of curvature change $\chi_{22} D/P$ for large $x/\sqrt{Rh}$.

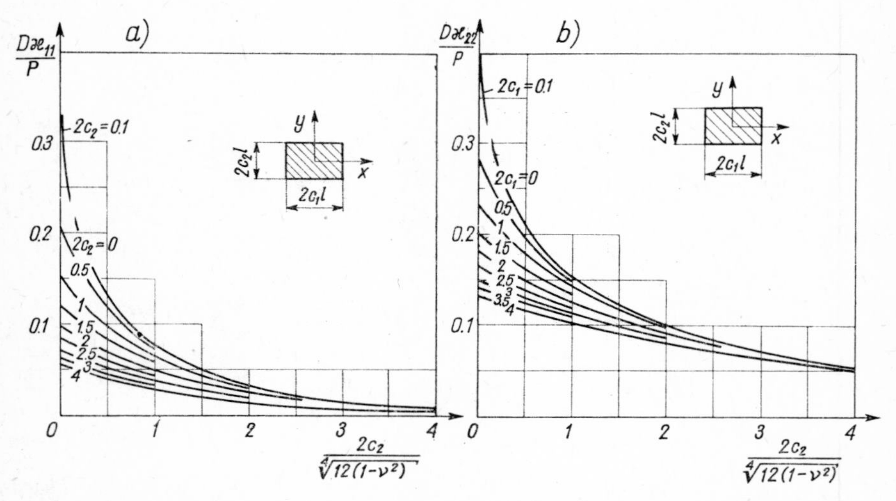

Fig. 116. Curvature changes $\varkappa_{11} D/P$ and $\varkappa_{22} D/P$ at the centre of the loaded area. The force is uniformly distributed over the surface $2c_1 l \times 2c_2 l$ (from J. Holand [10.21]).

the moments at $x = 0$, $y = 0$ can be found by integrating the influence functions $\varkappa_{11}$ and $\varkappa_{22}$. The moments M_{11} and M_{22} may be written as previously, (10.27). Here $D\varkappa_{11}$ and $D\varkappa_{22}$ depend also very mildly on ν. These quantities have been evaluated for $\nu = 0.3$ and are given in Figs. 116a and b.

10.4 Shallow cylindrical shells

10.4.1 *Singularities of shallow cylindrical shells*

A. Jahanshahi [10.27] examined singular solutions of the eighth order equation of a shallow cylindrical shell and found that the force singularities were expressible in terms of integrals of the products of the cylindrical and circular functions.

Let us consider a shallow circular cylindrical shell of a radius R and adopt the Cartesian system of non-dimensional coordinates $x = \bar{x}/R$, $y = R\varphi/R = \varphi$ having its origin 0 in the middle surface of the shell and at the point of application of the load. As was shown in § 1.21, in the case of the cylindrical shell the two equations (1.115) can be brought into the one eighth order differential equation (1.119). Since we consider the shallow shell, we neglect the term $1/R^2$ in (1.119). Let us examine the homogeneous equation

$$\Delta^4 w + 4\varkappa^4 \frac{\partial^4 w}{\partial x^4} = 0, \tag{10.28}$$

where $4\varkappa^4 = 12(1-\nu^2)\dfrac{R^2}{h^2}$.

Factoring the differential operator, we obtain the equation

$$L_1 L_2 L_3 L_4 w = 0,$$

where

$$L_i = \Delta + \beta_i \frac{\partial}{\partial x}, \qquad i = 1, \ldots, 4,$$

$$\beta_1 = -\beta_4 = (1+i)\varkappa, \qquad \beta_2 = -\beta_3 = (1-i)\varkappa.$$

It is evident that the solution of the equation

$$\Delta w + \beta_i \frac{\partial w}{\partial x} = 0 \tag{10.29}$$

also satisfies Eq. (10.28). It can be shown by direct substitution that

$$w = \frac{1 \pm i}{2} \varkappa \int_0^x e^{\pm i(1\pm i)\frac{\varkappa}{2} t}\, Y_0\left[(1 \pm i)\frac{\varkappa}{2}\sqrt{t^2+y^2}\right] \mathrm{d}t \pm$$

$$\pm i(1 \pm i)^2 \frac{\varkappa^2}{4} \int_0^y \int_0^t Y_0[(1 \pm i)s]\, \mathrm{d}s\, \mathrm{d}t$$

where $Y_0(\cdot)$ is the Bessel function of zero order and second kind. It follows that

$$\chi = \frac{1-i}{2}\varkappa\int_0^x \sin(1-i)\frac{\varkappa}{2}t\cdot Y_0\left[(1-i)\frac{\varkappa}{2}\sqrt{t^2+y^2}\right]\mathrm{d}t+$$

$$+i(1-i)^2\frac{\varkappa}{4}\int_0^y \mathrm{d}t\int_0^t Y_0[(1-i)s]\,\mathrm{d}s$$

is also the solution of Eq. (10.28). Examination of these results indicates that the singular solution corresponding to the action of the concentrated normal force takes the form

$$w = C\,\mathrm{im}\,\chi\left[x^2, y^2, (1-i)\frac{\varkappa}{2}\right].$$

Here, C is a constant determined from the condition of equilibrium of the force P with the resultant of the shearing forces for small r.

$$P = \lim_{\varepsilon\to 0}\left\{D\int_0^{2\pi}\frac{\partial}{\partial r}(\Delta w)_{r=\varepsilon}\,\varepsilon\,\mathrm{d}\vartheta\right\}, \qquad r = x^2+y^2.$$

Finally, we obtain

$$w = -\frac{PR^2}{4D\varkappa^2}\,\mathrm{im}\,\chi\left[x^2, y^2, (1-i)\frac{\varkappa}{2}\right] \tag{10.30}$$

where im denotes the imaginary part of the function χ. For the stress function we find a similar singular solution. The above solution can be presented in the form of real functions

$$w = -\frac{P}{8\pi D}\{\psi_k+\psi_l\}, \qquad \Phi = -\frac{2PR\varkappa^2}{8\pi}\{\psi_k-\psi_l\}, \tag{10.31}$$

where

$$\left.\begin{matrix}\psi_k\\ \psi_l\end{matrix}\right\} = \frac{2}{\varkappa}\int_0^x\left[\sinh\left(\frac{\varkappa}{2}x\right)\cos\left(\frac{\varkappa}{2}x\right)\mathrm{ker}\left(\frac{\varkappa}{\sqrt{2}}r\right)\mp\cosh\left(\frac{\varkappa}{2}x\right)\times\right.$$

$$\left.\times\sin\left(\frac{\varkappa}{2}x\right)\mathrm{kei}\left(\frac{\varkappa}{\sqrt{2}}r\right)\right]\mathrm{d}x+\int_0^{|y|}\int_0^t\left[\mathrm{ker}\left(\frac{\varkappa}{\sqrt{2}}s\right)\mp\mathrm{kei}\left(\frac{\varkappa}{\sqrt{2}}s\right)\right]\mathrm{d}s\,\mathrm{d}t.$$

In order to be sure that this solution is pure, i.e. that it is not contaminated by self-equilibrating singularities of high order, one must consider them as the singular part of Green's function for the concentrated force, integrated over an arbitrary area of the shell for a distributed load $Z(x, y)$ and show that:

(1) the solution obtained satisfies the non-homogeneous shell equations;

(2) w, $\partial w/\partial x$ and $\partial w/\partial y$ are continuous everywhere;

(3) the stress resultants N_{xx}, N_{yy}, N_{xy}, M_{xx}, M_{yy}, M_{xy}, Q_x, and Q_y are continuous everywhere.

This proof was carried out in detail in [10.31].

In the vicinity of the origin, w and Φ behave like

$$w = \frac{PR^2}{8\pi D} r^2 \ln r$$

and

$$\Phi = -\frac{PR}{48\pi} \varkappa^4 \left(\frac{r^4}{4} \ln r + x^2 r^2 \ln r\right).$$

The above solution can easily be generalized to the case of the force moving along the generator of the shell [10.27]. R. Doré and W. Flügge [10.30] obtained in a similar way the solutions for the displacement dislocations. The stress function and the components of the displacement vector are given for all these solutions.

10.4.2 *Shallow cylindrical shell subjected to hot spot*

We assume that the shell is free of external loads. Then we obtain the following set of equations for the shallow cylindrical shell.

$$\begin{aligned} D\Delta^2 w - R\frac{\partial^2 \Phi}{\partial x^2} &= -\frac{1+\nu}{h} DR^2 \Delta(\alpha_t T_2), \\ \frac{1}{Eh}\Delta^2 \Phi + R\frac{\partial^2 w}{\partial x^2} &= -R^2 \Delta(\alpha_t T_1), \end{aligned} \tag{10.32}$$

where T_1 is the average temperature of the shell at the arbitrary point (x, y), T_2 defines the difference of the temperature of the external and internal surface of the shell. Eliminating the stress function and introducing the intensities: T_{s1} of the plane hot spot and T_{s2} — intensity of the bending hot spot according to (3.114), we find

$$\Delta^4 w + 4\varkappa^4 \frac{\partial^4 w}{\partial x^4} = -\left[\frac{1+\nu}{h} R^2 T_{s2} \Delta^3 + 4\varkappa^4 R T_{s1} \Delta \frac{\partial^2}{\partial x^2}\right] \delta(x, y, 0, 0). \tag{10.33}$$

The hot spot is assumed to be located at the origin of coordinates. The particular solution of Eq. (10.33) which constitutes the singularities for the hot spot is

$$w = -\left[T_{s2} \frac{1+\nu}{4h} \operatorname{re} + T_{s1} \frac{\varkappa^2}{2R} \operatorname{im}\right] \cos(1-i)\frac{\varkappa}{2} x Y_0\left[(1-i)\frac{\varkappa}{2} r\right] \tag{10.34}$$

where re and im denote the real and imaginary parts of the complex function

$$\cos\left[(1-i)\frac{\varkappa}{2}x\right] Y_0\left[(1-i)\frac{\varkappa}{2}r\right].$$

Let us note that in the vicinity of the origin the deflection and the stress functions can be presented in the form

$$\begin{aligned} w &\cong -\left(\frac{1+\nu}{h}\right)\frac{T_{s2}}{2\pi}\ln r - \frac{\varkappa^4}{4\pi R} T_{s1}\left(\frac{r^2}{2}+x^2\right)\ln r, \\ \Phi &\cong \frac{\varkappa^4}{4R^2} D\left(\frac{1+\nu}{h}\right)\frac{T_{s2}}{\pi}\left(\frac{r^2}{2}+x^2\right)\ln r - \frac{Eh T_{s1}}{2\pi}\ln r. \end{aligned} \tag{10.35}$$

10.5 Effect of boundary conditions

10.5.1 *General solution of homogeneous equations*

The calculations performed in the previous sections concerned the infinitely long shell. We have found that the deflection and the stresses in cylindrical and nearly cylindrical shells are very slowly damped in the x-axis direction. The influence of the finite dimensions of the shell is therefore important and often must be taken into account.

Let us consider now a shell of finite dimensions. We assume that at its edges certain boundary conditions, expressed by the displacements or internal forces, should be satisfied:

$$w = w_b, \qquad u = u_b, \qquad v = v_b, \qquad \frac{\partial w}{\partial x} = \left(\frac{\partial w}{\partial x}\right)_b,$$

or (10.36)

$$N_{xx} = N_{xx_b}, \quad \overline{N}_{xy} = \overline{N}_{xy_b}, \quad \overline{Q}_x = \overline{Q}_{x_b}, \quad M_{xx} = M_{xx_b},$$

where the index b denotes boundary displacements and forces.

In order to satisfy these conditions let us assume that a certain system of edge forces acts at the shell edge. These additional edge forces are determined by the functions Φ, w which satisfy the homogeneous shell equations. Let us express the functions Φ and w by the Fourier series

$$w(x, \varphi) = \sum_n^{\infty} w_n e^{i\alpha x} \cos n\varphi,$$
$$\Phi(x, \varphi) = \sum_n^{\infty} \Phi_n e^{i\alpha x} \cos n\varphi, \tag{10.37}$$

Here α is a parameter.

Substitution of the homogeneous equations (9.1) yields the following set of algebraic equations:

$$\frac{D}{R_2^4}(\alpha^2+n^2-\chi_R)^2 w_n + \frac{1}{R_2^3}(\alpha^2+\lambda n^2-\varphi_R)\Phi_n = 0,$$
$$\frac{1}{Eh R_2^4}(\alpha^2+n^2-\mu_R)^2\Phi_n - \frac{1}{R_2^3}(\alpha^2+\lambda n^2-\varphi_R) w_n = 0. \tag{10.38}$$

Solving this set of equations, we find a relation between the stress function Φ_n and deflection w_n:

$$\Phi_n \cong \pm 2i\varkappa^2 \frac{D}{R_2} w_n, \tag{10.39}$$

and the same characteristic equation (9.5) as before. This equation has eight complex roots

$$\left.\begin{aligned}\alpha_1 &= -\alpha_5\\ \alpha_4 &= -\alpha_8\end{aligned}\right\} = a \pm bi, \qquad \left.\begin{aligned}\alpha_2 &= -\alpha_6\\ \alpha_3 &= -\alpha_7\end{aligned}\right\} = c \pm di,$$

which can be obtained from Eq. (9.6) by introducing new coefficients (10.4).

In expression (10.39) the positive sign is for the odd roots, the minus sign for even ones. The solution (10.40) contains eight arbitrary constants which can be obtained from the boundary conditions. We get the following general solution:

$$w = \sum_{j=1}^{8} \sum_{n=0}^{\infty} C_{jn} e^{i\alpha_j x} \cos n\varphi,$$

$$\Phi = 2\varkappa^2 \sum_{j=1}^{8} \sum_{n=0}^{\infty} i \frac{D}{R_2} (-1)^{j+1} C_{jn} e^{i\alpha_j x} \cos n\varphi. \tag{10.40}$$

On changing to real functions, this solution takes the form

$$\text{(a)} \quad w = \sum_{n=0}^{\infty} [C_{1,n} e^{-bx} \cos ax + C_{2,n} e^{-bx} \sin ax + \\ + C_{3,n} e^{-dx} \cos cx + C_{4,n} e^{-dx} \sin cx + \\ + C_{5,n} e^{+bx} \cos ax + C_{6,n} e^{+bx} \sin ax + \\ + C_{7,n} e^{+dx} \cos cx + C_{8,n} e^{+dx} \sin cx] \cos n\varphi, \tag{10.41}$$

$$\text{(b)} \quad \Phi = 2\varkappa^2 \frac{D}{R_2} \sum_{n=0}^{\infty} [C_{2,n} e^{-bx} \cos ax - C_{1,n} e^{-bx} \sin ax - \\ - C_{4,n} e^{-dx} \cos cx + C_{3,n} e^{-dx} \sin cx - \\ - C_{6,n} e^{+bx} \cos ax + C_{5,n} e^{bx} \sin ax + \\ + C_{8,n} e^{dx} \cos cx - C_{7,n} e^{dx} \sin cx] \cos n\varphi.$$

It can also be represented in the following way:

$$w = \sum_{n=0}^{\infty} [(C'_{1,n} \cosh bx + C'_{2,n} \sinh bx) \cos ax + \\ + (C'_{3,n} \cosh bx + C'_{4,n} \sinh bx) \sin ax + \\ + (C'_{5,n} \cosh dx + C'_{6,n} \sinh dx) \cos cx + \\ + (C'_{7,n} \cosh dx + C'_{8,n} \sinh dx) \sin cx] \cos n\varphi, \tag{10.42}$$

$$\Phi = 2\varkappa^2 \frac{D}{R_2} \sum_{n=0}^{\infty} [(C'_{1,n} \sinh bx + C'_{2,n} \cosh bx) \sin ax - \\ - (C'_{3,n} \sinh bx + C'_{4,n} \cosh bx) \cos ax - \\ - (C'_{5,n} \sinh dx + C'_{6,n} \cosh dx) \sin cx + \\ + (C'_{7,n} \sinh dx + C'_{8,n} \cosh dx) \cos cx] \cos n\varphi.$$

If the shell and the load are symmetrical with respect to a certain plane, the solution should also include only the functions symmetrical with respect to the some plane. The constants must then vanish: $C'_{2,n} = C'_{3,n} = C'_{6,n} = C'_{7,n} = 0$. We have still to find four constants for every n. In order to form the equations of the boundary conditions let us express Eq. (10.36) by means of deflection and stress function. For this purpose we calculate the internal forces resulting from the deflection w and stress function Φ. On substituting the deflection (10.42) in (9.7) we obtain the bending moment M_{xx} in the form (for $l = R_2$)

$$
\begin{aligned}
M_{xx} = \sum_{n=0}^{\infty} \{ & [(C_{1,n} m_x^a - C_{4,n} m_x^b) \cosh x + \\
& + (C_{2,n} m_x^a - C_{3,n} m_x^b) \sinh bx] \cos ax + \\
& + [(C_{3,n} m_x^a + C_{2,n} m_x^b) \cosh bx + (C_{4,n} m_x^a + \\
& + C_{1,n} m_x^b) \sinh bx] \sin ax + [(C_5 m_x^c - C_8 m_x^d) \cosh dx + \\
& + (C_6 m_x^c - C_7 m_x^d) \sinh dx] \cos cx + + [(C_7 m_x^c + \\
& + C_6 m_x^d) \cosh dx + (C_8 m_x^c + C_5 m_x^d) \sinh dx] \sin cx \} \cos n\varphi ,
\end{aligned}
\tag{10.43}
$$

where

$$
\begin{aligned}
m_x^a &= [-n^2(1-\nu) + \eta_1 - \nu] \frac{D}{R_2^2} , \\
m_x^b &= (\eta_2 - \varkappa^2) \frac{D}{R_2^2} , \\
m_x^c &= [-n^2(1-\nu) - \eta_1 - \nu] \frac{D}{R_2^2} , \\
m_x^d &= (\eta_2 + \varkappa^2) \frac{D}{R_2^2} ,
\end{aligned}
\tag{10.44}
$$

where η_1, η_2 are given by (10.4).

Other internal forces can be calculated in the similar way. For example, to obtain the bending moment M_{yy}, it is necessary to substitute in the place of coefficients m_x^a, m_x^b, ... respectively m_y^a, m_y^b, ..., etc. These expressions are collected below. For the moment M_{yy} we have

$$m_y^a = [n^2(1-\nu)+\nu\eta_1-1]\frac{D}{R_2^2},$$

$$m_y^c = [n^2(1-\nu)-\nu\eta_1-1]\frac{D}{R_2^2},$$

$$m_y^b = \nu(\eta_2-\varkappa^2)\frac{D}{R_2^2}, \tag{10.45}$$

$$m_y^d = \nu(\eta_2+\varkappa^2)\frac{D}{R_2^2}.$$

For the normal force N_{xx}

$$n_x^a = 0, \quad n_x^b = n^2\frac{D}{R_2^3}2\varkappa^2, \quad n_x^c = 0, \quad n_x^d = -n^2\frac{D}{R_2^3}2\varkappa^2. \tag{10.46}$$

For the force N_{yy}

$$n_y^a = (\eta_2-\varkappa^2)\frac{D}{R_2^3}2\varkappa^2,$$

$$n_y^b = -(-n^2+\eta_1)\frac{D}{R_2^3}2\varkappa^2,$$

$$n_y^c = -(\eta_2+\varkappa^2)\frac{D}{R_2^3}2\varkappa^2,$$

$$n_y^d = (-n^2-\eta_1)\frac{D}{R_2^3}2\varkappa^2.$$

The displacements u and v can be calculated also from an expression similar to (10.43). For the displacement v we have the following coefficients:

$$v^a = (\eta_2-\varkappa^2)\frac{1}{2\varkappa^2 n},$$

$$v^b = -[-(1+\nu)n^2+\eta_1]\frac{1}{2\varkappa^2 n},$$

$$v^c = (-\eta_2-\varkappa^2)\frac{1}{2\varkappa^2 n},$$

$$v^d = -[(1+\nu)n^2+\eta_2]\frac{1}{2\varkappa^2 n}.$$

The displacement u can be represented in a similar way.

10.5.2 *Shell with simply supported edges*

In this case we require the final total displacements w and v as well as the moment M_{xx} and normal force N_{xx} to be equal to zero at the edge of the shell. Let us denote the displacements and the internal forces produced at the edge of the shell by the force P which acts at the point $x = 0$, $y = 0$ by v_0 and w_0, M_{xx_0} and N_{xx_0}, respectively.

Requiring these displacements and forces together with the boundary effects for every n to be equal to zero when $x = \pm L$, we obtain the set of $4 \cdot n$ equations for the boundary conditions for one edge. The deflection of the shell and all other distortions at the edge caused by the force P depend on the rapidly decreasing function $e^{i\alpha_1 x} = e^{-(b-ai)x}$. With increase of n, also the magnitude of b increases (see Table 10) and all distortions for $x = L > R$ vanish rapidly. Therefore, it is sufficient to take into account only the first few terms of the series (10.43). So we solve the set of equations for the general load only for the first few integers n. Knowing the constants $C_{i,n}$, we can calculate all displacements and internal forces from (10.43)–(10.46).

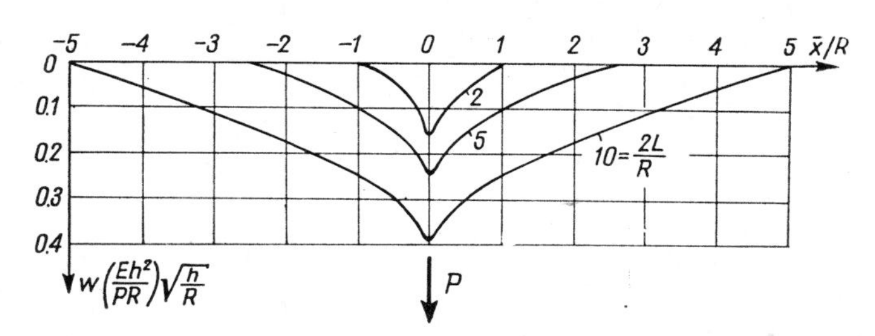

Fig. 117. Deflection of a cylindrical shell of the length L and simply supported at edges for various ratios L/R.

The deflection calculated for the cylindrical shell with simply supported edges is presented graphically in Fig. 117. The same results can easily be deduced from the diagram for an infinitely long shell by the method of images. This method was first used by I. Nadai [2.1] in the theory of plates (see § 3.12.1).

The problem of concentrated loads on a cylindrical shell was introduced by Yuan and Ting [10.7]. The procedure is illustrated in Fig. 118 for the case of the load P acting on a simply supported shell with the length L. The stresses in the shell can be found by loading an infinitely long shell with alternative positive and negative fictitious forces which are images of the real load with respect to the ends of the real shell and fictitious ones. In view of the adjacent symmetry the conditions at the dashed lines are

$$w = M_{xx} = N_{xx} = v = 0,$$

which correspond to those of a simple supported edge.

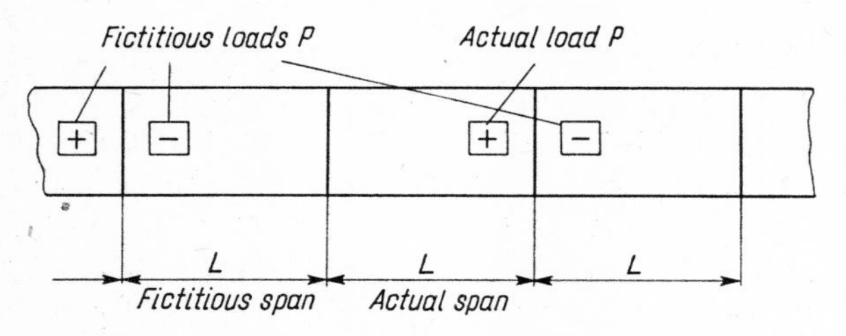

Fig. 118. Method of images applied to a cylindrical shell.

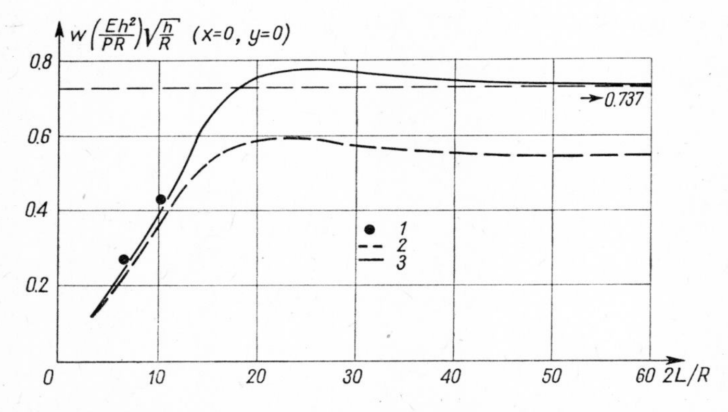

Fig. 119. Deflection of cylindrical shell at the point of application of the load for various ratios 2L/R. Dots 1 denote experiment results, line 2 results of the Donnell–Vlasov equations; line 3 results from the Flügge and Morley equations.

The numerical results given in Fig. 117 are equal to those achieved by Yuan and Ting [10.7] who used Flügge's equations. It should be mentioned here that the calculation of the roots of the characteristic Flügge's equation is troublesome. Equations (10.25), whose roots can be obtained from the closed formulae, give the same results with accuracy of the order of less than one per cent.

The relation between the maximum deflection of the shell at the point of application of the load and the length of the shell is shown in Fig. 119.

We observe that for short shells the differences between the results from the improved Eqs. (1.119) and Vlasov's ones are slight, about 5 per cent for $2L/R < 10$. These differences increase quickly with the increase of the length of the shell, reaching for long shells the magnitude of about 25 per cent.

In the case where the load is distributed over a small area, $2c_1 \times 2c_2$, the method of images can also be used. The boundary forces are practically independent of the extent of the area on which acts the load. Thus the influence surfaces depend only on the distance from the actual to the fictitious loads.

10.5.3 *The other variant of the solution*

The solution for the cylindrical shell of length L with free supported edges loaded by a concentrated normal force can be obtained in another way. Namely, we can present the solution in the form of the series

$$w = \sum A e^{p\varphi} \cos(nx/R), \qquad \Phi = \sum B e^{p\varphi} \cos(nx/R),$$

where $n = m\pi R/L$. If we apply Donnell's equations, we obtain the characteristic equation

$$(p^2 - n^2)^4 + 4\varkappa^4 n^4 = 0,$$

which has the roots

$$p = \pm\alpha_1 \pm i\beta_1, \qquad p = \pm\alpha_2 \pm i\beta_2,$$

where

$$\left.\begin{matrix}\alpha_1\\ \alpha_2\end{matrix}\right\} = \left[\frac{n}{2}\{(n \pm \varkappa) + [(n \pm \varkappa)^2 + \varkappa^2]^{1/2}\}\right]^{1/2}, \qquad \beta_i = \frac{n\varkappa}{2\alpha_i}. \tag{10.47}$$

The solution takes the form

$$\begin{aligned} w = \cos\frac{nx}{R}[&e^{-\alpha\varphi}(A_1 \cos\beta_1\varphi + A_2 \sin\beta_1\varphi) + \\ &+ e^{-\alpha_2\varphi}(A_3 \cos\beta_2\varphi + A_4 \sin\beta_2\varphi) + \\ &+ e^{\alpha_1\varphi}(A_5 \cos\beta_1\varphi + A_6 \sin\beta_1\varphi) + \\ &+ e^{\alpha_2\varphi}(A_7 \cos\beta_2\varphi + A_8 \sin\beta_2\varphi)]. \end{aligned}$$

For the stress function Φ we have a similar expression.

This solution can be interpreted in the following way. Let us imagine that the cylindrical shell consists of two infinitely long panels with the common edge along the generator $\varphi = 0$, each wound around an infinite number of times, with a radius of curvature R, but one wound in one direction and the other in the opposite direction. Each panel may then be considered to be semi-infinite in extent in the φ direction. Then only the negative exponential terms are retained so that the evaluation of only four constants is required. Since the panels are wound continuously through an infinite number of revolutions, the final deflection w for any angular position φ is given by the series

$$w = w_0 + w_1 + w_2 + w_3 + \ldots \tag{10.48}$$

where w_0 is the deflection for φ and w_1, w_2, w_3 are given by replacing φ by $(2\pi - \varphi)$, $(2\pi + \varphi)$, $(4\pi - \varphi)$, ..., etc. If the radial line load $q = q_0 \cos(nx/R)$ is applied along the generator $\varphi = 0$, this is equivalent to a transverse shear force $Q_\varphi = \frac{1}{2} q_0 \cos(nx/R)$ applied along the edge $\varphi = 0$ of each of the two imaginary panels. The constants A_1 can be calculated from the boundary conditions for the edge $\varphi = 0$. In the considered case we have, if $\varphi = 0$,

$$\frac{\partial w}{R \partial \varphi} = 0, \quad v = 0, \quad N_{x\varphi} = 0, \quad Q_\varphi = \frac{q_0}{2} \cos \frac{nx}{R}.$$

This solution may now be used in considering a line load uniformly distributed along the part of the generator $\varphi = 0$ if this load is expressed in the Fourier series form. Next the result obtained for analysing the effect of loads applied along parts of generators may be used to obtain stresses and deflections in the region of rectangular areas of loading by direct integration. If the shell is loaded by the moment distributed along the generator $\varphi = 0$, the solution can be obtained by expressing the moment in Fourier series form with the terms

$$M_{\varphi\varphi} = M(n) \cos \frac{m\pi x}{L}.$$

Because of the symmetry about $x = 0$ the same solution is applicable (10.47). Because of the antisymmetry about $\varphi = 0$ Eq. (10.48) is replaced by

$$w = w_0 - w_1 + w_2 - w_3 + \ldots$$

The boundary conditions take the form

$$\text{for} \quad \varphi = 0 \quad w = u = N_{\varphi\varphi} = 0$$

and

$$M_{\varphi\varphi} = -M(n)\cos\frac{m\pi x}{L}.$$

The stress resultants and deflections can be determined by summing the series. J. Kempner, I. Sheng and F. V. Pohle [10.6] obtained in this way the local loaded cylindrical shell based on the results of the paper by N. J. Hoff, J. Kempner, and F. V. Pohle [10.5].

10.5.4 *Solution in double trigonometric series*

The problem of the shell with simply supported edges can also be solved by means of double Fourier series. In this case the boundary conditions are identically satisfied and the coefficients of the series are determined by the simple formulae. Such a treatment in the case of a concentrated force was applied for the first time in the book by Vlasov [1.12]. Solving the problem by means of double series many more coefficients have to be calculated to obtain the same accuracy as in the case of the solution presented by a single series. But for computing only the deflection, the amount of work is not very large.

Let us consider a shallow or nearly cylindrical shell simply supported at the edges and loaded by a concentrated normal force P at the arbitrary point (x_0, y_0). The dimensions of the shell are a and b (Fig. 120). The coordinate

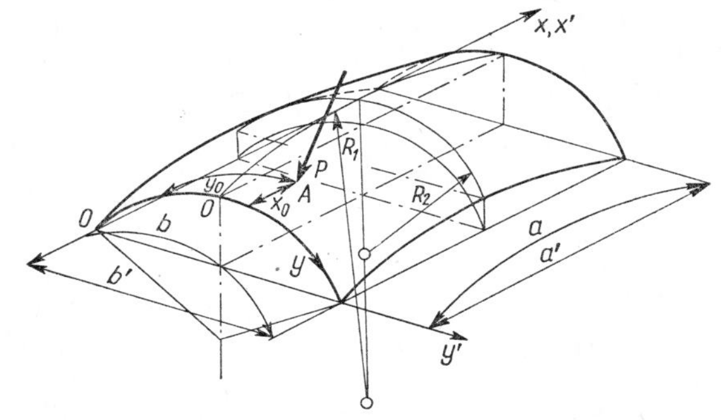

Fig. 120. Notations for the segment of double-curvature shell.

y is taken on the shell surface. The force P can be expressed by means of the double Fourier series (3.24). Introducing it into Eq. (9.1), we obtain the expressions for w and Φ analogous to those derived previously by means of Fourier integrals (9.2):

$$w = \frac{1}{D}\sum_{m=1}^{\infty}\sum_{n=1}^{\infty}\frac{q_{mn}}{D_{mn}}\left[\left(\frac{m\pi}{a}\right)^2+\left(\frac{n\pi}{b}\right)^2-\frac{1}{R_1R_2}\right]^2\sin\frac{m\pi x}{a}\sin\frac{n\pi y}{b},$$

$$\Phi = 4\varkappa^4R_2\sum_{m=1}^{\infty}\sum_{n=1}^{\infty}\frac{q_{mn}}{D_{mn}}\left[\left(\frac{m\pi}{a}\right)^2+\lambda_R\left(\frac{n\pi}{b}\right)^2-\right. \tag{10.49}$$

$$\left.-\frac{1}{R_1}\left(\frac{1}{R_1}+\frac{1}{R_2}\right)\right]\sin\frac{m\pi x}{a}\sin\frac{n\pi y}{b},$$

where

$$D_{mn} = \left[\left(\frac{m\pi}{a}\right)^2+\left(\frac{n\pi}{b}\right)^2-\frac{1}{R_1^2}+\frac{1-\nu}{2R_1R_2}-\frac{1}{R_2^2}\right]^2\left[\left(\frac{m\pi}{a}\right)^2+\left(\frac{n\pi}{b}\right)^2-\right.$$

$$\left.-\frac{1}{R_1R_2}\right]^2+4\varkappa^4\left[\left(\frac{m\pi}{a}\right)^2+\lambda_R\left(\frac{n\pi}{b}\right)^2-\frac{1}{R_1}\left(\frac{1}{R_1}+\frac{1}{R_2}\right)\right]^2,$$

$$4\varkappa^4 = Eh/DR^2;$$

q_{mn} is given by formula (3.26).

The internal forces and moments can be obtained from the above expressions by formulae (9.7). We see that the above solutions satisfy identically the boundary conditions for simply supported edges. At the edges we have

for $x = 0$ and $x = a$

$$w = 0, \quad M_{xx} = 0,$$

for $y = 0$ and $y = b$

$$w = 0, \quad M_{yy} = 0.$$

The conditions are satisfied since

$$w = \frac{\partial^2 w}{\partial x^2} = \frac{\partial^2 w}{\partial y^2} = 0$$

when $x = 0$ and $x = a$ and

$$w = \frac{\partial^2 w}{\partial y^2} = \frac{\partial^2 w}{\partial x^2} = 0$$

when $y = 0$ and $y = b$. The differentiation of the stress function yields the membrane forces. We have for $x = 0$ and $x = a$

$$\Phi = \frac{\partial^2 \Phi}{\partial x^2} = \frac{\partial^2 \Phi}{\partial y^2} = 0, \quad \sigma_{xxN} = \sigma_{yyN} = 0,$$

and for $y = 0$ and $y = b$

$$\Phi = \frac{\partial^2 \Phi}{\partial x^2} = \frac{\partial^2 \Phi}{\partial y^2} = 0, \qquad \sigma_{xxN} = \sigma_{yyN} = 0.$$

That gives also $u = 0$ for $y = 0$ and $y = b$ and $v = 0$ for $x = 0$ and $x = a$.
The bending moment M_{xx} calculated from (10.49) takes the form

$$M_{xx} = \sum_{m=1}^{\infty} \sum_{n=1}^{\infty} \frac{q_{mn}}{D_{mn}} \left[\left(\frac{m\pi}{a}\right)^2 + \left(\frac{n\pi}{b}\right)^2\right]^2 \left[\left(\frac{m\pi}{a}\right)^2 + \nu \left(\frac{n\pi}{b}\right)^2\right] \times$$

$$\times \sin \frac{m\pi x}{a} \sin \frac{n\pi y}{b}. \tag{10.50}$$

Similar expressions can be written for other internal forces and moments. By comparison of the above formulae with those for plates we observe that the bending moments in the shell are always smaller than the moments in the plate with the same dimensions, boundary conditions, and loadings. The shell carries the loads in a different way than the plate. The membrane forces in the shell play a significant role, whereas in the plate they do not appear. The shell represents a space structure in which the internal forces are distributed in a more reasonable way. The solution presented can also be applied to the cylindrical shell of a closed cross-section. It is then necessary to substitute $b = 2\pi R$ and $y = \varphi R$.

In the case where the load is distributed over a rectangular surface of dimensions $2c_1 \times 2c_2$ the solution can easily be obtained by expressing the load by means of Fourier series (3.24). Bijlaard (1954–1955) [10.4, 10.13, 10.14] calculated in this way the stresses in the cylindrical shell simply supported at the edges under the action of forces and bending moments distributed over rectangular surface of various dimensions. The numerical results were obtained by means of an IBM computer for various dimensions of the shell and loaded surface. For the bending moments the largest number of terms calculated are 7000. The results of these computations are presented graphically in Fig. 130. An analogous computation was performed in [10.32] for a large range of changes of parameters. In these calculations the smallest length of the side of the rectangle $2c_1$ was taken to be $(1/96)R$ and more than a million terms were calculated for the moments M_{xx} and M_{yy}. The exactness of the results obtained was examined by the calculation of the rest of the series. In every case the error did not exceed eight per cent. In the case where the bending moment was the load and the length of the loaded area was $(1/96)R$, the error was twenty per cent.

The method of the double trigonometric series was also applied for isotropic shallow shells in papers [10.82], [10.87], [10.91] and for orthotropic shells in [10.89], [10.92]. The method of finite differences was applied in [10.88] and the Bubnov-Galerkin method in [10.83].

Elling [11.2] obtained a singular solution in the form of a single trigonometric series. The coefficients of this series were obtained as the sum of the exponential series with respect to the polar radius.

Sanders considered the problem of arbitrary shallow shells subjected to the action of concentrated forces and moments and obtained the singular solutions solving the complex equations (1.123) (a). He considered also the displacement dislocations [9.15].

10.5.5 *Cylindrical shell with free edges*

If the ends of the cylindrical shell are free, the loading presented in Fig. 121 produces deformations consisting principally of bending. In such a case the magnitude of the deflection can be obtained approximately by neglecting the strain in the middle surface of the shell [3.1].

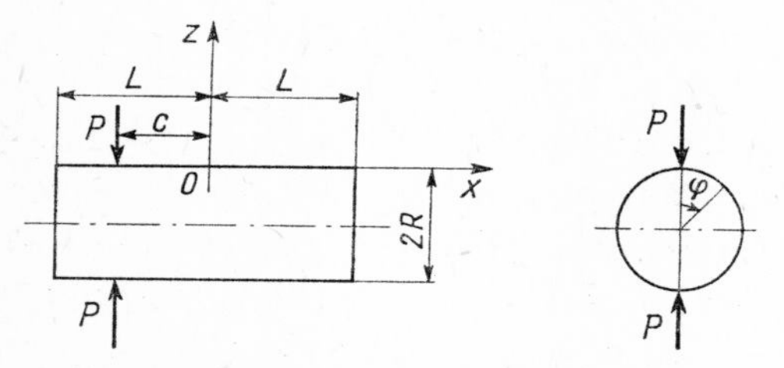

Fig. 121. Cylindrical shell with free edges.

If we assume that $\Phi = 0$ and express the deflection w by the following series:

$$w = R \sum_{n=1}^{\infty} w_n^{(1)} \cos n\varphi + x \sum_{n=1}^{\infty} w_n^{(2)} \cos n\varphi, \tag{10.51}$$

we can calculate the total strain energy of the shell from (1.139).

By using the principle of virtual displacements, the equations for calculating the coefficients $w_n^{(1)}$ and $w_n^{(2)}$ can be found. As a result we find the series

$$w = \frac{PR^3}{\pi DL} \sum_{n=2,4,\dots}^{\infty} \left\{ \frac{1}{(n^2-1)^2} + \frac{n^2 cx}{(n^2-1)^2 [\frac{1}{3} n^2 L^2 + 2(1-\nu) R^2]} \right\} \cos n\varphi . \tag{10.52}$$

If the force P acts at the middle, $c = 0$ and the displacement for $\varphi = 0$ is

$$w = \frac{PR^3}{\pi DL} \sum_{n=2,4,\dots}^{\infty} \frac{1}{(n^2-1)^2} = 0.0745 \frac{PR^3}{2DL} . \tag{10.53}$$

For $\varphi = \pi/2$

$$w = -\frac{PR^3}{\pi DL} \sum_{n=2,4,\dots}^{\infty} \frac{(-1)^{n/2+1}}{(n^2-1)^2} = -0.0685 \frac{PR^3}{2DL} .$$

The above solution does not satisfy all conditions at the free edge of the shell and neglects the local effects near the loading point. In Eq. (10.51) it has been assumed that the generator remains a straight line. But the local deflection is in this case relatively small and does not substantially effect the total deflections of the shell. The results (10.53) are in satisfactory agreement with experiment.

The method described above can be used in calculating the deflection of other shells of inextensional deformation, for example a portion of a cylindrical shell which is cut from a complete cylinder and with movable supports.

10.6 Cylindrical orthotropic shell

10.6.1 *Shell with simply supported edges*

The deflection of the orthotropic cylindrical shell with simply supported edges loaded by a normal force can be obtained by means of Fourier series. We assume that the deflection is represented by the series

$$w = \sum_{m}^{\infty} \sum_{n}^{\infty} w_{mn} \sin \frac{m\pi x}{L} \cos n\varphi .$$

Introducing it into Eqs. (1.155), we find the deflection and the stress function in the form $(L(\ ) = 0)$

$$E_1 w = 12(1-\nu_1\nu_2)\frac{R^4}{h^3}\sum_m^\infty\sum_n^\infty\left[\left(\frac{m\pi R}{L}\right)^4+2\lambda^2\mu_2\left(\frac{m\pi R}{L}\right)^2 n^2+\right.$$

$$\left.+\lambda^4 n^4\right]\frac{q_{m,n}}{D(m,n)}\sin\frac{m\pi x}{L}\cos n\varphi,$$

$$\Phi = 4\varkappa^4 R^3\sum_m^\infty\sum_n^\infty\left(\frac{m\pi R}{L}\right)^2\frac{q_{m,n}}{D(m,n)}\sin\frac{m\pi x}{L}\cos n\varphi, \tag{10.54}$$

where

$$D(m,n) = \left[\left(\frac{m\pi R}{L}\right)^4+2\lambda^2\mu_1\left(\frac{m\pi R}{L}\right)^2 n^2+\lambda^4(n^2-1)^2-\right.$$

$$\left.-2\lambda^2\mu_3\left(\frac{m\pi R}{L}\right)^2\right]\left[\left(\frac{m\pi R}{L}\right)^4+2\lambda^2\mu_2\left(\frac{m\pi R}{L}\right)^2 n^2+\lambda^4 n^4\right]+$$

$$+4\varkappa^4\left(\frac{m\pi R}{L}\right)^4 \tag{10.55}$$

and

$$4\varkappa^4 = 12(1-\nu_1\nu_2)\frac{E_2}{E_1}\frac{R^2}{h^2}, \qquad \lambda^4 = \frac{E_2}{E_1} = \frac{D_2}{D_1},$$

$$\mu_1 = \frac{H}{\sqrt{D_1 D_2}} = \nu_2\sqrt{\frac{E_1}{E_2}}+(1-\nu_1\nu_2)\frac{2G}{\sqrt{E_1 E_2}},$$

$$\mu_2 = \frac{\sqrt{E_1 E_2}}{2G}-\nu_2\sqrt{\frac{E_2}{E_1}}, \tag{10.56}$$

$$\mu_3 = \nu_2\sqrt{\frac{E_1}{E_2}}+(1-\nu_1\nu_2)\frac{G}{\sqrt{E_1 E_2}}.$$

In the case where the concentrated force P acts on the shell at the point $x = x_0$, $\varphi = 0$, the series representing the load takes the form

$$Z(x,\varphi) = \frac{P}{\pi R^2}\sum_{m=1}^\infty\left[\frac{1}{2}+\sum_{n=0}^\infty\cos n\varphi\right]\sin\frac{m\pi x_0}{L}\sin\frac{m\pi x}{L}, \tag{10.57}$$

then

$$q_{m,0} = \frac{P}{2\pi R^2}\sin\frac{m\pi x_0}{L}, \qquad q_{m,n} = \frac{P}{\pi^2 R^2}\sin\frac{m\pi x_0}{L}.$$

The series (10.54) is rapidly convergent for the calculation of the deflection w. Iu. P. Artiukhin [10.29] calculated in this way the deflection of the orthotropic shell loaded at the central cross-section ($x = L/2$) taking into account about 100 terms. The last term taken into account was smaller than 0.1 per cent of the sum of the series. The computation was performed with the assumption that

$$G = \frac{\sqrt{E_1 E_2}}{2(1+\sqrt{\nu_1 \nu_2})}. \tag{10.58}$$

which agrees well with the properties of resins reinforced by the glass fibre. The results of the computation are presented in Fig. 122. We observe that the maximum deflection $w_{(0,\,0)}$ decreases with the increase of λ.

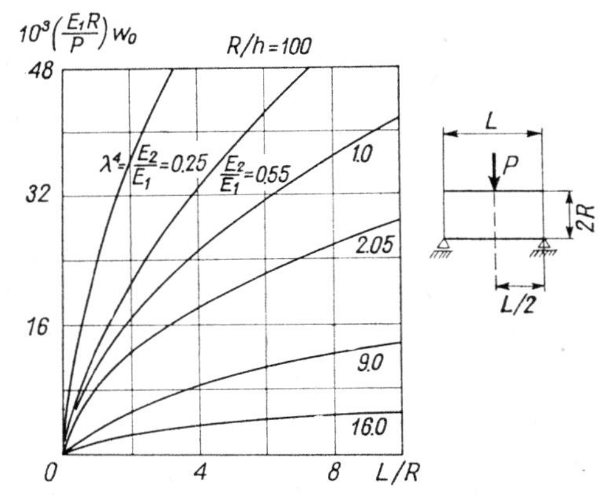

Fig. 122. Deflection of orthotropic cylindrical shell at the point of application of a load. The shell is simply supported at edges (from Iu. P. Artiukhin [10.29]).

If we neglect in the relation (10.55) for $D(m,n)$ the term $\lambda^2\mu_3(m\pi R/L)^2$ and unity in the expression $\lambda^4(n^2-1)^2$, we obtain the same results as if we used the Donnell–Vlasov equations for shallow shells. The differences between the results of the improved equations and those of Donnell–Vlasov are greater for larger values of λ. Artiukhin found that for $R/L = 10$ and $E_2/E_1 = 0.09$ the error is $\sim$ 5 per cent, but for $E_2/E_1 = 25$ the error is 20 per cent.

In a similar way Iu. R. Jigalko [10.26] considered the orthotropic cylindrical shell loaded by a normal force. The shell was simply supported at the edges. The solution was obtained by means of the equations of cylindrical orthotropic shells due to S. A. Ambartsumian [1.29]. The maximum deflection was computed for various values of the geometrical parameters. It was found that the deflection decreased considerably if the larger modulus E_2 corresponded to the circumferential direction. In the same paper the sandwich shell was considered.

10.6.2 *Infinitely long shell*

Let us consider now an infinitely long orthotropic cylindrical shell loaded by two radial forces (Fig. 123). The solution of this problem can be found by means of a Fourier integral in the same way as previously for the isotropic shell. We introduce the non-dimensional coordinate x referred to the radius R. Solving the set of equations (1.155), we find the deflection of the shell in the following form:

$$w_{(x,\varphi)} = \frac{2PR^2}{\pi^2 D_1}\left\{\frac{1}{2}\int_0^{\infty}\frac{\cos\alpha x\,\mathrm{d}\alpha}{\alpha^4-2\lambda^2\mu_3\alpha^2+\lambda^4+4\varkappa^4}+\right.$$

$$\left.+\sum_{n=2,4,\dots}^{\infty}\int_0^{\infty}[\alpha^4+2\lambda^2\mu_2\alpha^2n^2+\lambda^4n^4]\frac{\cos\alpha x}{D(\alpha,n)}\mathrm{d}\alpha\cos n\varphi\right\}, \tag{10.59}$$

where

$$D(\alpha,n) = [\alpha^4+2\lambda^2\mu_2\alpha^2n^2+\lambda^4n^4][\alpha^4+2\lambda^2\mu_1\alpha^2n^2+$$

$$+\lambda^4(n^2-1)^2-2\lambda^2\mu_3\alpha^2]+4\varkappa^4\alpha^4. \tag{10.60}$$

In order to facilitate numerical computations, the denominator can be simplified to the form

$$D(\alpha,n) \cong \{\alpha^4+2\lambda^2[(\mu_1+\mu_2)n^2-\mu_3]\alpha^2+\lambda^4n^2(n^2-1)\}^2+4\varkappa^4\alpha^4. \tag{10.61}$$

Integrating by means of calculus of residues we have the deflection

$$w = \mathrm{im}\frac{PR^2}{8\pi D_1\varkappa^2}\left\{\sum_{j=1}^{2}\frac{1}{2}\frac{\alpha_j e^{i\alpha_j x}}{\varkappa^2\pm i(\alpha_j^2-\lambda^2\mu_3)}+\right.$$

$$\left.+\sum_{j=1}^{2}\sum_{n=2,4,\dots}^{\infty}\frac{[\alpha_j^4+2\lambda^2\mu_2\alpha_j^2n^2+\lambda^4n^4]e^{i\alpha_j x}\cos n\varphi}{\alpha_j^3[\varkappa^2\pm i(\alpha_j^2+\lambda^2(\mu_1+\mu_2)n^2-\lambda^2\mu_3]}\right\}, \tag{10.62}$$

where α_1, α_2 are the roots of the characteristic equation

$$D(\alpha,n) = 0. \tag{10.63}$$

Artiukhin [10.29] computed the value of the deflection at the point of application of the load by means of a similar Fourier series. The computations were performed with the accuracy of 0.01 per cent, 150 terms being taken into account. The results are given in Fig. 123. As we see, with the increasing stiffeness in the circumferential direction, the deflection $w_{(0,0)}$ is more rapidly damped in the longitudinal direction. From previous considerations we re-

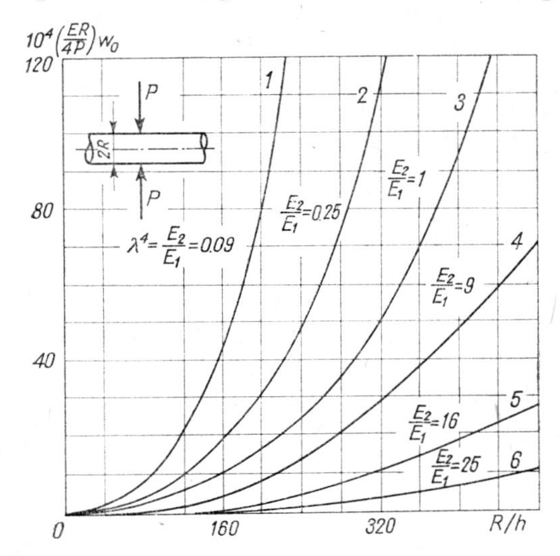

Fig. 123. Deflection of an infinitely long orthotropic cylindrical shell at the point of application of the load (Iu. P. Artiukhin [10.29]).

member that the shell in the vicinity of the point of application of the load behaves like a plate. Therefore, we conclude that the asymptotic formulae for the singular internal forces and displacements are for the orthotropic shell the same as for the orthotropic plate and can be obtained from Eqs. (3.159), (3.160), and (3.162). In the particular case where $H = \sqrt{D_1 D_2}$ and $\mu_1 = 1$ the following asymptotic formulae (introducing the polar coordinates r, ϑ) are proposed in paper [10.29]:

$$w = \frac{PR^2}{8\pi D_1 \lambda^3} (\lambda^2 \cos^2\vartheta + \sin^2\vartheta) r^2 \ln r.$$

The bending moments are

$$M_{rr} = \frac{P}{4\pi\lambda^3} [(\lambda^2 + \nu_2)\cos^2\vartheta + (1 + \nu_2 \lambda^2)\sin^2\vartheta] \ln r, \qquad (10.64)_1$$

$$M_{\vartheta\vartheta} = \frac{P\lambda}{4\pi}\,[(1+\nu_1\lambda^2)\cos^2\vartheta + (\lambda^2+\nu_1)\sin^2\vartheta]\ln r, \tag{10.64$_2$}$$

$$M_{r\vartheta} = \frac{P(\lambda^2-\nu_2)(1-\lambda^2)}{4\pi\lambda^3}\sin 2\vartheta \ln r. \tag{10.64$_3$}$$

The cylindrical orthotropic shell was considered in the paper by S. A. Khrustenko [10.17]. He has found that in the case where the shell is loaded by a concentrated normal force the asymptotic formulae are the same as for the isotropic shell provided that the shear modulus G is given by Eq. (10.58). Then we have the following asymptotic formulae for the bending moments:

$$\begin{aligned} M_{rr} &= \frac{P(1+\sqrt{\nu_1\nu_2})}{4\pi}\left(\frac{E_1}{E_2}\right)^{1/4}\ln\frac{r}{R}, \\ M_{\vartheta\vartheta} &= \frac{P(1+\sqrt{\nu_1\nu_2})}{4\pi}\left(\frac{E_2}{E_1}\right)^{1/4}\ln\frac{r}{R}. \end{aligned} \tag{10.65}$$

The shear modulus G, Eq. (10.58), is a hypothetical value. The problems of orthotropic cylindrical shells subjected to local loads were considered in papers [10.29], [10.41], [10.49], [10.74]. The sandwich shells were considered in [10.29], [10.45], [10.65], [10.66], [10.68], [10.79].

References 10

[10.1] S. ŁUKASIEWICZ: The influence of the double curvature on the rigidity of the shell of revolution, *Arch. Bud. Masz. 14.1* (1967).

[10.2] L. S. D. MORLEY: The thin-walled circular cylinder subjected to concentrated radial loads, *Quart. Mech. Appl. Math. 13.1* (1960) 24–37.

[10.3] SHAO WEN YUAN: Thin cylindrical shells subjected to concentrated loads, *Quart. Appl. Math. 4* (1946) 13.

[10.4] P. P. BIJLAARD: Stresses from radial loads in cylindrical pressure vessels, *Weld J. 3* (1954) 615.

[10.5] N. J. HOFF, J. KEMPNER, F. V. POHLE: Line load applied along generators of thin-walled circular cylindrical shells of finite length, *Quart. Appl. Math. 11.4* (1954) 411.

[10.6] J. KEMPNER, J. SHENG and W. V. POHLE: Tables and curves for deformation and stresses in circular cylindrical shells under localized loadings, *J. Aero. Sci. 2* (1957) 119.

[10.7] S. W. YUAN, L. TING, On radial deflections of a cylinder subjected to equal and opposite concentrated radial loads, *J. Appl. Mech. 24* (1957) 278.

[10.8] L. TING, S. W. YUAN, On radial deflection of a cylinder of finite length with various end conditions, *J. Aero. Sci. 25* (1958) 230.

[10.9] N. J. HOFF, The accuracy of Donnell's approximation for thin-valled circular cylinders, *J. Mech. Appl. Math. 12* (1959) 89.

[10.10] R. F. HOFFMANN, T. ARIMAN, Thermal and mechanical stresses in nuclear reactor vessels, *Symp. Nuclear. React. Containment Buildings and Pressure Vessels*, Berlin 1971.

[10.11] N. J. HOFF: Boundary-value problems of the thin-walled circular cylinder, *J. Appl. Mech. 21* (1954) 343.

[10.12] J. HOLAND: Characteristic equations in the theory of circular cylindrical shells, *Aero. Quart. 13* (1962).

[10.13] P. P. BIJLAARD: Stresses from local loadings in cylindrical pressure vessels, *Trans. ASME 77.6* (1955).

[10.14] P. P. BIJLAARD: Stresses from radial loads and external moments in cylindrical pressure vessels, *Weld. J. 34. 12* (1955).

[10.15] NAUBEREIT: Einflussflächen der Kreiszylinder Schale belastet durch Einzelmomente, *Wiss. Zeitsch. der Techn. Universität Dresden, Heft 1* (1963) 146–156.

[10.16] K. W. BIEGER: Die Kreiszylinderschale unter konzentrierten Belastungen, *Ing.-Arch. 30* (1961) 57–62; Dissertation, *Einflussflächen der Kreiszylinderschalen*, T. U., Berlin 1959.

[10.17] S. A. KHRUSTENKO: On the action of concentrated loads on orthotropic cylindrical shells, *Trudy II Vses. Konf. po Teorii Plastin i Obolochek*, Kiev 1962.

[10.18] K. SHESTOPAL: Equilibrium of closed cylindrical shells under action of concentrated forces, *Ukr. Mat. Journ. 15.1* (1963), 106–114 (in Russian).

[10.19] V. ABONINA: Application of trigonometric series to the solution of cylindrical shells on concentrated loads, *Sb. Trudov Leningradsk. U-ta Inj. Transporta 190* (1962) (in Russian).

[10.20] B. KLEIN: Effect of local loadings in pressurized circular cylindrical shells, *Aircraft Eng. 30* (1958) 356–361.

[10.21] J. HOLAND: Influence surfaces for bending moments in circular cylindrical shells or curved plates, *Int. Assoc. Bridge and Struct. Eng. Public. 21* (1961).

[10.22] T. E. HALS: *Circular cylindrical shells subjected to concentrated loads*, Institutt for Statick, Norges Tekniske Hogeskole, Trondheim, February 1966.

[10.23] IU. V. LITOVTSEV: Stability of visco-elastic shells under action of local loads, *M.T.T. 5* (1968) (in Russian).

[10.24] V. A. RODIONOVA: Solution for cylindrical shells loaded by concentrated forces of front surface, *Vestnik Leningradsk. U-ta 13* (1968) (in Russian).

[10.25] IU. IA. SHEVLIAKOV, V. P. SHEVCHENKO: Local stresses in cylindrical shell in the region surrounding concentrated loads, *Gidroaeromekh. i teoria uprugosti 6 (1)* (1968) (in Russian).

[10.26] IU. P. JIGALKO: Action of concentrated forces on the orthotropic and bimetallic cylindrical shells, *VI Vses. Konf. po Teorii Plastin i Obolochek*, 1966 Baku (in Russian).

[10.27] A. JAHANSHAHI: Forces singularities of shallow cylindrical shells, *J. Appl. Mech. 30.3* (1963) 343.

[10.28] P. V. VELICHKO, T. A. SHEVLIAKOV, V. P. SHEVCHENKO: *Proceedings of VII All Union Conf. on Theory of Shells and Plates*, Dnepropetrovsk 1969, Izd. "Nauka", Moscow 1970 (in Russian).

[10.29] IU. P. ARTIUKHIN: Solution for one-layer and many-layers orthotropic shells under local loads, *S. Issledovania po Teorii Plastin i Obolochek, 4*, Izd. Kazanskogo U-ta, 1966 (in Russian).

[10.30] R. DORÉ and W. FLÜGGE: Singular solutions for shallow cylindrical shells, *J. Appl. Mech. 40.1* (1973), 215–220.

[10.31] R. DORÉ: *Singular solutions for shallow cylindrical shells*, Stanford University, Doctoral dissertation 1969.

[10.32] W. M. DAREVSKII: Contact problems in the theory of shells (Action of local loads on shells), *Sb. Tr. VI Vses. Konf. po Teorii Obolochek i Plastinok*, Moscow, "Nauka" 1966 (in Russian).

[10.33] F. GARBRECHT: Belastungsversuch einer H. P.-Schale mit Einzellasten, *Bauingenieur 42.10* (1967)

[10.34] M. BADSTUBE: Ortliche Krafteinleitungen in Rotationsschalen, *IFL Mitt. 6.2.* (1967).

[10.35] E. I. GRIGOLIUK, V. M. TOLKACHEV: Equilibium of shells of revolution loaded on a meridian, *Izv. AN SSSR, MTT 3* (1970) (in Russian).

[10.36] I. PATSELT: Development of influence functions for thin toroidal shell under axisymmetrical load, *Izv. AN SSSR*, (*TT 2* V1970) (in Russian).

[10.37] IA. M. GRIGORENKO, G. K. SUDAVTSOVA: Spherical shells of revolution under local loads in the pole, *Prikladnaia Mekhanika 9.6* (1973) (in Russian).

[10.38] IA. M. GRIGORENKO: *Isotropic and anisotropic sandwich shells of revolution of variable rigidity*, Kiev, "Naukovaia Dumka", 1973, gl. V. (in Russian).

[10.39] A. A. FONINA: On a solution for shells of revolution under local loads, *Sudost. i Morsk. Sooruzh. Resp. Mezved. Temat. Nauch.-Tekhn. sb. 20* (1973) (in Russian).

[10.40] I. ICHINO, H. TAKAHASHI: Theory of non-symmetrical bending state for cylindrical shell. Stresses in a cylindrical shell with a local radial load, *Bull. JSME 8.30* (1965).

[10.41] M. KOZAROV: *The effect of concentrated loads on orthotropic cylindrical shells, Theory of Plates nad Shells*, Bratislava, 1966.

[10.42] R. O. STAFFORD, K. RIM: A general method of improving the accuracy of approximate solutions of circular cylindrical shells, *SIAM J. Appl. Math. 14.2* (1966).

[10.43] V. T. BUCHWALD: Some problems of thin circular cylindrical shells, I. The equations *J. Math. and Phys. 46.3* (1967).

[10.44] K. MIZOGUCHI, H. SHIOTA, K. SHIRAKAWA: Deformation and stress in a cylindrical shell under concentrated loading, *First rept. Radial loading. Bull. JSME 11. 45* (1968).

[10.45] A. A. ANTIPOV, V. G. POPOV: Local strength of three-layer panel under the action of concentrated loads, *Tr. Nikolaevsk. korablestroit. In-ta 32* (1969) (in Russian).

[10.46] V. G. NEMIROV: On fundamental solutions in the theory of circular cylindrical shell, *PMM 33.6* (1969) (in Russian).

[10.47] P. SEIDE: Effect of internal pressure on an infinite cylindrical shell subjected to concentrated radial loads, *AIAA J. 7.10* (1969).

[10.48] J. L. SANDERS, J. G. SIMMONDS: Concentrated forces on shallow cylindrical shells, *Trans. ASME, E. 37. 2* (1970).

[10.49] L. I. MOGILEVSKII: On influence of the factor of nonshallowness on the deflection while solving orthotropic cylindrical shell under radial concentrated force, *Tr. seminara po teor. obolochek, KFTI AN SSSR 2* (1971) (in Russian).

[10.50] L. I. MOGILEVSKII: On influence of the boundary conditions on edges of cylindrical shell on deflection in the region of application of concentrated force, *Tr. seminara po teor. obolochek, KFTI AN SSSR 2* (1971) (in Russian).

[10.51] K. MIZOGUCHI, K. SHIRAKAWA, K. HASEGAWA: Deformation and stress in cylindrical

shells under concentrated loading, *Second Rept. Circumferential Loading. Bull. JSME 14.75* (1971).
[10.52] A. G. GIRCHENKO, V. S. KRAVCHUK, A. O. RASSKAZOV, IU. K. CHEKUSHKIN, S. S. ELMANOVICH: Investigations of cylindrical shells loaded by radial concentrated force, *V sb. Prostranstvennye konstruktsii v Krasnoiarsk. krae*, Krasnoiarsk 1972 (in Russian).
[10.53] V. I. LEONOV, KH. S. KHAZANOV: Fundamental solution of equations of shallow cylindrical shells in polar coordinates, *Tr. Kuibysh. Aviats. In-ta 63* (1972) (in Russian).
[10.54] V. V. KLIMOV, V. I. GRIDASOV, L V. MAZANOWA: Application of generalized Fourier series to solve closed circular cylindrical shells under concentrated loads, *V sb. Differentsialnye uravnenia i vychislitelnaia matematika*, Saratov, 1972, vyp. 1 (in Russian).
[10.55] N. S. KONDRASHOV, On harmonic influence functions of cylindrical shell, *Prikladnaia Mekhanika 8.5* (1972) (in Russian).
[10.56] SH. KH. TUBEEV: Cylindrical shells of variable thickness under the action of concentrated moments, *Tr. Mosk. Energ. In-ta 120* (1972) (in Russian).
[10.57] IU. P. VINOGRADOV, IU. I. KLIEV: State of stress and deformation of cylindrical shell under concentrated load, *Izv. vuzov. Mashinostroenie 11* (*1973*) (in Russian).
[10.58] D. V. SINGH, R. SINHASAN, D. P. MITTAL: Surface stresses on cylindrical shell with arbitrary internal loading, *J. Inst. Eng.* (*India*). *Mech. Eng. Div. 53.6* (1973).
[10.59] A. S. ARYA, S. K. AGARWAL: Analysis of cylindrical shells subjected to line loads, *Indian Concr. J. 42.3* (1968).
[10.60] A. K. NAGHDI: Bending of a simply supported circular cylindrical shell subjected to uniform line load along a generator, *Intern. J. Solids and Struct. 4.11* (1968).
[10.61] I. SARKADI-SZABO: Circular cylindrical shell supported along a generator, *Trans. ASME, E 36.4* (1969).
[10.62] A. K. NAGHDI: Cylindrical shell subjected to longitudinal line load, *J. Eng. Mech. Div. Proc. ASCE 96.5* (1970).
[10.63] A. K. NAGHDI, J. M. GERSTING: The effect of a transverse shear acting on the edge of a circular cutout in a simply supported circular cylindrical shell, *Ing.-Arch., 42.2* (1973).
[10.64] J. G. SIMMONDS, Influence coefficients for semi-infinite and infinite circular cylindrical elastic shells, *J. Math. and Phys. 5.2* (1966).
[10.65] C. V. YOGANANDA, Ring loading of a long, thin, circular cylindrical shell enclosing a soft, solid core: a recalculation by the Love function method of elasticity, *Internat. J. Mech. Sci. 8.12* (1966).
[10.66] H. S. LEVINE, J. M. KLOSNER, Transversally isotropic cylinders under band loads, *J. Eng. Mech. Div. Proc. ASCE 93.3* (1967).
[10.67] C. V. YOGANANDA, S. RANGANATH; Symmetric solution for a corefilled cylindrical shell under a band of pressure, *J. Inst. Engr.* (*India*). *Mech. Engng Div. 48.5* (1968), Part 3.
[10.68] C. V. YOGANANDA: Influence lines for bending under a ring load of a free shell a shell embedded in a soft medium and a shell containing a soft core, *Acta mech. 7.4* (1969).
[10.69] C. V. YOGANANDA: Transversly isotropic cylinders under shrink fits, *ZAMM 50.11* (1970).

[10.70] SH. KH. TUBEEV: Bending of cylindrical shell of variable thickness under the action of concentrated loads, *Tr. Moskovs. Energ. In-ta 104* (1972) (in Russian).

[10.71] K. T. IYENGAR, R. SUNDARA, V. K. SEBASTIAN: Comparison of elasticity and shell-theory solutions for finite circular cylindrical shells, *Nucl. Eng. and Des. 21-1* (1972).

[10.72] V. I. LITVINENKO, Calculation for long cylindrical shell under the action of local uniformly distributed circular load, *Tr. Leningr. In-ta Vod. Transp. 140* (1973). (in Russian).

[10.73] C. V. YOGANANDA, S. K. GOYAL: Analytical investigation of a cylindrical shell embedded in a soft medium, *Rev. Roum. sci. techn. Ser. mec. appl. 5* (1972).

[10.74] S. N. SUKHININ: Action of local loads on orthotropic cylindrical shell, *Raschet prostr. konstr. 12* (1969) (in Russian).

[10.75] J. C. YAO, Long cylindrical tube subjected to two diametrically opposite loads, *Aeronaut. Quart. 20.4* (1969).

[10.76] B. V. NERUBAILO, V. A. SIBIRIAKOV, On calculation of cylindrical shell under local load, *Izv. Vuzov. Str-voi arkhitek. 6* (1970) (in Russian).

[10.77] B. V. NERUBAILO, V. A. SIBIRIAKOV: Nomograms for determination of stresses in locally loaded circular cylindrical shell, *Izv. Vuzov. Aviats. Tekhn. 3* (1970) (in Russian).

[10.78] B. V. NERUBAJLO, V. A. SIBIRIAKOV: Calculation for cylindrical shell under the action of radial local load, *Izv. vuzov. Mashinostr. 1* (1971) (in Russian).

[10.79] A. D. GRUCKOV: Stress-deformation state of sandwich shell under local load, *Stroit. Mekh. i raschet sooruzh. 2* (1973) (in Russian).

[10.80] R. V. MODESTOVA, A. M. SIMAKIN, E. N. SAMOILENKO, G. N. STEPANOVA: Determination of state of stresses of circular cylindrical shell changing *suddenly* its wall thickness under local load, *Probl. Prochnosti 2* (1973 (in Russian).

[10.81] F. WAN: Laterally loaded elastic shells of revolution, *Ing.-Arch. 42-4* (1973).

[10.82] T. N. GORSHUNOVA, E. I. MIKHAILOVA, V. JA. PAVILAINEN: Calculations for doubly curved shallow shells using electronic computers, *Sb. ECVM v stroit. mekhanike.* L 1966. (in Russian).

[10.83] P. I. KOKHREIDZE: Calculations for shallow shells under local loads by various boundary conditions, *Tr. Grusinskogo politechn. in-ta 3* (1967) (in Georgian).

[10.84] D. D. RABOTIAGOV: Problems of doubly curved shallow shells under linear loads, *Sb. Tr. Leningradskogo in-ta inż. z-d transporta 267* (1967).

[10.85] S. P. GAVELIA: On the state of stress of shallow shells under concentrated loads, *DAN URSR, A, 4* (1968) (in Ukrainian).

[10.86] I. S. DERIABIN: Simplified solution for doubly curved shallow shells clamped at the contour and at two parallel edges, *Sb. Tr. Leningradskogo inzh.-stroit. in-ta 57* (1968) (in Russian).

[10.87] L. V. CHIRADZE: Shallow shell under the action of concentrated forces factors, *Tr. Gruzinsk. Politekhn. In-ta 2* (1969) (in Russian).

[10.88] V. S. REKSHINSKII, B. M. MIZIN; Solution for shallow shells under the action of local loads, *Izv. vuzov. Str-vo i arkhitektura 3* (1970) (in Russian).

[10.89] A. S. KHRISTENKO, A. E. KALKO: Solution for rectangular in projection shallow shell with the help of double trigonometric series under local load, *Tr. Nikolaevskogo Korablestr. In-ta 46* (1971) (in Russian).

[10.90] V. P. Shevchenko: On the solution of the problem of statics of shallow shells under the action of local loads, *Prikladnaia Mekhanika* 7.6 (1971) (in Russian).

[10.91] L. V. Chiradze: On the solution for shallow shells under local loads, *Soobshch. AN Gruz. SSR 67.1* (1972) (in Russian).

[10.92] A. K. Kalko, A. S. Khristenko: *State of stress and displacement of shallow orthotropic and isotropic shells under local loads,* V sb. *Teoria obolochek i plastin,* M. "Nauka", 1973 (in Russian).

11

Shells under various concentrated loads

11.1 Shell loaded by a concentrated bending moment

We found out in Section 8.2 devoted to spherical shells that the solution for the concentrated bending moment M_i could be obtained by the differentiation of the solution for the concentrated normal force $P_z = P$. However, the solution for the concentrated tangential force $P_i = P$ should be subtracted from this result, according to the formula

$$w(M_i) = \frac{M_i}{P}\left\{\frac{1}{A_i}\frac{\partial}{\partial \alpha_i}[w(P_z)] - \frac{1}{R_i}w(P_x)\right\} = w_1(M_i) + w_2(M_i);$$

M_i means the moment whose vector representation is tangential to the α_j-axis (i not summed).

Therefore this case will not be considered in detail. We present only the bending moments resulting from local loads, i.e. for the integrals between the limits B and ∞ corresponding to $w_1(M_1)$.

$$\left.\begin{matrix} M^B_{xx} \\ M^B_{yy} \end{matrix}\right\} = \frac{M}{4\pi l}\int_B^\infty \{[\pm(1-\nu)S_1(k_1x)\beta] - \\ - [(1+\nu)\pm(1-\nu)]C_1(k_1x)\}\mathrm{e}^{-\beta x}\cos\beta y\,\mathrm{d}\beta, \tag{11.1}$$

$$M^B_{xy} = (1-\nu)\frac{M}{4\pi l}\int_B^\infty [S_1(k_1x)\beta - C_1(k_1x)]\mathrm{e}^{-\beta x}\sin\beta y\,\mathrm{d}\beta.$$

The integrals above can easily be evaluated by using the table of integrals (Table 16). If R_1 and R_2 increase simultaneously to infinity, then $k_1 \to 0$, and we obtain results identical with those for the flat plate. The membrane forces in the shell are multiplied by the ratio $R_2/4\pi l^3$ and decrease to zero as $R_2 \to \infty$. In this way we infer the relations obtained previously for a plate. When the shell, at the point of application of the load, does not differ much from the sphere then, using result (9.25), we obtain the following relation for the normal displacement w_1:

$$w_1 = -\frac{Ml}{2\pi D}\Bigg\{\operatorname{kei}' r\cos\vartheta + \\ + \frac{1-\lambda_R}{2}\Bigg[\left(\frac{1}{2}\operatorname{kei}' r - \frac{1}{4} r\operatorname{ker} r\right)\cos\vartheta + \\ + \left(\frac{8}{r^3} + \frac{8}{r^2}\operatorname{ker}' r + \frac{4}{r}\operatorname{kei} r - \frac{3}{2}\operatorname{kei}' r + \frac{1}{4} r\operatorname{ker} r\right)\cos 3\vartheta\Bigg]\Bigg\}. \quad (11.2)$$

11.2 The cylindrical shell loaded by a moment distributed over a small surface

The cylindrical shell subjected to the moment M_0 distributed along the distance $2c$ and over the surface of the rectangle $2b \times 2c$ has been considered by T. E. Hals [10.22]. He obtained the internal forces and moments in the shell, using the solution given by Holand [10.21] for the normal force acting on the segment of the cylindrical surface. The solution for the moment given as a linearly variable normal load $q(x)$ can be obtained by taking the solution for the unit force as the Green function for the problem considered. The line load $q(x)$ can be represented as the sum of the concentrated forces $q(x)\,dz$ acting along the straight line $y = 0$. The intensity of these forces varies linearly. The resultant of this load is the moment M_0. We find the relation between M_0 and $q(x)$ from the following equation: Taking $q(x) = q_0 \dfrac{x}{c}$, we have

$$M_0 = 2\int_0^c q_0 \frac{x}{c}\, x\,dx = \frac{2}{3} q_0 c^2.$$

We obtain the internal forces and moments in the shell integrating in a similar way as for the formulae obtained for the unit concentrated load. For example, the internal moment M_{xx} produced by the line load $q(x)$ is given by

$$M_{xx} = \int_{-c}^{c} q_0 \frac{x}{c} m_{xx} \mathrm{d}x,$$

where m_{xx} is the influence function for the moment M_{xx}, i.e. m_{xx} is the moment M_{xx} produced in the shell by a concentrated unit normal force. Integrating the above relation along y between the limits $-b$ and b, we obtain the solution

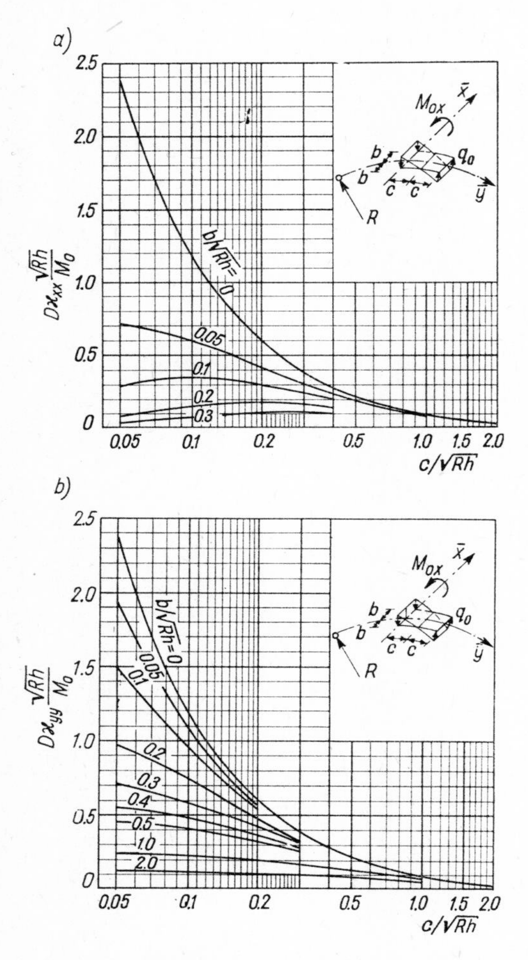

Fig. 124. a) Curvature change $D\varkappa_1$ at the point $(-c, 0)$ at the edge of the loaded area produced by the bending moment M_{oy}; b) curvature change $D\varkappa_2$ produced by the same moment in the point $(-c, 0)$ (from T. E. Hals [10.22]).

for the case of the shell loaded by a moment distributed over the surface of the rectangle $2b \times 2c$. For example, we have

$$M_{xx} = \int_{-b}^{b} \left[\int_{-c}^{c} q_0 \frac{x}{c} m_{xx} \,\mathrm{d}x \right] \mathrm{d}y,$$

where now the relation between M_0 and q_0 is

$$M_0 = \tfrac{4}{3} q_0 c^2 b .$$

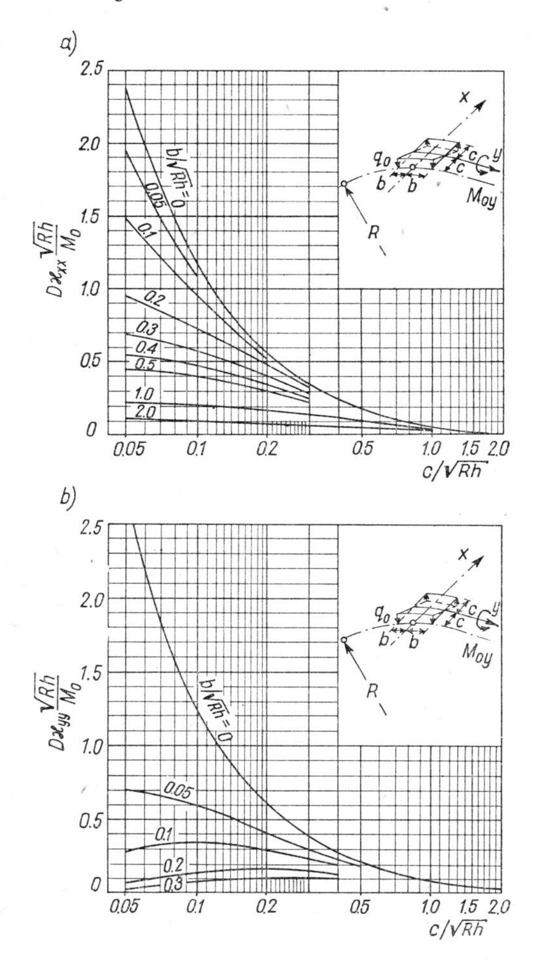

Fig. 125 a), b). Curvature changes $D\varkappa_1$ and $D\varkappa_2$ at the point $(-c, 0)$ at the edge of the loaded area produced by the distributed bending moment M_{0x} ([10.22]).

The results of the calculations by T. E. Hals are presented in the Fig. 124. Hals [10.22] considered the segment of an infinitely long cylindrical shell with simply supported edges. The width of the segment was $6\sqrt{Rh}$. It appears, however, that the boundary conditions have little influence on the magnitude of the internal forces in the vicinity of the loaded region if this region is not very close to the edge. The figures present the changes of the curvature $D\varkappa_{xx}$ and $D\varkappa_{yy}$ at the point lying on the boundary of the loaded region for different values of b/c. Figures 125a, b present the results for the case where the vector of the moment M_0 is directed perpendicularly to the axis of the cylinder. In Figs. 125c and d this moment is parallel to the axis. The bending moments in the shell are given by the formulae

$$M_{xx} = D(\varkappa_{xx}+\nu\varkappa_{yy}),$$

$$M_{yy} = D(\varkappa_{yy}+\nu\varkappa_{xx}).$$

Figure 126 presents the angles of the rotations $\partial w/\partial x$, $\partial w/\partial y$ in the middle of the loaded region. We obtain the real values of the angles by multiplication of the values taken from the diagram by the coefficient M_0/Eh^3. As a result of calculations, we find that the effect of the finite lenght of the shell is negligible if the distance of the loaded region from the boundary is larger than $2\sqrt{Rh}$. If this distance is smaller and the shell has simply supported edges, this effect can easily be taken into account by means of the method of images.

Comparing the results given here with the moments obtained previously in § 3.9 for the infinite plate loaded by a moment introduced by a rigid element, we find that the moments are of the same order. The solutions presented in this section are obtained taking into account the displacement $w_1(M)$ only.

11.3 Shell loaded by a force tangential to its surface

Let us consider now a shell of double curvature subjected to a force tangential to its surface. This force is directed along the x-axis lying in the plane of the larger of the principle radii of curvature. Let us assume that the shell extends infinitely in all directions. We neglect therefore the boundary effects. Let us assume further that the radii of curvature are constant in the area near the loading point. Expressing the force as a Fourier integral on substitution in Eq. (7.57), we have

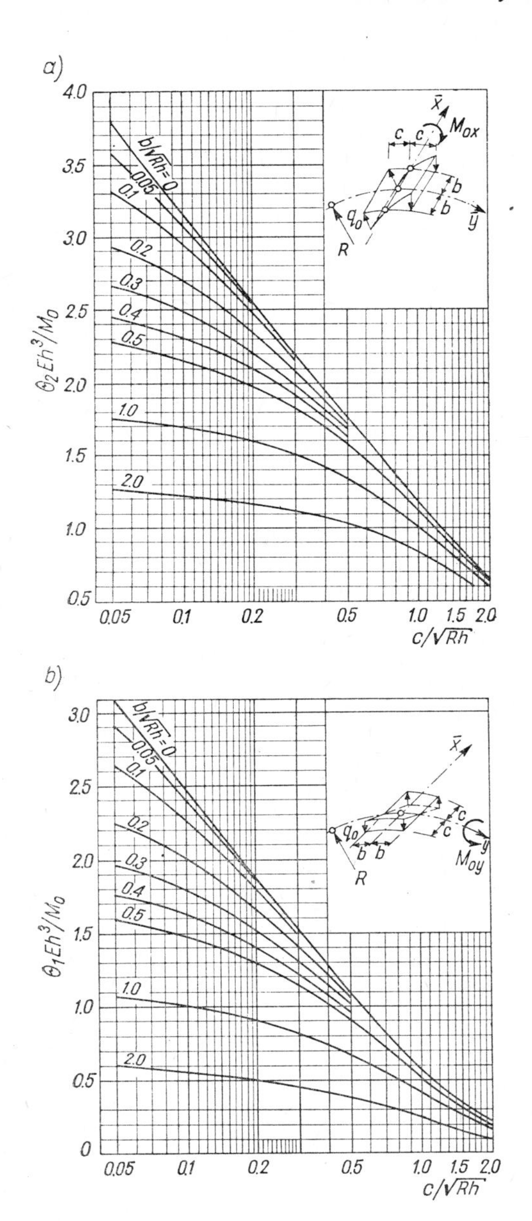

Fig. 126. Angles of rotation θ_1 and θ_2 at the centre of the loaded area point (0, 0) *produced by the moments M_{0x} and M_{0y}* ([10.22]).

$$\frac{D}{l^4}[\Delta+\chi_R]^2 w-\frac{1}{l^2}\Delta_k\Phi = \frac{T}{R_1\pi^2 l}\int_0^\infty\int_0^\infty \frac{\sin\alpha x}{\alpha}\cos\beta y\,d\alpha\,d\beta,$$

$$\frac{1}{Ehl^4}[\Delta+\mu_R]^2\Phi+\frac{1}{l^2}\Delta_k w = \frac{T}{Eh\pi^2 l^3}\int_0^\infty\int_0^\infty \alpha\left(\frac{\beta^2}{\alpha^2}-\nu\right)\sin\alpha x\cos\beta y\,d\alpha\,d\beta. \tag{11.3}$$

Solving Eqs. (11.3), we find

$$w = \frac{Tl^3}{\pi^2 R_2 D}\int_0^\infty\int_0^\infty [\lambda_R(\alpha^2+\beta^2-\mu_R)^2-(\beta^2-\nu\alpha^2)(\alpha^2+\lambda_R\beta^2-\varphi_R)]\times \times\frac{\sin\alpha x\cos\beta y\,d\alpha\,d\beta}{\alpha D(\alpha,\beta)},$$

$$\Phi = \frac{Tl}{\pi^2}\int_0^\infty\int_0^\infty [\lambda_R(\alpha^2+\lambda_R\beta^2-\varphi_R)+(\beta^2-\nu\alpha^2)(\alpha^2+\beta^2-\chi_R)^2]\times \times\frac{\sin\alpha x\cos\beta y\,d\alpha\,d\beta}{\alpha D(\alpha,\beta)}, \tag{11.4}$$

where

$$D(\alpha,\beta) = (\alpha^2+\beta^2-\chi_R)^2(\alpha^2+\beta^2-\mu_R)^2+(\alpha^2+\lambda_R\beta^2-\varphi_R)^2.$$

Integrating these expressions by means of the calculus of residues, we obtain

$$w = \frac{Tl^3}{4\pi R_2 D}\times \times \operatorname{im}\int_0^\infty \sum_{k=1}^{2}\frac{i[\lambda_R+(-1)^k(\beta^2-\nu\alpha_k^2)]e^{-i\alpha_k x}\cos\beta y\,d\beta}{\alpha_k^2(\alpha_k^2+\beta^2+\frac{1}{2}i(-1)^k-\frac{1}{2}(\chi_R+\mu_R))}. \tag{11.5}$$

The bending moments and the membrane forces can easily be obtained by differentiation under the integrals for the deflection w and the stress function Φ.

11.3.1 *Local loads*

In order to obtain stresses it is necessary to calculate the above integrals numerically. It can be proved that in this case the integrals for the bending moments converge relatively fast, whereas the integrals for the membrane forces converge very slowly. In order to accelerate numerical calculations, let us divide the load in a local and general part. For the local load, i.e. for $\beta > B$, we can omit the coefficients χ_R, μ_R; φ_R. Changing the lower limit of integration from zero to B and substituting the simplified expressions (9.33) for the roots of characteristic equation, we obtain, on transformation, the following integrals determining the bending moments

$$\left.\begin{matrix} M_{xx}^B \\ M_{yy}^B \end{matrix}\right\} = \frac{T}{4\pi R_2} \int\limits_B^\infty \{\mp(1-\nu^2)S_2(k_1 x) - \\ - [(1+\nu)^2 \mp (1-\nu)^2] S_3(k_1 x)\} e^{-\beta x} \cos\beta y \, d\beta . \tag{11.6}$$

For membrane forces we have

$$\left.\begin{matrix} N_{xx}^B \\ N_{yy}^B \end{matrix}\right\} = -\frac{T}{4\pi l} \int\limits_B^\infty [(1+\nu)S_1(k_1 x) + \\ + [(1-\nu) \pm (1+\nu)] 2C_1(k_1 x)] e^{-\beta x} \cos\beta y \, d\beta ,$$

$$N_{xy}^B = -\frac{T}{4\pi l} \int\limits_B^\infty [(1+\nu)S_1(k_1 x)\beta + \\ + (1-\nu)C_1(k_1 x)] e^{-\beta x} \sin\beta y \, d\beta . \tag{11.7}$$

The integrals above can easily be calculated (see Table 16, p. 552)

Similar expressions can be written for the moments M_{yy}, M_{xy}. For $\bar{r} \ll \bar{l}$ the moments are proportional to $x^3 \cdot r^2$, which gives zero stresses at the loading point. For small $x \ll 1$, the functions $S_1(k_1 x)$, $C_1(k_1 x)$ can be replaced by $S_1(k_1 x) = x$, $C_1(k_1 x) = 1$. Then the membrane forces are determined by the expressions identical with those calculated previously for the flat plate. In conclusion—the shell in the area near the loading point behaves in the same way as the flat plate (see § 3.5). All singularities of the stresses are identical. The solution for a force uniformly distributed along a certain distance $2c$ can be derived very easily by multiplying the integrands by $(\sin\beta c)/\beta c$.

For $x = 0$ the membrane stresses in the shell are identical with those in the flat sheet. We have

$$-c < y < c, \qquad \sigma_{xx} = -\frac{T}{4hc}, \qquad \sigma_{yy} = -\nu\frac{T}{4hc},$$
$$y < -c \quad \text{and} \quad y > c, \qquad \sigma_{xx} = \sigma_{yy} = 0. \tag{11.8}$$

11.3.2 *Solution in polar coordinates*

The problem considered can be solved in a similar way as that used in Section 9.3 introducing the system of polar coordinates. Then the deflection and stress function can be presented in the form

$$w = (1+\nu)\frac{Tl^3}{\pi^2 R_2 D}\int_0^{\pi/2}\int_0^{\infty}\left[\frac{\lambda_R+\nu}{1+\nu}-(1-\lambda_R)\sin^2\varphi\right]\times$$
$$\times\cos\varphi\sin(\gamma x\cos\varphi)\cos(\gamma y\sin\varphi)\,\frac{d\gamma\, d\varphi}{D(\gamma)},$$
$$\Phi = \frac{Tl}{\pi^2}\int_0^{\pi/2}\int_0^{\infty}[\lambda_R(\cos^2\varphi+\lambda_R\sin^2\varphi)+\gamma^4(\sin^2\varphi-\nu\cos^2\varphi)]\times \tag{11.9}$$
$$\times\sin(\gamma x\cos\varphi)\cos(\gamma y\sin\varphi)\,\frac{d\gamma\, d\varphi}{D(\gamma)\gamma^2\cos\varphi},$$

where

$$D(\gamma) = \gamma^4+(\cos^2\varphi+\lambda_R\sin^2\varphi)^2.$$

The first integral (11.9) can be evaluated by the same method as the integrals in the previous article, transforming the denominator to $1/[(\gamma^4+1)(1-R)]$, where R is determined by (9.13). Expressing the fraction $1/(1-R)$ in power series, we get, on integration,

$$w = (1+\nu)\frac{Tl^3}{2\pi R_2 D}\sum_{n=0,1,\dots}^{\infty}\ \sum_{m=1,3,\dots}^{4n+3}[2(1-\lambda_R)]^n a_{m,n} U^0_{m,n}, \tag{11.10}$$

where

$$U^k_{m,n} = \int_0^{\infty}\frac{\gamma^k J_m(\gamma_r)\, d\gamma}{(\gamma^4+1)^{n+1}}. \tag{11.11}$$

The solutions of the above integrals are presented in Table 15.

$$a_{m,0} = \frac{\lambda_R+\nu}{1+\nu}\nu^{0,m}-(1-\lambda_R)\nu^{2,m},$$

$$a_{m,n} = a_{m,n-1} \bar{x} \left(\nu^{2;0} - \frac{1-\lambda_R}{2} \nu^{4,0} \right)$$

where

$$\nu^{2n,2n+1} = \frac{1}{2^{2n}} \cos(2n+1)\vartheta,$$

$$\nu^{2n,2n-m} = \frac{1}{2^{2n}} \frac{2n(2n-1)\ldots(2n-(m-3)/2)}{((m+1)/2)!} (2n-m)\cos(2n-m)\vartheta.$$

For small values of $1-\lambda_R$ we have

$$w = (\lambda_R+\nu) \frac{Tl^3}{2\pi R_2 D} \times$$
$$\times \left\{ \left[1 - \frac{(1-\lambda_R)(1-\nu-2\lambda_R)}{4(\lambda_R+\nu)} \right] \left(\frac{1}{r} + \mathrm{ker}' r \right) \cos\vartheta + \right.$$
$$+ \frac{1-\lambda_R}{4} \left[\frac{r}{2} \mathrm{kei}\, r \cos\vartheta - \right.$$
$$\left. \left. - \frac{1+\nu}{\lambda_R+\nu} \left(\frac{1}{r} + \frac{4}{r} \mathrm{ker}\, r - \frac{8}{r^2} \mathrm{kei}' r - \mathrm{ker}' r \right) \cos 3\vartheta \right] \right\}. \tag{11.12}$$

The bending moment takes the form of the integral

$$M_{xx} = (1+\nu) \frac{Tl}{\pi^2 R_2} \int_0^{\pi/2} \int_0^{\infty} \gamma^2 \left[\frac{\lambda_R+\nu}{1+\nu} - (1-\lambda_R)\sin^2\varphi \right] \times$$
$$\times [1-(1-\nu)\sin^2\varphi] \cos\varphi \sin(\gamma x \cos\varphi) \cos(\gamma y \sin\varphi) \frac{\mathrm{d}\gamma\, \mathrm{d}\varphi}{D(\gamma)}. \tag{11.13}$$

On integration we obtain the series

$$M_{xx} = (1+\nu) \frac{Tl}{2\pi R_2} \sum_{n=0,1,\ldots}^{\infty} \sum_{m=1,3,\ldots}^{4n+5} [2(1-\lambda_R)]^n c_{m,n} U^2_{m,n}, \tag{11.14}$$

where

$$c_{m,0} = \frac{\lambda_R+\nu}{1+\nu} \nu^{0,m} - \frac{1+2\nu(1-\lambda_R)-\nu^2}{1+\gamma} \nu^{2,m} + (1-\lambda_R)(1-\nu)\nu^{4,m}, \quad \text{etc.}$$

In a similar way we obtain the other internal moments. For the moment M_{yy} we have

$$c_{m,0} = \nu \frac{\lambda_R+\nu}{1+\nu} \nu^{0,m} - \frac{\lambda_R+\lambda_R\nu^2-2\nu^2}{1+\nu} \nu^{2,m} - (1-\lambda_R)(1-\nu)\nu^{4,m}, \quad \text{etc.}$$

For the moment M_{xy} we find

$$c_{m,0} = \frac{\lambda_R+\nu}{1+\nu}\eta^{0,m} - \frac{1+2\nu-\lambda\nu}{1+\nu}\eta^{2,m} - (1-\lambda_R)\eta^{4,m}, \qquad \text{etc.}$$

Now we calculate the membrane forces which can be obtained by using the stress function Φ. We see, however, that the integral $(11.9)_2$ for the stress function differs from all other integrals, for it contains $\cos\varphi$ in the denominator. This makes certain difficulties in integration. But we can divide this integral into two parts, namely, we can write

$$\Phi = \Phi_1 + \Phi_2,$$

where

$$\Phi_1 = \frac{Tl}{\pi^2}\int_0^{\pi/2}\int_0^{\infty}\frac{1}{\gamma^2\cos\varphi}\sin(\gamma x\cos\varphi)\cos(\gamma y\sin\varphi)\,\mathrm{d}\gamma\,\mathrm{d}\varphi,$$

$$\Phi_2 = -\frac{T}{\pi^2}\int_0^{\pi/2}\int_0^{\infty}\{(1+\nu)\gamma^4+(1-\lambda_R)[1-$$

$$-(1-\lambda_R)\sin^2\varphi]\}\cos\varphi\sin(\gamma x\cos\varphi)\cos(\gamma y\sin\varphi)\frac{\mathrm{d}\gamma\,\mathrm{d}\varphi}{\gamma^2 D(\gamma)}. \tag{11.15}$$

We see now that only the first integral of $(11.15)_1$ contains $\cos\varphi$ in the denominator. The second one can be evaluated in the same way as previously. It is easy to find a closed solution for the first integral. Namely, we have

$$\Phi_1 = -\frac{Tl}{2\pi}(r\ln r\cos\varphi - r\vartheta\sin\vartheta) + \text{const.} \tag{11.16}$$

Then the membrane forces resulting from Φ_1 are

$$N_{xx_1} = -\frac{T}{2\pi l}\frac{\cos\vartheta}{r}, \qquad N_{yy_1} = \frac{T}{2\pi l}\frac{\cos\vartheta}{r}, \qquad N_{xy_1} = -\frac{T}{2\pi l}\frac{\sin\vartheta}{r}.$$

The membrane forces resulting from the second part of the stress function can be obtained in the form of series as previously. On integration we have the following expression:

$$N_{xx_2} = -\frac{T}{2\pi l}\sum_{n=0,1,\dots}^{\infty}\sum_{m=1,3,\dots}^{4n+5}[2(1-\lambda_R)]^n[(1+\nu)b_{m,n}U^4_{m,n} +$$

$$+(1-\lambda_R)d_{m,n}U^0_{m,n}]. \tag{11.17}$$

The coefficients $b_{m,n}$ and $d_{m,n}$ are defined by the following recurrent formulae:

$$b_{m,0} = \nu^{2,m}, \quad b_{m,n} = b_{m,n-1} \times \left(\nu^{2,0} - \frac{1-\lambda_R}{2} \nu^{4,0}\right)$$

and

$$d_{m,n} = \nu^{2,m} - (1-\lambda_R)\nu^{4,m}, \quad d_{m,n} = d_{m,n-1} \times \left(\nu^{2,0} - \frac{1-\lambda_R}{2} \nu^{4,0}\right).$$

For the force N_{yy_2} we have the same formula (11.17) but the coefficients are

$$b_{m,0} = \nu^{0,m} - \nu^{2,m}, \qquad d_{m,0} = \nu^{0,m} - (2-\lambda_R)\nu^{2,m} + (1-\lambda_R)\nu^{4,m}.$$

For the force N_{xy_2} we obtain an analogous series. The coefficients $b_{m,0}$ and $d_{m,0}$ are given by the following recurrent formulae

$$b_{m,0} = \eta^{0,m} - \eta^{2,m}$$

and

$$d_{m,0} = \eta^{0,m} - (2-\lambda_R)\eta^{2,m} + (1-\lambda_R)\eta^{4,m},$$

$$b_{m,n} = 0 \quad \text{if} \quad m < 4n+3,$$

$$d_{m,n} = 0 \quad \text{if} \quad m < 4n+5.$$

It can be proved on the basis of numerical computations that the series obtained above converge sufficiently rapidly for the calculation of the stresses in the shell of double positive Gaussian curvature in the range $0.1 \leqslant \lambda_R \leqslant 1$. The series usually converge more rapidly than the series

$$\sum_{n=1}^{\infty} (1-\lambda_R)^n / 2^{2n-1}.$$

However, for large n the number of coefficients which must be calculated increases quickly, although they are calculated from the simple recurrent formulae. The series developed give exact results only in the case where the concentrated loads act at a certain distance of about several l from the edge. We know that all distortions in the shell of positive curvature are very quickly damped. The influence of the edge (except for the case of free or movable edges) does not spread far and does not disturb the distribution of stresses and displacements in the vicinity of the point of application of the load. So, these results should be correct not only for shallow shells, but also for arbitrary ones.

11.3.3 *The cylindrical shell loaded by a tangential force*

The solution for the case of the infinite cylindrical shell loaded by the set of m tangential forces T acting in the x-axis direction can be obtained representing the load by means of the series of Fourier integrals (10.1). If m is the number of forces applied at equal distances along the circumference, we have the following expression for the load X:

$$X(x, \varphi) = \frac{mT}{\pi^2 R^2} \int_0^\infty \left(\frac{1}{2} + \sum_{n=1}^\infty \cos mn\varphi \right) \cos \alpha x \, d\alpha .$$

Solving the set of equations (1.127) and assuming Donnell's approximation, we obtain the following result:

$$w = \frac{mTR^2}{\pi^2 D} \left[-\frac{\nu}{2} \int_0^\infty \frac{\sin \alpha x \, d\alpha}{\alpha(\alpha^4 + 4\varkappa^4)} + \right.$$

$$\left. + \int_0^\infty \sum_{m=1}^\infty \frac{\alpha(m^2 n^2 - \nu\alpha^2) \sin \alpha x \cos mn\varphi \, d\alpha}{(\alpha^2 + m^2 n^2)^4 + 4\varkappa^4 \alpha^4} \right],$$

$$\Phi = \frac{mTR}{\pi^2} \left[-\frac{\nu}{2} \int_0^\infty \frac{\alpha \sin \alpha x \, d\alpha}{\alpha^4 + 4\varkappa^4} + \right.$$

$$\left. + \int_0^\infty \sum_{n=1}^\infty \frac{(m^2 n^2 - \nu\alpha^2)(\alpha^2 + m^2 n^2)^2 \sin \alpha x \cos mn\varphi \, d\alpha}{\alpha[(\alpha^2 + m^2 n^2)^4 + 4\varkappa^4 \alpha^4]} \right].$$

The integrals above can easily be evaluated by means of the calculus of residues. Since the above solution is antisymmetrical with respect to the cross-section $x = 0$, the evaluation can be performed only for $x > 0$ (see Table 16).

The case of the tangential force acting in the circumferential direction can be solved in a similar way.

A cylindrical segment under the action of tangential force T (Fig. 127) was considered by T. E. Hals [10.22] using the shallow shells equations. The solution was obtained by the method proposed by J. Holand [10.21], i.e. by dividing the infinitely long shell into two parts in the plane of the loaded cross-section $x = 0$. Each part was loaded by half of the concentrated force which was presented by a Fourier series. Next the boundary conditions resulting from the conditions of symmetry with respect to the plane $x = 0$ were taken into account. The results of the computations are presented in Fig. 128. The diagrams present the influence surfaces for the membrane forces for the tangential load $T = 1$. The figures give only one quarter of the influence surface. The forces in the remaining quarters of the area can easily be defined

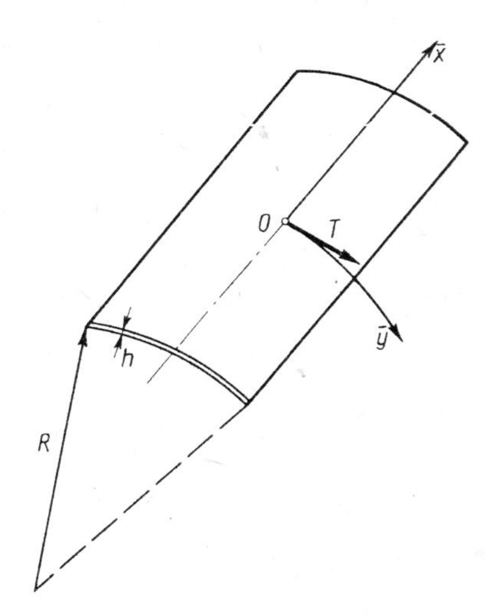

Fig. 127. Segment of a cylindrical shell subjected to the concentrated tangential force T.

taking into account the fact that the stress field is symmetrical with respect to the plane $x = 0$ and antisymmetrical with respect to the plane $y = 0$. The membrane forces and bending moments are zero at the loaded point. In the vicinity of this point the membrane forces are defined by formulae identical with that for a flat infinite sheet loaded on an interior point by a concentrated tangential force. In order to obtain the true value of membrane forces the value taken from the diagram should be devided by $\sqrt{Rh}$.

The influence surfaces presented here and in the previous sections permit the calculation of stresses in the circular cylindrical shell under an arbitrary load. The effects of the finite dimensions of the shell with simply supported edges can be taken into account by means of the method of images.

11.4 Asymptotic relations

In order to facilitate the approximate computation of the stresses and displacements in the shell we can isolate from the previously obtained formulae the

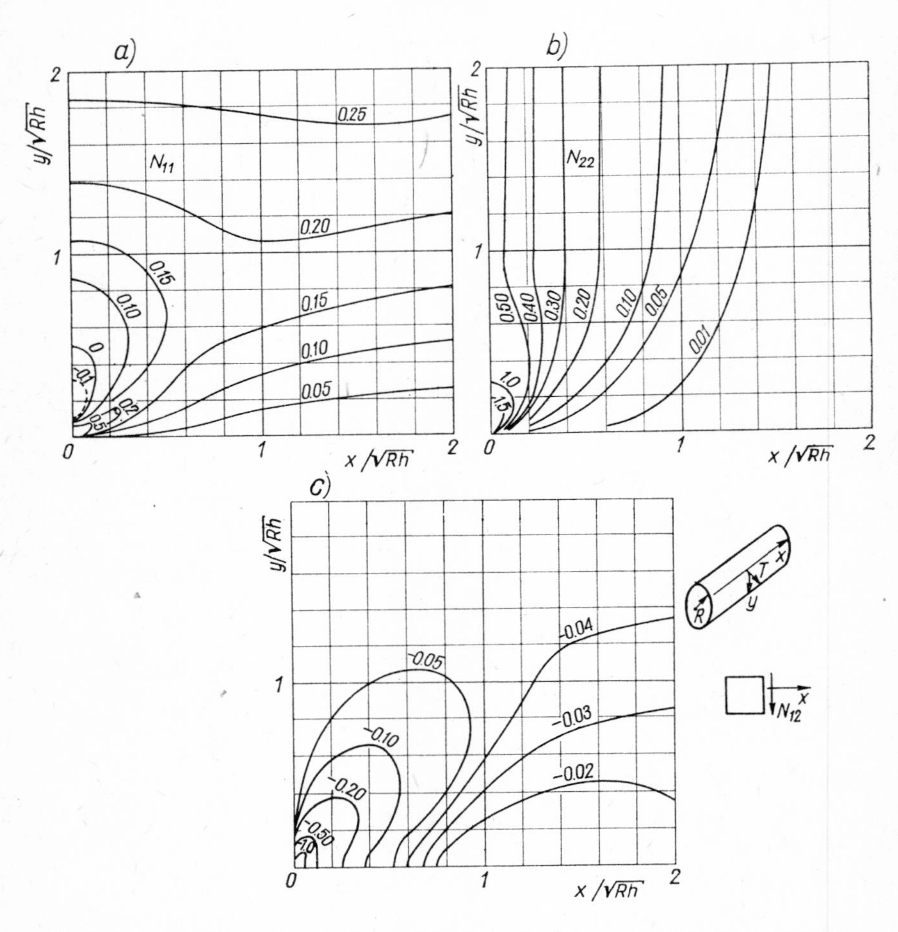

Fig. 128. Influence surfaces for membrane forces produced by tangential concentrated force applied to cylindrical segment. a) N_{11}, *b)* N_{22}, *c)* N_{12}. *Surfaces for* N_{11} *and* N_{22} *are symmetrical with respect to y-axis and antisymmetrical with respect to x-axis. The influence surface for the tangential force* N_{12} *is symmetrical to the x-axis and antisymmetrical to y-axis* ([10.22]).

singular terms, obtaining in this way the asymptotic relations for these internal forces and moments which are very large in the vicinity of the singular points. Namely: near the points of application of the loads, at the ends of the loaded segment or at the corner points of the loaded rectangle, etc. Such formulae are derived in the works by Novozhilov and Chernykh [11.1] by dividing the series into two parts, slowly and rapidly divergent. It is possible to express

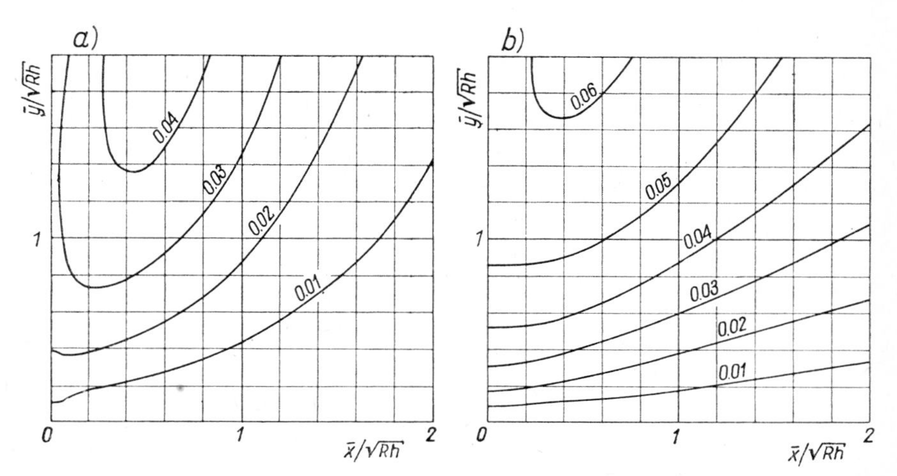

Fig. 129 a), b). Influence surfaces for bending moments M_{xx} and M_{yy} produced in a cylindrical segment by a tangential force T ([10.22]).

the sum of the first slowly divergent series in the form of a finite combination of elementary functions. The second, fast divergent series, can be neglected in the computation of the large displacements or internal forces in the vicinity of the loaded point. The asymptotic formulae obtained in this way have an identical form with those for plates and become more exact the smaller the distance between the loading and the point where the stresses are calculated. However, the asymptotic formulae are insufficiently accurate for the calculation of stresses at points remoted from the singular points.

For an arbitrary shell loaded by a concentrated force P the following asymptotic formulae exist for the bending moments:

$$M_{xx} \simeq M_{yy} \simeq -\frac{1+\nu}{4\pi} P \ln \frac{r}{r_0}. \tag{11.18}$$

For the shell loaded by a concentrated bending moment M_1 whose vector has the direction of the coordinate x we have

$$\begin{aligned} M_{xx} &\simeq -\frac{M_1}{4\pi r} \sin\vartheta[2(1-\nu)\cos^2\vartheta - 1 - \nu], \\ M_{yy} &\simeq +\frac{M_1}{4\pi r} \sin\vartheta[2(1-\nu)\cos^2\vartheta + 1 + \nu], \\ M_{xy} &\simeq -\frac{M_1}{4\pi r} (1+\nu)\cos\vartheta(2\sin^2\vartheta - 1). \end{aligned} \tag{11.19}$$

We obtain similar relations for the bending moment M_2 acting in the direction of the second coordinate y by introducing $\vartheta_1 = \pi/2-\vartheta$.

For the shell loaded by a tangential force T acting in the direction of the coordinate x we find

$$
\begin{aligned}
u &= -\frac{(3-\nu)(1+\nu)}{4Eh} T \ln\frac{r}{r_0}, \\
N_{xx} &\simeq \frac{T}{4\pi r}\cos\vartheta[2(1+\nu)\sin^2\vartheta-3-\nu], \\
N_{yy} &\simeq -\frac{T}{4\pi r}\cos\vartheta[2(1+\nu)\sin^2\vartheta-1+\nu], \\
N_{xy} &\simeq -\frac{T}{4\pi r}\sin\vartheta[2(1+\nu)\cos^2\vartheta+1-\nu].
\end{aligned}
\tag{11.20}
$$

When the force T acts in the y-direction, we have

$$
\begin{aligned}
v &= -\frac{(3-\nu)(1+\nu)}{4Eh} T \ln\frac{r}{r_0}, \\
N_{xx} &\simeq -\frac{T}{4\pi r}\sin\vartheta[2(1+\nu)\cos^2\vartheta-1+\nu], \\
N_{yy} &\simeq \frac{T}{4\pi r}\sin\vartheta[2(1+\nu)\cos^2\vartheta-3-\nu]. \\
N_{xy} &\simeq -\frac{T}{4\pi r}\cos\vartheta[2(1+\nu)\sin^2\vartheta+1-\nu].
\end{aligned}
\tag{11.21}
$$

As we have said before, in reality the concentrated forces do not exist. They are the idealization of the loads distributed over a small surface and enable us to simplify the calculation of these quantities which remain limited by the concentrated loads (for example, the deflection under the concentrated normal force). But there still remains the question of the determination of these quantities which are unbounded when the real local load is replaced by a load concentrated only at one point of the shell (for example, moments M_{xx} and M_{yy} resulting from the concentrated normal force).

The following method seems to be most appropriate here. We suppose that the maximum values of these quantities approximately equal the values resulting from the asymptotic formulae calculated on the boundary of the loaded region when the distributed load is replaced by the force concentrated at one point at the centre of this region. The question is: how small should

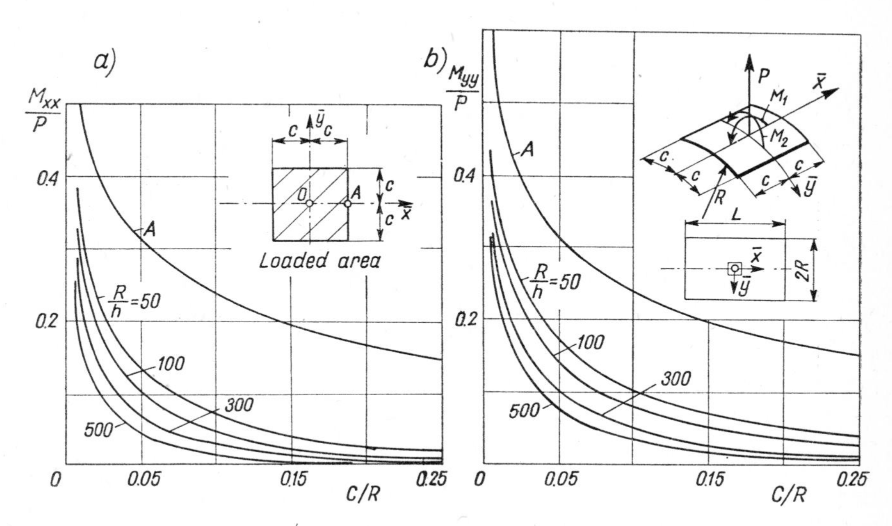

Fig. 130. Comparison of results of asymptotic formulae and results of exact computations for cylindrical shell subjected to normal force distributed over the surface $2c \times 2c$.

this area be to have sufficient accuracy? The answer to this question can be obtained by comparison of the results from the asymptotic formulae and respective results of exact computations, for example, the results obtained by Bijlaard [10.13], or Holand [10.21]. The comparison for the cylindrical shell taken from the paper by Darevskii [10.32] is shown in Figs. 130, 131. The lines A in these figures correspond to the asymptotic formulae, other lines result from the calculation performed by means of the double Fourier series for the quadratic form of the loaded area. For the shell considered the ratio $\pi R/L = 1$ (where L is the length of the shell). With decrease of the ratio $\pi R/L$, the curves change slightly.

The bending moments resulting from the action of the force P distributed over the surface $2c \times 2c$ were calculated by means of Fourier series at the central point 0 of the loaded surface. The values resulting from asymptotic formulae were calculated at the point A on the boundary of the loaded surface. In the case of the shell loaded by concentrated bending moments the internal moments were calculated by both the methods at the point A. As shown in the Figs. 130, 131 the asymptotic formulae give much larger values of stresses for the isotropic cylindrical shell. It may be expected that the same principle concerns also the orthotropic cylindrical shell or an arbitrary shell. Therefore one should take this into consideration when using them.

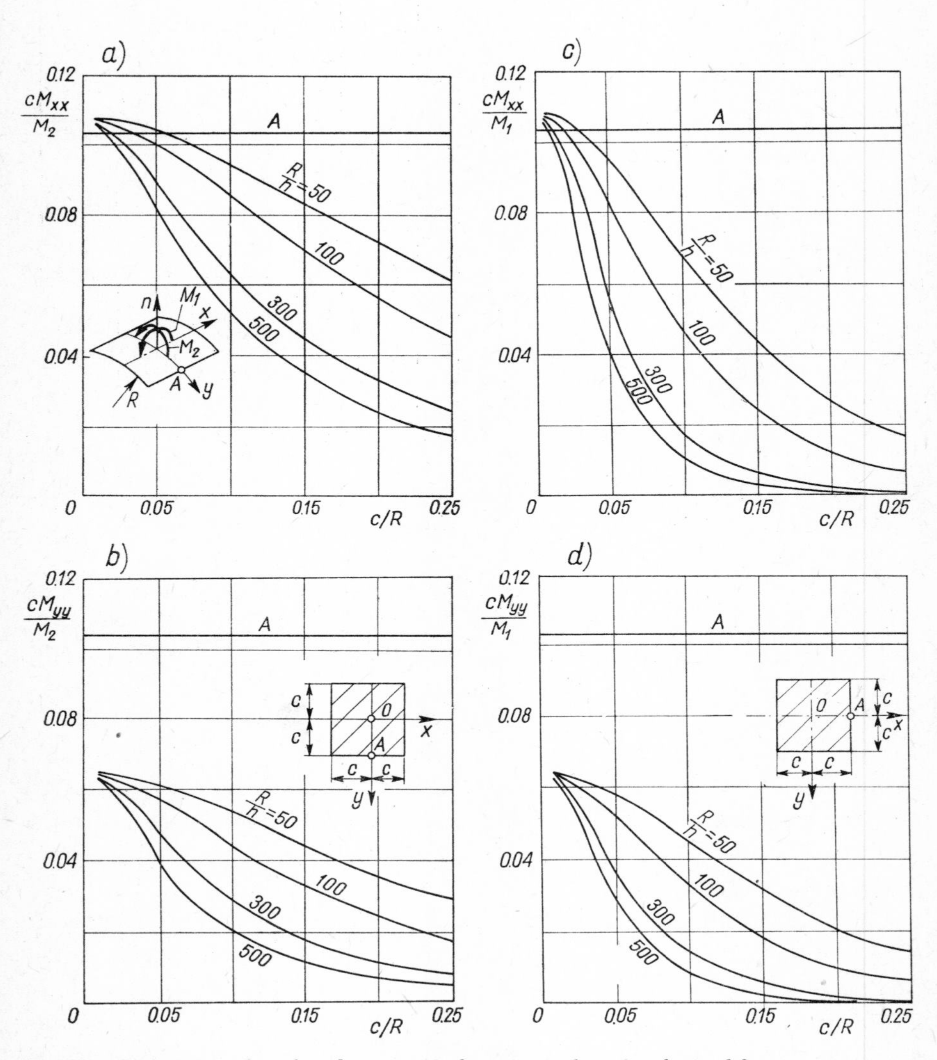

Fig. 131. Comparison of results of asymptotic formulae and results obtained from exact computations for a cylindrical shell subjected to bending moments M_1 and M_2 distributed over the surface $2c \times 2c$. Internal moments M_{xx} and M_{yy} are calculated from asymptotic formulae at the point A ([10.32]).

References 11

[11.1] V. V. Novozhilov, K. F. Chernykh: On calculation of shells under local loads, in: *Issledovania po uprugosti i plastichnosti, No. 2*, Leningradskii Universitet, 1963 (in Russian).

[11.2] R. Elling: Concentrated loads applied to shallow shells, *J. Eng. Mech. Div. Proc. ASCE 99. 12* (1973).

[11.3] K. Mizoguchi, K. Shirakawa: Deformation and strength of a cylindrical shell subjected to a concentrated twisting couple, *Bull. JSME 15.82* (1972).

[11.4] P. M. Velichko, Iu. A. Shevliakov, V. P. Shevchenko: Deformation of shells of positive curvature under the action of arbitrary concentrated forces, in: *Kontsentratsia napriazhenii 3* (1971) (in Russian).

[11.5] N. A. Kil: On the action of local loads on shells, *Izv. Vuzov. Str-vo i Arkhitektura 3* (1973) (in Russian).

[11.6] I. G. Teregulov, L. I. Mogilevckii: Asymptotic formulae for concentrated loads taking into account physical nonlinearity in thin shells, *Prikladnaia Mekhanika 9.12* (1973), (in Russian).

12

Edge loads

12.1 Shell loaded at free edge by a lateral force

12.1.1 *Solution in Fourier integrals*

Let us consider an arbitrary shell loaded by a concentrated force acting at the free edge. In order to simplify the calculations let us assume that the shell spreads infinitely in the x and y-directions. In this way we neglect the effect

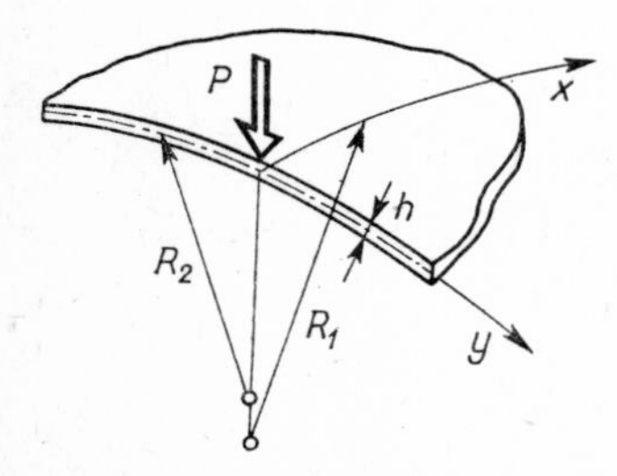

Fig. 132. Shell loaded at its free edge by a concentrated shear force.

of the other edges as we are most interested in the distribution of the stresses caused by the concentrated force in the vicinity of the loading point. Let us assume also that in this area the main radii of curvature are constant. We perform the calculation in the non-dimensional system of coordinates (x, y) referred to the characteristic length l. Representing the deflection and the stress function by Fourier integrals and substituting these functions in the homogeneous equations (9.1), we find the following set of algebraic equations (neglecting the effect of transverse normal and shear stresses):

$$\begin{aligned} &\frac{D}{l^4}(\alpha^2+\beta^2-\chi_R)^2 w_\beta + \frac{1}{R_2 l^2}(\alpha^2+\lambda_R\beta^2-\varphi_R)\Phi_\beta = 0, \\ &\frac{1}{Ehl^4}(\alpha^2+\beta^2-\mu_R)^2\Phi_\beta - \frac{1}{R_2 l^2}(\alpha^2+\lambda_R\beta^2-\varphi_R)w_\beta = 0, \end{aligned} \tag{12.1}$$

where χ_R, μ_R, φ_R are coefficients depending on the ratio h/R_2 and defined by Eqs. (10.3). Solving this set, we have

$$\Phi_\beta = \pm i\frac{DR_2}{l^2} w_\beta \frac{\alpha^2+\beta^2-\chi_R}{\alpha^2+\beta^2-\mu_R} \approx \pm i\frac{DR_2}{l^2} w_\beta. \tag{12.2}$$

Eliminating the function Φ_β from Eq. $(12.1)_1$, we find the characteristic equation identical as previously (9.5). The roots of this equation take the form

$$\alpha_1 = a+bi, \qquad \alpha_2 = c+di, \tag{12.3}$$

where the coefficients a, b, c, d are defined by (9.6). The general solution is

$$\begin{aligned} w = \int_0^\infty [&C_1 e^{-bx}\cos\alpha x + C_2 e^{-bx}\sin\alpha x + C_3 e^{-dx}\cos cx + \\ &+ C_4 e^{-dx}\sin cx + C_5 e^{bx}\cos ax + C_6 e^{bx}\sin ax + \\ &+ C_7 e^{dx}\cos cx + C_8 e^{dx}\sin cx]\cos\beta y\, d\beta, \\ \Phi = \frac{DR_2}{l^2}\int_0^\infty [&C_2 e^{-bx}\cos ax - C_1 e^{-bx}\sin ax - \\ &- C_4 e^{-dx}\cos cx + C_3 e^{-dx}\sin cx - C_6 e^{bx}\cos ax + \\ &+ C_5 e^{-dx}\sin ax + C_8 e^{dx}\cos cx - C_7 e^{bx}\sin cx]\cos\beta y\, d\beta. \end{aligned} \tag{12.4}$$

In order to solve the given problem we have to find the value of the eight constants from the boundary conditions. For the free edge we have

$$x = 0, \qquad M_{xx} = 0, \qquad N_{xx} = 0;$$

$$\bar{N}_{xy} = N_{xy} + \frac{M_{xy}}{R_2} = 0, \qquad Q_x + \frac{\partial M_{xy}}{\partial y} = P\delta(y).$$

With increasing distance from the loaded edge the displacements and the stresses should decrease to zero. Then we have $x \to \infty$; $w \to 0$, $\Phi \to 0$. This condition is satisfied if we neglect the expressions containing a real positive part of the power $e^{\alpha x}$. In this way four constants disappear from Eq. (12.4). For the free edge we find the following conditions; $x = 0$:

$$
\begin{aligned}
M_{xx} &= D\left[\frac{\partial^2 w}{\partial x^2}+\nu\frac{\partial^2 w}{\partial y^2}+k_R(\lambda_R^2+\nu)w\right]=0,\\
N_{xx} &= -\left[\frac{\partial^2 \Phi}{\partial y^2}+\lambda_R k_R \Phi-\lambda_R\frac{D}{R_2}\Delta w\right]=0,\\
\bar{Q}_x &= -\frac{D}{l^3}\left\{\frac{\partial^3 w}{\partial x^3}+(2-\nu)\frac{\partial^3 w}{\partial x\,\partial y^2}+k_R[1+\lambda_R^2]w\right\}_{x=0}=P\delta(y),\\
\bar{N}_{xy} &= \frac{\partial^2 \Phi}{\partial x\,\partial y}+(1-\nu)\frac{D}{R_2}\frac{\partial^2 w}{\partial x\,\partial y}=0.
\end{aligned}
\tag{12.5}
$$

As first of all we are interested in the determination of the state of stress in the small area near the point of application of the load, we can neglect the coefficients k_R, μ_R, etc., in solving the problem by means of shallow shell equations. On substituting in the simplified Eqs. (12.5) the functions w and Φ, from Eqs. (12.4) we obtain the following set of algebraic equations:

$$
\begin{matrix} N_{xx}\\ \bar{N}_{xy}\\ M_{xx}\\ \bar{Q}_x \end{matrix}
\quad
\begin{bmatrix} a_{11} & a_{12} & a_{13} & a_{14}\\ a_{21} & a_{22} & a_{23} & a_{24}\\ a_{31} & a_{32} & a_{33} & a_{34}\\ a_{41} & a_{42} & a_{43} & a_{44} \end{bmatrix}
\cdot
\begin{bmatrix} C_1\\ C_2\\ C_3\\ C_4 \end{bmatrix}
=
\begin{bmatrix} 0\\ 0\\ 0\\ P_0 \end{bmatrix},
\tag{12.6}
$$

where

$$
\begin{aligned}
P_0 &= Pl^2/\pi D,\\
a_{11} &= 0, \quad a_{12}=1, \quad a_{13}=0, \quad a_{14}=-1,\\
a_{21} &= a, \quad a_{22}=b, \quad a_{23}=-c, \quad a_{24}=-d,\\
a_{31} &= -\beta^2(1-\nu)+\eta_1-\nu,\\
a_{32} &= \eta_2-\tfrac{1}{2},\\
a_{33} &= -\beta^2(1-\nu)-\eta_1-\nu;\\
a_{34} &= \eta_2+\tfrac{1}{2},\\
a_{41} &= -\beta^2(1-\nu)b-\eta_1 b-(\eta_2-\tfrac{1}{2})a,\\
a_{42} &= \beta^2(1-\nu)a+\eta_1 a-(\eta_2-\tfrac{1}{2})b,\\
a_{43} &= -\beta^2(1-\nu)d+\eta_1 d-(\eta_2+\tfrac{1}{2})c,\\
a_{44} &= \beta^2(1-\nu)c-\eta_1 c-(\eta_2+\tfrac{1}{2})d.
\end{aligned}
\tag{12.7}
$$

The coefficients a, b, c, d, η_1, η_2 result from the characteristic equation (9.5) and are given by Eqs. (9.6). Solving Eqs. (12.6), we find the following magnitudes of the constants C_i:

$$C_1 = \{[\beta^2(1-\nu)+\eta_1](b-d)-2c\eta_2\}\frac{Pl^2}{D(\beta)\pi D},$$

$$C_2 = C_4 = -[\beta^2(1-\nu)(a+c)-\eta_1(a-c)]\frac{Pl^2}{D(\beta)\pi D}, \tag{12.8}$$

$$C_3 = -\{[\beta^2(1-\nu)-\eta_1](b-d)+2a\eta_2\}\frac{Pl^2}{D(\beta)\pi D};$$

$D(\beta)$ is here the determinant of the matrix of the coefficients a_{ij} and has the form

$$\begin{aligned} D(\beta) = {} & \{[\beta^2(1-\nu)+\eta_1]b+(\eta_2-\tfrac{1}{2})a\}\{2c\eta_1- \\ & -[\beta^2(1-\nu)+\eta_1](b-d)\}+\{(\beta^2(1-\nu)(a+c)- \\ & -\eta_1(a-c)-\eta_2(b+d)+\tfrac{1}{2}(b-d)\}\{\beta^2(1-\nu)(a+c)+ \\ & +\eta_1(a-c)\}+\{[\beta^2(1-\nu)-\eta_1]d+(\eta_2+\tfrac{1}{2})c\}\{2a\eta_2+ \\ & +[\beta^2(1-\nu)-\eta_1](b-d)\}. \end{aligned} \tag{12.9}$$

As we know, the coefficients χ_R, μ_R, φ_R are multiplied by the ratio l^2/R_2^2 whose magnitude is of order h/R_2 and are, therefore, very small. But we can easily be convinced that when $\beta = 0$ the coefficients C_i (12.8) become indeterminate as 0/0. In this situation even very small magnitudes can considerably affect the integration for small β, and the same is true for the value of the entire integral. However, as the integrals for the internal forces converge more slowly than those for the deflection, these coefficients have a smaller influence on the internal forces than on the deflection. Holand [10.21] proved that in the case of a concentrated load acting on a cylindrical shell the internal forces can be evaluated with an accuracy of one per cent from the Donnell equation, i.e. by neglecting the coefficients k_R, χ_R, μ_R, φ_R. Only the displacement w, which converges rapidly, should be found using them. However, this treatment is not worth doing as the displacement w cannot be evaluated exactly by means of the integral (12.4).

The integral (12.4) for the displacement was evaluated numerically for a few values of the parameter λ_R (Fig. 133). From this computation the shell loaded at its free edge deflects more and the deflected area is greater than in the case of the shell loaded at some distance from the edge. For example, when the deflection of an infinitely long cylindrical shell calculated by means of the integral (9.1) with $R_2/h = 100$ is $0.709 \cdot \frac{PR}{Eh^2}\sqrt{\frac{R_2}{h}}$ whereas the deflection of the semi-infinite shell loaded at its free edge, calculated by means of the integral (12.4) is $5.03 \cdot \dfrac{PR}{Eh^2}\sqrt{\dfrac{R^2}{h}}$, i.e. about seven times more. For the spherical shell the difference is even larger. Also the deflected area increases, for a cylinder the deflections vanish practically only at a distance of about several hundred l measured in the x-axis direction.

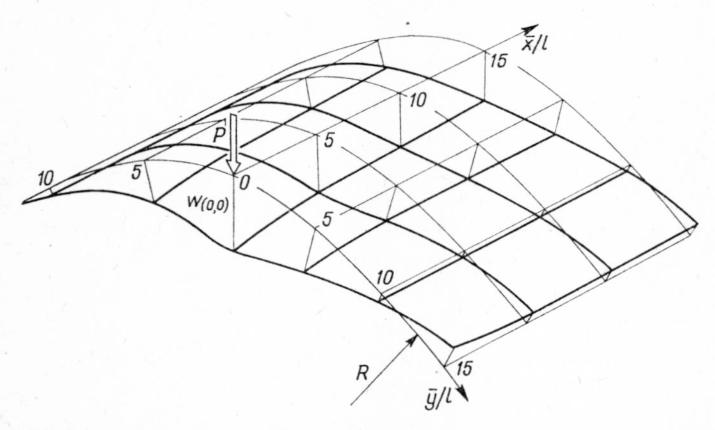

Fig. 133. Deflection of a cylindrical shell loaded at its free edge by a concentrated shear force.

This phenomenon affects the accuracy of the results obtained from integrals (12.4) by neglecting some boundary conditions. Moreover, the deflection of the shell of double curvature is calculated on the assumption that the radii of curvature are constant. Of course, this assumption is correct only in the vicinity of the point considered. At a greater distance the constant value of one of the radii of curvature requires the change of the other one. The above leads to certain errors if the reasons for the deflection are also the strains and displacements appearing at the distant points.

However, the internal forces decrease more rapidly with increasing distance from the point of application of the load and depend to a smaller degree on the boundary conditions. Therefore the magnitude of the stresses obtained by means of the double integrals will be more exact than the value of the deflection. Calculating the bending moment M_{xx}, we obtain the following result:

$$M_{xx} = \frac{D}{l^2}\int_0^\infty [e^{-bx}\{C_1[-\beta^2(1-\nu)+\eta_1]+C_2(\eta_2-\tfrac{1}{2})\}\cos ax -$$

$$-\mathrm{e}^{-bx}\{C_1(\eta_2-\tfrac{1}{2})-C_2[-\beta^2(1-\nu)+\eta_1]\}\sin ax+$$
$$+\mathrm{e}^{-dx}\{C_3[-\beta^2(1-\nu)-\eta_1]+C_4(\eta_2+\tfrac{1}{2})\}\cos cx-$$
$$-\mathrm{e}^{-dx}\{C_3(\eta_2+\tfrac{1}{2})-C_4[-\beta^2(1-\nu)-\eta_1]\}\sin cx]\cos\beta y\,\mathrm{d}\beta. \qquad (12.10)$$

The moment M_{yy} can be obtained by means of Eq. (9.8). The moment M_{xy} is

$$M_{xy} = (1-\nu)\frac{D}{l^2}\int_0^\infty [(-C_1 b+C_2 d)\mathrm{e}^{-bx}\cos ax-$$
$$-(C_1 a+C_2 d)\mathrm{e}^{-bx}\sin ax+(-C_3 d+C_4 c)\mathrm{e}^{-dx}\cos cx-$$
$$-(C_3 c+C_4 d)\mathrm{e}^{-dx}\sin cx]\beta\sin\beta x\,\mathrm{d}\beta. \qquad (12.11)$$

The membrane forces are as follows

$$N_{xx} = \frac{DR_2}{l^3}\int_0^\infty [-C_1\mathrm{e}^{-bx}\sin ax+C_2 3^{-bx}\cos ax+$$
$$+C_3\mathrm{e}^{-dx}\sin cx-C_4\mathrm{e}^{-dx}\cos cx]\beta^2\cos\beta y\,\mathrm{d}\beta,$$

$$N_{yy} = \frac{DR_2}{l^3}\int_0^\infty \{[-C_1(\eta_2-\tfrac{1}{2})+C_2(-\beta^2+\eta_1)]\mathrm{e}^{-bx}\cos ax-$$
$$-[C_1(-\beta^2+\eta_1)+C_2(\eta_2-\tfrac{1}{2})]\mathrm{e}^{-bx}\sin ax+$$
$$+[C_3(\eta_2+\tfrac{1}{2})-C_4(-\beta^2-\eta_2)]\mathrm{e}^{-dx}\cos cx+ \qquad (12.12)$$
$$+[C_3(-\beta^2-\eta_2)+C_4(\eta_2+\tfrac{1}{2})]\mathrm{e}^{-dx}\sin cx\}\cos\beta y\,\mathrm{d}\beta,$$

$$N_{xy} = \frac{DR_2}{l^3}\int_0^\infty [(C_1 a+C_2 b)\mathrm{e}^{-bx}\cos ax-$$
$$-(C_1 b-C_2 a)\mathrm{e}^{-bx}\sin ax-(C_3 c+C_4 d)\mathrm{e}^{-dx}\cos cx+$$
$$+(C_3 d-C_4 c)\mathrm{e}^{-dx}\sin cx]\beta\sin\beta\,\mathrm{d}\beta.$$

The above integrals can be evaluated only by numerical methods. This computation is, however, difficult, because of the complicated form of the integrands and their slow convergence. The integrands for the bending moments converge especially slowly and require many computations. Near the edge the numerical integration does not give any result and even long computations with the use of digital computers are not successful. Therefore in this case the computations must be made in two stages: numerical in the portion 0–B and analytical from B–∞, using the simplifications proposed previously. First, we calculate the integrals in the portion B–∞ corresponding to the local load.

12.1.2 *Local load*

If we substitute approximate values for the roots of the characteristic equation (9.33) in Eqs. (12.8) and then develop them in power series with respect to $1/\beta$, taking into consideration only the two first terms, we obtain the following simplified formulae:

$$\begin{aligned} D_{(\beta)} &= (1-\lambda_R)(3+\nu)(1-\nu)\beta^4, \\ C_1 &= -[2\beta^2(1-\nu)k_1+8k_1^2\beta]\frac{P_0}{D(\beta)}, \\ C_2 = C_4 &= -[2\beta^2(1-\nu)k_1-\tfrac{1}{2}k_1(1+\lambda_R)]\frac{P_0}{D(\beta)}, \\ C_3 &= -[-2\beta^2(1-\nu)k_1+8k_1^2\beta]\frac{P_0}{D(\beta)}. \end{aligned} \tag{12.13}$$

The above formulae are accurate enough when $\beta > \beta \lim = B$. The differences between the exact and simplified values of $D(\beta)$ for $\lambda_R = 0$ and $\nu = 0.3$ are indicated in Table 8.

Table 8.

β	$D(\beta)$ exact	$D(\beta)$ simplified	error %
0.655	0.4967	0.4251	14.4
1.53	13.064	12.658	3.1
3.73	452.36	447.15	1.5

We observe that the denominator $D(\beta)$ is identical with that corresponding to the plate on an elastic foundation. On substitution of approximate values of the coefficients C_i we find the following bending moments:

$$M_{xx}^B = \frac{P}{\pi}\frac{1-\nu}{3+\nu} S_1(k_1 x) \int_B^\infty e^{-\beta x} \cos \beta y \, d\beta,$$

$$M_{yy}^B = -\frac{P}{\pi}\frac{1-\nu}{3+\nu} \int_B^\infty \left[2\frac{1+\nu}{1-\nu} C_1(k_1 x) + S_1(k_1 x)\beta \right] \frac{e^{-\beta x}}{\beta} \cos \beta y \, d\beta, \tag{12.14}$$

$$M_{xy}^B = \frac{1-\nu}{3+\nu}\frac{P}{\pi} \int_B^\infty \left[\frac{1+\nu}{1-\nu} C_1(k_1 x) + S_1(k_1 x)\beta \right] \frac{e^{-\beta x}}{\beta} \sin \beta y \, d\beta.$$

It may be noticed that the above formulae are similar to those for the bending moments in the plate loaded by the force P at its free edge. The expressions for the plate can be obtained from those for the shell by substituting $R_1 = R_2 = \infty$ (then $k_1 = 0$) and decreasing the lower limit of integration to zero ($B \to 0$).

For $k_1 = 0$,

$$S_1(k_1 x) = x, \qquad C_1(k_1 x) = 1.$$

and we derive the formulae identical with those obtained previously in Section 12.5.

Integrals (12.14) can easily be evaluated by the use of Table 16 given at the end of this book. We obtain then the following bending moments for $r \ll 1$:

$$M_{xx}^B = \frac{1-\nu}{3+\nu}\frac{P}{\pi} S_1(k_1 x) \frac{e^{-Bx}}{r} \cos(By+\vartheta),$$

$$M_{xx}^B = -\frac{1-\nu}{3+\nu}\frac{P}{\pi} \left[-2\frac{1+\nu}{1-\nu} C_1(k_1 x) (\ln r + \gamma_0 + \ln B) + \right.$$
$$\left. + S_1(k_1 x) \frac{e^{-Bx}}{r} \cos(By+\vartheta) \right], \tag{12.15}$$

$$M_{xy}^B = \frac{1-\nu}{3+\nu}\frac{P}{\pi} \left[\frac{1+\nu}{1-\nu} C_1(k_1 x)\vartheta + S_1(k_1 x) \frac{e^{-Bx}}{r} \sin(By+\vartheta) \right],$$

For large magnitudes of r the complete expression for the integrals (12.14) should be used.

The membrane forces in the shell produced by the self-equilibrating local load are

$$\left.\begin{matrix} N^B_{xx} \\ N^B_{yy} \end{matrix}\right\} = -\frac{1}{(3+\nu)(1-\nu)}\frac{PR_2}{\pi l^4}\int\limits_B^\infty \Bigg[(1-\nu)S_2(k_1 x)-$$

$$-(1+\nu\pm(1-\nu))S_3(k_1 x)\frac{1}{\beta}\Bigg]e^{-\beta x}\cos\beta y\,d\beta,$$

$$N^B_{xy} = -\frac{1}{(3+\nu)(1-\nu)}\frac{PR_2}{\pi l^4}\int\limits_B^\infty \Bigg[(1-\nu)S_2(k_1 x)- \tag{12.16}$$

$$-(1-\nu)S_3(k_1 x)\frac{1}{\beta}\Bigg]e^{-\beta x}\sin\beta y\,d\beta.$$

The integrals for normal forces can be evaluated in a similar way. For the force N_{xx} we have the expression for $r \ll 1$

$$N^B_{xy} = -\frac{1}{(3+\nu)(1-\nu)}\frac{PR_2}{\pi l^4}\Bigg[(1-\nu)S_2(k_1 x)\frac{e^{-Bx}}{r}\cos(By+\vartheta)-$$

$$-2S_3(k_1 x)(\ln r+\gamma_0+\ln B)\Bigg]. \tag{12.17}$$

Similar expressions exist for the forces N_{xy} and N_{yy}. We see that for the point $r = 0$ only the moments have singularities of the type $1/r$ or $\ln r$. The membrane forces have finite values at this point.

For a very small x, the moments can be expressed by means of the simplified asymptotic expressions:

$$M_{xx} = \frac{1-\nu}{3+\nu}\frac{P}{\pi}\frac{x^2}{r^2},$$

$$M_{yy} = -\frac{1-\nu}{3+\nu}\frac{P}{\pi}\left[-\frac{2(1+\nu)}{1-\nu}\ln r+\frac{x^2}{r^2}\right]+\text{const}, \tag{12.18}$$

$$M_{xy} = \frac{1-\nu}{3+\nu}\frac{P}{\pi}\left[\frac{1+\nu}{1-\nu}\vartheta+\frac{xy}{r^2}\right],$$

which are identical with those obtained previously for a plate (4.9), (4.12).

At the point $x = 0$, the forces N_{xx}, N_{yy}, and N_{xy} resulting from formulae (12.16) are equal to zero and only give small corrections for the area near the loading point. These expressions give much smaller values, caused by the more rapid convergence of the integrals for the membrane forces than for the moments.

12.1.3 *Solution by means of Fourier series*

As we found before, the displacement defined by the integral (11.4) includes the effect of a free edge but does not include the effect of the boundary conditions of the remaining edges of the shell. In the case of a tube of closed cross-section the continuity of the displacements of points of the circular cross-section is not ensured. Therefore the displacement (12.4) can be treated only as approximate furnishing some information about the shape of the shell near the loading point. In order to calculate the exact value of the displacement all boundary conditions of the shell should be considered. For example, while investigating the closed cylindrical shell, Fourier series should be used. Then the condition of continuity of the circular cross-section is identically fulfilled. These series converge quickly for the displacements.

The respective formulae can easily be obtained from the previous results (Section 4.1). The sign of the integral should be replaced by the sign of the sum and the variable βy by the variable $(n\varphi)$. Since now $l = R_2$, $4\varkappa^4 \neq 1$; the roots a, b, c, d, and the coefficients η_1, η_2 should be calculated here from the characteristic equation (10.3).

P. Seide [12.3], taking as a starting point Donnell's equations, calculated in this way the displacement of a semi-infinite cylinder loaded at its free edge by a system of concentrated and distributed radially forces (Fig. 134). From this calculation it follows that the radial deflection at the edge $x = 0$ is given by

$$w_{(x=0)} = \left[1+16\sum_{n=1}^{\infty}\frac{[(\psi_n^2+1)^{1/2}+\psi_n]^{1/2}}{\psi_n \varDelta_n}\cos mn\varphi\right] w_p,$$

where

$$\begin{aligned} w_p &= \sqrt[4]{3(1-\nu^2)}\,\frac{m}{\pi}\frac{PR}{Eh^2}\sqrt{\frac{h}{R}}, \\ \psi_n &= 4n^2\frac{m^2h}{\sqrt{12(1-\gamma^2)}\,R}, \\ \varDelta_n &= (1+\nu)^2\psi_n+2[(1+\psi_n^2)^{1/2}-\psi_n], \end{aligned} \tag{12.19}$$

and m is the number of equally spaced concentrated forces.

The slope $\mathrm{d}w/\mathrm{d}x$ at the edge is given by

$$\left.\frac{\mathrm{d}w}{\mathrm{d}x}\right|_{(x=0)} = -\left[1+8(1+\nu)\sum_{n=1}^{\infty}\frac{1}{\varDelta_n}\cos nm\varphi\right]\frac{3(1-\nu^2)}{\pi}\frac{mP}{Eh^2}.$$

If the radial forces are uniformly distributed symmetrically over angular segments of $\pm\ \varepsilon/2$ (Fig. 134) the above result can easily be modified by multiplying the terms of the series by the quantity

$$\sin\left(n\pi\frac{m\varepsilon}{2\pi}\right)\Big/ n\pi\frac{m\varepsilon}{2\pi}.$$

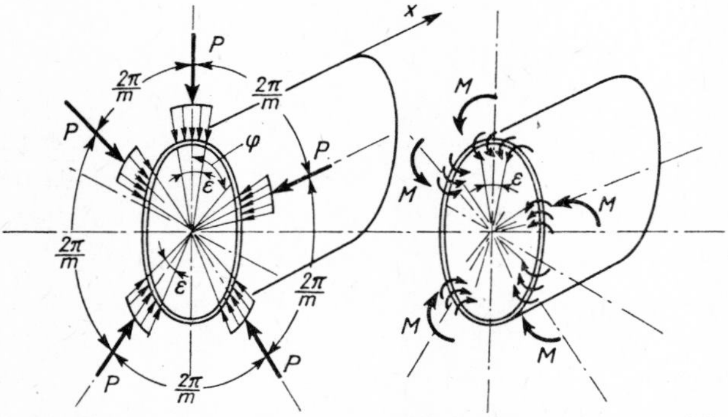

Fig. 134. Cylindrical shell subjected at its free edge to a system of forces and moments distributed along angular segment εR (from P. Seide [12.3]).

Then the deflection at the edge takes the form

$$w = \left\{1+16\sum\frac{[(\psi_n^2+1)^{1/2}+\psi_n]^{1/2}\sin\left(n\pi\frac{m\varepsilon}{2\pi}\right)}{\psi_n\Delta_n n\pi\frac{m\varepsilon}{2\pi}}\cos nm\varphi\right\} w_p. \qquad (12.20)$$

The results of computation of the sum of the above series are shown in Fig. 135. We see that the diagram for the concentrated force is approximately a straight line within a large range of the parameter $m^2h/R\sqrt{12(1-\nu^2)} < 2\cdot 10^{-1}$. This dependence enables us to write the following approximate formula for the displacement of the cylindrical shell at the loading point.

$$\frac{Eh^2}{PR}\sqrt{\frac{h}{R}}\, w_{(0,0)} = \frac{2.16}{m}[3(1-\nu^2)]^{3/4}. \qquad (12.21)$$

For $R/h = 100$, $\nu = 0.3$, $m = 2$ we obtain the displacement $\sim 300\ w_p$ which corresponds to

$$w_{(0,0)} = 2.29\frac{PR}{Eh^2}\sqrt{\frac{R}{h}}.$$

This value differs sharply from the displacement calculated by means of the integral (12.4).

The membrane stresses calculated by means of the division of the load into a local and general parts are here identical with those obtained by calculation by means of Fourier series for $n = 80$. The first eight terms of the development into series have been regarded as the result of the general load. Figure 136 shows the distribution of the membrane and bending stresses in

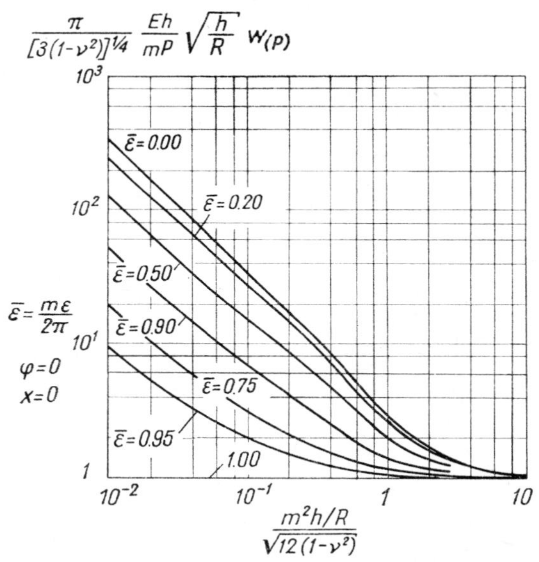

Fig. 135. Deflection of a cylindrical shell subjected to a system of distributed forces at the centre of the loaded segments ([12.3]).

the cylindrical shell in the direction of x-axis. The bending stresses were calculated by the method of the division of the load. The first eight terms of the Fourier series were also taken as a general part. The diagram shows the resultant stresses. We see that the largest stresses are caused by the bending moment M_{yy}. These stresses vanish rather slowly. Thus, the shell works essentially as a ring loaded in its plane.

12.2 Shell loaded by a force normal to the edge and tangential to the middle surface

The next problem is to define the stresses in a shell loaded by a force N normal to the edge of the shell and tangential to the x-axis (Fig. 137). We solve

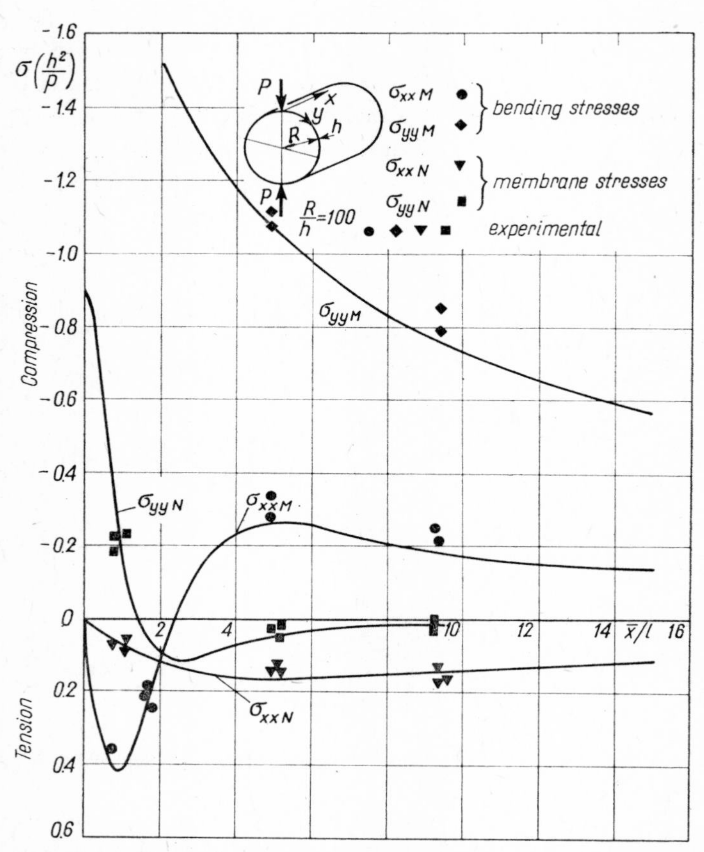

Fig. 136. Stresses in a cylindrical shell loaded at its free edge by two concentrated shear forces.

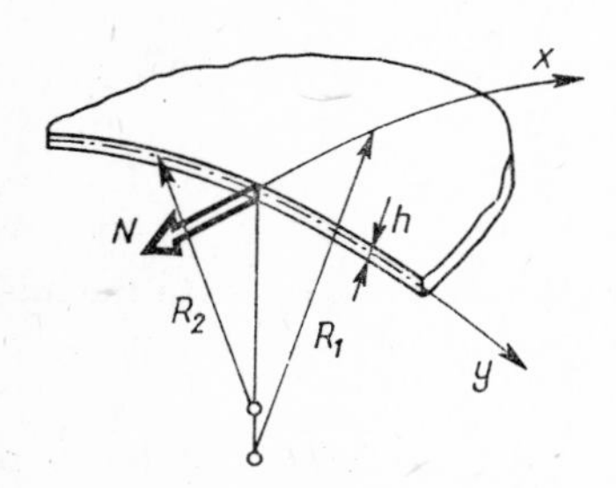

Fig. 137. Shell subjected at the free edge by a concentrated normal force.

this problem in the same way as previously. The boundary conditions to be fulfilled are the following: at the free adge:

$$x = 0, \quad M_{xx} = 0, \quad N_{xx} = N\delta(y), \quad \overline{N}_{xy} = N_{xy} + \frac{M_{xy}}{R_2} = 0,$$

$$\overline{Q}_x = Q_x + \frac{\partial M_{xy}}{\partial y} = 0$$

and the condition for $x \to \infty$

$$\Phi \to 0 \quad \text{and} \quad w \to 0.$$

We obtain then a set of equations identical with those found previously, only the right-hand side has a different form. We have

$$[a_{ij}] \times \begin{bmatrix} C_1 \\ C_2 \\ C_3 \\ C_4 \end{bmatrix} = \begin{bmatrix} N_0 \\ 0 \\ 0 \\ 0 \end{bmatrix}, \tag{12.22}$$

where

$$N_0 = \frac{Nl}{\pi D R_2} \frac{1}{\beta^2}.$$

After solving we have

$$C_1 = -\frac{N_0}{D(\beta)} [b_4(c_3 d - c d_3) + c_4(d b_3 + b d_3) + d_4(c b_3 + b c_3)],$$

$$C_3 = \frac{N_0}{D(\beta)} [a_4(d b_3 + b d_3) + b_4(a_3 d + d_3 a) + d_4(a_3 b - b_3 a)], \tag{12.23}$$

$$C_4 = -\frac{N_0}{D(\beta)} [a_4(b_3 c + b c_3) + b_4(c_3 a + c a_3) + c_4(b_3 a - b a_3)],$$

$$C_2 = N_0 + C_4, \quad C_5 = C_6 = C_7 = C_8 = 0.$$

In case we know the value of constants C_1–C_8, the displacement and internal forces in the shell can be calculated from formulae (12.4). However, the analytical evaluation of these integrals is impossible. In order to accelerate the numerical calculations we divide the load into local and general parts. From Eq. (12.22) we see that the matrix of the coefficients with the unknown C_1–C_8 is here identical with that from the former problem. These equations differ from (12.6) only in the terms on the right-hand side. The value $D(\beta)$ is the determinant of the matrix of the coefficients corresponding to C_1–C_4 and is therefore the same now.

For the self-equilibrating local load we obtain the following simplified expressions by substituting in formulae (12.23) the simplified roots of the characteristic equation.

$$\begin{aligned} C_1 &= -\frac{N_0}{4}\frac{\beta}{k_1} = -C_3, \\ C_2 &= \frac{N_0}{4}\left[\frac{\beta}{k_1}+2\right], \\ C_4 &= \frac{N_0}{4}\left[\frac{\beta}{k_1}-2\right], \\ C_5 &= C_6 = C_7 = C_8 = 0. \end{aligned} \tag{12.24}$$

The internal forces in the shell can be obtained by substituting the above coefficients C_i into Eqs. (12.4). On differentiating we have

$$\begin{aligned} \left.\begin{matrix} N_{xx} \\ N_{yy} \end{matrix}\right\} &= \frac{N}{\pi l}\int_B^\infty \{\pm S_1(k_1 x)\beta + C_1(k_1 x)\}\mathrm{e}^{-\beta x}\cos\beta y\,\mathrm{d}\beta, \\ N_{xy}^B &= \frac{N}{\pi l}\int_B^\infty S_1(k_1 x)\beta\mathrm{e}^{-\beta x}\sin\beta y\,\mathrm{d}\beta. \end{aligned} \tag{12.25}$$

In the case where the shell becomes very flat, the radii of curvature tend simultaneously to infinity ($R_2/R_1 \to 1$) and the coefficient $k_1 = \sqrt{(1-\lambda_R)/8}$ tends to zero. Then the solution (12.25) becomes identical with that obtained in § 5.5.1 for the flat plate loaded by a concentrated force on its edge (Eqs. (5.33)).

For the local bending moments we have the following formulae:

$$\begin{aligned} \left.\begin{matrix} M_{xx}^B \\ M_{yy}^B \end{matrix}\right\} &= \frac{Nl}{\pi R_2}\int_B^\infty \{\pm(1-\nu)S_2(k_1 x)\beta + (1+\nu)S_3(k_1 x)\}\mathrm{e}^{-\beta x}\cos\beta y\,\mathrm{d}\beta, \\ M_{xy}^B &= (1-\nu)\frac{Nl}{\pi R_2}\int_B^\infty \left[S_2(k_1 x)\beta - \frac{\lambda_R}{4}S_3(k_1 x)\right]\mathrm{e}^{-\beta x}\sin\beta y\,\mathrm{d}\beta. \end{aligned} \tag{12.26}$$

When the radius of curvature of the shell R_2 tends to infinity, the shell becomes a plate and bending moments vanish. The coefficient $Nl/\pi R_2$ preceding the sign of the integrals is proportional to h/R_2 and when $R_2 \to \infty$, $M_{ij} \to 0$. The bending moments disappear also when the thickness of the shell tends to zero.

The functions $S_2(k_1 x)$, $S_3(k_1 x)$ occurring in formulae (12.26) are for small values of x of the order x^2. It happens that the bending caused by a force tangential to the shell does not increase infinitely at the loading point, but has a finite value in the whole area of the shell. Integrals (12.25) and (12.26) can easily be evaluated by the use of the table of integrals (p. 559). As in the area very near to the point of application of the load, the shell behaves like a flat plate and the asymptotic formulae for the singular forces are the same as for plates.

12.3 The shell loaded at its edge by a concentrated bending moment

This case can be treated in the same way as previous ones. The boundary conditions are the following:

At the free edge for $x = 0$

$$M_{xx} = M\delta(y), \qquad N_{xx} = 0,$$

$$N_{xy} + \frac{M_{xy}}{R_2} = 0,$$

$$Q_x + \frac{\partial M_{xy}}{\partial y} = 0,$$

and, for $x \to \infty$,

$$\Phi \to 0 \quad \text{and} \quad w \to 0.$$

The bending moment is expressed by the Fourier integral, then the set of equations for the boundary conditions is

$$[a_{ij}] \times \begin{bmatrix} C_1 \\ C_2 \\ C_3 \\ C_4 \end{bmatrix} = \begin{bmatrix} 0 \\ 0 \\ M_0 \\ 0 \end{bmatrix}, \tag{12.27}$$

where $M_0 = Ml/\pi D$. After solving we obtain the coefficients C_1–C_8 defined by the formulae

$$C_1 = \frac{M_0}{D(\beta)}[c(b_1+d_4)-c_4(b-d)], \qquad C_2 = C_4 = \frac{M_0}{D(\beta)}(ac_4+ca_4),$$

$$C_3 = \frac{M_0}{D(\beta)}[a(b_4+d_4)+a_4(b-d)], \qquad C_5 = C_6 = C_7 = C_8 = 0. \tag{12.28}$$

Knowing the constants of integration C_1–C_8, we can calculate the displacement and internal forces from formulae (12.4), (9.7). The calculation of these integrals is, however, complicated both by the slow convergence and complicated structure of the formulae. In order to permit effective numerical calculations we divide the load into local and general parts. For the local load we shall determine simplified expressions valid for $\beta > B_{\text{lim}}$.

Considering the roots of the characteristic equation according to Eqs. (9.33), we obtain on computation for a local load

$$C_1 = \frac{M_0}{D(\beta)}[2\beta^3(1-\nu)k_1-4\beta^2k_1^2(1+\nu)],$$

$$C_2 = \frac{M_0}{D(\beta)}[2\beta^3(1-\nu)k_1] = C_4, \tag{12.29}$$

$$C_3 = \frac{M_0}{D(\beta)}[-2\beta^3(1-\nu)k_1-4\beta^2k_1(1+\nu)],$$

$$C_5 = C_6 = C_7 = C_8 = 0.$$

The internal forces are as follows:

$$M_{xx}^B = \frac{M}{\pi l}\int_B^\infty \left[C_1(k_1x)-\frac{1-\nu}{3+\nu}S_1(k_1x)\beta\right]e^{-\beta x}\cos\beta y\,d\beta,$$

$$M_{yy}^B = \frac{M}{\pi l}\frac{1-\nu}{3+\nu}\int_B^\infty [C_1(k_1x)-S_1(k_1x)\beta]e^{-\beta x}\cos\beta y\,d\beta, \tag{12.30}$$

$$M_{xy}^B = \frac{M}{\pi l}\frac{1-\nu}{3+\nu}\int_B^\infty \left[\frac{2}{1-\nu}C_1(k_1x)-S_1(k_1x)\beta\right]e^{-\beta x}\sin\beta y\,d\beta.$$

The displacement of the cylindrical shell subjected to two concentrated bending moments was calculated numerically by replacing the integral (12.4) by a single Fourier series. This series was found to be well convergent. The results of the computation are given in Fig. 138. This figure shows the shape of the shell in the vicinity of the loading point. The displacement decreases quickly with increase of the distance from the point subjected to the moment both in the x and y-axis directions. The maximum displacement of the shell is equal to $\sim 2.9Ml/\pi D$ which was obtained by means of Eqs. (10.19) neglecting the effect of transverse shear deformations.

The displacement of the cylindrical shell subjected to the bending moments, both concentrated and distributed over equal and equally spaced

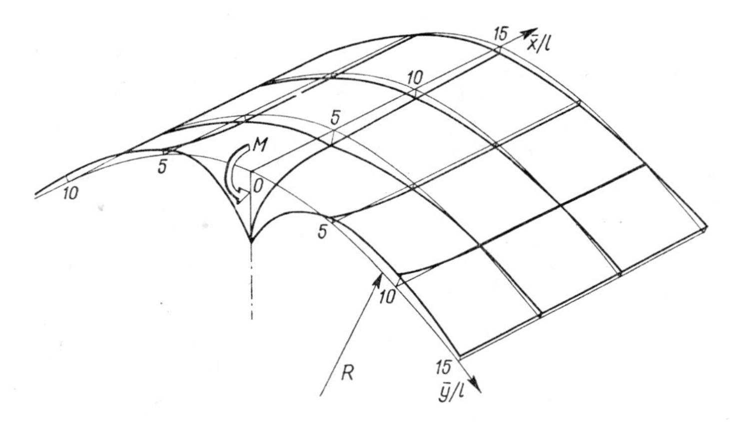

Fig. 138. Deflections of a shell subjected at the free edge to the concentrated bending moment.

angular segments, was calculated by P. Seide [12.3] by means of a single Fourier series. Taking Donnell's equation as the starting point, the following series was obtained for the maximum radial deflection at the origin:

$$w(M)_{(x=0)} = \left[1+8(1+\nu)\sum_{n=1}^{\infty}\frac{1}{\Delta_n}\cos nm\varphi\right]\frac{\sqrt{3(1-\nu^2)}}{\pi}\frac{mM}{Eh^2}, \tag{12.31}$$

where Δ_n is given by Eqs. (12.19). The slope $\mathrm{d}w/\mathrm{d}x$ at the origin is given by

$$\left.\frac{\mathrm{d}w(M)}{\mathrm{d}x}\right|_{x=0} = -\left[1+4\sum_{n=1}^{\infty}\frac{[(\psi_n^2+1)^{1/2}+\psi_n]^{1/2}}{\Delta_n}\cos nm\varphi\right]\times$$

$$\times[3(1-\nu^2)]^{3/4}\frac{2mM\sqrt{h}}{Eh^3\sqrt{R}};$$

for ψ_n see Eq. (12.19).

If the moments are distributed over angular segments $\pm\varepsilon$, the analysis is very easily modified by multiplying the terms of the series by the quantity

$$\sin\left(n\pi\frac{m\varepsilon}{2\pi}\right)\Big/ n\pi\frac{m\varepsilon}{2\pi}.$$

The results of the numerical computation of the series are given in Figs. 139, 140.

The membrane and bending stresses in the shell were calculated by the method of the division of the load into a local and general parts. The first 12

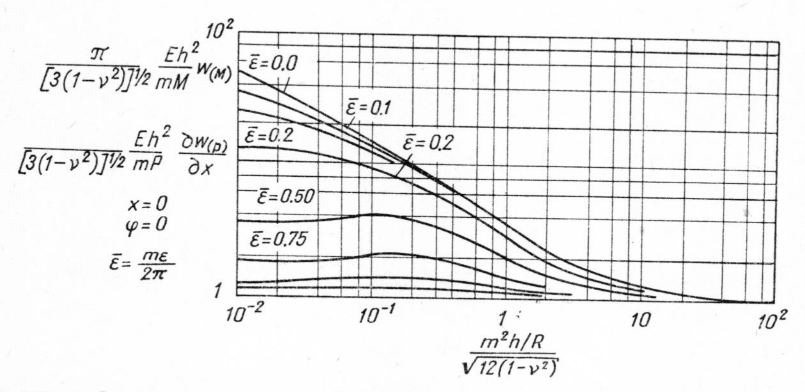

Fig. 139. Deflection of a cylindrical shell subjected at the free edge to a system of moments distributed along angular segments according to Fig. 134 ([12.3]).

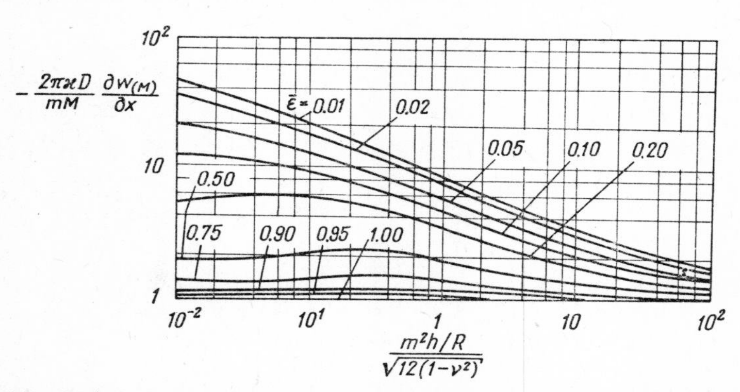

Fig. 140. Angle $\partial w/\partial x$ at the edge of the cylindrical shell subjected to a bending moment distributed along the angular segment εR ([12.3]).

terms of the development into Fourier series were adopted as a general part. The values calculated from formulae (12.30) for the local load were added to the results obtained. The results of the calculation are presented graphically in Fig. 141. It reveals that at the loading point both the internal bending moments in the shell are infinitely large, but their signs are different. Similar

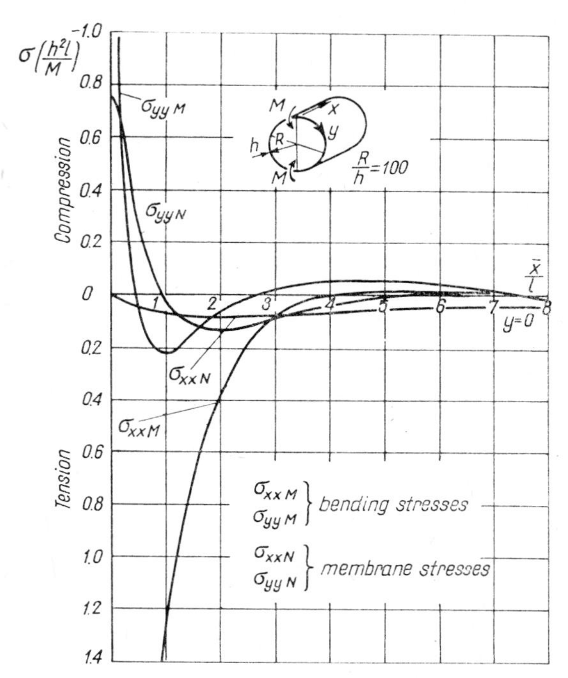

Fig. 141a. Membrane and bending stress variations in a cylindrical and spherical shell subjected at the free edge to concentrated bending moments.

results were obtained in § 4.2, where we considered the plate loaded by a concentrated moment at its edge. The singularities of the moments had the same character. The membrane stresses have at the loading point a finite value and vanish rather quickly with increase of the distance from this point. We see that already at the distance of the order of 6–7 l we have negligible stresses in comparison to those at the distance of about 1 l. The distribution of the stresses in the spherical shell is given in Fig. 141b.

The stresses in the shell loaded by the moment uniformly distributed along the edge can easily be obtained. It is adequate to multiply the integrands for the local load by $(\sin \beta c)/\beta c$ and next to evaluate them. We obtain then the expressions similar to those given in Section 4.2 for the plate.

Cylindrical shells subjected to concentrated loads at the free edge were considered by I. L. Sharinov [12.4, 12.8] and T. V. Galoian [12.11, 12.12]; A. S. Khristenko studied the same problem in the case of orthotropic circular cylindrical shells [12.6, 12.7, 12.9].

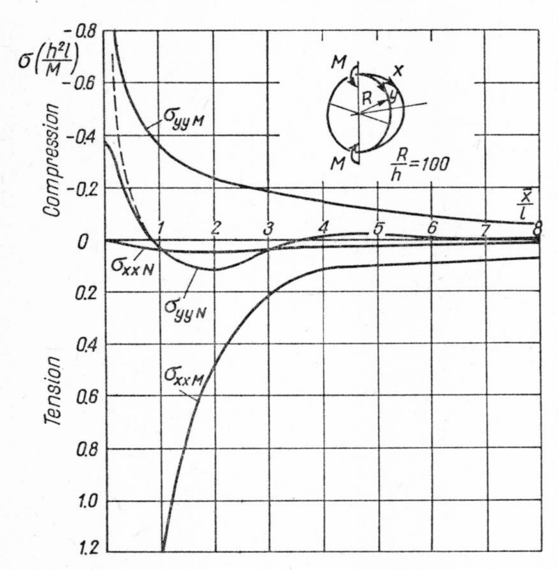

Fig. 141b. Membrane and bending stress variations in a cylindrical and spherical shells subjected at the free edge to concentrated bending moments.

References 12

[12.1] S. Łukasiewicz: On shell loaded on its free edge, *Arch. Bud. Masz. 16.3* (1969).

[12.2] J. G. Berry: On thin hemisherical shells subjected to concentrated edge moments and forces, *Proc. 3rd Midwest Conf. on Solid Mechanics.*

[12.3] P. Seide: On the bending of circular cylindrical shells by equal and equally spaced end radial shear forces and moments, *J. Appl. Mech. 28.1* (1961).

[12.4] I. L. Sharinov: The action of the concentrated load at free edge on cylindrical shell, *MTT 3* (1967) (in Russian).

[12.5] D. R. Westbrook: Applications of asymptotic integration to potential problems in shells. I. The flexure problem of St. Venant, *SIAM J. Appl. Math. 15.4* (1967).

[12.6] A. S. Khristenko: Action of the concentrated load applied at a free edge of orthotropic cylindrical shell, *Tr. Nikolaievskogo Korablestroit. In-ta 25* (1968) (in Russian).

[12.7] A. S. Khristenko: On the action of concentrated forces and moments applied at free edge on orthotropic circular cylindrical shell, *MTT 4* (1968) (in Russian).

[12.8] I. L. Sharinov: On the action of a concentrated edge load on a cylindrical shell, *MTT 4* (1968) (in Russian).

[12.9] A. S. Khristenko, T. A. Iurchenko: On the problem of evaluation of asymptotic formulae, *Tr. Nikolaevskogo Korablestroit. In-ta 32* (1969) (in Russian).

[12.10] A. Kildegaard: *Bending of a cylindrical shell subject to axial loading, theory of thin shells, Proceedings of second IUTAM Symposium on the theory of thin shells,* Copenhagen, 1967, Springer-Verlag, Berlin-Heidelberg-New York 1969.

[12.11] T. V. GALOIAN: State of stress of circular cylindrical shell loaded by edge radial forces, *Sb. Tr. Moskovsk. Inzh. Stroit. In-ta 84* (1970) (in Russian).

[12.12] T. V. GALOIAN: State of stress of circular cylindrical shell loaded by edge tangential forces, *Sb. Tr. Moskovsk. Inzh. Stroit. In-ta 84* (1970) (in Russian).

[12.13] I. M. PIROGOV: I. P. DIMITRIENKO, F. I. SELITSKII: Bending of orthotropic cantilever cylindrical shell by a shear force applied at free edge, *Sb. Tr. Vses. Zaochn. Politekhn. 59* (1970) (in Russian).

13

Large deflections of shells

13.1 Spherical shell

Experiments with the spherical shell loaded by a central concentrated normal force proved that the deflections calculated from (8.8) are compatible with those found in experiments only for small initial values of the force. A *circular dimple*, i.e. a region of reversed curvature, appears in the shell and spreads outwards as the load increases. This particular non-linear behaviour of the shell aroused the interest of many scientists. C. B. Biezeno [13.1] (1935) solved this problem by reducing it to two simultaneous non-linear differential equations for two unknown functions: v, the displacement normal to the axis of symmetry, and $\varphi = \mathrm{d}w/\mathrm{d}r$, where w is the displacement in the direction of the axis. Assuming the function

$$\varphi = C_1 \frac{r}{R} + C_2 \frac{r}{R} \ln \frac{b}{r} \tag{13.1}$$

similar to that obtained for the flat plate, the first equation was solved for v which together with φ was substituted into the second equation. As a result the relationship between the applied load and the deflection of the shell w was obtained. (See Fig. 142.)

Chien [13.2] treated the case of a spherical cup with a line load uniformly distributed along a circle concentric with the apex and of radius αc. Putting $\alpha = 0$ gives the case of a concentrated load. Also two simultaneous non-linear differential equations were found for w and N_{rr}. To obtain the solution the following form was assumed for w:

$$w = w_0 \left(1 - \frac{r^2}{b^2}\right)\left(1 - \frac{1+\nu}{5+\nu}\frac{r^2}{b^2}\right), \tag{13.2}$$

where w_0 is the deflection at the apex. From one of the equations the corresponding membrane stresses were obtained. Calculating the strain energy U of the shell and putting $dU/dw_0 = 0$, an equation relating r, α, and w_0 was

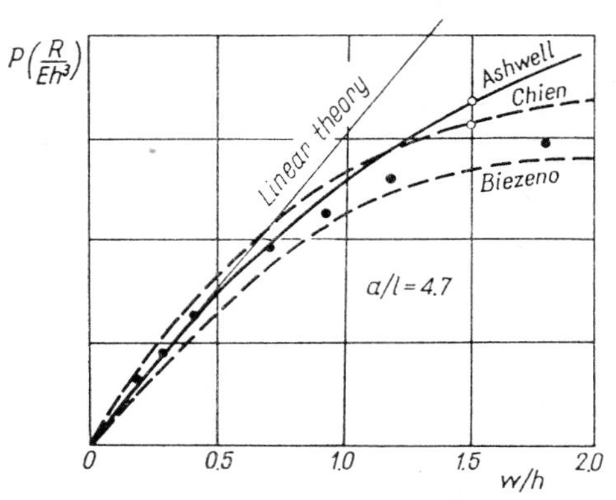

Fig. 142. Load-deflection curves for a spherical shell loaded by a concentrated radial force. Points from Ashwell's experiment.

found. The results of both calculations are presented graphically in Fig. 142. The dotted lines are those given by Biezeno [13.1] and Chien [13.2] for $b/l = 4.7$.

The same problem was considered in work [13.5] by D. G. Ashwell who solved the non-linear problem of large displacement of a shell by means of the combination of two linear solutions. Namely, observing the behaviour of the shell during deformation, he noticed that the dimple appearing on the shell surface has a spherical shape of negative reversed curvature in comparison with the initial shape (Fig. 143). Only on the boundary of the two regions does a larger deviation from the sphere appear. It can be assumed that the dimple inner region and outer region undergo small displacements from that of the spherical shell. Then for both regions linear equations can be adopted. In this way the large deflection and non-linear behaviour of the shell can be described in terms of two sets of equations of small deflections only. The compatibility and equilibrium conditions should be satisfied on the boundary between the two regions. The results of numerical calculations obtained by this method are presented graphically in Fig. 142. The full line is that predicted by D. G. Ashwell. The thin full lines result from the linear theory.

R. R. Archer [13.7] reduced the problem of the spherical shell loaded by a concentrated force to the solution of a set of three algebraic equations by use of the non-linear Reissner equations and the method of finite differences. A simplified method of calculation of large deflections proposed recently by Pogorelov [13.11] is also worth mentioning here. This method

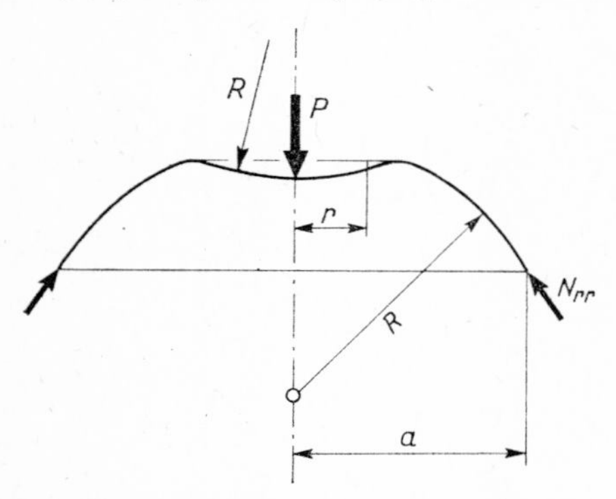

Fig. 143. Shape of a deflected spherical shell.

is similar to that of Ashwell, noting that the shell deflects elastically, and takes a form resembling one of the forms of its isometric transformation. This means that geometrical bending plays the most important role in deformation of the shell, while the strains in its middle surface are of smaller importance. Assuming that the shell is made of steel of modulus of elasticity $E = 2 \cdot 10^6$ KG/cm^2 and yield stress $\sigma = 4 \cdot 10^3$ KG/cm^2, the corresponding strain in the middle surface does not exceed $\varepsilon = \sigma/E = 2 \cdot 10^{-3}$ which is a sufficiently small value. The deformation of the shell by such a change of metric of middle surface does not differ much from the isometric transformation, thus the metric does not change at all. This means that it is possible to predict, to a certain degree, the shape of the deflected shell, by looking for it among the isometric transformations. Considering a spherical shell, we come to a conclusion that the simplest isometric transformation corresponds to a mirror-like reflection (Fig. 144). We can assume therefore that the deflection produced by a concentrated force has approximately the shape represented in this figure and consists of the growth of the dimple together with a ridge where the shell is rapidly bent.

The relation between the deflection and the load can be found by means of the calculus of variations. One of the fundamental variational principles is: *Under the action of the given load, the shell takes the form for which the vari-*

ation of the functional $W = U - L$ is equal to zero; U here being the total strain energy caused by the change of the shape of the shell and L is the work of the external load. Thus: $\delta(U-L) = 0$.

In our case it is easy to calculate the work of the external load, viz.:

$$L = P \cdot 2f \tag{13.3}$$

where $2f$ is the deflection of the shell.

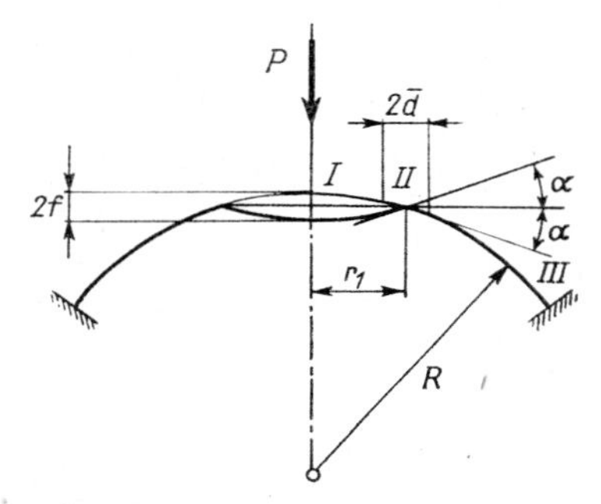

Fig. 144. Isometric transformation of spherical shell.

The calculation of the strain energy is more troublesome. Let us do it in approximation. Let us divide the area of the shell into three portions: I — central, II — area near the ridge, III — external area. The strain energy is defined by expressions (1.135) and (1.136). Let us calculate the energy separately in these three portions. Let us consider, at first, the external area. The changes of the main curvatures $\varkappa_1$, $\varkappa_2$ are here equal to zero, which yields the bending energy $U_b^{III} = 0$. The strain energy of the middle surface for this portion is also very small because of small strains, and can be neglected. As a result we can assume that the entire strain energy in this area is equal to zero.

In the central portion, I, the shape of the shell resembles the shape of the mirror-like reflection. The energy appears mainly as a result of changes of curvature:

$$\varkappa_1 = -\frac{2}{R}, \qquad \varkappa_2 = -\frac{2}{R}.$$

Then we have

$$U_b^I = \frac{D}{2} \iint\limits_{I} (\varkappa_1^2 + \varkappa_2^2 + 2\nu\varkappa_1\varkappa_2)\,dS = \frac{4D}{R^2}(1+\nu)S_1, \tag{13.4}$$

where S_1 is the surface of the area I. The strain energy in the middle surface is neglected here. Now we have to calculate the energy of the area II near the ridge. Both the bending stresses and the stresses in the middle surface are large in this area. Assuming that the deflected area near the ridge is relatively narrow, we have

$$\varepsilon_2 = \frac{u}{r_1},$$

u being the displacement in the radial direction, r_1 the radius of the circle ranging outward from the ridge, and ε_2 the circumference strain.

The strain ε_1 in the meridian direction is small and can be neglected. Thus we obtain the energy produced by the strains in the middle surface of this portion to be equal to

$$U = \frac{Eh}{2} \iint_{\text{II}} \left(\frac{u}{r_1}\right)^2 \mathrm{d}S. \tag{13.5}$$

We evaluate this integral over the entire area II. The bending energy in this area depends on the change of the shape of the middle surface. This energy will be calculated separately for the external and internal strip contiguous to the ridge.

By the assumption that the shell near the point of application of the load is a shallow shell, the change of the curvature for the external strip is

$$\varkappa_1 = \frac{\partial^2 w}{\partial r^2} = w''. \tag{13.6}$$

The change of the second principal curvature can be determined by means of Fig. 144 and is

$$\varkappa_2 = \frac{1}{R}\left(1 + \frac{w'}{\alpha}\right) - \frac{1}{R} = \frac{w'}{\alpha R}. \tag{13.7}$$

The positive values of w are here measured in the inward direction. Substituting the above into (13.4), we obtain the bending energy for the internal strip to be equal to

$$U_b^{\text{II}} = \pi r_1 D \int_0^{\bar{d}} \left[w''^2 + \left(\frac{w'}{\alpha R^2}\right)^2 + 2\nu w'' \frac{w'}{\alpha R} \right] \mathrm{d}s, \tag{13.8}$$

where $2\bar{d}$ is the width of the strip. The middle term in the above integrand is considerably smaller than the others and is neglected. The third expression can be integrated. As $w'(0) = -\alpha$, $w'(d) \simeq 0$ and we obtain the following formula:

$$U_b^{II} = \pi r_1 D \int_0^{\bar{d}} w''^2 ds - \pi r_1 D\nu \frac{\alpha}{R}. \tag{13.9}$$

The complete strain energy contained in the external strip, contiguous to the ridge, is

$$U_1^{II} = \pi r_1 D \int_0^{\bar{d}} w''^2 ds + \pi r_1 \frac{Eh}{2} \int_0^{\bar{d}} \frac{u^2}{r_1^2} ds - \pi r_1 D\nu \frac{\alpha}{R}. \tag{13.10}$$

For the internal strip we have

$$\varkappa_1 = -\frac{1}{R} + w'' - \frac{1}{R} = w'' - \frac{2}{R}, \qquad \varkappa_2 = -\frac{w'}{\alpha R} - \frac{2}{R}.$$

Calculating the energy of the internal strip in a similar way, we obtain

$$U_2^{II} = \pi r_1 D \int_0^{\bar{d}} w''^2 ds + \pi r_1 \frac{Eh}{2} \int_0^{\bar{d}} \frac{u^2}{r_1^2} ds - \frac{4(1+\nu)\pi r_1 D}{R} + \pi r_1 D\nu \frac{\alpha}{R}. \tag{13.11}$$

We observe that the calculation of the full strain energy of the shell would be possible if we know the functions u and w describing the deformation of the shell near the ridge. In order to determine these functions let us develop an additional equation for u and w resulting from the assumption that the strains in the generator direction are very close to zero. We can then write the equation

$$ds^2 = (dr + du)^2 + (dz + dw)^2. \tag{13.12}$$

Thus

$$dr\,du + dz\,dw + \frac{1}{2}(dw^2 + du^2) = 0.$$

When the width $2\bar{d}$ of the strip in which the deformations appear is small, this equation can be simplified and written as

$$u' + \alpha w' + \frac{1}{2} w'^2 = 0. \tag{13.13}$$

The functions u and w can be found by means of the calculus of variation, demanding the satisfaction of the condition of the minimum of the strain energy U_1^{II} and U_2^{II} in the area II contiguous to the ridge. As the expressions for the energy U_1^{II} and U_2^{II} differ from one another only by constant terms, this condition is reduced to finding the minimum of the functional J:

$$J = \int_0^{\bar{d}} \left[D(w'')^2 + \frac{Eh}{2} \frac{u^2}{r_1^2} \right] \mathrm{d}s \tag{13.14}$$

and satisfying simultaneously Eq. (13.13) and the boundary conditions.

Introducing new non-dimensional variables

$$\bar{u} = \frac{u}{dr_1 \alpha^2}, \quad \bar{w} = \frac{w'}{\alpha}, \quad \bar{s} = \frac{s}{r_1 d}; \quad d^4 = \frac{h^2}{12 r_1^2 \alpha^2},$$

we obtain

$$U^{\mathrm{II}} = \frac{\pi r_1^{1/2} E h^{5/2} \alpha^{5/2}}{12^{3/4}(1-\nu^2)} \cdot J, \tag{13.15}$$

where

$$J = \int_0^{d^*} [(\bar{w}''^2) + \bar{u}^2] \mathrm{d}s, \quad d^* = \frac{\bar{d} \cdot 12^{1/4} \alpha^{1/2}}{r_1^{1/2} h^{1/2}}.$$

The boundary conditions for the functions u and w are assumed to be the following: $w'(0) = -\alpha$, $u(0) = 0$, which by substituting non-dimensional variables yields $\bar{w}'(0) = -1$, $\bar{u}(0) = 0$, $\bar{w}'(d^*) = 0$, $\bar{u}(d^*) = 0$. These conditions mean that the displacements $\bar{u}$ and $\bar{w}$ are of a local character and vanish at a certain distance from the ridge.

Using the calculus of variations, we can find the minimum of the functional to be 1.15. If now J is substituted into the expression for the full strain energy of the shell, we find the following simple result:

$$U = 2\pi r_1 c \frac{E h^{5/2} \alpha^{5/2}}{r_1^{1/2}}, \tag{13.16}$$

where c is a constant equal to $c = J/12^{3/4}(1-\nu^2) \simeq 0.19$. For the shallow shell we have

$$\alpha = r_1/R, \quad 2f = \frac{r_1^2}{R}.$$

On substitution we have

$$U = 2\pi c E (2f)^{3/2} h^{5/2} \frac{1}{R}. \tag{13.17}$$

The functional W which is the subject of our investigation has the following simple form:

$$W = 2\pi c E(2f)^{3/2} h^{5/2} \frac{1}{R} - 2fP. \tag{13.18}$$

Equating the variation of the functional W with respect to the deflection f to zero, we have $\mathrm{d}W/\mathrm{d}f = 0$ which results in the following formula for the deflection of the shell.

$$2f = \frac{R^2 P^2}{9\pi^2 c^2 E^2 h^5}. \tag{13.19}$$

The dependence (13.19) between the deflection $2f$ and the load P is presented graphically in Fig. 145.

The results of the experiments by F. A. Penning [13.9] are also shown there. These experiments were performed with a number of shallow spherical shells with clamped edges made of aluminium and loaded at the apex through a small circular contact area. This revealed that the behaviour of the shell is depended on the ratio of the radius of the clamped edge b to the characteristic length l. Different deflection patterns were observed during the testing. The thicker shells suffered plastic yielding in the vicinity of the load and did not buckle. Thus, permanent deflections remained at the apex.

The thinner shells buckled and did not show any evidence of permanent deflections after having been loaded to their highest value of load. Deflections calculated from small-deflection theory showed good agreement for the initial shape of the load deflection curves.

The deflection given by formula (13.19) corresponds quite well to the curve resulting from experiments for small values of the load and for smaller magnitudes of b/l. When the shell is relatively thin (b/l large) and asymmetric buckling modes appear, Pogorelov's curve deviates considerably from experimental values and does not give the lowest buckling load.

The experiments proved that the deflection patterns are symmetrical only when the force is smaller than a certain critical value. After surpassing this value the shape of the deflection undergoes a change and its magnitude grows rapidly. The deformed area takes the form similar to the triangle, quadrilateral, etc. The problem of determination of the critical force can be solved by analysing the stability of the shape of the ridge appearing on the shell surface. The shape can be expressed in the form of the equation

$$r_1 = \varrho(1 + p\cos k\vartheta) \tag{13.20}$$

where p is a certain small parameter.

Finding the strain energy in the area limited by the curve (13.20) and the energy in the area near the ridge, the force corresponding to the stability of the assumed form of deflection can be obtained by means of the calculus of variations. For $k = 3$ we obtain the deflection in the form of a star with three arms. The critical force for this shape is

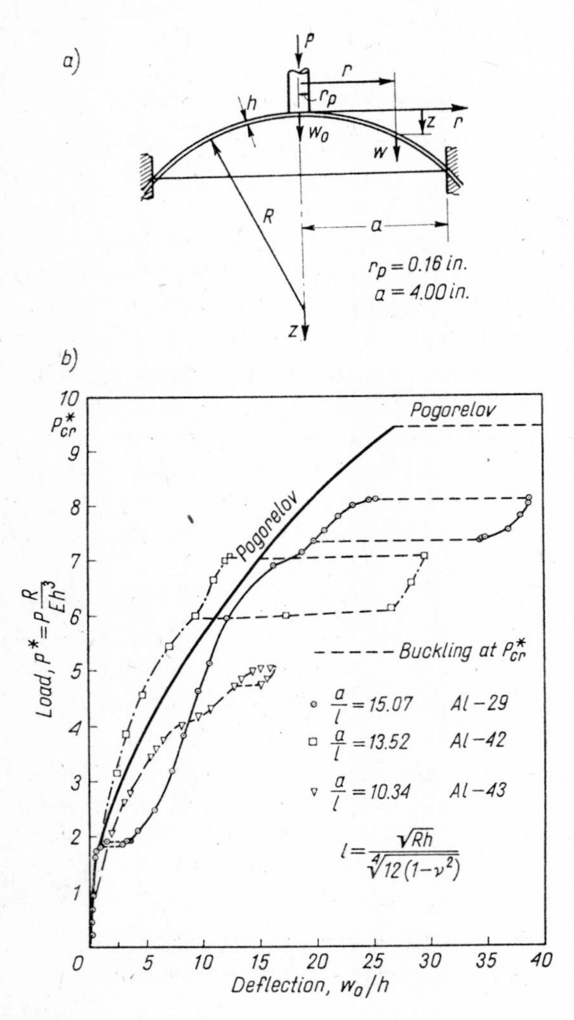

Fig. 145. Comparison of results by Pogorelov's theory with results by F. A. Penning's [13.9] *experiments.*

$$P_{cr} = 3\pi \frac{Eh^3}{R}. \tag{13.21}$$

In Fig. 145 this force is given by the horizontal straight line

$$\frac{PR}{Eh^3} = 3\pi.$$

I. G. Teregulov and L .I. Mogilevskii [13.12] also considered the non-linear effects of large deflections in the case of shallow spherical shells.

13.2 Arbitrary shell of positive double curvature

The non-linear displacements of a shell of positive double curvature ($R_1 \neq R_2$) can be calculated in a manner similar to that for the sphere. An elliptical dimple, which has the form of the original surface but of negative reversed curvature, can be observed on the shell surface loaded by a concentrated normal force. Pogorelov calculated the complete energy as being the effect of this dimple, adding the bending energy of the isometrically deformed area and the energy stored in the area along the ridge. On integration he obtained the following simplified expression for the energy U contained in the shell and being the result of the action of the concentrated normal force.

$$U = \pi c E h^{5/2} (2f)^{3/2} \left(\frac{1}{R_1} + \frac{1}{R_2} \right) \quad \text{where} \quad c = 0.19. \tag{13.22}$$

Calculating the work of the external force $L = P \cdot 2f$ and equating the variation of the functional $W = U - L$ to zero, we find the relation between the deflection of the shell and the load.

$$P = \frac{3\pi c}{2} E h^{5/2} \left(\frac{1}{R_1} + \frac{1}{R_2} \right) \sqrt{2f}. \tag{13.23}$$

The maximum stresses appear near the rib and are the effect of bending in the direction perpendicular to the rib. These stresses can be calculated approximately from

$$\sigma_b = \frac{Eh}{2} \max(w'') = 0.9E \left(\frac{2fh}{R_1 R_2} \right)^{1/2}. \tag{13.24}$$

If the concentrated force acts on the shell through the rigid insert, Pogorelov gives the following formula for the smallest critical buckling force of the shell:

$$P_{cr} = \frac{2Eh^2 S}{\sqrt{3(1-\nu^2)} R_1 R_2}; \tag{13.25}$$

S is here the surface of the ellipse generated on the contour of the insert. For the sphere loaded by a force through the circular insert of radius r_1, the surface is $S = \pi r_1^2$. If a concentrated bending moment acts through the rigid insert, Pogorelov (13.11) gives the following value:

$$M_{cr} = \frac{2Eh^2 Sa}{\sqrt{3(1-\nu^2)}\, R_1 R_2};$$

a is here half of an axis of the ellipse. For the circular insert the following condition is satified:

$$Sa = \pi r_1^3.$$

References 13

[13.1] C. B. BIEZENO: Über die Bestimmung der Durchschlagkraft einer schwachgekrummten kreisformigen Platte, *Z. Angew. Math. Mech. 15.10* (1935); C. B. Biezeno and R. Grammel, *Engineering Dynamics*, 484, London 1965; 2. Blackie, London, 484, 1956.

[13.2] CHIEN WEI-ZANG and HU HAI-CHANG: On the snapping of a thin spherical cap, *9th Int. Congr. Appl. Mech., 6*, University of Brussels, 309, 1957.

[13.3] R. M. EVAN-IVANOWSKI, H. S. CHENG and T. C. LOO: Experimental investigation of deformation and stability of spherical shells subjected to concentrated loads at the apex, *Proc. of the Fourth U. S. Nat. Congr. Appl. Mech. ASME* (1962) 563–575.

[13.4] W. C. VOSS, D. PEABODY, H. R. STALEY and A. G. H. DIETZ: Thin shallow domes loaded excentrically, *Proc. ASCE 73* (1947) 1173–1195.

[13.5] D. G. ASHWELL: On large deflection of a spherical shell with an inward point load, *Proc. of IUTAM Symp. Theory Thin Elastic Shells Delft, 1959, North Holland Publish., Amsterdam 43–63*, 1960.

[13.6] G. A. THURSTON and F. A. PENNING: Effect of initial imperfections on the stability of shallow spherical shells, *AFOSR, Sci. Report No. 64, 1627*, August 1964.

[13.7] R. R. ARCHER: On the numerical solutions of the nonlinear equations for shells of revolution, *J. of Math. Phys. 41.3* (1962) 165–178.

[13.8] F. A. PENNING AND G. A. THURSTON: The stability of shallow spherical shells under concentrated load, *NASA CR-265*, July 1965.

[13.9] F. A. PENNING: Experimental buckling modes of clamped shallow shells under concentrated load, *J. Appl. Mech. 33. 2* (1966) 297–304.

[13.10] D. BUSHNELL: Bifurcation phenomena in spherical shells under concentrated and ring loads, *AIAA J. 5. 11* (1967) 2035.

[13.11] A. V. POGORELOV: *Geometrical methods in nonlinear theory of elastic shells*, Izd. Nauka, Moscow 1967 (in Russian).

[13.12] I. G. TEREGULOV and L. I. MOGILEVSKII: The influence of the geometrical nonlinearity in asymptotic formulae for circular plates and shallow spherical dome loaded by a concentrated force, *Tr. seminara po teorii obolochek. KFTI AN SSSR, Kazan. 3* (1973) (in Russian).

14

Design of plates and shells under concentrated loads

The results obtained in the previous chapters proved that concentrated loads cause a concentration of stresses in thin-walled structure. When the force acts at one point of the surface, there always appear infinitely large stresses. In a real structure the stresses are limited by the plasticity of the material and in the vicinity of the loading point the plasticized areas appear. Sometimes these areas have a local character and do not determine the strength of the whole structure. This takes place, for example, in the cantilever plate (Fig. 49) loaded by a concentrated force at the free edge. At the loading point the moment M_{yy}, bending the plate in the plane parallel to the edge, is infinitely large, while the moment M_{xx} has a finite value. In a real, metal plate the stresses σ_{yy} never become infinitely large, as they can only reach the yield point. It cannot, however, destroy the whole plate as here its bending strength in the direction of the x-axis perpendicular to the edge has a larger significance. Only the reaching of the yield point by the stress σ_{xx} at the built-in edge can be dangerous for the strength of the plate.

We see therefore that not always is the local concentration of stresses a direct reason for damage. Of course, the structures submitted to fatigue loads should be considered differently. There the local concentration can be the reason for the local crack, which, widening with the lapse of time, leads to damage. In this case we always try to design a structure having the smallest concentration of stresses. We have then at our disposal two ways. The first consists in introducing the load in the thin-walled structure in such a way that it acts on the greatest portion of this surface and changes slowly. For example, from the diagram of stresses (Fig. 25) we see that the distribution of the force over a certain surface affects to a great degree the distribution of the stresses in the circular plate. A similar effect is noted for the distribution of load along a certain length of the edge.

The other way consists of designing a local increase of the rigidity of the structure, at the loaded place, and shaping the structure in such a way that it can carry the concentrated loads without or with a small concentration of stresses.

For example, at the loaded place we apply a bar or a rib whose task is to distribute the load over the whole structure in the form of tangential and normal stresses applied in a continuous way. Then, however, a certain concentration of stresses occur in the wall of the structure. We consider such a problem in detail in the next section.

When we assume that in the whole structure the reduced stresses reach a certain constant value, then we design it for a uniform strength. Such problems consist of defining the shape, thickness or the distribution of strengthening elements in concrete structures and are usually complicated. For concentrated loads they have not been widely worked yet. We will cite here a few basic solutions.

14.1 Infinitely large plate jointed to a long bar and loaded at an interior point

Let us investigate a case where the concentrated force is introduced in the plate by means of a bar. We shall consider a very long bar of constant cross-section bonded with an infinitely large plate of the thickness h (Fig. 146). The force P is applied at one cross-section of the bar. The main object of our investigation is the manner in which the force is transferred from the bar to the plate. The bar is placed symmetrically with respect to the plate, bending being excluded. Let us assume that the forces acting between the plate and the bar consist of the distributed tangential line load $q(x)$. The plate is in a plane state of stress.

The above assumptions are equivalent to the assumption that the bar is a one-dimensional body with zero bending rigidity. The distributed load $q(x)$ acting between the plate and the bar can be represented by means of the Fourier integral

$$q(x) = \frac{1}{\pi}\int_0^\infty q(\alpha)\cos\alpha x\,\mathrm{d}\alpha. \tag{14.1}$$

Performing the calculations similar to those in Section 4.2, we obtain the stresses in the plate as

$$\begin{aligned}
\sigma_{xx} &= \frac{1}{4\pi h}\int_0^\infty [3+\nu-\alpha y(1+\nu)]\mathrm{e}^{-\alpha y}q(\alpha)\sin\alpha x\,\mathrm{d}\alpha,\\
\sigma_{yy} &= -\frac{1}{4\pi h}\int_0^\infty [1-\nu-\alpha y(1+\nu)]\mathrm{e}^{-\alpha y}q(\alpha)\sin\alpha x\,\mathrm{d}\alpha,\\
\sigma_{xy} &= -\frac{1}{4\pi h}\int_0^\infty [-2+\alpha y(1+\nu)]\mathrm{e}^{-\alpha y}q(\alpha)\cos\alpha x\,\mathrm{d}\alpha.
\end{aligned} \tag{14.2}$$

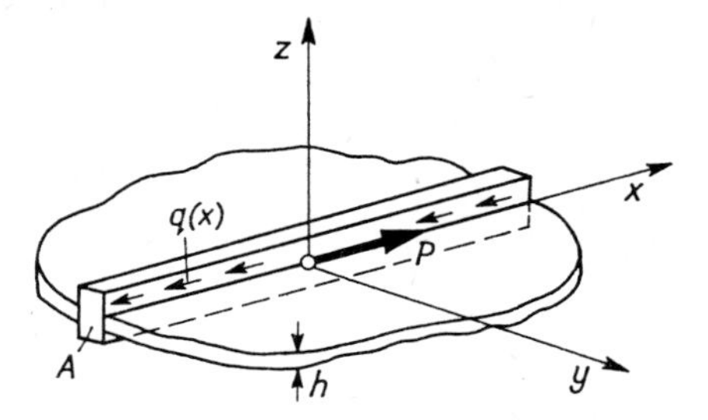

Fig. 146. Infinite plate jointed with a bar and loaded by a concentrated force P applied to the cross-section of the bar.

The strains in the plate and the bar are equal along the joint because they are bonded together. The strains in the plate for $y = 0$ resulting from the above stresses are

$$\varepsilon_p = -\frac{(3-\nu)(1+\nu)}{4\pi hE}\int_0^\infty q(\alpha)\sin\alpha x\,\mathrm{d}\alpha. \tag{14.3}$$

The strains in the bar depend on the force $N(x)$, which is the force acting in the cross-section of the bar, and the stiffness $E_b A$ of the bar. The force N results from the simultaneous action of the load P and the distributed force $q(x)$. The value of the force N in an arbitrary cross-section can be calculated from the equilibrium condition of the bar and we have:

$$\int_{-\infty}^{x} q(x)\,\mathrm{d}x - P = N.$$

Representing the load P by means of the Fourier integral, we obtain the following strain in the bar:

$$\varepsilon_b = \frac{N}{E_b A} = -\frac{1}{E_b A\pi} \int_{-\infty}^{x} \left[\int_{0}^{\infty} (P - q(x)) \cos \alpha \mathrm{d}\alpha \right] \mathrm{d}x. \tag{14.4}$$

Since $\varepsilon_p = \varepsilon_b$, we have

$$\frac{\partial \varepsilon_p}{\partial x} = \frac{\partial \varepsilon_b}{\partial x}. \tag{14.5}$$

On solving Eq. (14.5) we obtain the force $q(x)$ in the form

$$q(x) = \frac{P}{\pi} \int_{0}^{\infty} \frac{\cos \alpha x \,\mathrm{d}\alpha}{k\alpha + 1} \tag{14.6}$$

where

$$k = (3-\nu)(1+\nu)\frac{AE_b}{4hE}.$$

On integration we have

$$q(x) = -\frac{P}{k}\left[\cos\frac{x}{k}\,\mathrm{Ci}\,\frac{x}{k} + \sin\frac{x}{k}\,\mathrm{Si}\,\frac{x}{k}\right], \tag{14.7}$$

where $\mathrm{Ci}\,\frac{x}{k}$ and $\mathrm{Si}\,\frac{x}{k}$ are the sine and cosine integrals defined by

$$\mathrm{Si}\,px = -\int_{x}^{\infty} \frac{\sin pt}{t}\,\mathrm{d}t, \qquad \mathrm{Ci}\,px = -\int_{x}^{\infty} \frac{\cos pt}{t}\,\mathrm{d}t.$$

The numerical values of the functions $\mathrm{Si}\,x$ and $\mathrm{Ci}\,x$ can be found in tables of special functions. The force $N(x)$ takes the form

$$N(x) = \frac{P}{\pi}\left[\sin\frac{x}{k}\,\mathrm{Ci}\,\frac{x}{k} - \cos\frac{x}{k}\,\mathrm{Si}\,\frac{x}{k}\right], \tag{14.8}$$

$$0 < x < \infty.$$

From the above results we have:

$$N(x) = \frac{P}{2} \quad \text{if} \quad x \to 0, \qquad N(x) = \frac{Pk}{\pi}\frac{1}{x} \quad \text{if} \quad x \to \infty, \tag{14.9}$$
$$q(x) = \frac{P}{k}\left[\ln\frac{x}{k} + \gamma_0\right] \quad \text{if} \quad x \to 0.$$

If we know the value $q(\alpha) = P/(k\alpha+1)$ the stresses in the plate can be obtained simply by integration, according to (14.2). The distribution of the force $q(x)$ is presented graphically in Fig. 147. At the point $x = 0$ the force

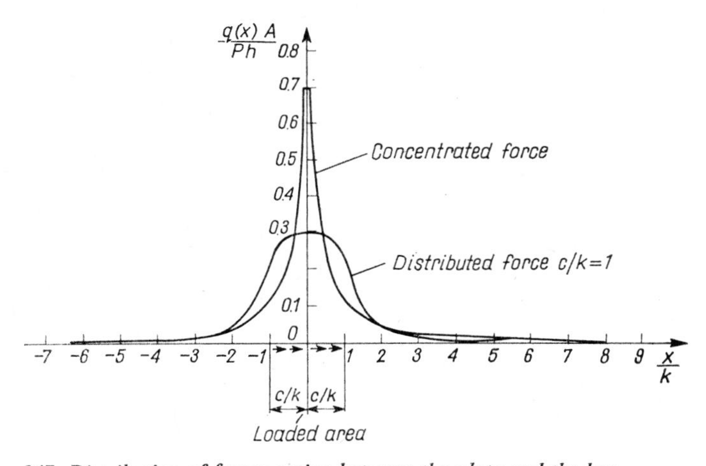

Fig. 147. Distribution of forces acting between the plate and the bar.

$q(x)$ is infinitely large. The above solution was obtained by Melan [14.6] in 1932. The solution in the case where the force P is uniformly distributed along the distance $2c$ can simply be obtained by multiplication of the integrand (14.6) by $(\sin\alpha c)/\alpha c$. Then we find

$$q(x) = \frac{P}{\pi}\int_0^\infty \frac{\sin\alpha c\cos\alpha x\,d\alpha}{\alpha c(1+k\alpha)}, \tag{14.10}$$

On integrating we have

$$q(x) = \frac{P}{2\pi c}\left\{\mu + \cos\frac{c-x}{k}\,\mathrm{Si}\,\frac{c-x}{k} - \sin\frac{c-x}{k}\,\mathrm{Ci}\,\frac{c-x}{k} + \right.$$
$$\left. + \cos\frac{c+x}{k}\,\mathrm{Si}\,\frac{c+x}{k} - \sin\frac{c+x}{k}\,\mathrm{Ci}\,\frac{c+x}{k}\right\}$$

where

$$\mu = 2\int_0^\infty \frac{\sin \alpha c \cos \alpha x}{\alpha} \, d\alpha .$$

For $-c < x < c$ we have $\mu = \pi$, for $x < -c$ and $x > c$, $\mu = 0$.

The function $q(x)$ is presented graphically in Fig. 147 for $c/k = l$. This corresponds to the following value of the bar cross-section:

$$A = \frac{4hc}{(3-\nu)(1+\nu)} .$$

We see that the force P acting on the bar is distributed along a certain distance, the force $q(x)$ appearing between the bar and the plate is finite for every x.

The maximum value of the force $q(x)$ is for $x = 0$

$$q(0) = \frac{P}{2\pi c}\left\{\pi + 2\left((\cos\frac{c}{k}\,\mathrm{Si}\,\frac{c}{k} - \sin\frac{c}{k}\,\mathrm{Ci}\,\frac{c}{k}\right)\right\}. \tag{14.11}$$

For $c \to 0$, $q(x) \to \infty$.

The problem of an infinite bar attached to a semi-infinite plate at the edge and loaded by a concentrated force can be solved in the same way.

If we assume that the bending rigidity of the bar is equal to zero, the forces between the bar and the plate consist only of the distributed line load $q(x)$. The solution takes a similar form, and the coefficient k is

$$k = \frac{2AE_b}{hE} . \tag{14.12}$$

The above problems can be presented mathematically also in another way, namely by means of integral-differential equations.

Let us calculate the strain in the plate produced by the distributed line load $q(x)$ by means of the superposition of strains produced by a single unit tangential force acting at the points $(\xi, 0)$ along the line of the joint with intensity $q(\xi)\,d\xi$. Then we have

$$\varepsilon(x) = \int_{-\infty}^{\infty} q(\xi)\varepsilon_0(x-\xi)\,d\xi \tag{14.13}$$

where $\varepsilon_0(x)$ is the strain produced in the plate at the origin of coordinates by a unit concentrated force acting at the point $(x, 0)$ and is the Green function of our problem. This function can easily be obtained from the solution (14.3) and is given by the formula

$$\varepsilon_0(x) = \frac{(3-\nu)(1+\nu)}{4\pi hE} K(x), \quad \text{where} \quad K(x) = \frac{1}{x}. \tag{14.14}$$

Let us introduce the function $p(x)$ which represents the external load acting on the bar. Then the differential equation of equilibrium of the bar takes the form

$$N'(x)+q(x)+p(x) = 0, \quad \text{where} \quad N'(x) = \frac{dN(x)}{dx}, \tag{14.15}$$

representing the equation of compatibility $\varepsilon(x)_p = \varepsilon(x)_b$ along the joint in the form of an integral-differential equation we have

$$\frac{1}{k} N(x)+\frac{1}{\pi} \int_{-\infty}^{\infty} \frac{N'(\xi)+p(x)}{x-\xi} d\xi = 0. \tag{14.16}$$

The solution of the above equation can be obtained by means of the Fourier transform and takes the form (14.8) if $p(x)$ is given by $p(x) = P\delta(x)$.

The above-described problems were considered by R. Muki and E. Sternberg [14.39] taking into account the bending rigidity of the bar. In the particular case of the bar with rectangular cross-section, they obtained a rigorous solution applying both for the plate and the bar the equations of the plane state of stress.

The differences in the distributions of the normal force $N(x)$ obtained in this way are not essential (of the order 15 per cent) and are smaller the larger the coefficient $h_b E_b/hE$. More significant differences were observed between the stresses in the plate in the immediate vicinity of the loaded place as obtained by means of the above-mentioned methods. The problem of the transfer of the load to a sheet from a rivet-attached stiffener was considered by B. Budiansky and T. T. Wu [14.43]. More information about problem discussed in this section can be found in the paper by E. Sternberg [14.42].

14.2 Strip plate jointed with a stringer of finite length

We consider now the case presented in Fig. 148 in which the semi-infinite sheet is loaded through a bar of constant cross-section. Such a problem was considered by E. Reissner [14.2], S. U. Benscoter [14.18], and F. C. Monge [14.44]. See also the analysis due to J. N. Goodier and C. S. Hsu [14.1].

Before we start to solve the given problem let us consider the results obtained from the more general case indicated by Fig. 149. A thin flat plate D_1

overlaps another thin plate D_2, the two being bonded over the whole of the common surface D. The assemblage is in equilibrium under some set of in-plane forces applied to the non-overlapping parts of D_1 and D_2. Bending is excluded here. The differential equations for the elastic displacements in plane stress can be obtained by substituting into the equilibrium equations the

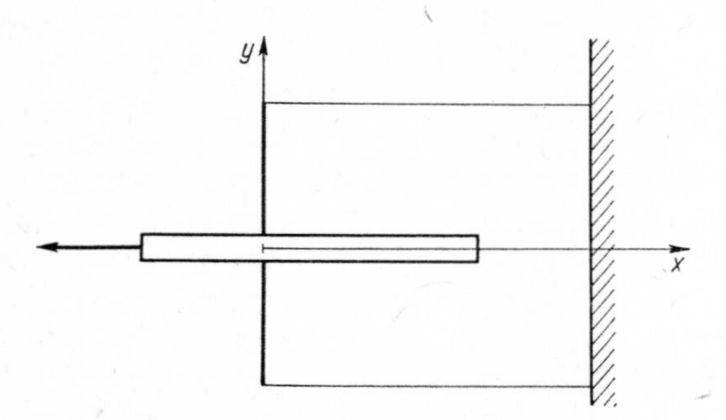

Fig. 148. Strip plate loaded through a bar.

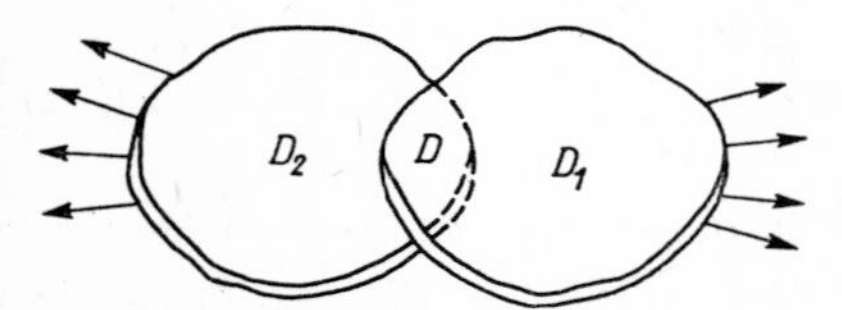

Fig. 149. Two plates D_1 and D_2 jointed over the surface D.

stresses in terms of strains and later in terms of displacements. Then we have two equations of the type

$$\Delta u + \frac{1+\nu}{1-\nu}\frac{\partial}{\partial x}\left(\frac{\partial u}{\partial x} + \frac{\partial v}{\partial y}\right) + \frac{X}{hG} = 0,$$

$$\Delta v + \frac{1+\nu}{1-\nu}\frac{\partial}{\partial y}\left(\frac{\partial u}{\partial x} + \frac{\partial v}{\partial y}\right) + \frac{Y}{hG} = 0, \tag{14.17}$$

where X, Y are body forces per unit area. The displacements in the common part D of the plates D_1 and D_2 are the same throughout because the plates are bonded together, but the body forces are equal and opposite.

Introducing this fact into the above equations, we see that they can be satisfied only in the case where $X = 0$, $Y = 0$ provided Poisson's ratios are equal for both plates. It results from the above that no force is transmitted by the bond to the interior of D. If the Poisson ratios are the same, the force transmitted by the bond at the common surface D can be only a line distribution confined to the periphery. The uniform thickness of the plates and their Young's moduli need not be identified.

In any case it is possible to deduce the line distribution of the force around the periphery of D when the stress is known just outside the periphery. Considering the equilibrium of a small rectangular element of D_1 at the periphery (Fig. 150), we obtain

$$h_1(\sigma_{n_1}-\sigma'_{n_1}) = F_n, \quad h_1(\tau_{ns_1}-\tau'_{ns_1}) = F_s, \tag{14.18}$$

where h_1 is the thickness of the plate D_1. σ_{n_1}, τ_{ns_1} and σ'_{n_1}, τ'_{ns_1} are the stress components in D_1 on the inside and outside of the periphery. $F_s\,\mathrm{d}s$ and $F_n\,\mathrm{d}s$

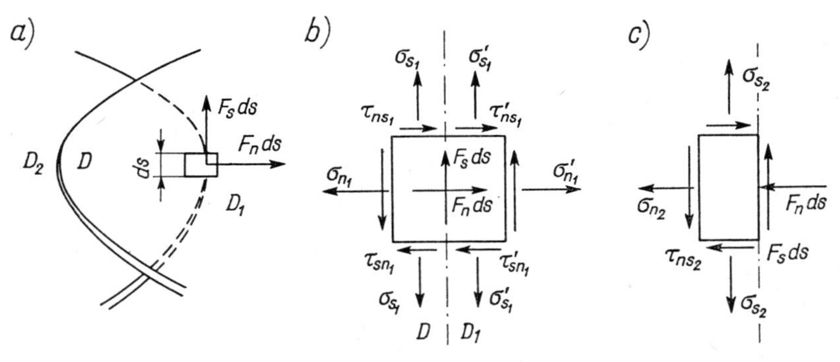

Fig. 150. Stress components at the periphery of D.

are the line forces along and normal to the periphery. The stresses σ_{n_2}, σ_{s_2}, τ_{ns_2} and forces $F_s\,\mathrm{d}s$, $F_n\,\mathrm{d}s$ act on the underlying element of D_2. Equilibrium yields

$$h_2\sigma_{n_2} = -F_n, \quad h_2\tau_{ns_2} = -F_s \tag{14.19}$$

where h_2 is the thickness of the plate D_2. The strains in both plates throughout D are the same. Then

$$\varepsilon_{n_2} = \varepsilon_{n_1}, \quad \varepsilon_{s_2} = \varepsilon_{s_1}, \quad \gamma_{ns_2} = \gamma_{ns_1}. \tag{14.20}$$

Moreover, the extensional strain ε_{s_1} along and just inside the periphery is the same just outside in D_1 denoted by ε'_{s_1}; $\varepsilon_{s_1} = \varepsilon'_{s_1}$. We obtain from the above equations, taking into account the stress–strain relations, the following relations between bond forces F_n, F_s and stresses just outside the periphery. For $\nu_1 = \nu_2$ we have

$$F_n = -\frac{h_1 \sigma'_{n_1}}{1+E_1 h_1/E_2 h_2}, \qquad F_s = -\frac{h_1 \tau'_{ns_1}}{1+E_1 h_1/E_2 h_2}. \tag{14.21}$$

If $h_1 = h_2$ and $E_1 = E_2$,

$$F_n = -\tfrac{1}{2} h_1 \sigma'_{n_1}, \quad F_s = -\tfrac{1}{2} h_1 \tau'_{ns_1},$$

which indicates that half of each type of loading in D_1 is immediately transmitted to D_2.

Equations (14.21) were obtained by J. N. Goodier and C. S. Hsu [14.1]. The case where the Poisson ratios in both plates are not equal was studied by R. Muki and E. Sternberg [14.40]. However, it results from this paper that the influence of the internal surface forces X and Y appearing if $\nu_1 \neq \nu_2$ in the common surface D is not important and only slightly changes the distribution of stresses in the plates.

Coming to our problem of a bar jointed to a plate and transmitting tension to it, we can expect that this transmission is not effected by a smooth distribution of forces along the lap joint. If the surface of the point has a form of an elongated rectangle, the line distribution of the transmitting forces along the periphery can be replaced by one line distribution and two concentrated forces, one where the bar meets the plate and a second at the end of the bar. Taking into account the equilibrium equation

$$N'(x)+q(x) = 0$$

and the equation of compatibility $\varepsilon(x)_p = \varepsilon(x)_b$, we find the integral equation

$$\frac{N(x)}{AE_b} = \frac{1}{\pi Eh}\left[Q_1 K(x,0)+Q_2 K(x,0) - \int_0^a N'(\xi) K(x,\xi)\,d\xi\right]; \tag{14.22}$$

$K(x, \xi)$ is here the Green function of the problem and has the form [5.12]

$$4K(x,\xi) = \frac{(3-\nu)(1+\nu)}{x-\xi} + \frac{5-2\nu+\nu^2}{x+\xi} + 2(1+\nu)^2 \frac{(x-\xi)\xi}{(x+\xi)^3}, \tag{14.23}$$

and represents the strain at the point $(x, 0)$ due to a single force unity acting at the point $(\xi, 0)$ in the negative x-axis direction

$$\varepsilon_0(x,\xi) = \frac{1}{\pi hE} K(x,\xi).$$

If we consider the bar with a rectangle cross-section of the dimensions abh_b, the concentrated forces Q_1 and Q_2 can be calculated as the resultants of the distributed forces acting along the shorter edges b of the rectangle surface of the joint. Treating the bar loaded by the force P as a sheet, we obtain the force Q_1 from (14.21)

$$Q_1 = F_n b = \frac{P}{1+E_b h_b/Eh} \,. \tag{14.24}$$

The second force is $Q_2 = N(a)$.

If we assume that the load transmitted from the bar to a plate consists only of a single line load $q(x)$, we assume simultaneously that the bar is narrow and $b \to 0$. Then, if $A = \text{const}$, $h_b \to \infty$. Introducing this into (14.21), we find that $Q_1 \to 0$, $Q_2 \to 0$. The same result was obtained by Reissner [14.2] based on physical considerations. The question if the concentrated forces appear or not in the plate has been widely discussed in the literature. See, for example, the references given in [14.41]. The following fact testifies against the appearance of the concentrated forces. The concentrated force acting on the plate produces at the point of application infinitely large strains, while the force acting on the bar brings about in this place only finite strains. As the strains of the plate and the bar should be equal along the line of the joint, the concentrated forces should be zero. Thus Eq. (14.22) takes the form

$$\frac{N(x)}{AE_b} + \frac{1}{\pi Eh} \int_0^a N'(x) K(x, \xi) \, d\xi = 0. \tag{14.25}$$

Then we have the differential-integral equation with the singular kernel at $x = \xi$. The solution of this equation can be carried out only numerically by reducing it to algebraic equations. The integral in Eq. (14.25) is an improper one. This integral can be replaced for calculation by an ordinary integral in the following way:

$$\int_0^a \frac{N'(\xi)\,d\xi}{x-\xi} = -\int_0^a \frac{N'(x)-N'(\xi)}{x-\xi}\,d\xi + N'(x)\int_0^a \frac{d\xi}{x-\xi} \,,$$

where $N'(\xi) = dN(\xi)/d\xi$. The last integral can be evaluated as

$$\int_0^a \frac{d\xi}{x-\xi} = \ln\left(\frac{x}{a-x}\right).$$

From the asymptotic solutions presented in [14.41] it results that the distributed line forces $q(x)$ are singular at the ends of the segment of the joint:

$$q(x) = O(x^{-\alpha_*}), \quad x \to 0,$$
$$q(x) = O(1/\sqrt{x-a}), \quad x \to a, \tag{14.26}$$

where α_* is the real root of the transcendental equation in the interval (0, 1)

$$\cos[(1-\alpha)\pi] - \frac{2(1+\nu)}{3-\nu}(1-\alpha)^2 + \frac{8-(3-\nu)(1+\nu)}{(3-\nu)(1+\nu)} = 0, \qquad 0 < \nu < \frac{1}{2};$$

α_* changes from $\alpha_* = 0$ for $\nu = 0$ to $\alpha_* \cong 0.3$ for $\nu = 0.5$.

The above results are in agreement with the results obtained by A. M. Hens [14.11] and W. T. Koiter [14.7].

The problem of the bar with the rectangular cross-section was considered by Goodier and Hsu [14.1] based on equation (14.22). Assuming that $\nu_b = \nu$, $E_b = E$ and $h_b = h$, they obtained from (14.24) $Q_1 = \frac{1}{2}P$. Reducing Eq. (14.22) to eight simultaneous algebraic equations for the eight values $N_1, N_2, \ldots, N_8$ of $N(x)$ at eight selected points of the interval $0 < x < a$, the results presented in Fig. 151 were obtained. In this figure there are also given the results of

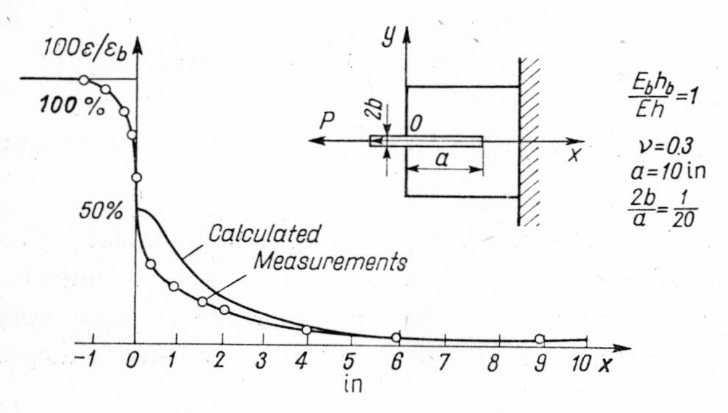

Fig. 151. Comparison of the results obtained by J. N. Goodier and C.S. Hsu [14.1], *and the experiment in the case of the transmission of a load to the strip plate through a bar.*

measurements made on a model by means of strain gages. The bar consisting of two equal strips was attached on each side of a plate. The thickness of each bar was half that of the plate. The points in Fig. 151 show the tensile strain in the strip. The sharp drop of tension is observed at the entry to the plate exceeding the predicted 50 per cent. The above results suggest that the fatigue strength of such a joint depends more on the detailed local character of the joint where the bar meets the plate, than on the length of the joint. The same problem was solved by E. Krahn [14.45] assuming that the surface forces in the common region are not zero and are uniformly distributed along the width of the bar. The results were similar and differed from [14.1] by less than 8 per cent.

14.3 The bar of finite length jointed with an infinite plate

The forces acting between the infinitely large plate and a bar of finite length (Fig. 152) can be found in the same way. All previous equations conserve their validity, only the kernel $K(x, \xi)$ here has the form (14.14).

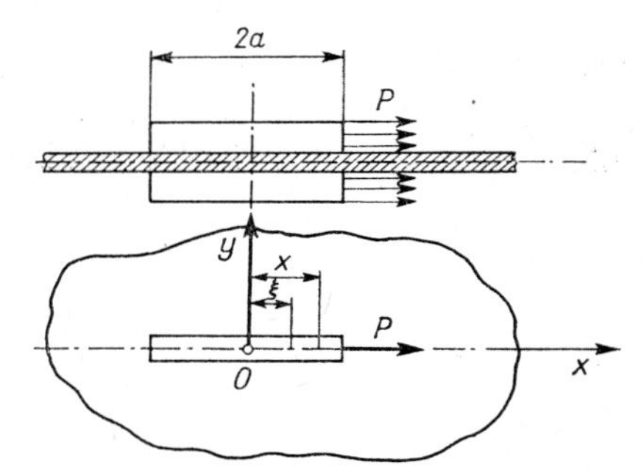

Fig. 152. Transmission of a force to the plate through a bar of finite length.

There integral equation of the problem is

$$N(x)+\frac{k}{\pi}\int_{0}^{a}\frac{N'(x)-P(x)}{x-\xi}\,d\xi = 0. \tag{14.27}$$

The closed solution of the above equation is unknown, but the approximate solution can be obtained by reducing Eq. (14.27) to a set of algebraic equations. The line forces $q(x)$ are singular at the ends of the bar:

$$q(x) = O\left(\frac{1}{\sqrt{x}}\right) \text{ if } x \to 0 \quad \text{and} \quad q(x) = O\left(\frac{1}{\sqrt{x-a}}\right) \text{ if } x \to a.$$

The above problem was the subject of the papers by G. A. Morar and G. Ja. Popov.

14.4 Optimum design of elements introducing the load

Let us consider two elastic elements, for example two sheets I and II, jointed along the line AB. The forces acting on each sheet should be equal and oppositely directed. The equations of compatibility of strains for both sheets along the line AB can be written in the form of integral equations. For example, equating in both plates the strain ε_{xx} along the line AB, we find (Fig. 153).

$$\begin{aligned}\varepsilon_{xx} &= \int_A^B [K_1^{\mathrm{I}}(x,\xi)q_n(\xi)+K_2^{\mathrm{I}}(x,\xi)q_t(\xi)]\,\mathrm{d}\xi \\ &= \int_A^B [K_1^{\mathrm{II}}(x,\xi)q_n(\xi)+K_2^{\mathrm{II}}(x,\xi)q_t(\xi)]\,\mathrm{d}\xi,\end{aligned} \tag{14.28}$$

where K_1^{I}, K_2^{I} and K_1^{II}, K_2^{II} are the influence functions for both plates, respectively. These functions can be find applying the unite forces tangential and normal to the considered edge of the plates at the point ξ. Similar equations are for the strains ε_{yy}, ε_{xy}.

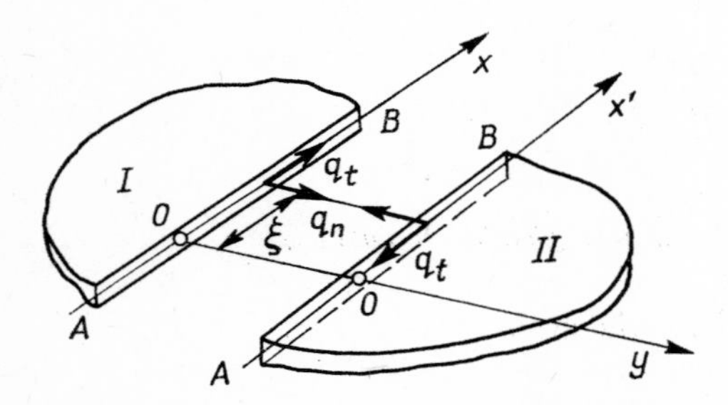

Fig. 153. Forces acting between two jointed elastic elements.

If we know the influence functions K_i^{I}, K_i^{II}, then, solving the equations (14.28) together with the equations of equilibrium, we can define the distribution of forces q_n and q_t acting between the jointed elastic elements. These equations can be used to design the proper shape of both elements. If the influence functions, or certain parameter of the influence functions are considered as unknowns, then solving the equations of compatibility we obtain the geometry of the structure. We know that the influence functions (Green's functions) depend on the shape of the structure and on the boundary conditions. Assuming proper and desirable distribution of the reacting forces q_n and q_t, we obtain the optimal shape of the structure. This problem reduces to the solution of the integral equations. However, if we look only for certain parameters of the influence functions, these equations become ordinary algebraic equations. Let us describe this problem by the following example.

14.4.1 *Bar of variable cross-section*

Considering the plate jointed with a bar of finite length we find that at the ends of the bar appears a singular distribution of the tangential forces q_t. These forces produce infinitely large stresses in the plate. However, this can be avoided by a proper design of the bar. If we demand, for example, that between the plate and the bar appear only uniformly distributed tangential forces $q(x)$, the cross-section of the bar at the ends should decrease to zero. Let us solve such a problem and define the shape of the bar from a condition that the forces acting between the bar and the plate have a constant value q_0. The strains in the bar of variable cross-section are, $x < a$ (Fig. 154),

$$\varepsilon_0(x) = \frac{q_0(x+a)}{E_1 A(x)} = \frac{P(x+a)}{2aE_1 A(x)} \,. \tag{14.29}$$

The strains in the plate are

$$\varepsilon(x) = -\int_{-a}^{a} q(\xi)\varepsilon_0(x-\xi)\,\mathrm{d}\xi = -\frac{(3-\nu)(1+\nu)q_0}{4\pi hE} \ln\left|\frac{x-a}{x+a}\right| \,. \tag{14.30}$$

Equating the strains in the plate and in the bar, we have

$$A(x) = A_0 \frac{x+a}{a\ln\left|\dfrac{x-a}{x+a}\right|} \quad \text{where} \quad A_0 = \frac{2\pi haE}{(3-\nu)(1+\nu)E_b} \,. \tag{14.31}$$

It results from the above formula that for $x = 0$ the cross-section of the bar increases to infinity and for $x > 0$ it becomes negative, which is impossible. But the real cross-section can be obtained if the concentrated force is applied at the point $x = 0$. Then for $x > 0$ the strains in the plate are negative and the cross-section becomes positive. If the cross-section of the bar is circular, its radius changes as given in Fig. 154. For $x = 0$ the radius is infinitely large. But if the load is distributed along a certain distance $2c$, we obtain a finite radius of the bar for $x = 0$. In this case the force N is equal to zero in the central cross-section of the bar, $N(x) = 0$ for $x = 0$.

Equating the strains in the plate and in the bar for $-c \leqslant x \leqslant c$, we obtain

$$k(x)\frac{q_0 a}{\pi}\int_{-a}^{a}\frac{\mathrm{d}\xi}{x-\xi} = \frac{P(a-c)}{2c}\,x, \tag{14.32}$$

where

$$k(x) = (3-\nu)(1+\nu)\frac{E_b A(x)}{4hE} \,.$$

Then we have

$$A(x) = A_0 \frac{a-c}{ac} \frac{x}{-\ln\left|\frac{x-a}{x+a}\right|}.$$

Calculating the $\lim_{x\to 0} A(x)$ we find the value of the central cross-section of the bar

$$A_{(x=0)} = \frac{1}{2}\frac{a-c}{c} A_0 .$$

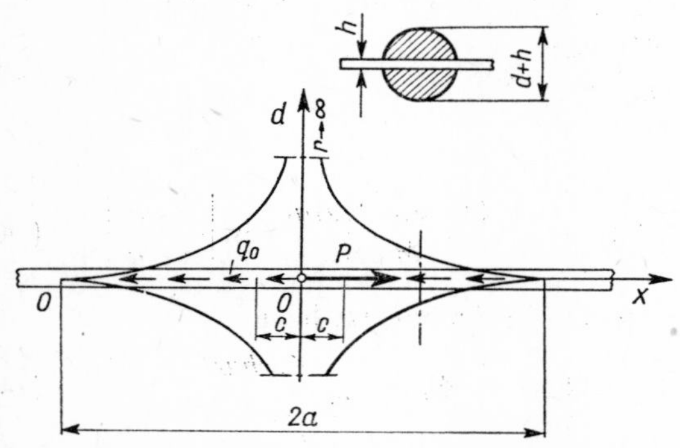

Fig. 154. Radius of the cross-section of the designed circular bar introducing the tangentia force into a sheat.

If we assume that the bar has a circular cross-section, we obtain the value for the radius of the bar $r = (A/\pi)^{-1/2}$ and if, for example, $E_b = E$, $h = 0.01a$, $c = a/2$, then $r = 0.053a$ for $x = 0$.

We observe that in order to avoid the concentration of stresses in the plate at the ends of the bar, we have to change the shape of the bar so that its cross-section decreases to zero at the ends. Then only distributed forces $q(x)$ appear between the plate and the bar. But, even if the forces $q(x)$ are constant, the stresses in the plate increase to infinity at the ends of the bar. Thus, if we want the plate to be free of infinitely large stresses, the ends of the bar should get thinner more rapidly than the formulae indicate.

Let us assume that the interaction between the bar and the plate $q(x)$ is given by the equation

$$q(x) = q_0 \left(1 - \left(\frac{x}{a}\right)^2\right)^2 . \tag{14.33}$$

Let us calculate now the shape of the bar. As previously, the force P is distributed along the distance $2c$. We find the strain in the interval $a \leqslant x \leqslant c$ and $0 \leqslant x \leqslant a$. For $c \leqslant x \leqslant a$ the strain in the bar is

$$\varepsilon(x)_b = -\frac{1}{A(x)E_b}\int_x^a q(x)\,\mathrm{d}x.$$

Introducing the force $q(x)$ from (14.33), we find

$$\varepsilon(x)_b = -\frac{q_0 a}{A(x)E_b}\left[-\frac{1}{5}\left(\frac{x}{a}\right)^5+\frac{2}{3}\left(\frac{x}{a}\right)^3-\frac{x}{a}+\frac{8}{15}\right].$$

The strain in the plate is

$$\varepsilon(x)_p = -\frac{(3-\nu)(1+\nu)}{4\pi E}\left\{\left[1-\left(\frac{x}{a}\right)^2\right]^2 \ln\left|\frac{x-a}{x+a}\right|-\frac{10}{3}\frac{x}{a}+2\left(\frac{x}{a}\right)^3\right\}.$$

Equating the strains in the plate and in the bar, we find the cross-section of the bar. For $c \leqslant x \leqslant a$ we have

$$A(x) = 2A_0\frac{\frac{1}{5}\left(\frac{x}{a}\right)^5-\frac{2}{3}\left(\frac{x}{a}\right)^3+\frac{x}{a}-\frac{8}{15}}{\left[1-\left(\frac{x}{a}\right)^2\right]^2 \ln\left|\frac{x-a}{x+a}\right|-\frac{10}{3}\frac{x}{a}+2\left(\frac{x}{a}\right)^3}. \tag{14.34}$$

In the interval $0 \leqslant x \leqslant c$ the strain in the bar is

$$\varepsilon(x)_b = -\frac{1}{A(x)E_b}\left[\int_x^a q(x)\,\mathrm{d}x-\frac{c-x}{2c}P\right].$$

The following equation results from the equilibrium of the bar:

$$q_0\int_{-a}^{a}\left[1-\left(\frac{x}{a}\right)^2\right]^2\mathrm{d}x = P \tag{14.35}$$

which gives

$$q_0 = \frac{15}{16}\frac{P}{a}.$$

Finally, we obtain the cross-section of the bar in the segment $0 \leqslant x \leqslant c$ to be:

$$A(x) = 2A_0 \frac{\frac{1}{5}\left(\frac{x}{a}\right)^5 - \frac{2}{3}\left(\frac{x}{a}\right)^3 + \frac{x}{a} - \frac{8}{15}\frac{x}{c}}{\left[1-\left(\frac{x}{a}\right)^2\right]^2 \ln\left|\frac{x-a}{x+a}\right| - \frac{10}{3}\frac{x}{a} + 2\left(\frac{x}{a}\right)^3} \,. \tag{14.36}$$

If $x \to 0$, the numerator and the denominator of the expression (14.36) decrease to zero and the value of the cross-section becomes indeterminate of form 0/0. Calculating the $\lim A(x)$ for $x \to 0$, we find the value of the central cross-section of the bar:

$$\underset{x \to 0}{A(x)} = A_0 \frac{3}{8}\left(\frac{8}{15}\frac{a}{c} - 1\right).$$

Since the cross-section can be a positive value only, the following condition must be satisfied:

$$\frac{8}{15}\frac{a}{c} > 1.$$

Then

$$c < \frac{8}{15} a.$$

The condition $c < 8a/15$ results from the assumed distribution of the force P and the load $q(x)$ and consists of the condition demanding the strains in the bar have the sign compatible with the sign of the strains in the plate. The cross-section of the bar $A(x)$ calculated for $c = 0.5a$ is presented graphically in Fig. 155. The thinner line represents the value of $A(x)$ resulting from Eq. (14.32), the other line from Eqs. (14.34), (14.36). If the force $q(x)$ is constant, the cross-section increases more rapidly than in the second case considered.

The values of the stresses σ_{xx}, σ_{yy} and the reduced stresses $\sigma_{re}^2 = \sigma_{xx}^2 - \sigma_{xx} \cdot \sigma_{yy} + 3\tau_{xy}^2$ in the plate along the joint ($y = 0$) are given in Fig. 156. We see that the reduced stresses σ_{re} do not change rapidly and are almost constant in the interval in which the bar is bonded with the plate. It can be proved [14.48] that it is not necessary to assume the function (14.33) as the most appropriate to present the distribution of the forces $q(x)$ in the joint in order to avoid the concentration of stresses in the plate. The same result can be obtained if, for example, $q(x) = q_0 (1 - (x/a)^2)^{1/2}$. The reduced stresses obtained in this way are even more uniform in the joint and their maximum value is smaller.

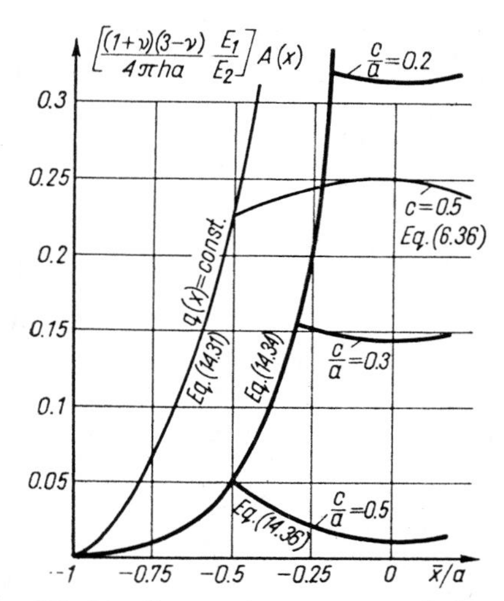

Fig. 155. Distribution of cross-section of a bar.

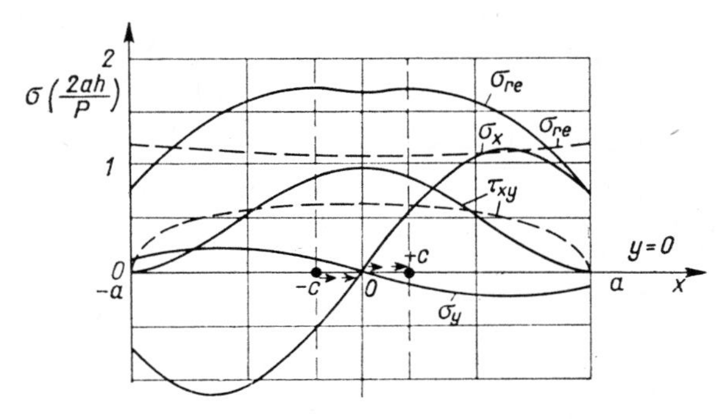

Fig. 156. Stress variation in a loaded segment. The line ------ presents the results obtained by assuming $q(x) = q_0[1-(x/a)^2]^{1/2}$.

14.5 Plate strengthened at the edge by a bar with bending rigidity

Let us consider a semi-infinite plate jointed at the edge $x = 0$ with an infinitely long bar with bending rigidity. At the point $y = 0$ the bar is loaded by a concentrated force P perpendicular to the edge (Fig. 157). Let us assume that the bar transfers the load to the plate only by means of the pressure $p(y)$ and no forces of friction appear between the plate and the bar. The aim of our considerations is the definition of the distribution of the pressure $p(y)$.

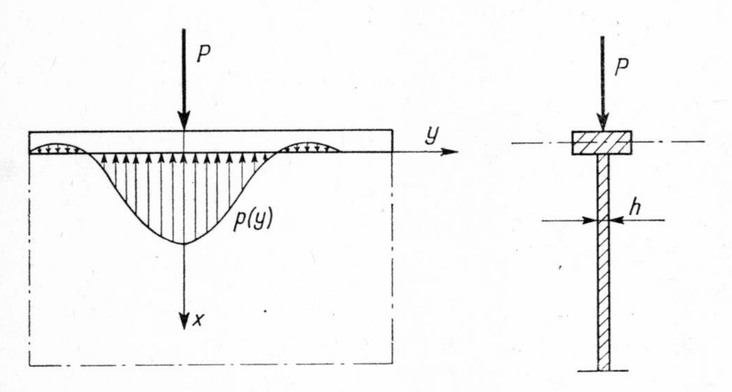

Fig. 157. Plate loaded through a bar applied at the edge.

Let us denote by E_b the modulus of elasticity of the bar, J_b the moment of inertia of the cross-section of the bar with respect to its neutral axis. The plate has thickness h and modulus of elasticity E. Let us represent the action between the plate and the bar and also the force P by means of Fourier integrals.

The equation of equilibrium for the bar bent by the force P and the distributed load $p(y)$ takes the form

$$E_b J_b \frac{\partial^4 u_p}{\partial y^4} = \int_0^\infty \left[\frac{P}{\pi} - p(\beta)\right] \cos \beta y \, d\beta, \tag{14.37}$$

where u_p means the deflection of the bar in the direction normal to its axis Since the bar is jointed with the plate along the edge, the displacements of the bar and the plate must be equal along this line. We can calculate the displacement u_p of the plate produced by the load $p(y)$ using the solution (5.32). Taking into account Eq. (5.4), we find the following relation for the deflection u_p of the plate at the edge:

$$E \frac{\partial u^4}{\partial y^4} = \frac{2}{h} \int_0^\infty p(\beta) \beta^5 \cos \beta y \, d\beta, \qquad x = 0. \tag{14.38}$$

Since at the edge $u = u_b$, we have $\partial u^4/\partial y^4 = \partial u_b^4/\partial y^4$. Introducing into the above equation the values (14.37) and (14.38), we obtain the following integral for the normal load $p(y)$:

$$p(y) = \frac{P}{\pi} \int_0^\infty \frac{\cos \beta y \, d\beta}{1+2\beta^3 k^3} \tag{14.39}$$

where

$$k^3 = \frac{E_b J_b}{Eh}.$$

The largest value of the pressure between the plate and the bar appears at the point $y = 0$. It is possible to find for this point the closed solution of the integral (14.39). We have

$$p_{\max} = \frac{P}{\pi} \int_0^\infty \frac{d\beta}{1+2\beta^3 k^3} = \frac{2P}{3\sqrt{3}\sqrt[3]{2}k} = 0.30549P \sqrt[3]{\frac{Eh}{E_b J_b}}.$$

If the force P acting on the bar is distributed along the distance $2c$, we obtain the solution by multiplying the integrand (14.39) by $(\sin \beta c)/\beta c$. The above problem was solved by Marguerre [3.48] and K. Girkmann [3.49]. A similar problem of the bar with finite length was solved by E. Reissner [3.51].

14.6 An infinite rigid body

A plate subject to the action of an infinitely rigid body of width $2c$ (Fig. 158) pressing the edge was solved by M. Sadowski [3.54], H. Borowicka [3.52], and G. Schubert [3.53], and in another way by Girkmann [1.23]. These authors obtained the distribution of pressure neglecting the friction between the body and the plate and assuming the plane state of strain in the plate to be

$$p(y) = \frac{P}{\pi\sqrt{c^2-y^2}}. \tag{14.40}$$

The above expression gives infinitely large pressures at the points $y = \pm c$. As in reality the stresses in the plate can have only finite values, a different distribution of stresses appear, and plasticized areas occur near the points $y = \pm c$. The solution of this elastoplastic problem and the solutions for the cases of the plates made of ideally plastic material or hardened plastic material can be found in the books concerning the theory of plasticity such as V. V. Sokolovskii [14.47], A. Nádai [14.46], and others.

The orthotropic plate whose one direction of orthotropy is parallel to the edge was considered in paper [3.46]. Expression (14.40) is valid also in this case.

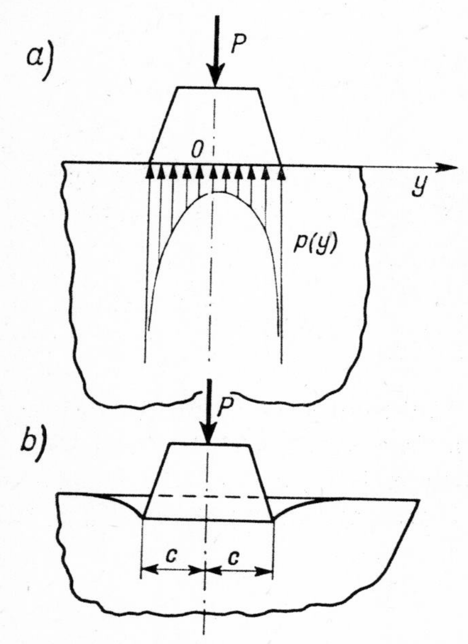

Fig. 158. Plate under the force applied to the edge through an infinitely rigid body.

14.7 A bar of variable rigidity

From the previous sections, the forces transmitted from the bar to the plate are singular if the bar has a finite length. The concentration of stress which is produced in the plate can be avoided if we properly design the element introducing the load.

Let us assume that between the bar and the plate there acts a distributed pressure $p(y)$. The displacement of the bar u_p is defined by the equation

$$(E_b J_b u_p'')'' = p(y). \tag{14.41}$$

Let us equate the displacement of the bar with the displacement of the plate at the edge produced by the load $p(y)$. The displacement produced by a unit load $P = 1$ acting at the point ξ for $x = 0$ is

$$Eu = -\frac{2P}{\pi h}\ln\frac{y-\xi}{a}. \tag{14.42}$$

On differentiation of (14.42) with respect to y we find

$$u'' = \frac{2P}{Eh\pi}\frac{1}{(y-\xi)^2}. \tag{14.43}$$

The displacement produced by the load $p(y)$ is defined by

$$\frac{d^2u}{dy^2} = \frac{2}{\pi Eh}\int_{-\infty}^{\infty}\frac{p(\xi)\,d\xi}{(y-\xi)^2}. \tag{14.44}$$

Equating the displacements of the bar $u_b'' = M/E_bJ_b$ and that of the plate along the edge, we find for the infinitely long bar the equation

$$\frac{2}{\pi Eh}\int_{-\infty}^{\infty}\frac{p(\xi)\,d\xi}{(y-\xi)^2} = \frac{M(y)}{E_bJ_b} \tag{14.45}$$

where $M(y)$ is the moment in the bar produced by the load $p(y)$. If the load $p(y)$ is an unknown function, the above equation is a first order Fredholm integral equation with the singular kernel at $y = \xi$.

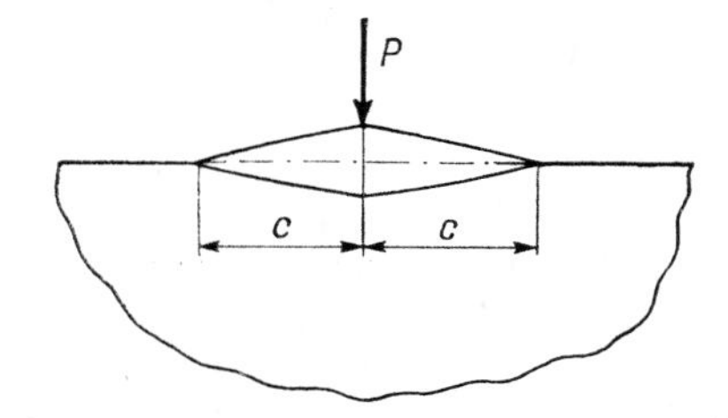

Fig. 159. Shape of a designed bar.

For the bar of constant bending rigidity E_bJ_b the solution of Eq. (14.45) takes the form (14.39). Let us assume that the bar is of finite length $2c$, and the distribution of force $p(y)$ acting between the plate and the bar is uniform, $p(y) = P/2c = \text{const}$. Calculating the stiffness E_bJ_b of the bar from Eq. (14.45), which becomes now an ordinary algebraic equation, we obtain

$$E_bJ_b = \frac{\pi Eh}{8c}(c-y)^3(c+y). \tag{14.46}$$

The shape of the bar of constant width has been presented in Fig. 159. The stresses in the plate in spite of the rapidly varying loading are not singular at the points $y = \pm c$, $x = 0$. The problem of the bar transferring the load into the infinite plate can be solved in the same way.

14.8 A plate strengthened by a semi-infinite bar

Consider a semi-infinite plate jointed at the edge with a bar of constant cross-section. A force P is applied at the end of the bar. The problem of determining the line tangential forces $q(x)$ acting between the bar and the plate by the assumption that the bar has zero bending rigidity was solved by W. T. Koiter

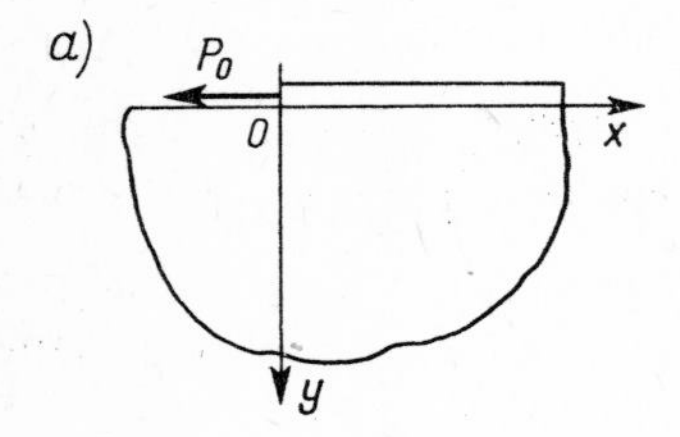

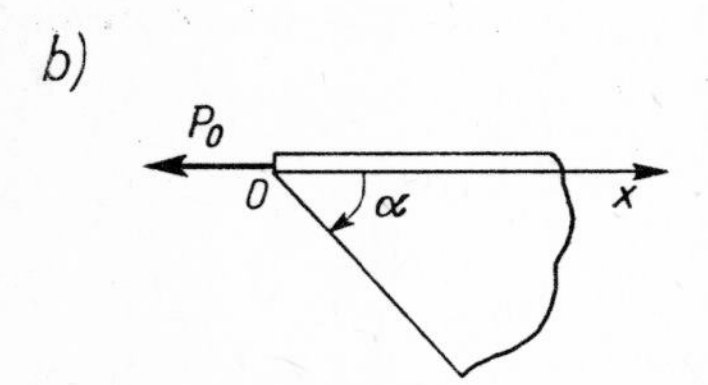

Fig. 160. Semi-infinite plate jointed at the edge with a bar of constant cross-section.

[14.7] by means of the Mellin transform. The result indicated that the forces $q(x)$ are singular at the end of the bar. We have for small x (Fig. 160a)

$$q(x) = \frac{\sqrt{2}}{\pi}\frac{PEh}{E_b A_b} x^{-1/2}. \tag{14.47}$$

The normal force in the bar is represented by the series

$$\frac{N(x)}{P} = 1 - \frac{\sqrt{2x}}{\pi}\,[1 - x(0.25425 - 0.1061 \ln x) + \ldots].$$

The above solution is valid also for an infinite plate jointed with the semi-infinite bar. The integral equations in both cases are identical. Here x denotes the non-dimensional coordinate which in the case of an infinite plate is

$$\bar{x} = \frac{1}{8}(3-\nu)(1+\nu)\frac{E_b A}{Eh}x, \quad \bar{x}\ (\text{cm}). \tag{14.48}$$

For the semi-infinite plate loaded at the edge

$$\bar{x} = \frac{E_b A}{Eh}x,$$

where E_b is the modulus of elasticity of the bar, A its cross-section, E, h the modulus of elasticity and thickness of the plate, respectively.

From the above relations, the force $q(x)$ at the end of the bar is infinitely large and produces infinitely large stresses in the plate. Similar results were obtained by Benthem [14.8] for the wedge plate (Fig. 160b). For the apex angle $\alpha = 135°$ the tangential forces are singular $q(x) = q_0 x^{-0.09166} + \text{constant}$ where q_0 is a certain constant value. Let us notice that the order of singularity is here smaller which means that the manner of introduction of the load in the plate is more advantageous here. It is possible to avoid the singularity of $q(x)$ by a proper design of the shape of the plate or by changing the cross-section of the bar. The singularities of the stresses at corners of plates were discussed by M. L. Williams [14.9].

14.9 Optimum design of plates under concentrated forces acting in its middle plane

14.9.1 *Elastic design*

We consider a plate loaded by a concentrated force at its edge. The force acts in the middle plane of the plate. Let us define the thickness of the plate (Fig. 161) in such a way that the stresses produced by the concentrated force fulfil the condition of uniform strength at each point of the plate. If the thickness of the plate is small compared to the other dimensions of the plate and changes slowly, the state of stress in the plate can be considered approximately plane. The Huber–Mises yield condition for the plane state of stress takes the form (14.55).

In order to find the shape of the plate we use the equation

$$\Delta\left(\frac{1}{h}, \Delta\Phi\right) - (1+\nu)L\left(\frac{1}{h}, \Phi\right) = 0. \tag{14.49}$$

Here $L(\,)$ is the operator given previously in (1.104). The stresses in the plate of variable thickness are expressed through the stress function:

$$\sigma^{ij} = \frac{1}{h} d^{i\alpha} d^{j\beta} \Phi_{,\alpha\beta},$$

where $h = h(x, y)$ is the variable thickness of the plate. The functions $h(x, y)$ and $\Phi(x, y)$ must fulfil the differential equation (14.49) together with the boundary conditions. We assume that the stress function Φ has the same form as the function Φ for the plate of constant thickness, Eq. (5.32),

$$\Phi = -\frac{P}{\pi} r\vartheta \sin\vartheta.$$

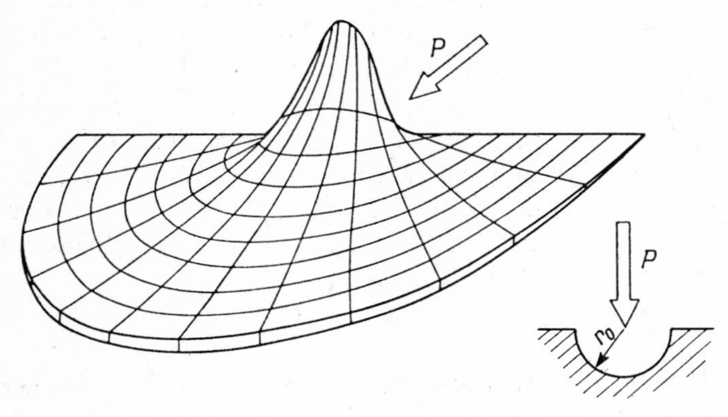

Fig. 161. Thickness of a plate of a uniform strength loaded at its edge (from J. Kapkowski [14.15]).

Then Eqs. (14.49) and $\sigma_{red} = \sigma_0$ are satisfied when the thickness of the plate is defined by

$$h = \frac{\Delta\Phi}{\sigma_0} = \frac{2P}{\sigma_0} \frac{\cos\vartheta}{r}. \tag{14.50}$$

The shape of the plate is given in Fig. 161. We see from formula (14.50) that if $r \to 0$, the thickness of the plate tends to infinity. This result is in contradiction with the assumption that the thickness of the plate is small and the state of stresses is plane. However, if we reject the neighborhood of the point of application of the force, assuming that the force is distributed along a certain radius r in the form of distributed radial pressure $n_r = -\frac{2P}{\sigma_0} \frac{\cos\vartheta}{r_0}$, the thickness of the plate can be defined from formula (14.50) with sufficient accuracy.

In a similar way we can calculate the shape of the wedge loaded at the apex by a concentrated force (Fig. 162). Assuming the stress function Φ identical with that for the flat wedge of constant thickness, we obtain the thickness of the wedge plate of constant strength

$$h = \frac{P}{\sigma_0 r}\left[\frac{\cos\varepsilon\cos\vartheta}{\alpha_0+\frac{1}{2}\sin 2\alpha_0}+\frac{\sin\varepsilon\sin\vartheta}{\alpha_0-\frac{1}{2}\sin 2\alpha_0}\right]. \tag{14.51}$$

It should be noticed that the exact solution can be obtained in this way only when the clamped edge gives rise to the stresses in the radial direction. In

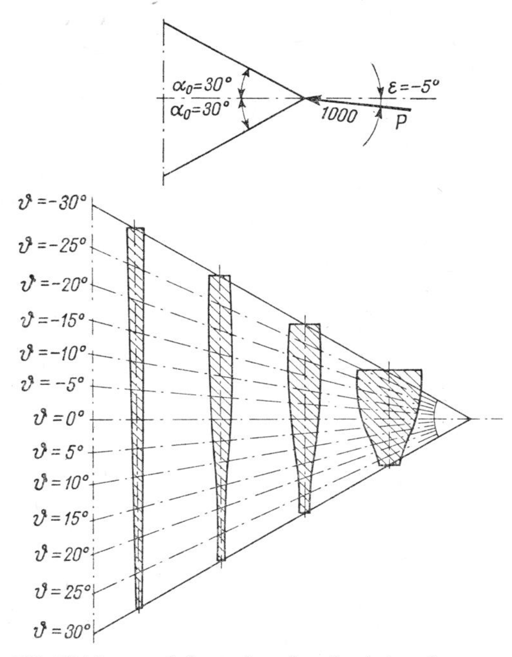

Fig. 162. Thickness of the wedge plate loaded at the apex ([14.15]).

other cases formula (14.51) is exact for points lying some distance from the clamped place. Also, the area in the immediate vicinity of the loading point should be excluded from consideration. There is still one limit on the validity of the solution resulting from the condition that the expression for the thickness of the plate cannot change sign in the whole area of the plate. Otherwise Eq. (14.51) is not reasonable for real materials. The numerical results for the wedge of angle $2\alpha_0 = 60°$ loaded at the angle $\varepsilon = -5°$ by the force $P = 1000$ kG are given in Fig. 162.

The above solution satisfies the conditions of equilibrium, compatibility of strains and is correct both for the plate being in the elastic state and also in the plastic state. In the case of the plate being in the elastic state this solution satisfies the condition that the reduced stresses are in a constant ratio with the yield stress of the material in the entire area of the plate. The above treatment can be called the *elastic design*.

The above solution was obtained by J. Kapkowski [14.15].

14.9.2 *Plastic design of plates*

The plate loaded in its plane can also be designed in another way. We can require that under a certain load every point of the plate simultaneously reaches the plastic state. We can assume here that the plate is made from a perfectly rigid—plastic material. Then, the equations of the theory of plasticity govern the state of stress in the plate: namely, the equations of internal equilibrium, the condition of plasticity, and the relations between the stresses and the strain velocities. Let us mention here the basic equations without deriving them. Readers anxious to study the problem more thoroughly should refer to special monographs concerning the mathematical theory of plasticity.

Let us consider two basic states: a plane state of stress and a plane state of strain; $\dot{\varepsilon}_z = 0$. The plane state of strain is determined by four components σ_{xx}, σ_{yy}, τ_{xy}, σ_{zz}. Then the above-mentioned equations take the following form:

The Huber–Mises yield condition. The assumption that the material becomes plastic when the intensity of the tangential stress reaches a certain critical value, gives a simple relation

$$\sigma_1 - \sigma_2 = 2k, \tag{14.52}$$

where σ_1, σ_2 are the principal stresses, and $k = \sigma_{pl}/\sqrt{3}$.

The same relation is valid for the *Tresca yield condition*, which assumes that the material becomes plastic if the maximum tangential stress reaches a certain critical value. In this case $k = \sigma_{pl}/2$. Expressing condition (14.52) by the components σ_{xx}, σ_{yy} τ_{xy}, we obtain

$$(\sigma_{xx} - \sigma_{yy})^2 + 4\tau_{xy}^2 = 4k^2.$$

The relations between the stresses and the strain velocities (*Levy–Mises relations*) take the form

$$\frac{\dot{\varepsilon}_x}{\sigma_{xx}-\sigma_s} = \frac{\dot{\varepsilon}_y}{\sigma_{yy}-\sigma_s} = \frac{\dot{\varepsilon}_{xy}}{\tau_{xy}}, \tag{14.53}$$

where $\sigma_s = \frac{1}{3}(\sigma_{xx}+\sigma_{yy}+\sigma_{zz})$ is the mean stress, and $\dot{\varepsilon}_x = \mathrm{d}\varepsilon_x/\mathrm{d}t$ strain velocity. As $\dot{\varepsilon}_z = 0$, the above condition (with the assumption of incompressibility of the material) gives the following relation between the stresses σ_{xx}, σ_{yy}, σ_{zz}:

$$\sigma_{zz} = \tfrac{1}{2}(\sigma_{xx}+\sigma_{yy}) \tag{14.54}$$

or

$$\sigma_3 = \tfrac{1}{2}(\sigma_1+\sigma_2).$$

The plane state of strain is met in thick or long elements of constant cross section towards the z-axis if the edge surfaces are undeformable.

On the contrary, the plane state of stress can be met in thin flat elements loaded at their edges by forces acting in their middle planes. We assumed that the coordinate axes x, y lie in the middle plane and the axis z is perpen-

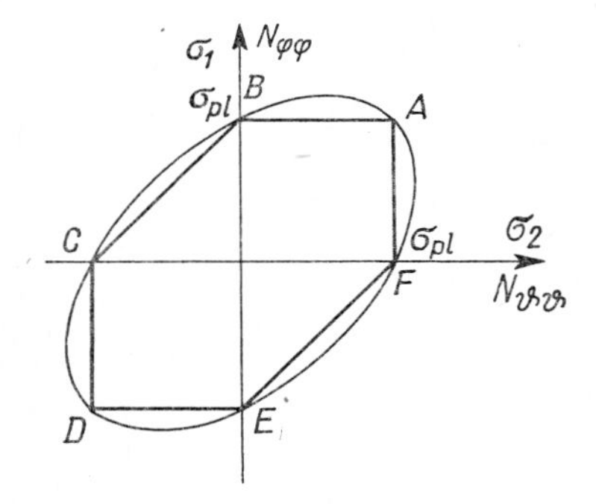

Fig. 163. Huber–Mises yield condition and Tresca yield condition for a plain state of stress.

dicular to them. Then $\sigma_{zz} = \sigma_3 = 0$. The Huber–Mises yield condition for the plane state of stress takes the following form

$$\sigma_1^2-\sigma_1\sigma_2+\sigma_2^2 = \sigma_{\mathrm{pl}}^2. \tag{14.55}$$

The ellipse presented in Fig. 163 is the representation of this condition on the plane of principal stresses. For the Tresca condition, two cases should be considered. If both stresses have opossite signs, $\sigma_1\sigma_2 \leqslant 0$, the condition takes the form

$$|\sigma_1-\sigma_2| = 2k, \quad \text{where} \quad k = \sigma_{\mathrm{pl}}/2. \tag{14.56}$$

In the case where the principal stresses have the same signs, i.e. when $\sigma_1\sigma_2 > 0$, we have at $\sigma_3 = 0$

$$|\sigma_1| = 2k \quad \text{for} \quad |\sigma_1| > |\sigma_2|,$$
$$|\sigma_2| = 2k \quad \text{for} \quad |\sigma_2| > |\sigma_1|.$$

These conditions are obvious if we use the representation of the state of stress given by Mohr circle.

Usually it is difficult to obtain the exact solution of all above-mentioned equations and we often limit ourselves to simplified solutions consisting of the determination of the statically admissible stress field or kinematically admissible strain velocity field. The statically admissible stress field has to satisfy the boundary conditions at the free and loaded edges of the plate and the conditions of internal equilibrium. Besides, the yield condition cannot be exceeded at any point of the field. The kinematic conditions may not be satisfied, i.e. conditions (14.53).

The velocity field is kinematically admissible if it represents a possible flow in the body and if all kinematic boundary conditions are satisfied. We conclude on the grounds of the extremal theorems of the theory of plasticity that we obtain a safe estimation of the dimensions of the structure by adopting the statically admissible stress field. By requiring the volume of the body to be minimum, we obtain the most economical shape. The real collapse load of an element designed in this way will be at least equal to the required collapse load. Designing the structure on the grounds of the kinematically admissible strain velocity fields, we obtain the real collapse load of the element to be at most equal to the required collapse load. This can result in too optimistic estimation of the collapse load of the structure. However, having the dimensions obtained by assuming the admissible field of velocities, the excess of the dimensions of the structure designed according to the statically admissible stress field can be estimated. If, for the assumed statically admissible stress field, the kinematic velocity field associated with it can be determined, the problem is solved exactly and the collapse load of the structure is equal to the applied limit load. However, minimum volume is not the only criterion determining the shape of the structure. Sometimes, manufacture reasons or the condition of minimum costs can be more important.

14.9.3 *Design for minimum weight*

The weight of the homogeneous plate depends on its volume. The volume of the plate is given by the integral

$$V = \iint_S h \,\mathrm{d}x\,\mathrm{d}y \tag{14.57}$$

where S is the surface of the plate. Let us determine the thickness h in such a way that the above integral attains its minimum. Then we have:

$$\delta V = 0.$$

The state of stress in the plate must satisfy the equations of equilibrium:

$$F = \frac{\partial N_{xx}}{\partial x} + \frac{\partial N_{xy}}{\partial y} + X = 0,$$
$$G = \frac{N_{yy}}{\partial y} + \frac{\partial N_{xy}}{\partial x} + Y = 0. \tag{14.58}$$

Besides, the yield condition should be satisfied at every point of the surface S. We apply the Tresca yield condition, which can be written in the following form:

$$\text{(a)}\ N_{xx}N_{yy} < 0, \qquad H = (N_{xx}-N_{yy})^2+4N_{xy}^2-4k^2h^2 = 0,$$
$$\text{(b)}\ N_{xx}N_{yy} > 0, \tag{14.59}$$
$$H = (N_{xx}-N_{yy})^2+4N_{xy}^2-[4kh-\varkappa(N_{xx}+N_{yy})]^2 = 0,$$

where $\varkappa = \operatorname{sgn}(N_{xx}+N_{yy})$.

Let us consider the first case (a). The problem of determining the thickness of the plate consists in determination of the minimum of the integral (14.57) with the additional conditions (14.58), (14.59). According to the rules of the variational calculus we introduce the function

$$K = h+\lambda F+\mu G+\nu H, \tag{14.60}$$

where λ, μ, ν are the Lagrange multipliers.

Now we solve the problem of the minimum of the integral for the function K:

$$\iint_S K \mathrm{d}x\,\mathrm{d}y = \min. \tag{14.61}$$

The Euler–Lagrange equations

$$\frac{\partial}{\partial x}\left(\frac{\partial K}{\partial(\partial h/\partial x)}\right)+\frac{\partial}{\partial y}\left(\frac{\partial K}{\partial(\partial h/\partial y)}\right)-\frac{\partial K}{\partial h} = 0,$$

. . . .

give the following set of differential equations for the functions λ, μ, ν, h, N_{xx}, N_{yy}, N_{xy}:

$$
\begin{aligned}
&1-8k^2\nu h = 0,\\
&\frac{\partial\lambda}{\partial x} = 2\nu(N_{xx}-N_{yy}),\\
&\frac{\partial\lambda}{\partial y}+\frac{\partial\mu}{\partial x} = 8\nu N_{xy},\\
&\frac{\partial\mu}{\partial y} = 2\nu(N_{yy}-N_{xx}).
\end{aligned}
\tag{14.62}
$$

Equations (14.62) together with conditions (14.59) form a complete set of equations for the seven unknown functions N_{xx}, N_{yy}, N_{xy}, λ, μ, ν, h. From Eq. $(14.62)_1$ the function ν is inversely proportional to the thickness h. Excluding ν from Eqs. (14.62) we find the equations

$$
\begin{aligned}
&\frac{\partial\mu/\partial y-\partial\lambda/\partial x}{\partial\mu/\partial x+\partial\lambda/\partial y} = \frac{N_{yy}-N_{xx}}{2N_{xy}} = \frac{\sigma_{yy}-\sigma_{xx}}{2\tau_{xy}},\\
&\frac{\partial\lambda}{\partial x}+\frac{\partial\mu}{\partial y} = 0.
\end{aligned}
\tag{14.63}
$$

Comparing the above relations with Eq. (14.53), we find that λ and μ represent the velocities $\dot{\varepsilon}_x$ and $\dot{\varepsilon}_y$ of the points of the plate. From Eq. (14.62) we find that

$$
\frac{\partial\mu}{\partial y}-\frac{\partial\lambda}{\partial x} = 4\nu h(\sigma_{yy}-\sigma_{xx}),
$$

and also

$$
\dot{\varepsilon}_y-\dot{\varepsilon}_x = \frac{D}{8k^2}(\sigma_{yy}-\sigma_{xx}),
$$

where

$$
D = \dot{\varepsilon}_x\sigma_{xx}+\dot{\varepsilon}_y\sigma_{yy}+\ \dots = \dot{\varepsilon}_1\sigma_1+\dot{\varepsilon}_2\sigma_2
$$

is the rate of dissipation of energy per unit volume of the material due to plastic action. Comparing the above equations, we find that

$$
D = \text{constant}. \tag{14.64}
$$

We see that if a plate is at collapse under the given loads and the plate has a collapse mode such that the rate D of dissipation of energy per unit volume is constant, then the plate is of minimum volume. The above condition was obtained in another way by D. C. Drucker and R. T. Shield [14.22], [14.12]. The solution of the problem, i.e. designing the plate for minimum weight, consists in solving the above equations. Eliminating the velocities from Eqs. (14.62), (14.63), the above set can be reduced to four quasi-linear differential equations. The solution of these equations can be found by integration along the characteristics similar to the case of plane problems of theory of plasticity. We mention below some results only from paper [14.14]. The reader who wants to study the problem more thoroughly should refer to [14.12], [14.13], etc. Many authors recently considered the problem of design of plates for minimum weight. The following authors should be mentioned: D. C. Drucker and R. T. Shield [14.12], W. Prager [14.13], M. Sh. Mikeladze [14.37], and M.I. Estrin [14.38] who used the above treatment.

14.9.4 *Support for a single force*

We suppose that the material of the plate can be considered to be in the semi-infinite region $x \geqslant 0$ and that the boundary $x = 0$ is a motion free support (Fig. 164). Let a single force P be applied anywhere in the right half plane. If the force makes an angle $\psi \leqslant \pi/4$ with the x-axis, then the minimum volume

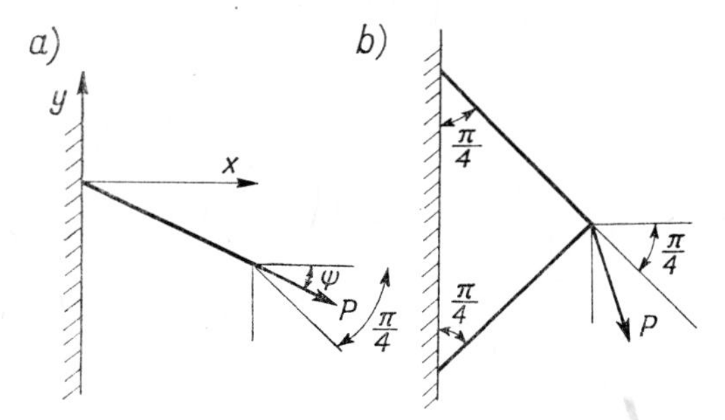

Fig. 164. Bar introducing single force into a semi-infinite region (from Te-Chiang Hu, R .T. Shield [14.14]).

design consists of a single bar in the direction of the force and connected to the supporting line as shown in Fig. 164a. The bar has a finite cross-sectional area but zero width. The structure is in the regime given by the point B at Fig. 163. For $\psi > \pi/4$ the minimum volume design consists of two bars from the point of application of the force to the support, each inclined at an angle $\pi/4$ with the x-axis. The structure is in the regime BC, Fig. 163.

Let us consider now a rigid circular support of radius a (Fig. 165). A single force P is applied at the point A in the plane of the circle at the distance R

from the centre O. If the force is vertically downward $\psi = \pi/2$ (Fig. 165a), the minimum volume design is then a plate ABC bounded by two similar equiangular spirals of angle $\pi/4$ which intersect orthogonally at the point A. The plate is in the regime BC and the lines of principal stress are equiangular spirals. The spirals which form the edges of the disc are two ribs, carrying a tensile and compressive force $P/\sqrt{2}$.

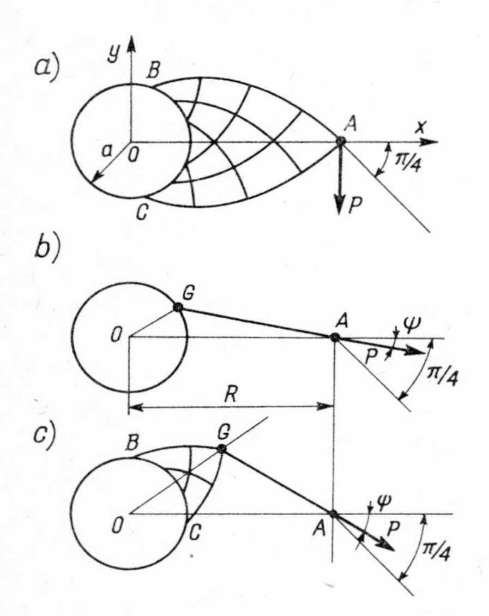

Fig. 165. Bar introducing single force into a circular support ([14.14]).

When the inclination ψ of the force to the horizontal line is less than $\pi/4$ the form of the design is presented in Fig. 165b or in Fig. 165c depending on whether $\sin\psi$ is smaller or greater than $a/\sqrt{2R}$. In the first case, the minimum volume design is the single bar AG. For $\sin\psi > a/\sqrt{2}R$ the design consists of a bar AG in tension and a plate GBC bounded by equiangular spirals.

The thickness of the plate in the problems presented in Fig. 165a and Fig. 165c can be found from the equilibrium equations. Assuming that the curvilinear coordinates coincide with the lines of principal stress, we have

$$\frac{\partial N_1}{\partial s_1} + \frac{N_1 - N_2}{\varrho_2} = 0, \qquad \frac{\partial N_2}{\partial s_2} + \frac{N_2 - N_1}{\varrho_1} = 0, \tag{14.65}$$

where ϱ_1, ϱ_2 are the curvatures of the lines of the principal stress. It is necessary to integrate the hyperbolic system of equations (14.65) in conjunction with the appropriate boundary conditions. The integration is simplified by the fact that the principal stress lines, which are characteristics of Eqs. (14.65), are a system of equiangular spirals. We obtain the following result for the case shown in Fig. 165a:

$$\sigma_0 rh = f(\xi, \eta) = 2PJ_0(2\sqrt{\xi\eta}) - P\left(\sqrt{\frac{\xi}{\eta}} - \sqrt{\frac{\eta}{\xi}}\right)J_1(2\sqrt{\xi\eta})$$

where the variables ξ, η are given by the relations

$$\vartheta = \xi+\eta, \qquad r = Re^{\xi-\eta},$$

and (r, ϑ) are polar coordinates, OA being the line $\vartheta = 0$.

The kinematically admissible velocity fields can be found for all above-mentioned problems.

14.10 Design of circular plates under lateral loads

The design of the plate under lateral loads consists of the determination of the variable thickness so that certain additional conditions regarding the magnitude of stresses, minimum of volume, etc. are satisfied.

Two ways of solving the problem are available. One of them consists in shaping the plate in such a way that it is just at the point of collapse under the given loads. The material is assumed to be perfectly plastic and at the point of collapse yield stresses appear in the entire plate. This treatment can be called *plastic design.*

The second way consists in assuming that the entire plate is in the elastic state and only the maximum reduced stresses in the external surfaces are equal to the yield stress or are less than yield. Such a treatment can be called *elastic design.*

14.10.1 *Plastic design*

Let us discuss first the "plastic design", which was investigated by H. G. Hopkins and W. Prager [14.19], and W. Freiberger and B. Tekinalp [14.20]. The problem under consideration is the design of a circular plate for minimum weight. The plate is loaded by a force P distributed over the surface $q = P/\pi c^2$, where c is the radius measured from the plate centre. The edge $r = R$ of the plate is simply supported or built-in. We follow here the results due to E.T.Onat, W. Schumann, and R. T. Shield [14.21], and assume that under the given load the entire plate is in the plastic state. As has been shown [14.22], a plate designed to collapse in a mode such that the rate of dissipation of energy D_A over a unit volume is constant on the surface of the plate, is the minimum volume design. In the case of solid plates this provides a relative minimum of the functional for the volume of the plate. Then

$$\frac{D_A}{h} = \text{const.} > 0.$$

Here, h is the thickness of the plate. Since we ignore the membrane stresses, the rate of dissipation of energy D_A due to plastic action per unit area of the middle surface of the plate is given by

$$D_A = M_{rr}\varkappa_{rr} + M_{\vartheta\vartheta}\varkappa_{\vartheta\vartheta}. \tag{14.66}$$

For equilibrium the moments satisfy the equation

$$\frac{d^2}{dr^2}(rM_{rr}) - \frac{d}{dr}(M_{\vartheta\vartheta}) + rq = 0. \tag{14.67}$$

The curvatures $\varkappa_{rr}$, $\varkappa_{\vartheta\vartheta}$ in the radial and circumferential directions are given by

$$\varkappa_{rr} = -\frac{d^2w}{dr^2}, \qquad \varkappa_{\vartheta\vartheta} = -\frac{1}{r}\frac{dw}{dr}.$$

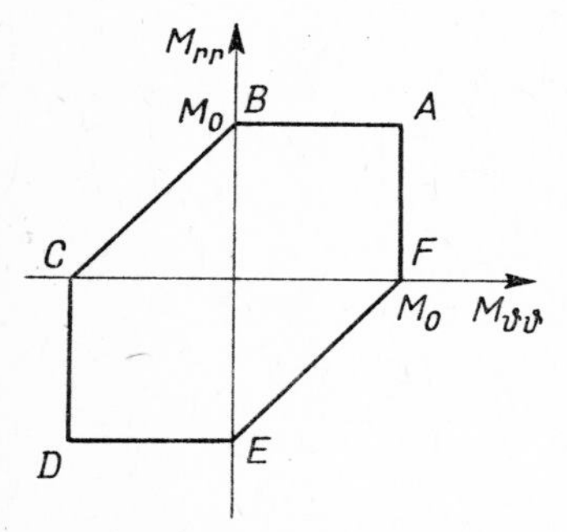

Fig. 166. Tresca's yield condition for a circular plate.

Moreover, it is assumed that the elastic—perfectly plastic material obeys Tresca's yield condition and the associated flow rule. Plastic yielding can occur when one or both of the moments M_{rr} and $M_{\vartheta\vartheta}$ is equal to the fully plastic moment M_0 or when the magnitude of $(M_{rr} - M_{\vartheta\vartheta})$ equals M_0. This condition defines a yield curve (Fig. 166). For a point on a side of the hexagon, the flow associated with the yield condition requires that during collapse the curvature vector ($\varkappa_{rr}$, $\varkappa_{\vartheta\vartheta}$) be parallel to the outwards directed normal to the sides. At the corner of the hexagon the vector must lie between the normals to the sides which meet at the corner. For a plate the fully plastic moment M_0 is given by

$$M_0 = \frac{1}{4}\sigma_{pl}h^2 \tag{14.68}$$

where σ_{pl} is the yield stress of the material.

Considering the simply supported plate we can assume that

$$M_{rr} = M_{\vartheta\vartheta} = M_0 \tag{14.69}$$

(point A in Fig. 166). This assumption and the equilibrium equation (14.67) together with the conditions at the centre and the edge of the plate lead to the equation

$$\int_r^R \frac{1}{\xi}\,\mathrm{d}\xi \int_0^\xi \varrho q(\varrho)\,\mathrm{d}\varrho = M_0. \tag{14.70}$$

For q = constant over the area $0 \leqslant r \leqslant c$ the distribution of the thickness for a simply supported plate is obtained by direct substitution of Eq. (16.69) into Eq. (14.66)

$$h = \begin{cases} \left(\dfrac{q}{\sigma_{pl}}\right)^{1/2}\left(c^2 - r^2 + 2c^2 \ln\dfrac{R}{c}\right)^{1/2}, & 0 \leqslant r \leqslant c, \\[2ex] \left(\dfrac{q}{\sigma_{pl}}\right)^{1/2} rc \ln\dfrac{R}{r}, & c \leqslant r \leqslant R. \end{cases} \tag{14.71}$$

If the collapse mode is associated with the point A, $\varkappa_{rr}$ and $\varkappa_{\vartheta\vartheta}$ must be non-positive. From Eq. (14.65) it follows that this condition is satisfied. When the concentrated force acts on the plate, the relative minimum volume design is

$$h = \left(\frac{P}{\pi\sigma_{pl}}\right)^{1/2} \ln\frac{R}{r} \tag{14.72}$$

with volume

$$V = \frac{1}{2}\left(\frac{P}{\sigma_{pl}}\right)^{1/2} \pi R^2.$$

For the built-in plate under the action of the force P we assume that point A applies for $0 \leqslant r \leqslant b$. Since near the edge $\varkappa_{rr} \leqslant 0$ and $\varkappa_{\vartheta\vartheta} \geqslant 0$, therefore regime C for $b \leqslant r \leqslant R$ is applied; b is a certain radius which can be found from the conditions for w and dw/dr at the junction $r = b$ of the two regions. We obtain the thickness from (14.67) as

$$h = \begin{cases} \left(\dfrac{P}{\pi\sigma_{pl}}\right)^{1/2} \ln\dfrac{b}{r}, & 0 \leqslant r \leqslant b, \\ \left(\dfrac{2P}{\pi\sigma_{pl}} \dfrac{r-b}{r}\right)^{1/2}, & b \leqslant r \leqslant R, \end{cases} \tag{14.73}$$

and $b = 0.74R$. The volume of the plate is found to be

$$V = 0.407 \left(\frac{P}{\sigma_{pl}}\right)^{1/2} \pi R^2. \tag{14.74}$$

For a circular plate in the fully plastic state and for both simply supported and built-in edges, the limit value of the central force P is $2\pi\sigma_{pl}h^2$. Then the plate of uniform thickness which is just at collapse under the force P has the volume

$$V = 0.798 \left(\frac{P}{\sigma_{pl}}\right)^{1/2} \pi R^2$$

for both built-in and simply supported edges. Comparing the above results with these for the constant thickness design, we see that the savings are 37 per cent (Eq. (14.72)) and 49 per cent (Eq. (14.73)).

Sandwich plates can be treated in a similar way. In the case of a plate built of a light-weight core of constant thickness H and two identical face sheets of variable thickness $h \ll H$, we require the minimum volume of the sheets. The core carries no bending stresses and the fully plastic moment M_0 has the value

$$M_0 = \sigma_{pl} H h. \tag{14.75}$$

The condition D_A/h is independent of h because of the linear dependences (14.66) and (14.75). We can directly obtain the deflection w using point A or C of Fig. 163. As the result we obtain the minimum volume design. In the case of a concentrated force P at the centre of the plate, the following thickness is obtained from the equilibrium equation (14.67) assuming that $M_{rr} = M_{\vartheta\vartheta} = M_0$:

$$h = \frac{P}{2\pi\sigma_{pl} H} \ln\frac{R}{r}. \tag{14.76}$$

The volume of the face sheets is

$$V = \frac{1}{2} \frac{PR^2}{\sigma_{pl} H}. \tag{14.77}$$

In the case of built-in edges the thickness is given by

$$h = \begin{cases} \dfrac{P}{2\pi\sigma_{pl} H} \ln \dfrac{2}{3} \dfrac{R}{r} & \text{if} \quad 0 \leqslant r \leqslant \dfrac{2}{3} R, \\[2ex] \dfrac{P}{2\pi\sigma_{pl} H} \left(1 - \dfrac{2}{3} \dfrac{R}{r}\right) & \text{if} \quad \dfrac{2}{3} R \leqslant r \leqslant R. \end{cases} \tag{14.78}$$

The volume of the face sheets

$$V = \frac{1}{3} \frac{PR^2}{\sigma_{pl} H} .$$

As was shown by H. G. Hopkins and W. Prager [14.19], the limit value of the concentrated central force P for a sandwich plate with constant fully plastic bending moment M_0 is $P = 2\pi M_0$ for both face sheets of constant thickness of the sandwich plate having the same load carrying capacity is

$$V = \frac{PR^2}{\sigma_{pl} H} .$$

This means that the volume of the facings of constant thickness is twice larger than the volume of facings of a simply supported plate and three times larger than that of a built-in plate, designed for minimum weight.

14.10.2. *Circular plate of uniform strength loaded by a lateral concentrated force*

Let us take for a second example the elastic design of a circular plate loaded at its centre by a concentrated force [14.17]. We determine the thickness in such a way that the maximum stresses in the plate are constant and equal to the yield stress.

In order to find the shape of the plate we have to solve the following set of equations:

(1) The equation of equilibrium of the plate (14.67) which in the case of axial symmetry can be represented in the form

$$M_{rr} + r \frac{dM_{rr}}{dr} - M_{\vartheta\vartheta} + Q_r r = 0 \tag{14.79}$$

where

$$Q_r = \frac{1}{r} \int_0^r q r \, dr$$

is the shear force.

(2) The condition of uniform strength.

(3) The equation of compatibility of strains.

The stresses in the external layers of the plate exceed by many times the tangential stresses which are largest in the middle plane. Therefore we shall consider the plate material only in the external layer. Neglecting the stresses σ_{zz} perpendicular to the middle surface, we have the two-dimensional state of stress. Let us apply the Huber–Mises yield condition which for the symmetrical case in a system of polar coordinates takes the form

$$\sigma_0^2 = \sigma_{rr}^2 - \sigma_{rr}\sigma_{\vartheta\vartheta} + \sigma_{\vartheta\vartheta}^2 \qquad (z = \pm h/2),$$

This condition is fulfilled identically if the stresses in the external layers are defined by

$$\left.\begin{matrix}\sigma_{rr}\\ \sigma_{\vartheta\vartheta}\end{matrix}\right\} = 2\frac{\sigma_0}{\sqrt{3}}\cos\left(\omega \pm \frac{\pi}{6}\right),$$

where $\omega = \omega(r)$ is an arbitrary function of variable r.

The bending moments in the plate are defined by the formulae

$$\left.\begin{matrix}M_{rr}\\ M_{\vartheta\vartheta}\end{matrix}\right\} = \frac{\sigma_0 h^2}{3\sqrt{3}}\cos\left(\omega \pm \frac{\pi}{6}\right). \tag{14.80}$$

Eliminating the thickness h, we obtain the set of equations, in which there are only two unknown functions φ and ω:

$$\begin{aligned}&\frac{d\varphi}{dr} - \frac{\varphi}{r}\frac{\sin(\mu-\omega)}{\sin(\mu+\omega)} = 0,\\ &r\frac{d\omega}{dr} = -\left[\frac{3\sqrt{3}k^2}{\sigma_0}\frac{Q_r}{r}\varphi^2 - f_1(\omega)\right]\frac{1}{f_2(\omega)},\end{aligned} \tag{14.81}$$

where

$$\mu = \arcsin\frac{\sqrt{3}}{2}\frac{1-\nu}{\sqrt{1-\nu+\nu^2}}, \qquad k = \frac{\sqrt{3}E}{4\sigma_0\sqrt{1-\nu+\nu^2}},$$

$$f_1(\omega) = \sin\omega\sin(\omega+\mu)[\sin(\omega+\mu) - 4\cos\mu\cos(\omega+\pi/6)],$$

$$f_2(\omega) = \frac{1}{2}\sin(\omega+\mu)[\cos(\mu-\pi/6) + 3\cos(2\omega+\mu+\pi/6)].$$

In order to define the stresses and the thickness of the plate it is adequate to find the solutions $\omega(r)$ and $\varphi(r)$ holomorphical in the vicinity of the singular point $r = 0$. Knowing the function $\omega(r)$, we obtain the thickness of the plate from the equation

$$h = \frac{r}{k\varphi}\sin(\mu+\omega). \tag{14.82}$$

Let us solve the above equations for the case of the plate loaded by a force distributed uniformly over the circular part of its surface defined by $0 \leqslant r \leqslant c$. For the loaded region $r < c$, the shear force is

$$Q_r = \frac{q_0 r}{2}.$$

For the unloaded region $r \geqslant c$

$$Q_r = \frac{q_0 c^2}{2r}.$$

At the radius $r = c$ where the load suddenly changes the function ω describing the state of stress should be continuous. Then the following conditions should be satisfied:

$$\omega^{(1)}(c) = \omega^{(2)}(c); \qquad \left(\frac{d\omega^{(1)}}{dr}\right)_{(r=c)} = \left(\frac{d\omega^{(2)}}{dr}\right)_{(r=c)}.$$

Numerical results can be obtained by a numerical integration of Eq. $(14.81)_1$. The shape of the plate is given in Fig. 167. We see that in the limiting case where $c \to 0$ and the load becomes a concentrated force, the thickness of the plate for $r = 0$ increases to infinity.

Knowing the solution of Eq. $(14.81)_1$, the deflection of the plate can easily be determined. To obtain the deflection it is sufficient to integrate the function $\varphi(r) = -dw/dr$ once:

$$w = -\int_0^r \varphi\, dr.$$

The maximum values of the deflection are at the centre of the plate they are given in Table 9 in the first row.

Table 9.

No.	c/a	1.0	0.5	0.2	0.01
1	w^*	0.139	0.353	1.01	3.09
2	h_0	0.867	0.860	0.79	0.77
3	$w^*_{h=\text{const.}}$	0.0887	0.202	0.305	0.344

In the table

$$w^* = w \frac{\pi D_0}{Pa^2},$$

D_0 is the rigidity of the plate of uniform strength at the point $r = 0$.

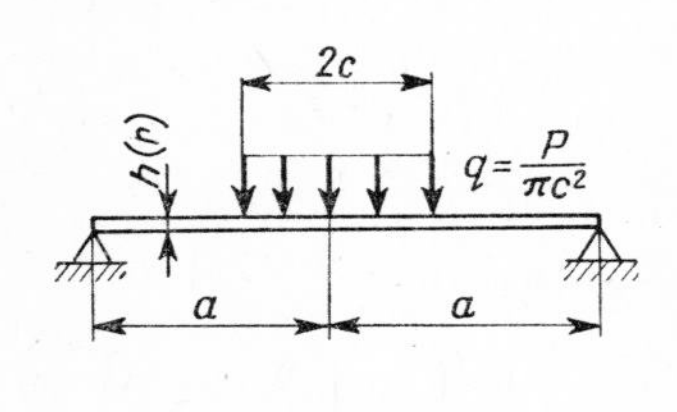

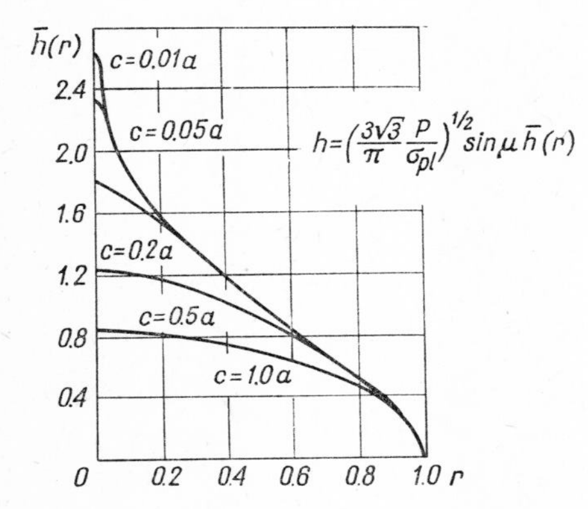

Fig. 167. Circular plate of a uniform strength.

Comparing the deflections of the designed plates of uniform strength with maximum deflections of the plates of uniform thickness calculated on the assumption that the maximum resistance of the material is identical with that for plates of uniform strength, one should say that these deflection are much larger. The thickness of the plate of uniform strength may be found from

$$h_{\text{const}} = h_0 \sqrt{\frac{M_{\text{constant}}}{M_0}}$$

where M_{constant} is the moment at $r = 0$ in the plate of constant thickness and M_0 is the moment in the plate of uniform strength. For the cases of loading considered, the thicknesses are given in the second row of Table 9.

The corresponding deflections of the plate of uniform thickness are given in the third row of Table 9.

A particularly large increase of the deflection is observed in the case where the force is distributed over a small area, for example $c = 0.01$. The deflection of the plate of uniform strength is in this case nearly ten times larger than that of the plate of constant thickness.

It should be pointed out here that formulae (14.80) and (14.81) defining the thickness of the plate are not valid near the loading point as they have been developed by simplified assumptions. Namely, Eqs. (14.80) are developed on the assumption that the plane state of stress governs in the entire plate. This condition is satisfied approximately when the thickness of the plate changes in a smooth way and to a small degree. However, the thickness defined by Eq. (14.82) increases infinitely at the loading point. Therefore, a three-dimensional state of stresses appears in the vicinity of this point and the solution described is only a simplified one.

14.11 Optimum design of shells

As we have seen from previous chapters, concentrated loads cause a strong concentration of stresses in a shell. Also, when the loads act on only a small portion of the shell surface, there appears a concentration of stresses making the influence of this area large in comparison to the other parts of the structure. Therefore, designers usually strengthen (locally) the loaded places by welding strips, ribs, or other strengthening elements. During the analysis a reinforced shell may be regarded as a shell whose thickness is given by the distributions.

We can also achieve an optimum shell by proper forming of its thickness and middle surface. The problem of the optimum design can be formulated, for example, in the following way. Let us assume that the loads acting on the structure should be defined to satisfy the respective condition of optimization. Simultaneously, the stresses in the structure should not exceed the admissible stresses. Each of the following conditions can be assumed as the optimization condition, namely: the condition of minimum weight, minimum volume, minimum surface, minimum or maximum stiffeness, maximum stability, etc. Analysing the problem of the optimum design of shells, we usually assume the condition of minimum weight as the governing criterion.

The problem of designing of a shell of uniform strength can be formulated in the following way. Let us define the shape of the middle surface in such a way that the condition of constant strength is satisfied at every point of the structure. The theoretical solutions of these problems are usually mathematically complicated and they are known only for shells of revolution and shallow shells. We mention here the papers by F. Tölke [14.24], G. Megareus [14.25], K. Federhoffer [14.29], and H. Ziegler [14.26] devoted to the problem of shells subjected to variable or constant pressure. The case of simultaneous action of the concentrated load and the pressure was discussed in [14.31]. The solutions in all the above-mentioned papers were obtained based on the membrane theory of shells. It was assumed that the bending areas do not appear in the shell. This condition imposes additional limitation regarding the edge loads and displacements which cannot always be satisfied. For example, at the clamped edge of the shell certain displacements appear. In such a case the solutions should be completed by the analysis of the edge areas.

14.12 Basic equations of the problem

Assuming that the shell is in the membrane state of stresses, we obtain the following set of equations:

(a) *Equations of equilibrium*

$$\frac{\partial}{\partial \alpha_1}(A_2 N_{11}) - N_{22}\frac{\partial A_2}{\partial \alpha_1} + \frac{\partial}{\partial \alpha_2}(A_1 N_{12}) + \frac{\partial A_1}{\partial \alpha_2} N_{12} + A_1 A_2 X = 0,$$

$$\frac{\partial}{\partial \alpha_2}(A_1 N_{22}) - N_{11}\frac{\partial A_1}{\partial \alpha_2} + \frac{\partial}{\partial \alpha_1}(A_2 N_{12}) + \frac{\partial A_2}{\partial \alpha_1} N_{12} + A_1 A_2 Y = 0, \qquad (14.83)$$

$$\frac{N_{11}}{R_1} + \frac{N_{22}}{R_2} = Z.$$

(b) A *condition of uniform strength*, which is equivalent to the Huber–Mises yield condition. For the plane state of stress, we have

$$\sigma_0 = \sqrt{\sigma_1^2 - \sigma_1 \sigma_2 + \sigma_2^2}\,, \qquad (14.84)$$

where σ_i are the principal stresses. We can also apply the linearized Tresca yield condition presented by the hexagon in Fig. 163.

(c) A *condition of optimization*. This condition can be formulated in various ways. If we require the minimum volume, it takes the form

$$\delta V = \delta \iint_S h \mathrm{d}S, \tag{14.85}$$

where $\mathrm{d}S$ denotes the element of the middle surface of the shell, and h is the thickness.

The above equations contain five unknown functions: the shape of the middle surface, the variable thickness of the shell $h(\alpha_1 \alpha_2)$, and the internal membrane forces, N_{11}, N_{22}, N_{12}. This set is often solved by the assumption that the principle stresses

$$\sigma_1 = \sigma_2 = \sigma_0 = \text{constant}, \tag{14.86}$$

are equal at every point of the shell. Condition (14.86) satisfying Eqs. (14.84) introduces usually additional limitation in comparison with the general condition (14.84). Then the set of equations (14.83) together with Eq. (14.86) make a set enabling the determination of the shape and thickness of the structure. However, if we use condition (14.84) instead of (14.86), one equation is lacking. In order to find a solution these equations should be completed by an additional equation. Then we can demand for example that the shell is a shell of minimum volume, etc.

The problem of the designing the shell can also be solved in a different way. Uniform strength can be achieved by designing the thickness by changing it symmetrically with respect to the middle surface. In this case the form of the middle surface does not undergo a change. To ensure constant strength it is adequate to determine the proper thickness. The solution, which is rather difficult to obtain, in this case reduces to solving the problem of the shell which carries both direct and bending stresses. Certain problems of shells in limit state were solved in this way, for example by R. T. Shield [14.33]. The solutions were obtained by the assumption that the material of the complete shell was in the plastic state. The basic criterion resulting from the condition of minimum volume demands that the dissipation of the strain energy per unit surface of the shell is a constant value.

There is also a third method of designing shells. We can resign from the variable thickness and search only for the shape of the middle surface. If we assume that h = constant, it is impossible to satisfy the additional condition such as minimum volume or minimum costs, etc., because this assumption determines at once the thickness and shape of the shell. However, it can be proved that also in this case the shell designed by the condition (14.86) and loaded only at the edges satisfies the condition of minimum volume.

14.13 Shells of constant thickness

If we assume that the thickness of the shell is constant, the shape of its middle surface is defined by Eqs. (14.83) and (14.84). This shape depends on the applied loads and boundary conditions, which should be assumed in a way permitting the appearance of the membrane state of stress in the shell. Let us assume that the shell is subjected to the load $Z = P(\alpha_1, \alpha_2)$ and the boundary forces only. The set of equations (14.83) and (14.84) can be satisfied by the assumption that $N_{11} = N_{22} = h\sigma_0$, $N_{12} = 0$. This corresponds to the point A on the Tresca hexagon (Fig. 163). Then the first and second equation of the group (14.83) and the yield condition (14.85) are satisfied identically. The third equation (14.83) takes the form

$$\frac{1}{R_1} + \frac{1}{R_2} = 2H = \frac{P}{\sigma_0 h},$$

which means that the mean curvature is proportional to the load $p(\alpha_1, \alpha_2)$. If we assume that the surface of the shell is given by the function $z = z(x, y)$, the mean curvature satisfies the condition

$$2H = \frac{z''(1+z^{\cdot 2}) - 2z'z^{\cdot}z'^{\cdot} + z^{\cdot\cdot}(1+z'^2)}{(1+z'^2+z^{\cdot 2})^{3/2}} = \frac{p}{\sigma_0 h} \tag{14.87}$$

where $z' = \mathrm{d}z/\mathrm{d}x$, $z^{\cdot} = \mathrm{d}z/\mathrm{d}y$. The condition of minimum volume yields

$$\delta V = h\delta \iint_S \sqrt{1+z'^2+z^{\cdot 2}}\,\mathrm{d}x\,\mathrm{d}y = 0.$$

Euler's equation for the minimum of the above functional takes the form

$$z''(1+z^{\cdot 2}) - 2z'z^{\cdot}z'^{\cdot} + z^{\cdot\cdot}(1+z'^2) = 0.$$

We see that if $p = 0$, the shell designed by the condition of constant stresses (14.86) is simultaneously the shell of minimum volume. It is difficult to obtain the exact solution of Eq. (14.87). It is known only for shallow shells and shells of revolution.

Shallow shells. If we design a shallow shell, Eq. (14.87) can be simplified by neglecting the non-linear terms. We obtain the simple equation for the function $z(x, y)$

$$\Delta z = p/h\sigma_0,$$

where Δ is the Laplace operator.

Shells of revolution. The shape of the shell of revolution can be defined by the function $z = z(r)$. Then the principal radii of curvature are given by the formulae

$$\frac{1}{R_1} = \frac{z''}{(1+z'^2)^{3/2}}, \qquad \frac{1}{R_2} = \frac{z'}{r(1+z'^2)^{1/2}}, \qquad \sin\varphi = \frac{z'}{(1+z'^2)^{1/2}},$$

where

$$z' = \frac{\mathrm{d}z}{\mathrm{d}r}, \qquad z'' = \frac{\mathrm{d}^2 z}{\mathrm{d}r^2}, \qquad r = R_2 \sin\varphi.$$

The equilibrium equations are the following

$$\frac{\mathrm{d}}{\mathrm{d}\varphi}(rN_{\varphi\varphi}\sin\varphi) = (p_r - p_\varphi \tan\varphi)r\frac{\mathrm{d}r}{\mathrm{d}\varphi}, \qquad \frac{N_{\varphi\varphi}}{R_1} + \frac{N_{\vartheta\vartheta}}{R_2} = p_r \tag{14.88}$$

where $p_r = Z$, $p_\varphi = X$.

The first equation can be integrated. We obtain

$$rN_{\varphi\varphi}\sin\varphi = \int r(p_r - p_\varphi \tan\varphi)\,\mathrm{d}r + C.$$

If $N_{\varphi\varphi} = N_{\vartheta\vartheta} = h\sigma_0$, we have

$$\sin\varphi = \frac{1}{h\sigma_0 r}\int r(p_r - p_\varphi\tan\varphi)\,\mathrm{d}r + \frac{C}{h\sigma_0}\frac{1}{r}.$$

If the shell is subjected to the pressure p_r and axial load P we find

$$\sin\varphi = \frac{1}{h\sigma_0 r}\int_0^r p_r r\,\mathrm{d}r + \frac{P}{2\pi h\sigma_0}\frac{1}{r}.$$

The integral of this equation for $p = $ constant takes the form

$$z = \int \frac{\mathrm{d}r}{\sqrt{\dfrac{r_0^4}{\varrho_0^2} + \dfrac{r^2}{(r_0^2+r^2)^2} - 1}} + z_0.$$

where we write

$$\varrho_0 = \frac{P}{2\pi h\sigma_0}, \qquad r_0^2 = \frac{P}{\pi p}.$$

The above integral can be evaluated by reducing it to elliptic integrals. However, the graphical integration is here more convenient. If the shell is under a concentrated load P only and $p = 0$, the above integral can be evaluated in the elementary way. We obtain

$$z = \varrho_0 \ln\left(\sqrt{\left(\frac{r}{\varrho_0}\right)^2 - 1} + \frac{r}{\varrho_0}\right) + z_0.$$

This is the equation of the brachistochrone. We see that $\varrho_0 = P/2\pi h\sigma_0$ is the smallest possible radius of the shell. Assuming that the shell is loaded by the internal pressure p_0 = constant only, we find

$$z' = \frac{p_0}{2\sigma_0 h} \int \frac{r^2 \mathrm{d}r}{\sqrt{r^2 - \left(\frac{p_0 r^2}{2\sigma_0 h} + C_1\right)^2}},$$

where C_1 is the integration constant. Assuming $C_1 = 0$, we find

$$z = \sqrt{R^2 - r^2},$$

where

$$R = \frac{2\sigma_0 h}{p_0}.$$

This is the equation of the sphere of radius R. It should be mentioned here that we do not obtain a different shape of the shell taking into account the more general condition (14.84) instead of (14.86), and that the above solutions are unique. It can be proved by the introduction into the equilibrium equations the stresses satisfying the Tresca yield condition. Assuming, for example, that $\sigma_1 = \sigma_0$ and $\sigma_2 = \chi\sigma_0$ or that $\sigma_1 = \chi\sigma_0$, $\sigma_2 = \sigma_0$ where $0 \leqslant \chi \leqslant 1$, we obtain from the first equation (14.83) the shape of the shell. The second equation defines the coefficient χ. Solving it we find that χ is always equal to unity and that it corresponds to condition (14.86).

14.14 Design of a shell of variable thickness

14.14.1 *Shell loaded by a concentrated force and by a pressure*

Let us consider at first a case of a shell loaded by a central concentrated force P and by a constant external lateral pressure p_0 simultaneously. Let us design the shell in such a way as to avoid the appearance of the bending moments and to assure equilibrium by membrane forces only. Simultaneously, let us demand each point of the shell surface fulfils the condition of uniform strength (14.84). Also the condition of minimum weight of the shell (14.85) should be fulfilled. Such an optimum design is performed without regard to difficulties and costs that may arise in its manufacture. Nevertheless, it can be the basis of comparison for any proposed structure.

The four unknowns to be found are the following: the internal forces $N_{\varphi\varphi}$, $N_{\vartheta\vartheta}$, the thickness of the shell h, and the shape of the shell determined by the function $z(r)$. We have at our disposal the following equations containing the unknown functions: two differential equations of equilibrium, the condition of constant strength. We assume that the material of the shell obeys the Tresca yield criterion (Fig. 163). The condition of minimum weight depends on the volume of the shell. The function $z(r)$ can be determined from the condition of minimum volume. Let us assume that $N_{\varphi\varphi} \leqslant N_{\vartheta\vartheta} \leqslant 0$. Then the stresses are represented by the straight line DE of the hexagon. We write

$$N_{\varphi\varphi} = -\sigma_0 h, \qquad N_{\vartheta\vartheta} = -\chi\sigma_0 h,$$

where χ is a coefficient which should be $0 \leqslant \chi \leqslant 1$, and σ_0 is the tensile yield stress of the material. On substitution into equilibrium equation (14.88) we obtain the thickness

$$h = \frac{p_0}{2\sigma_0}\,\frac{r_0^2+r^2}{r}\,\frac{(1+z'^2)^{1/2}}{z'}, \qquad r_0^2 = \frac{P}{\pi p_0}. \tag{14.89}$$

The volume of the shell is

$$V = \frac{\pi p_0}{\sigma_0}\int_0^{r_1} (r_0^2+r^2)\left(\frac{1}{z'}+z'\right)\mathrm{d}r.$$

The above functional has a minimum when the following Euler equation is satisfied

$$\frac{\mathrm{d}}{\mathrm{d}r}\left[(r_0^2+r^2)\left(-\frac{1}{z'^2}+1\right)\right] = 0,$$

which yields

$$z' = \pm\sqrt{\frac{r_0^2+r^2}{r_0^2+B+r^2}} \tag{14.90}$$

where B is an arbitrary constant. Equation (14.90) can be integrated. We obtain

$$z = -r_0^2 E\left[\mathrm{gd}\left(\operatorname{arcsinh}\frac{r_1}{\sqrt{r_0^2+B}}\right),\ \left(\frac{r_0^2+B}{r_0^2}\right)^{1/2}\right]+C \tag{14.91}$$

where

$$\mathrm{gd}(r) = \int_0^r \frac{\mathrm{d}t}{\cosh t}$$

and $E(\ )$ is an elliptic function. For $r = 0$ the slope depends on the constant B and is $z' = \pm[r_0^2/(r^2+B)]^{1/2}$. This angle has a real value when $(r^2+B) > 0$. The thickness takes the form

$$h = \frac{p_0}{2\sigma_0}\frac{1}{r}(2r_0^2+B+2r^2)^{1/2}(r_0^2+r^2)^{1/2}. \tag{14.92}$$

The above solution holds only when $0 \leqslant \chi \leqslant 1$. Therefore the value of χ should be evaluated; it can be obtained from the second equilibrium equation. On transformation we have

$$\chi = \frac{2r^2}{r^2+r_0^2} - \frac{rz''}{(1+z'^2)z'}.$$

Substitution of z'' and z' from (14.90) yields

$$\chi = \frac{r^2}{r_0^2+r^2}\,\frac{4r_0^2+B+4r^2}{2r_0^2+B+2r^2}; \tag{14.93}$$

χ satisfies the condition $0 \leqslant \chi \leqslant 1$ when $0 \leqslant r^4 \leqslant r_0^2(r_0^2-B/2)$. The constants B and C make it possible to determine the shape of the shell in accordance with the boundary conditions. Some examples have been given in Fig. 168.

Let us now assume that the height of the shell is not determined and its edge lies on the line $r = r_1$. The condition of minimum volume gives

$$(r_0^2+r^2)\left(-\frac{1}{z'^2}+1\right)_{r=r_1} = 0, \qquad z' = 1.$$

In conclusion we observe that minimum volume is obtained for the conical shape with the angle $\alpha = 45°$. The thickness should vary as

$$h = \frac{p_0}{\sqrt{2}\,\sigma_0}\,\frac{r_0^2+r^2}{r}. \tag{14.94}$$

When the shell is subjected only to one concentrated force P, we have

$$V = \frac{P}{\sigma_0}\int_0^{r_1}\left(\frac{1}{z'}+z'\right)dr. \tag{14.95}$$

The above functional is a minimum when $z' = \text{const}$. The absolute minimum of volume is for $z' = 1$, i.e. for the conical shell with the thickness $h = P/\sqrt{2}\pi\sigma_0 r$.

When the load is uniformly distributed in the horizontal plane, the components of the load are $Z = -p_0\cos^2\varphi$, $X = p_0\cos\varphi\sin\varphi$ and $Z - X\tan\varphi = -p_0$. The first equilibrium equation does not change and therefore the thickness of the shell and volume are expressed by means of the same formulae as previously. The only change is in the equation for χ. After some manipulation we obtain

$$\chi = r^2/(r_0^2+r^2). \tag{14.96}$$

We see that for arbitrary r the condition $0 \leqslant \chi \leqslant 1$ is fulfilled. The shape and the thickness of the shell are as previously.

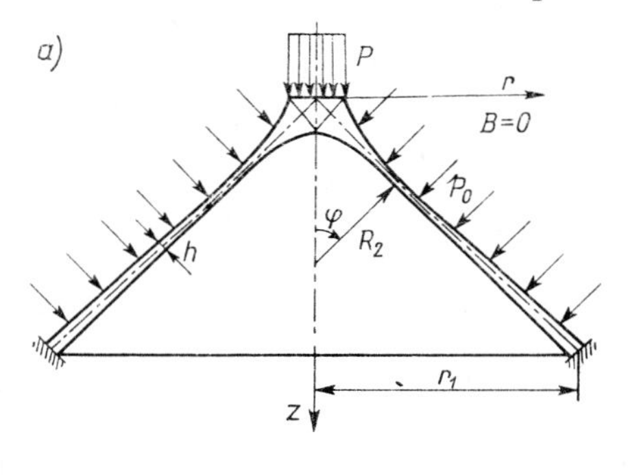

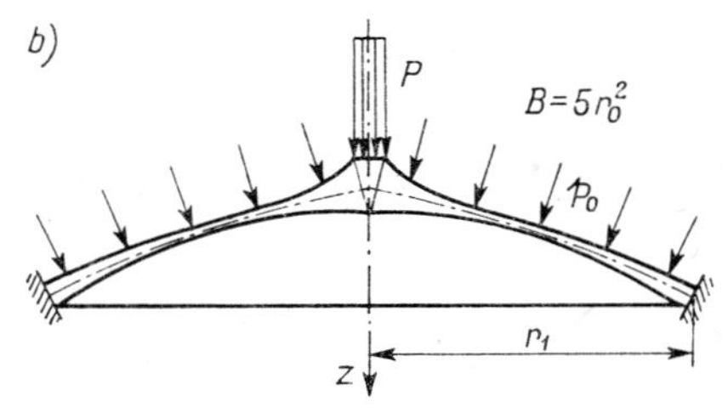

Fig. 168. Shape of a shell of minimum weight subjected to the single force P and the pressure p_0 *(from* [14.17]).

Now we consider the case where the concentrated force P acts in the direction opposite to the distributed load. We assume that the force $N_{\varphi\varphi}$ is a compressive force and $N_{\vartheta\vartheta}$ is a tensile force. Such a state of stress is represented by the straight line BC in Fig. 163. These forces can be defined in the following way:

$$N_{\varphi\varphi} = -\chi\sigma_0 h, \quad N_{\vartheta\vartheta} = (1-\chi)\sigma_0 h, \quad 0 \leqslant \chi \leqslant 1.$$

Then we obtain the thickness and the parameter χ by solving the equation of equilibrium. We have after some manipulation:

$$z' = \sqrt{\frac{2r_0^2-2r^2}{r_0^2+B+r^2}}\,. \tag{14.97}$$

The shape of the shell can be obtained from (14.97) in the form of elliptic integrals. The thickness of the shell takes the form

$$h = \frac{p_0}{2\sqrt{2}\,\sigma_0} \frac{1}{r} \frac{3r_0^4 + Br_0^2 - r^4}{(3r_0^2 + B - r^2)^{1/2}(r_0^2 - r^2)^{1/2}} \,. \tag{14.98}$$

The constants B, C enable one to obtain the shape of the shell required by the boundary conditions. When the height of the shell is not determined in advance, we obtain the minimum weight for $B = 0$. The parameter χ is determined by means of the second equilibrium equation (14.83). We obtain after some transformations

$$\chi = \frac{(r_0^2 - r^2)(3r_0^2 + B - r^2)}{3r_0^4 + r_0^2 B - r^4} \,; \tag{14.99}$$

for $r = 0$, $\chi = 1$, for $r = r_0$, $\chi = 0$.

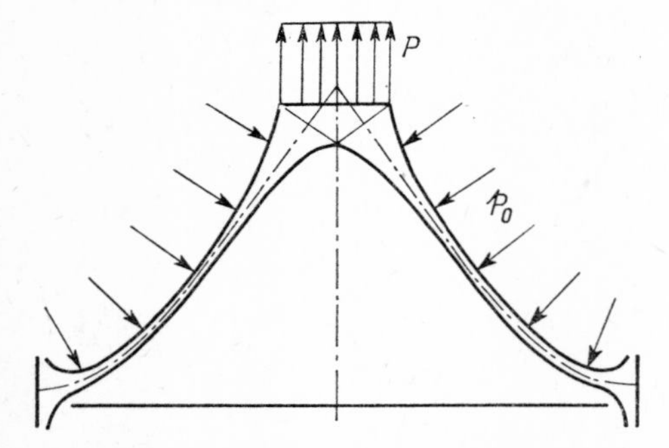

Fig. 169. Shape of a shell of minimum weight in the case where the load P is oppositely directed to the pressure p_0.

The above solution is valid only if $0 < r < r_0$ for $r = r_0$, $\chi = 0$, i.e. the force $N_{\varphi\varphi} = 0$ and the shell is in equilibrium under the action of a concentrated force and a distributed load. The shape of the shell is given in Fig. 169. We observe that the thickness of the shell tends to infinity as $r \to r_0$. The reason for this phenomenon is that the radii of curvature $R_1 \to 0$ and $R_2 \to \infty$ at this point. But the surface of this part of the shell is limited. We can imagine that the shell is strengthened there by a ring of certain cross-section, which carries the tension in the parallel circle. In the case of a uniformly distributed load in the horizontal plane we find the shape of the shell in a similar way.

$$z' = \pm \sqrt{\frac{2(r_0^2 - r^2)}{r_0^2 + B - r^2}} \,. \tag{14.100}$$

The thickness is

$$h = \frac{p_0}{2\sqrt{2}\,\sigma_0}\,\frac{1}{r}\left[\frac{(r_0^2-r^2)^2+(r_0^2+r^2)\,(r_0^2+B-r^2)}{(3r_0^2+B-3r^2)^{1/2}} - \right.$$

$$\left. - \frac{Br^4}{(r_0^2+B-r^2)^{1/2}}\right]\frac{1}{(r_0^2-r^2)^{1/2}}\,. \tag{14.101}$$

The absolute minimum occurs when $B = 0$ which gives $z' = \sqrt{2}$ = constant. It happens that the conical shape is also the structure of minimum weight. The angle of the shell is $\alpha = 54°44'$ and the thickness (Fig. 170) is

$$h = \frac{p_0}{2\sqrt{6}\,\sigma_0}\left(\frac{3r_0^2}{r}+r\right).$$

A shallow shell can be obtained if we assume large B.

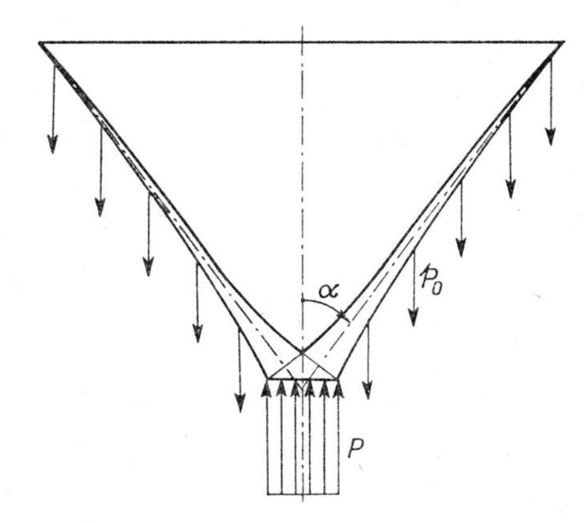

Fig. 170. Shell of minimum weight in the case of uniformly distributed load in a horizontal plane.

14.14.2 *Shell carrying its own weight*

Let us consider now a shell loaded by a single central concentrated normal force together with its own weight. If γ is the specific weight of the material, the components of load per unit area of the surface are

$$X = \gamma h \sin\varphi, \qquad Z = -\gamma h \cos\varphi.$$

Then

$$Z - X\tan\varphi = -\frac{\gamma h}{\cos\varphi}.$$

We assume that both internal forces are compressive and $N_{\varphi\varphi} \leqslant N_{\vartheta\vartheta} \leqslant 0$. Then this state of stress is expressed by the line *DE* in the hexagon of Fig. 163. Then $N_{\varphi\varphi} = -h\sigma_0$, $N_{\vartheta\vartheta} = -\chi h\sigma_0$ where $0 \leqslant \chi \leqslant 1$. On substitution into the first equilibrium condition (14.88) we have

$$\frac{\mathrm{d}}{\mathrm{d}r}\left(rh\frac{z'}{\sqrt{1+z'^2}}\right) = \frac{\gamma}{\sigma_0}rh\sqrt{1+z'^2}\,. \tag{14.102}$$

This equation can be integrated to yield:

$$h = C\frac{\sqrt{1+z'^2}}{rz'}\exp\left(\frac{\gamma}{\sigma_0}\int\left(\frac{1}{z'}+z'\right)\mathrm{d}r\right). \tag{14.103}$$

Here C is an arbitrary constant and can be found by comparing (14.84) and (14.99). For $\gamma = 0$ we should obtain the result for the shell loaded only by the concentrated force P. Then $C = P/2\pi\sigma_0$. The volume of the shell is

$$V = \frac{P}{\gamma}\int_0^{r_1}\left[\left(\frac{1}{z'}+z'\right)\exp\left(\frac{\gamma}{\sigma_0}\int\left(\frac{1}{z'}+z'\right)\mathrm{d}r\right)\right]\mathrm{d}r.$$

The above functional can be integrated to yield:

$$V = \frac{P}{\gamma}\left[\exp\left(\frac{\gamma}{\sigma_0}\int_0^{r_1}\left(\frac{1}{z'}+z'\right)\mathrm{d}r\right) - 1\right]. \tag{14.104}$$

If we assume that $z' > 0$ when $0 \leqslant r \leqslant r_1$, then the integral in the above expression is positive. The functional V has its minimum when this integral is minimum. That yields z = constant, i.e. the case of a conical shell. The absolute minimum is when $z' = 1$. The thickness of the shell $z' = 1$ is

$$h = \frac{P}{\sqrt{2}\pi\sigma_0}\frac{\mathrm{e}^{2\gamma r/\sigma_0}}{r}.$$

The parameter χ is determined by the equation

$$0 \leqslant \chi = \frac{\gamma}{\sigma_0}\frac{r}{z'} - \frac{rz''}{z'(1+z'^2)} \leqslant 1, \tag{14.105}$$

for $z' = 1$

$$r \leqslant \frac{\sigma_0}{\gamma}.$$

The shape of the shell is similar to those given in Fig. 168. When $P = 0$ and the shell is loaded only by its own weight, we obtain also the minimum weight for the conical shell. In order to compare the result obtained with that taken from [14.26] we can transform expression (14.103) observing Eq. (14.105). Then

$$\frac{\gamma}{\sigma_0}\frac{1}{z'} = \frac{\chi}{r} + \frac{z''}{z'(1+z'^2)} = \frac{\sqrt{1+z'^2}}{rz'}\frac{\mathrm{d}}{\mathrm{d}r}\left(\frac{z'r}{\sqrt{1+z'^2}}\right) + \frac{\chi-1}{r}.$$

Introducing this into Eq. (14.103), we have

$$h = h_0 \exp\left(\int\left[\frac{\gamma}{\sigma_0}z' - \frac{\chi-1}{r}\right]\mathrm{d}r\right).$$

Assuming $\chi = 1$, we find the known solution $h = h_0 \mathrm{e}^{\gamma z/\sigma_0}$. We see that only in the case where $P = 0$ the thickness of the shell is limited at the top and the shell is not conical.

Now we consider the case where the force P acts in the direction opposite to the weight. Assuming that $N_{\varphi\varphi} = \chi\sigma_0 h > 0$ and $N_{\vartheta\vartheta} = -(1-\chi)\sigma_0 h < 0$, we find on substitution into Eq. $(14.88)_1$ the thickness

$$h = C\frac{\sqrt{1+z'^2}}{\chi r z'}\exp\left[-\frac{\gamma}{\sigma_0}\int\left(\frac{1}{z'}+z'\right)\frac{\mathrm{d}r}{\chi}\right]. \tag{14.106}$$

On integration the volume of the shell is

$$V = 2\pi C\left\{1-\exp\left[-\frac{\gamma}{\sigma_0}\int_0^{r_1}\left(\frac{1}{z'}+z'\right)\frac{\mathrm{d}r}{\chi}\right]\right\}. \tag{14.107}$$

Assuming that $z' > 0$ when $0 \leqslant r \leqslant r_1$, we have the minimum of the functional V when the integral in Eq. (14.107) has its minimum.

The parameter χ can be evaluated from the equation

$$\frac{1}{\chi} = \frac{(z'-\gamma r/\sigma_0)(1+z'^2)}{z'(1+z'^2)+rz''}. \tag{14.108}$$

Introducing this into Eq. (14.107), we find the following differential equation:

$$z'^2\left[z'^2(1+B)-2(1+B)\frac{\gamma}{\sigma_0}r-2+B\left(\frac{\gamma}{\sigma_0}r^2\right)\right] = 0,$$

where B is an arbitrary constant. This equation has one solution $z'^2 = 0$ which corresponds to an infinitely thick flat sheet. Then there remains the algebraical equation of second order with respect to z'. Solving it and integrating, we find the shape of the shell.

Let us consider now the optimum shape for which the weight has absolute minimum value. From the variational conditions for the edge $r = r_1$ we find that $B = 0$. The solution takes the form

$$z' = \frac{\gamma r}{\sigma_0} \pm \sqrt{\frac{\gamma r}{\sigma_0} + 2}\,. \tag{14.109}$$

Since $z' > 0$, we take the positive sign. By introducing the non-dimensional coordinate $u = \gamma r/\sqrt{2}\,\sigma_0$, we have

$$z = \frac{\sigma_0}{\gamma}\left[u^2 + u\sqrt{u^2+1} + \ln\left(u + \sqrt{u^2+1}\right)\right] \tag{14.110}$$

and the thickness

$$h = \frac{P}{2\pi\sigma_0}\,\frac{e^{-f_1(u)}}{r}\,f_2(u), \tag{14.111}$$

where

$$f_1(u) = 2\left(u^2 + u\sqrt{u^2+1}\right) + \ln\left(u^2 + u\sqrt{u^2+1} + 1\right),$$

$$f_2(u) = \frac{(4u^2+3)\sqrt{u^2+1} + 4u^3 + 5u}{\sqrt{2}\,(u^2+1)\left(4u^2 + 4u\sqrt{u^2+1} + 3\right)^{1/2}}\,.$$

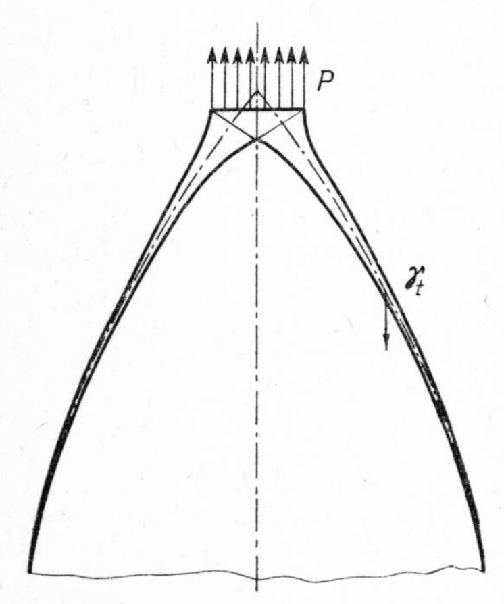

Fig. 171. Shell of minimum weight carrying its own weight (from [14.31]).

The shape of the shell is given in Fig. 171. Introducing z' from Eq. (14.109) into Eq. (14.108) we obtain the coefficient χ equal to 1 for $r = 0$ and $\chi = 0.4$ for $r = \infty$. It means that the condition $0 \leqslant \chi \leqslant 1$ is always satisfied.

The results presented have been obtained by means of the equations of the membrane theory of shells. These are exact enough only when the thickness of the shell does not change greatly and rapidly. Therefore, the results for the area near the top may not be quite accurate. However, they enable one to gain an idea of the optimum shape of the shell. Direct application of the results obtained may, in practice, lead to certain difficulties. At a large distance from the top the thickness of the shell determined by the given formulae is very small and the manufacture of such a shell would be impossible for technological reasons. In these areas the shell can buckle, which has not been considered above.

The review of the papers devoted to the optimum design of shells is given by M. Życzkowski [14.49].

14.15 Experimental design of the shell

14.15.1 *Designing with liquid analogy*

The exact solution of the equations for the problem of designing the shell under concentrated loads is rather complicated from the mathematical point of view. Therefore a simple, practical method of defining of the optimum shape structure would be very desirable. The first variant of the proposed experimental method consists in using the phenomenon of the surface tension appearing on the surface of two liquids not mixing together. Because of the surface tension, the external surface of the liquid behaves like an elastic membrane in which act uniform tensile forces constant at any point of the boundary surface. These forces depend only on the kind of liquids and their temperature. Inside the area bounded by the surface there occurs the pressure which can be defined by the equation

$$p = n\left(\frac{1}{R_1} + \frac{1}{R_2}\right) = 2Hn$$

where n is the surface tension and H the mean curvature of the shell.

If we place a drop of a liquid in a vessel filled by other kind of liquid, both not mixing one with another and having the same specific weight, the drop levitates freely taking the shape of a sphere. This shape corresponds to the solution § 14.13. The drop can be loaded by the boundary forces. For example, if we place the drop between two rings, Fig. 172, the drop sticks to the rings. When we strech them, it will take the shape of the shell under internal pressure p and tensile force P_x. The surface obtained is the rotationally symmetrical one.

The value of the ratio of the tensile force P_x and the internal pressure p can be found from the equilibrium condition

$$\frac{P_x}{\pi p} = \frac{(r\cos\alpha)_b}{H} - r_b^2 .$$

The index b denotes here the boundary values, r_b is the radius of the boundary ring, and H is the mean curvature at an arbitrary point of the surface.

Placing the drop between the frames of different shapes and loading them, we can obtain the shapes of shells subjected to internal pressure and boundary

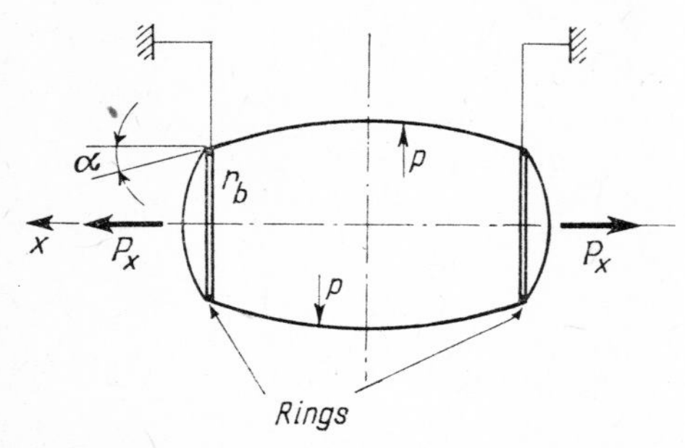

Fig. 172. Design of a shell created by a boundary surface of two liquids.

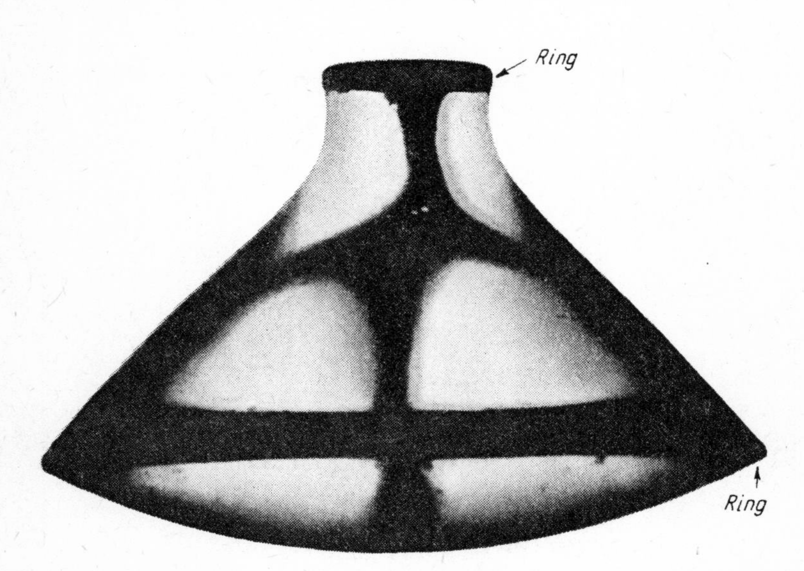

Fig. 173. Shape of a shell of revolution subjected to the action of the internal pressure and a tensile force uniformly distributed along the edges.

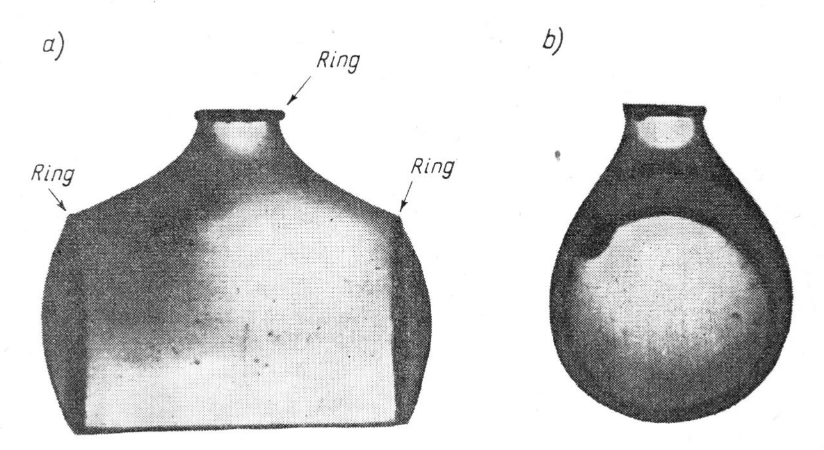

Fig. 174. Shape of a shell subjected to the internal pressure and the forces applied to three different diameters.

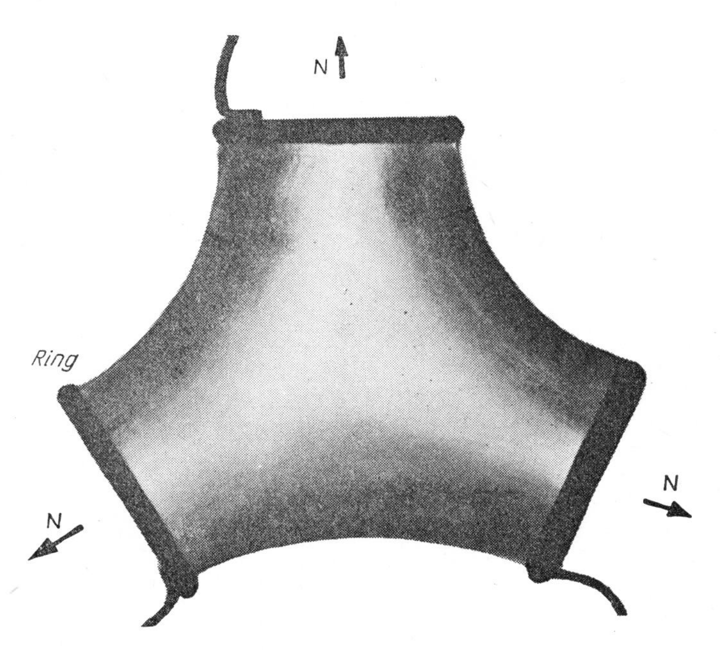

Fig. 175. Shape of a shell of uniform strength subjected to the action of an internal pressure and three tensile forces applied to the rings under the angle of 120°.

forces. In Figs. 173–176 several examples of shells designed in this way are presented. It should be stated that these shells satisfy the boundary conditions of supported edges which act on the support with the constant force. The calculation of the forces with which the shell acts on the supports consists

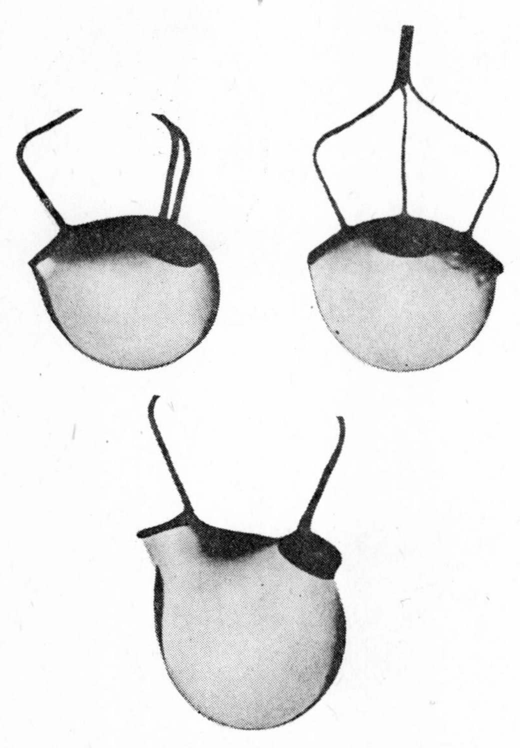

Fig. 176. Shape of a shell under variable internal pressure and forces applied to three rings. (The shape of a container under action of the weight of liquid was obtained by changing the specific weight of the liquid in which the drop was dipped.)

in the calculation of the resultants of the edge forces. For example, for Fig. 172 we have

$$P_x = \frac{p}{H} \oint \cos\alpha \, ds$$

where α denotes the angle between the tangential line to the surface and the direction of the x-axis.

14.15.2 *Designing in the plastic state*

The second version of the experimental method consists in designing the shell from a material which is in the plastic state during the experiment and which takes the form of the shell of constant strength under the action of applied loads. The method is based on the assumption that during the plastic process at each point of the shell there appears the state of stress which satisfies the yield condition and simultaneously the condition of constant strength. The shells presented in Figs. 177–180 are made in the following way. Plane or curved

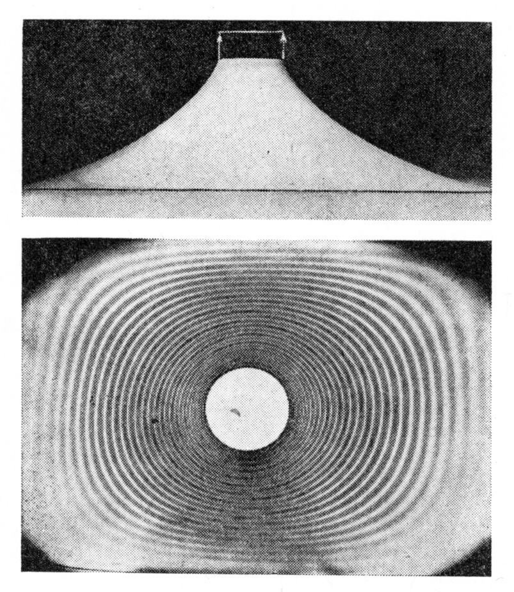

Fig. 177. Shape of a shell of uniform strength subjected to the force uniformly distributed along the circle. The contour lines are presented in the lower figure.

sheets of plexiglass were attached to the rigid frames and heated in the oven to the temperature of about 135°C at which the material became plastic. At this temperature the sheets were deformed under applied loads. After lowering the temperature the shape of the shell remained unchanged. Obviously, the shells formed in this way do not have constant wall thickness. Also the plastic flow does not take place by the constant yield stress. During the experiment a certain hardening of the material was observed. Therefore the shapes obtained did not correspond exactly to the shells of constant thickness designed

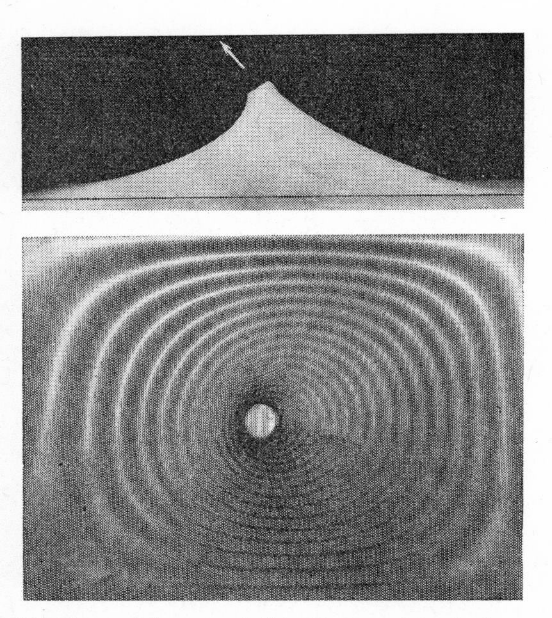

Fig. 178. Shape of the same shell in the case where the force acts with a certain angle in respect to the bottom.

Fig. 179. Shape of a shell subjected to internal pressure and force acting in the upper direction.

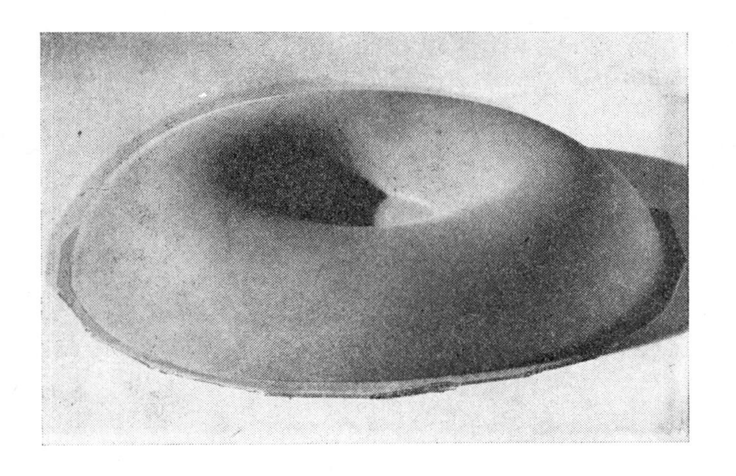

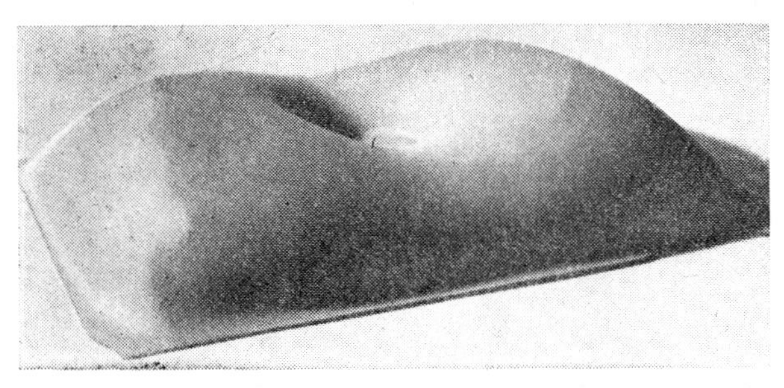

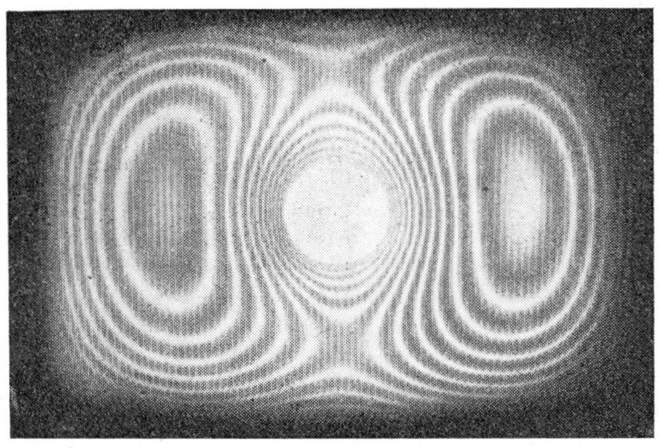

Fig. 180. Shape of a shell in the case where the force acts in down direction.

by the condition of constant strength. However, the equations of the optimum shells can be applied also to the shells for which the product $\sigma_0 h$ = constant. The change of the thickness gives the opposite effect to that caused by the hardening of the material. It can be expected that they cancel each other partly. In order to prove this hypothesis, the model of a shell of revolution was made of plexiglass. This shape corresponds to the optimum shell supported at the external edge and loaded by the line load distributed uniformly along the circle of the internal edge. The shape of the shell calculated theoretically was compared with that obtained from experiment.

It follows that the differences between the theoretical and experimental results are larger for non-shallow shells. The shape of the shell changes to a relatively high degree with the change of the thickness and the hardening of the material. However, it is possible to satisfy the condition $\sigma_0 h \cong$ constant by proper selection of the material and the temperature of the process of plastic flow.

References 14

[14.1] J. N. GOODIER, C. S. HSU: Transmission of tension from a bar to a plate, *J. Appl. Mech. Trans. ASME 21* (1954), 147.

[14.2] E. REISSNER: Note on the problem of the distribution of stress in a thin stiffened elastic sheet, *Proc. Nat. Sci.*, 26, April 1940, 300–305.

[14.3] W. SZCZEPIŃSKI: *Limit design of machine elements*, PWN, Warsaw, 1968 (in Polish).

[14.4] W. KOZŁOWSKI, Z. MRÓZ: Optimal design of discs subject to geometric constrains, *Int. J. Mech. Sci. 12* (1970) 1001–1021.

[14.5] I. M., RYŻYK, T. S. GRADSTEIN: *Tables of sums, products and integrals*, PWN, Warszawa 1964.

[14.6] E. MELAN: Ein Beitrag zur Theorie geschweister Verbindungen, *Ing.-Archiv. 3.2* (1932) 123.

[14.7] W. T. KOITER: On the diffusion of a load from a stiffener into a sheet, *Quart. J. Appl. Math. 3.2* (1955) 164.

[14.8] J. P. BENTHEM: Stress diffusion in wedge-shaped plates, *Simon Stevin, Wis. en Natuurkundig Tijdschrift 35, I; II, 31,* (1961).

[14.9] M. L. WILLIAMS: Stress singularities resulting from various boundary conditions in angular corners of plates in extension, *J. Appl. Mech. 19.4* (1954) 526.

[14.10] S. ŁUKASIEWICZ, W. BORAJKIEWICZ: Optimum design of a bar transmitting a load into a sheet, *Arch. Bud. Masz. 18.1* (1971) (in Polish).

[14.11] A.M. HENS: *De Inleding van een Kracht op een Verstiver in een half-oneindige Plaat, Report of the laboratorium voor Toegepaste Mechanica*, Technische Hogeschool, Delft, 1957.

[14.12] D. C. DRUCKER, R .T. SHIELD: Bounds on minimum weight design, *Quart. Appl. Math. 15* (1957) 269–281.

[14.13] W. PRAGER: On a problem of optimal design, *Proc. Symp. Non-Homogeneity in Elasticity and Plasticity*, Warsaw 1958.
[14.14] TE-CHIANG HU, R. T. SHIELD: Minimum volume design of discs, *ZAMP 12* (1961) 414–433.
[14.15] J. KAPKOWSKI: Introduction of concentrated force in a sheet taking into account the condition of uniform strength, *Arch. Bud. Masz. 7. 1* (1960) (in Polish).
[14.16] Z. BRZOSKA: Circular plates of uniform strength under symmetrical loads, *Arch. Bud. Masz. 1.* (in Polish).
[14.17] S. ŁUKASIEWICZ: Circular plates of uniform strength symmetrically loaded, *Arch. Bud. Masz. 7.1* (1960) (in Polish).
[14.18] S. U. BENSCOTER: Analysis of a single stiffener on an infinite sheet, *J. Appl. Mech. Trans. ASME 71* (1949) 242–246.
[14.19] H. G. HOPKINS, W. PRAGER: Limits of economy of material in plates, *J. Appl. Mech. 22* (1955) 372–374.
[14.20] W. FREIBERGER, B. TEKINALP: Minimum weight design of circular plates, *J. Mech. Phys. of Solids 4* (1959) 294–299.
[14.21] E. T. ONAT, W. SCHUMANN, R. T. SCHIELD: Design of circular plates for minimum weight, *ZAMP 8* (1957) 485–499.
[14.22] D. C. DRUCKER, R. T. SHIELD: Design for minimum weight, *Proc. 9th Int. Congr. Appl. Mech.*, Brussels 1956.
[14.23] A. E. GREEN: *Quart. Appl. Math. 7* (1949) 223.
[14.24] F. TÖLKE: Über Rotationsschalen gleicher Festigkeit für konstanten Innen- oder Aussendruck, *ZAMM 19* (1939) 338–343.
[14.25] G. MEGAREUS: Die Kuppel gleicher Festigkeit, *Bauing. 20* (1939) 232–234.
[14.26] H. ZIEGLER: Kuppeln gleicher Festigkeit, *Ing.-Arch. 26* (1958) 378–382.
[14.27] F. G. SHAMIEV: On the optimum design of shells of uniform weight, *Izv. AN SSSR, Seria Fiz. Mat. Tekhn. Nauk 5* (1963) (in Russian).
[14.28] V. DAĚEK: Zur Berechnung von Behälterböden gleicher Festigkeit, *Beton und Eisen 36. 5* (1937).
[14.29] K. FEDERHOFFER: Über Schalen gleicher Festigkeit, *Bauingenieur 20* (1939) 366.
[14.30] V. B. CHEREVADSKII: Cable net surfaces of revolution symmetrically loaded, *Sb. Issl. po Teorii Plastin i Obolochek*, Izd. Kazansk. U-ta, *5* (1967) (in Russian).
[14.31] S. ŁUKASIEWICZ: On the optimum design of shells loaded by concentrated forces, in: *Theory of Thin Shells, Proc. IUTAM Symp. Copenhagen 1967*, Springer Verlag 1969.
[14.32] W. READ: Equilibrium shapes for pressurised domes, *J. Appl. Mech. Trans. ASME, ser. B, II* (1963).
[14.33] R. T. SHIELD: On the optimum design of shells, *J. Appl. Mech. Trans. ASME* (1969) 316–322.
[14.34] W. PRAGER, R. SHIELD: A general theory of optimum plastic design, *J. Appl. Mech. Trans. ASME, E34* (1967) 184–186.
[14.35] W. FREIBERGER: Minimum weight design of cylindrical shells, *J. Appl. Mech. Trans. ASME 23. 14* (1956).
[14.36] M. A. SAVE, R .T. SHIELD: Minimum weight design of sandwich shells subjected to fixed and moving loads, *Proc. Eleventh Int. Congr. Appl. Mech.*, Munich 1964.

[14.37] M. Shch. Mikeladze: On anisotropic shells of minimum weight, *Soobshchenia Gruzinskoi AN SSR 19.1* (1957) (in Russian).

[14.38] M. I. Estrin: *On plates of minimum weight with plane state of stresses*, *Voprosy teorii plastichnosti i prochnosti stroitelnykh konstruktsii*, Moskva 1961 (in Russian).

[14.39] R. Muki, E. Sternberg: Transfer of load from an edge stiffener to a sheet — a reconsideration of Melan's problem, *J. Appl. Mech. 34.3* (1967) 679.

[14.40] R. Muki, E. Sternberg: On the stress analysis of overlapping bounded elastic sheets, *Internat. J. Solids and Structures 4. 1* (1968) 75.

[14.41] R. Muki, E. Sternberg: On the diffusion of a transverse tension-bar into a semi-infinite elastic sheet, *J. Appl. Mech. 35. 4* (1968) 737.

[14.42] E. Sternberg: Load transfer and load-diffusion in elastostatics, *Proceedings of the Sixth U. S. National Congress of Applied Mechanics* 1970, pp. 34–61.

[14.43] B. Budiansky, T. T. Wu: Transfer of load to a sheet from a rivet-attached stiffener, *J. Math. and Phys. 40. 2* (1961) 142.

[14.44] F. C. Monge: *Estudo de Barra Sometide a Esfuerzo Axial Soldata a una Placa Semiindefinida*, Instituto Tecnico de la Construccion y Edificacion, Publicacion No. 59.

[14.45] E. Krahn: Discussion of Ref. [14.1], *J. Appl. Mech. 22. 1* (1955) 139.

[14.46] A. Nádai: *Plasticity*, McGraw-Hill; *Theory of flow and fructure of solids*, McGraw-Hill Book Co., New York 1950.

[14.47] V. V. Sokolovskii: *Theory of plasticity*, Moskva-Leningrad 1950 (in Russian).

[14.48] J. Zwoliński: *Optimum design of joints*, Dissertation, Politechnika Warszawska, 1974 (in Polish).

[14.49] M. Życzkowski: Optimisation of shell structures, in: *Proceedings of the Symposium on Shell Structures, Theory and Applications*, Kraków 1974.

Appendix

Certain remarks on the problem of a concentrated force acting at the edge of the plate

In § 5.51 we considered the problem of a semiinfinite plate loaded at the edge by a concentrated force. Studying this case, we obtained a simple radial distribution of stress. However, this well-known solution is not complete. It can be proved considering the equilibrium of an arbitrary region surrounding the point of application of the load. Let us consider the equilibrium of an element presented in Fig. 181. Calculating the resultant of the stress σ_{rr} in

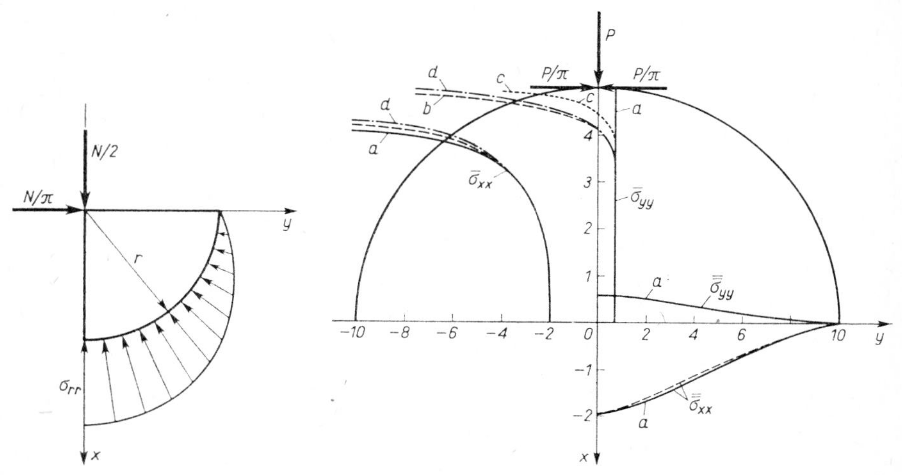

Fig. 181. Equilibrium of an element surrounding the point of application of the load.
Fig. 182. The stress distribution along the vertical diameter $\bar{\sigma}_{xx}\bar{\sigma}_{yy}$ *and the horizontal diameter* $\bar{\bar{\sigma}}_{xx}, \bar{\bar{\sigma}}_{yy}$ *divided by the quantity* P/hd, *a — classical theory, b — finite element analysis* [A3], *c — experimental method* [A2], *d — paper by Hondros* [A1].

the y-axis direction, we conclude that the equilibrium is possible only if the force N/π acts in the y-axis direction at the point of application of the load. Then the complete solution consists of the radial distribution of stress σ_{rr}

and a pair of oppositely directed forces N/π acting at the point of application of the load. Only in this case the conditions of the internal equilibrium of the plate and the boundary conditions are satisfied. The existence of these additional forces is usually neglected and left behind, because they do not produce the stresses. Similar forces appear in the second considered case, § 5.5.2. *A force tangential to the edge.* The equilibrium of the region surrounding the point of application of the load needs the existence of two concentrated forces T/π normal to the edge of the plate and acting in the opposite directions.

The problem of the action of two concentrated forces on a circular disc was considered by G. Hondros [A1] and experimentally by J. T. Pindera, S. B. Mazurkiewicz and M. A. Khattab [A2]. They called the attention to the discrepancies between the theory and the experimental results in the discussed case. The results of the experiments [A2], finite elements analysis [A3], the classical theory and the results given in paper [A1] by Hondros are presented in Fig. 182. We observe that the pair of concentrated forces predicted by the theory of elasticity is replaced in the reality by a distributed stress σ_{yy} which is singular at the point of application of the load. From the above comparison we can draw the following conclusion. If we apply a model of a concentrated force which does not exist in reality, the results obtained by this assumption can also be irrational. For example, we obtained also the infinitely large displacement at the point of application of the load, which has no physical meaning. These contradictions are caused by the very application of a notion of a concentrated force as well as by the initial assumptions of the linear theory of elasticity.

References

[A1] G. Hondros: The evaluation of Poisson's ratio and the modulus of materials of a low tensile resistance by the Brasilian test, *Australian Journal of Applied Sciences 10. 3* (1959) 243–268.

[A2] J. T. Pindera, S. B. Mazurkiewicz, M. A. Khattab: Stress field in circular disc loaded along diameter; Discrepancies between the analytical and experimental results. Experimental stress analysis, *Paper No. CR-10 presented at SESA Spring Meeting, Wichita, Kansas, May 14–19, 1978.*

[A3] Ken P. Chong, *Finite element and other analysis of split discs. Research report*, University of Wyoming, Loranie, 1978.

Tables

Table 10a. The coefficients for M_{max} in the square plate of simply supported edges under the load distributed over the surface $c \times d$, $b/a = 1$, $\nu = 0.3$.

β_1	c/a	0	0.2	0.4	0.6	0.8	1.0
	0	∞	0.251	0.180	0.141	0.112	0.092
	0.2	0.308	0.214	0.161	0.127	0.103	0.084
d/b	0.4	0.232	0.179	0.141	0.113	0.092	0.076
	0.6	0.188	0.150	0.121	0.099	0.091	0.067
	0.8	0.155	0.126	0.103	0.085	0.070	0.057
	1.0	0.127	0.105	0.086	0.071	0.058	0.048

Table 10b. The coefficients β_1 and β_2 for the rectangular plate of the dimensions a, b under the load distributed over the surface $c \times d$ for $\nu = 0.3$.

		β_1						β_2					
	d/a \ c/a	0	0,2	0,4	0,6	0,8	1,0	0	0,2	0,4	0,6	0,8	1,0
$b=1.4a$	0		0.276	0.208	0.163	0.134	0.110		0.299	0.230	0.183	0.151	0.125
	0.2	0.332	0.239	0.186	0.152	0.125	0.103	0.246	0.208	0.175	0.147	0.124	0.102
	0.4	0.261	0.207	0.168	0.138	0.115	0.095	0.177	0.157	0.138	0.119	0.101	0.083
	0.6	0.219	0.181	0.151	0.126	0.105	0.086	0.138	0.125	0.111	0.097	0.083	0.069
	0.8	0.187	0.158	0.134	0.112	0.094	0.078	0.112	0.102	0.091	0.080	0.069	0.058
	1.0	0.162	0.139	0.118	0.100	0.084	0.070	0.093	0.085	0.077	0.068	0.058	0.049
	1.2	0.141	0.122	0.104	0.089	0.075	0.062	0.079	0.072	0.065	0.058	0.050	0.042
	1.4	0.123	0.106	0.091	0.077	0.065	0.054	0.068	0.062	0.056	0.050	0.043	0.036
$b=2a$	0	∞	0.289	0.222	0.175	0.144	0.118	∞	0.294	0.225	0.179	0.148	0.122
	0.2	0.347	0.252	0.199	0.163	0.135	0.111	0.242	0.203	0.170	0.143	0.120	0.099
	0.4	0.275	0.221	0.181	0.150	0.125	0.103	0.172	0.152	0.133	0.114	0.097	0.081
	0.8	0.203	0.174	0.148	0.126	0.106	0.088	0.107	0.097	0.087	0.076	0.065	0.054
	1.2	0.161	0.141	0.122	0.105	0.089	0.074	0.074	0.068	0.061	0.054	0.046	0.039
	1.6	0.130	0.115	0.101	0.087	0.074	0.062	0.056	0.051	0.046	0.040	0.035	0.029
	2.0	1.07	0.094	0.083	0.072	0.061	0.051	0.044	0.041	0.037	0.032	0.028	0.023

Table 11. Functions ber(x), bei(x) and their derivatives.

x	ber(x)	bei(x)	ber′(x)	bei′(x)
0.00	1.000 00	0	0	0
01	000 00	0.000 025	−0.000 000 06	0.005 000
02	000 00	000 100	000 000 50	010 000
03	000 00	000 225	000 001 69	015 000
04	000 00	000 400	000 004 00	020 000

Table 11 (*cont.*).

x	ber(x)	bei(x)	ber′(x)	bei′(x)
0.05	1.000 00	0.000 625	−0.000 007 81	0.025 000
06	000 00	000 900	000 013 50	030 000
07	000 00	001 225	000 021 44	035 000
08	000 00	001 600	000 032 00	040 000
09	000 00	002 025	000 045 56	045 000
0.10	1.000 00	0.002 500	−0.000 062 50	0.050 000
0.2	0.999 98	010 000	000 500 0	099 999
3	999 87	022 500	001 687 5	149 99
4	999 60	039 998	004 000	199 97
0.5	0.999 02	0.062 493	−0.007 812	0.249 92
6	997 98	0.89 980	013 498	299 80
7	996 25	122 45	021 433	349 56
8	993 60	159 89	031 989	399 15
9	989 75	202 27	045 537	448 46
1.0	0.984 38	0.249 57	−0.062 446	0.497 40
1	977 14	301 73	083 082	545 81
2	967 63	358 70	107 81	593 52
3	955 43	420 41	136 97	640 34
4	940 08	486 73	170 93	686 01
1.5	0.921 07	0.557 56	−0.210 01	0.730 25
6	897 89	632 73	254 54	772 74
7	869 97	712 04	304 84	813 10
8	836 72	795 26	361 18	850 93
9	797 52	882 12	423 84	885 74
2.0	0.751 73	0.972 29	−0.493 07	0.917 01
1	698 69	1 065 4	569 06	944 18
2	637 69	161 0	652 00	966 61
3	568 05	258 5	742 02	983 61
4	489 05	357 5	839 20	994 43
2.5	0.399 97	1.457 2	−0.943 58	0.998 27
6	300 09	556 9	−1.055 1	994 26
7	188 71	655 7	173 8	981 49
8	065 112	752 9	299 3	958 97
9	−0.071 368	847 2	431 4	925 66
3.0	−0.221 38	1.937 6	−1.569 8	0.880 48
1	385 53	2.022 8	714 1	822 30
2	564 38	101 6	863 6	749 92

Table 11. (*cont.*)

x	ber(x)	bei(x)	ber′(x)	bei′(x)
3.3	−0.758 41	2.172 3	−2.017 7	0.662 14
4	968 04	233 4	175 5	557 69
3.5	−1.193 6	2.283 2	−2.336 1	0.435 30
6	435 3	319 9	498 3	293 66
7	693 3	341 3	660 8	131 49
8	967 4	345 4	822 2	−0.052 527
9	−2.257 6	330 0	980 7	259 65
4.0	−2.563 4	2.292 7	−3.134 7	−0.491 14
1	884 3	230 9	281 8	748 17
2	−3.219 5	142 2	420 0	−1.031 9
3	567 9	023 6	546 5	343 3
4	928 3	1.872 6	658 8	683 3
4.5	−4.299 1	1.686 0	−3.753 7	−2.052 6
6	678 4	461 0	828 0	452 0
7	−5 063 9	194 6	878 2	881 8
8	453 1	0.883 66	900 6	−3.342 2
9	842 9	525 15	891 1	833 1
5.0	−6.230 1	0.116 03	−3.845 3	−4 354 1
1	610 7	−0.346 66	758 9	904 6
2	980 3	−0.865 84	627 0	−5.483 5
3	−7.334 4	−1 444 3	444 5	−6.089 2
4	667 4	−2.084 5	206 4	719 9
5.5	−7.973 6	−2.789 0	−2.907 0	−7.372 9
6	−8.246 6	−3 559 7	−2.541 0	−8.045 4
7	479 4	−4.398 6	−2.102 4	−8.733 6
8	664 4	−5.306 8	−1.585 5	−9.433 3
9	793 7	−6.285 4	−0.984 38	−10.139
6.0	−8.858 3	−7.334 7	−0.293 08	−10.846
1	849 1	−8.454 5	0.494 29	−11.547
2	756 1	−9.643 7	1.383 5	−12.235
3	568 8	−10.901	2.380 2	−12.901
4	276 2	−12.223	3.489 9	−13.536
6.5	−7.866 9	−13.607	4.717 4	−14.129
6	328 7	−15.047	6.067.5	−14.670
7	−6.649 2	−16.538	7.544 2	−15.146
8	−5.815 5	−18.074	9.151 0	−15.543
9	−4.814 6	−19.644	10.891	−15.847

Table 11. (*cont.*)

x	ber(x)	bei(x)	ber′(x)	bei′(x)
7.0	−3.632 9	−21.239	12.765	−16.041
1	−2.257 1	−22.848	14.774	−16.109
2	−0.673 70	−24.456	16.918	−16.033
3	1.130 8	−26.049	19.194	−15.792
4	3.169 5	−27.609	21.600	−15.367
7.5	5.455 0	−29.116	24.130	−14.736
6	7.999 4	−30.548	26.777	−13.875
7	10.814	−31.882	29.532	−12.763
8	13.909	−33.092	32 382	−11.373
9	17.293	−34.147	35.314	−9.680 6
8.0	20.974	−35.017	38.311	−7.660 3
1	24.957	−35.667	41.353	−5.285 5
2	29.245	−36.061	44.415	−2.529 6
3	33.840	−36.159	47.472	0.634 1
4	38.738	−35.920	50.492	4.231 8
8.5	43.936	−35.298	53.442	8.289 5
6	49.423	−34.246	56.281	12.832
7	55.187	−32.714	58.967	17.883
8	61.210	−30.651	61.451	23.465
9	67.469	−28.003	63.682	29.598
9.0	73.936	−24.713	65.601	36.299
1	80.576	−20.724	67.145	43.583
2	87.350	−15.976	68.246	51.460
3	94.208	−10.412	68.813	59.936
4	101.10	−3.969 3	68.821	69.012
9.5	107.95	3.410 6	68 132	78.684
6	114.70	11.787	66.674	88.940
7	121.26	21.218	64.353	99.763
8	127.54	31.758	61.070	111.12
9	133.43	43.459	56.720	122.99
10.0	138.84	56.370	51.195	135.31
1	143.63	70.534	44.384	148.03
2	147.67	85.987	36.171	161.08
3	150.81	102.76	26.438	174.38
4	152.90	120.87	15.066	187.82
10.5	153.77	140.32	1.9344	201.30
6	153.23	161.12	−13.076	214.69

Table 11. (*cont.*)

x	ber(x)	bei(x)	bei′(x)	bei′(x)
7	151.09	183.25	−30.083	227.85
8	147.14	206.68	−49.202	240.59
9	141.17	231.35	−70.544	252.75
11.0	132.95	257.21	−94.212	264.12
1	122.25	284.14	−120.30	274.46
2	108.81	312.06	−148.90	283.54
3	92.383	340.80	−180.08	291.07
4	72.707	370.21	−213.89	296.76
11.5	49.517	400.08	−250.37	300.29
6	22.543	430.18	−289.55	301.32
7	−8.483 2	460.25	−331.41	299.48
8	−43.828	489.97	−375.92	294.37
9	−83.753	519.00	−423.01	285.58
12.0	−128.51	546.95	−472.57	272.67
1	−178.34	573.38	−524.46	255.18
2	−233.48	597.82	−578.51	232.62
3	−294.11	619.72	−634.46	204.50
4	−360.42	638.51	−692.03	170.30
12.5	−432.56	653.56	−750.87	129.49
6	−510.62	664.17	−810.58	81.534
7	−594.59	669.61	−870.67	25.889
8	−684.75	669.07	−930.59	−37.992
9	−780.78	661.72	−989.72	−110.65
13.0	−882.65	646.64	−1 047.3	−192.61
1	−990.17	622.87	−1 102.7	−284.38
2	−1 103.1	598.42	−1 154.8	−386.45
3	−1 221.0	545.22	−1 202.7	−499.28
4	−1 343.4	489.19	−1 245.3	−623.27
13.5	−1 469.8	420.18	−1 281.5	−758.77
6	−1 599.5	337.04	−1 309.9	−906.08
7	−1 731.5	238.57	−1 329.2	−1 065.4
8	−1 865.0	112.35	−1 337.7	−1 236.9
9	−1 998.7	−9.210	−1 333.9	−1 420.5
14.0	−2 131.3	−160.94	−1 316.1	−1 616.1
1	−2 261.3	−332.82	−1 282.3	−1 823.5
2	−2 387.1	−526.02	−1 230.7	−2 042.3
3	−2 506.8	−741.65	−1 159.1	−2 272.0
4	−2 618.2	−980.75	−1 065.4	−2 511.6

Table 11. (*cont.*)

x	ber(x)	bei(x)	ber′(x)	bei′(x)
14.5	−2 719.1	−1 244.3	−947.37	−2 760.4
6	−2 806.8	−1 533.1	−802.69	−3 016.9
7	−2 878.6	−1 847.9	−628.96	−3 279.9
8	−2 931.6	−2 189.2	−423.74	−3 547.4
9	−2 962.3	−2 557.4	−184.56	−3 817.5
15.0	−2 967.3	−2 952.7	91.056	−4 087.8

Table 12. Functions ker (x), kei (x) and their derivatives

x	ker (x)	kei (x)	ker′ (x)	kei′ (x)
0.00	∞	−0.785 398 2	−∞	0
01	4.721 121	785 255 1	−99.996 07	0.026 105 56
02	4.028 033	784 895 4	−49.992 15	045 279 94
03	3.622 666	784 358 1	−33.321 56	061 838 67
04	3.335 121	783 664 2	−24 984 31	076 699 29
0.05	3.112 154	−0.782 828 3	−19.980 40	0.090 297 73
06	2.930 048	781 861 6	−16.643 16	102 890 9
07	2.776 152	780 773 3	−14.258 31	114 648 5
08	2.642 914	779 571 0	−12.468 71	125 691 5
09	2.525 463	778 261 5	−11.075 94	136 110 2
0.10	2.430 474 0	−0.776 850 6	−9.960 959 3	0.145 974 8
2	1.733 142 7	758 124 9	−4.922 948 5	222 926 8
3	337 218 6	733 101 9	−3.219 865 2	274 292 1
4	062 623 9	703 800 2	−2.352 069 9	309 514 0
0.5	0.855 905 9	−0.671 581 7	−1.819 799 8	0.333 203 8
6	693 120 7	637 440 5	456 538 6	348 164 4
7	561 378 3	602 175 5	190 943 3	356 309 5
8	452 882 1	566 367 6	−0.987 335 1	359 042 5
9	362 514 8	530 511 1	825 868 7	357 443 2
1.0	0.286 706 2	−0.494 994 6	−0.694 603 9	0.352 369 9
1	222 844 5	460 129 5	585 905 3	344 521 0
2	168 945 6	426 163 6	494 643 2	334 473 9
3	123 455 4	393 291 8	417 227 4	322 711 8
4	085 126 0	361 664 8	351 055 1	309 641 6

Table 12. (*cont.*)

x	ker (x)	kei (x)	ker′ (x)	kei′ (x)
1.5	0.052 934 9	−0.331 395 6	−0.294 181 6	0.295 608 1
6	026 029 9	302 565 5	245 114 7	280 903 8
7	003 691 1	275 228 8	202 681 8	265 777 2
8	−0.014 696 1	249 417 1	165 942 4	250 438 5
9	029 661 4	225 142 2	134 128 2	235 065 7
2.0	−0.041 664 5	−0.202 400 1	−0.106 601 0	0.219 807 9
1	051 106 5	181 172 6	082 823 4	204 789 7
2	058 338 8	161 430 7	062 337 3	190 113 7
3	063 670 5	143 135 7	044 747 9	175 863 8
4	067 373 5	126 241 5	029 712 3	162 106 9
2.5	−0.069 688 0	−0.110 696 1	−0.016 929 8	0.148 895 4
6	070 825 7	096 442 9	006 135 8	136 268 9
7	070 973 6	083 421 9	0.002 904 3	124 255 8
8	070 296 3	071 570 7	010 399 0	112 874 8
9	068 939 0	060 825 5	016 534 2	102 136 2
3.0	−0.067 029 2	−0.051 121 9	0.021 476 2	0.092 043 1
1	064 678 6	042 395 5	025 373 8	082 592 2
2	061 984 8	034 582 3	028 360 3	073 775 2
3	059 032 9	027 619 7	030 555 4	065 579 4
4	055 896 6	021 446 3	032 066 2	057 988 1
3.5	−0.052 639 3	−0.016 002 6	0.032 988 6	0.050 982 1
6	049 315 6	011 231 1	033 408 7	044 539 4
7	045 971 7	007 076 7	033 403 0	038 636 4
8	042 646 9	003 486 7	033 040 0	033 248 0
9	039 373 61	000 410 81	032 380 46	028 348 32
4.0	−0.036 178 85	0.002 198 40	0.031 478 49	0.023 910 62
1	033 084 40	004 385 82	030 381 79	019 908 04
2	030 107 58	006 193 61	029 132 42	016 313 67
3	027 261 77	007 661 27	027 767 30	013 100 84
4	024 556 89	008 825 62	026 318 68	010 243 31
4.5	−0.021 999 88	0.009 720 92	0.024 814 54	0.007 715 43
6	019 595 03	010 378 86	023 279 08	005 492 26
7	017 344 41	010 828 72	021 733 00	003 549 67
8	015 248 19	011 097 40	020 193 91	001 864 78
9	013 304 90	011 209 51	018 676 61	000 415 22
5.0	−0.011 511 73	0.011 187 59	0.017 193 40	−0.000 819 98
1	009 864 74	011 052 01	015 754 36	001 860 79

Table 12. (*cont.*)

x	ker (x)	kei (x)	ker′ (x)	kei′ (x)
5.2	0.008 359 11	0.010 821 28	0.014 367 57	−0.002 726 05
3	006 989 28	010 512 06	013 039 35	003 433 49
4	005 749 13	010 139 29	011 774 46	003 999 69
5.5	−0.004 632 16	0.009 716 31	0.010 576 33	−0.004 440 16
6	003 531 56	009 254 96	009 447 17	004 769 28
7	002 740 38	008 765 72	008 388 18	005 000 41
8	001 951 58	008 257 74	007 399 67	005 145 84
9	001 258 12	007 739 02	006 481 21	005 216 89
6.0	−0.000 653 04	0.007 216 49	0.005 631 71	−0.005 223 92
1	000 129 53	006 696 06	004 849 57	005 176 37
2	0.000 319 05	006 182 75	004 132 75	005 082 93
3	000 699 12	005 680 77	003 478 86	004 951 05
4	001 016 83	005 193 58	002 885 23	004 788 03
6.5	0.001 278 080	0.004 723 992	0.002 348 995	−0.004 600 032
6	001 488 446	004 274 219	001 867 130	004 392 632
7	001 654 215	003 845 947	001 436 521	004 170 782
8	001 777 354	003 440 398	001 053 999	003 938 849
9	001 865 512	003 058 385	000 716 382	003 700 651
7.0	0.001 922 022	0.002 700 365	0.000 420 510	−0.003 459 509
1	001 950 901	002 366 486	000 163 267	003 218 285
2	001 955 861	002 056 629	−0.000 058 386	002 979 421
3	001 940 312	001 770 454	000 247 403	002 744 978
4	001 907 373	001 507 429	000 406 628	002 516 671
7.5	0.001 859 888	0.001 266 868	−0.000 538 787	−0.002 295 904
6	001 800 431	001 047 959	000 646 478	002 083 800
7	001 731 326	000 849 790	000 732 165	001 881 234
8	001 654 654	000 671 373	000 798 170	001 688 855
9	001 572 275	000 511 664	000 846 677	001 507 120
8.0	0.001 485 834	0.000 369 584	−0.000 879 724	−0.001 336 313
1	001 396 782	000 244 032	000 899 210	001 176 567
2	001 306 386	000 133 902	000 906 891	001 027 888
3	001 215 743	000 038 090	000 904 388	000 890 168
4	001 125 797	−0.000 044 491	000 893 190	000 763 209
8.5	0.001 037 349	−0.000 114 902	−0.000 874 656	−0.000 646 733
6	000 951 070	000 174 175	000 850 022	000 540 398
7	000 867 511	000 223 306	000 820 407	000 443 813
8	000 787 120	000 263 248	000 786 819	000 356 543
9	000 710 249	000 294 910	000 750 159	000 278 127

Table 12 (*cont.*)

x	ker(x)	kei(x)	ker$'(x)$	kei$'(x)$
9.0	0.000 637 164	−0.000 319 153	−0.000 711 231	−0.000 208 079
1	000 568 055	000 336 788	000 670 745	000 145 903
2	000 503 046	000 348 579	000 629 326	000 091 093
3	000 442 203	000 355 236	000 587 517	000 043 145
4	000 385 540	000 357 420	000 545 789	000 001 559
9.5	0.000 333 029	−0.000 355 743	−0.000 504 544	0.000 034 158
6	000 284 604	000 350 768	000 464 122	000 064 485
7	000 340 168	000 343 010	000 424 806	000 089 887
8	000 199 598	000 332 940	000 386 830	000 110 811
9	000 162 751	000 320 983	000 350 379	000 127 684
10.0	0.000 129 466	−0.000 307 524	−0.000 315 597	0.000 140 914

Table 13. Functions ber$_p(x)$, bei$_p(x)$ and their derivatives, $p = 1, 2, \ldots, 5$.

x	ber$_1(x)$	bei$_1(x)$	ber$_1'(x)$	bei$_1'(x)$
0	0	0	−0.353 553	0.353 553
1	−0.395 868	0.307 557	−0.476 664	0.212 036
2	−0.997 078	0.299 775	−0.720 532	−0.305 845
3	−1.732 64	−0.487 45	−0.635 99	−1.364 13
4	−1.869 25	−2.563 82	−0.658 74	−2.792 83
5	0.359 78	−5.797 91	4.251 33	−3.327 80
6	7.462 20	−7.876 68	10.206 52	0.235 45
7	20.368 9	−2.317 2	14.677 5	12.780 7
8	32.506 9	21.673 5	5.866 4	36.882 2
9	20.719 2	72.054 3	−37.108 0	61.749 0
10	−59.478	131.879	−132.087	45.127

x	ber$_2(x)$	bei$_2(x)$	ber$_2'(x)$	bei$_2'(x)$
0	0	0	0	0
1	0.010 411	0.124 675	0.041 623	−0.248 047
2	0.165 279	0.479 225	0.327 788	−0.437 789
3	0.808 37	0.891 02	1.030 93	−0.286 47
4	2.317 85	−0.725 36	1.975 73	0.853 82

Table 13. (*cont.*)

x	$ber_2(x)$	$bei_2(x)$	$ber'_2(x)$	$bei'_2(x)$
5	4.488 43	−1.422 10	2.049 97	3 785 30
6	5.242 91	−7.432 44	−1.454 56	8.368 74
7	−0.950 35	−17.592 4	−12.493 0	11.015 1
8	−22.889 0	−25.438 9	−32.589 1	1.300 6
9	−65.869 2	−10.134 8	−50.963 2	−38.551 6
10	−111.779	−66.610	−28.840	−121.987

x	$ber_3(x)$	$bei_3(x)$	$ber'_3(x)$	$bei'_3(x)$
0	0	0	0	0
1	0.013 788	0.015 629	0.039 433	0.048 634
2	0.085 612	0.144 210	0.093 575	0.239 418
3	0.130 44	0.565 38	0.072 00	0.636 27
4	−0.282 63	1.437 76	−0.914 09	1.073 55
5	−2.094 35	2.454 41	−2.922 76	0.695 57
6	−6.430 04	1.901 46	−5.747 81	−2.498 96
7	−12.876 5	4.407 2	−6.249 2	−11.222 9
8	−15.420 4	−22.575 0	3.979 6	−25.707 4
9	3.166 6	−54.538 7	38.354 6	−35.563 4
10	72.253	81.423	104.463	−7.513

x	$ber_4(x)$	$bei_4(x)$	$ber'_4(x)$	$bei'_4(x)$
0	0	0	0	0
1	−0.002 601	−0.000 130	−0.010 395	−0.000 781
2	−0.040 97	−0.008 30	−0.080 56	−0.024 83
3	−0.193 27	−0.093 02	−0.234 32	−0.183 52
4	−0.493 10	−0.499 85	−0.323 71	−0.716 65
5	−0.628 67	−1.727 62	0.248 34	−1.834 36
6	0.648 3	−4.230 2	2.770 0	−3.701 1
7	6.083 5	−7.116 9	8.745 2	−1.921 9
8	19.094 7	−5.288 8	17.319 5	7.703 5
9	38.667	14.082	19.140	34.545
10	46.579	70.500	−12.148	80.465

Table 13. (*cont.*)

x	$\mathrm{ber}_5'(x)$	$\mathrm{bei}_5'(x)$	$\mathrm{ber}_5'(x)$	$\mathrm{bei}_5'(x)$
0	0	0	0	0
1	0.000 192	−0.000 176	0.000 973	−0.000 866
2	0.006 80	−0.004 84	0.017 84	−0.011 00
3	0.058 59	−0.025 54	0.104 78	−0.028 32
4	0.273 08	−0.033 53	0.360 76	0.046 69
5	0.851 04	0.211 43	0.851 11	0.565 64
6	1.830 5	1.475 6	1.007 4	2.220 0
7	2.209 0	5.242 3	−0.847 2	5.589 6
8	−1.821 3	12.812 8	−8.623 9	9.233 7
9	−18.619	21.384	−26.955	5.504
10	−58.722	15.193	−53.427	−24.511

Table 14. Functions $\mathrm{ker}_p(x)$, $\mathrm{kei}_p(x)$ and their derivatives, $p = 1, 2, \ldots, 5$.

x	$\mathrm{ker}_1(x)$	$\mathrm{kei}_1(x)$	$\mathrm{ker}_1'(x)$	$\mathrm{kei}_1'(x)$
1	−0.740 32	−0.242 000	0.887 60	0.794 74
2	−0.230 81	0.080 05	0.287 98	0.073 632
3	−0.049 90	0.080 27	0.100 18	−0.038 005
4	0.005 351 3	0.039 166	0.022 690	−0.036 928
5	0.012 737	0.011 578	−0.002 318 3	−0.018 366
6	0.007 676 1	0.000 288 35	−0.005 920 4	−0.005 612 7
7	0.002 743 6	−0.002 148 9	−0.003 660 5	0.000 156 61
8	0.000 322 86	−0.001 567 0	−0.001 352 3	0.000 985 18
9	−0.000 355 78	−0.000 650 05	−0.000 185 34	0.000 748 45
10	−0.000 322 80	−0.000 123 52	0.000 158 19	0.000 321 35

x	$\mathrm{ker}_2(x)$	$\mathrm{kei}_2(x)$	$\mathrm{ker}_2'(x)$	$\mathrm{kei}_2'(x)$
1	0.418 03	1.884 2	−0.141 46	−4.120 8
2	0.261 47	0.309 00	−0.154 87	−0.528 81
3	0.128 39	0.036 80	−0.107 07	−0.116 58
4	0.048 134	−0.017 938	−0.055 546	−0.014 942
5	0.011 184	−0.018 065	−0.021 667	0.008 046 0
6	−0.001 088 3	−0.009 093 7	−0.005 268 9	0.008 255 2
7	−0.002 910 5	−0.002 820 5	0.000 411 05	0.004 265 4

Table 14. (*cont.*)

x	$\mathrm{ker}_2(x)$	$\mathrm{kei}_2(x)$	$\mathrm{ker}_2'(x)$	$\mathrm{kei}_2'(x)$
8	−0.001 819 9	−0.000 149 65	0.001 334 7	0.001 373 7
9	−0.000 683 40	0.000 477 20	0.000 863 10	0.000 102 03
10	−0.000 101 28	0.000 370 64	0.000 335 85	−0.000 215 04

x	$\mathrm{ker}_3(x)$	$\mathrm{kei}_3(x)$	$\mathrm{ker}_3'(x)$	$\mathrm{kei}_3'(x)$
1	4.887 3	−6.269 7	−16.290	17.772
2	0.298 02	−0.886 82	−0.850 42	1.296 6
3	−0.036 451	−0.236 02	−0.080 360	0.300 78
4	−0.052 071	−0.060 518	0.017 701	0.092 108
5	−0.029 283	−0.007 685 2	0.022 436	0.025 293
6	−0.011 450	0.004 511 5	0.012 925	0.003 405 0
7	−0.002 707 2	0.004 464 6	0.005 212 6	−0.001 977 0
8	0.000 267 67	0.002 263 3	0.001 292 3	−0.002 029 8
9	0.000 720 5	0.000 714 8	−0.000 094 4	−0.001 059 0
10	0.000 456 3	0.000 047 3	−0.000 327 3	−0.000 347 9

x	$\mathrm{ker}_4(x)$	$\mathrm{kei}_4(x)$	$\mathrm{ker}_4'(x)$	$\mathrm{kei}_4'(x)$
1	−47.753	3.981 0	19.199	−8.035
2	−2.774 9	0.940 03	5.966 1	−1.042 3
3	−0.410 62	0.348 52	0.740 16	−0.323 58
4	−0.057 09	0.137 36	0.136 71	−0.131 38
5	0.007 143	0.049 433	0.020 426	−0.054 819
6	0.012 375	0.014 000	−0.003 344	−0.020 620
7	0.007 257	0.001 780	−0.005 361	−0.006 088
8	0.002 878 3	−0.001 192 6	−0.003 228 8	−0.000 841 8
9	0.000 680 7	−0.001 153 8	−0.001 317 5	0.000 516 8
10	−0.000 072 2	−0.000 584 3	−0.000 327 2	0.000 522 9

x	$\mathrm{ker}_5(x)$	$\mathrm{kei}_5(x)$	$\mathrm{ker}_5'(x)$	$\mathrm{kei}_5'(x)$
1	287.76	253.88	−1407.9	−1306.0
2	10.209	6.076 6	−24.226	−17.818

Table 14. (*cont.*)

x	$\mathrm{ker}_5(x)$	$\mathrm{kei}_5(x)$	$\mathrm{ker}_5'(x)$	$\mathrm{kei}_5'(x)$
3	1.367 9	0.353 1	−2.402 6	−1.125 3
4	0.327 07	−0.052 99	−0.465 59	−0.071 26
5	0.077 13	−0.056 32	−0.117 13	0.026 42
6	0.012 982	−0.029 378	−0.029 468	0.023 332
7	−0.001 719	−0.011 767	−0.005 162	0.011 279
8	−0.003 146 2	−0.003 455 3	0.000 774 4	0.005 038 1
9	−0.001 873 6	−0.000 417 5	0.001 375 4	0.001 529 2
10	−0.000 746 0	0.000 324 1	0.000 837 2	0.000 200 1

Table 15. The roots of the characteristic equation $(\alpha^2+\beta^2)^4+(\alpha^2+\lambda_R\beta^2)^2 = 0$, $\alpha_1 = a+bi$, $\alpha_2 = c+di$.

$\lambda_R = 1.0$				
β	a	b	c	d
0.11	0.0000	0.0781	0.7032	0.7110
0.22	0.0000	0.2059	0.6905	0.7240
0.44	0.0000	0.4377	0.6397	0.7832
0.66	0.0000	0.6555	0.5725	0.8732
0.88	0.0000	0.8767	0.4953	1.0094
1.10	0.0000	1.0974	0.4241	1.1787
1.32	0.0000	1.3179	0.3611	1.3694
1.54	0.0000	1.5383	0.3179	1.5724
2.20	0.0000	2.1990	0.2260	2.2117

$\lambda_R = 0.8$				
β	a	b	c	d
0.11	0.0000	0.0732	0.7024	0.7117
0.22	0.0000	0.1855	0.6874	0.7274
0.44	0.0020	0.3882	0.6312	0.7933
0.66	0.0064	0.5879	0.5539	0.9094
0.88	0.0139	0.7887	0.4767	1.0718
1.10	0.0238	0.9927	0.4137	1.2654
1.32	0.0346	1.2005	0.3672	1.4748
1.54	0.0451	1.4115	0.3333	1.6911
2.20	0.0707	2.0569	0.2745	2.3510

Table 15 (*cont.*)

$\lambda = 0.6$

β	a	b	c	d
0.11	0.0001	0.0662	0.7017	0.7124
0.22	0.0012	0.1618	0.6843	0.7309
0.44	0.0087	0.3380	0.6215	0.8091
0.66	0.0258	0.5178	0.5429	0.9455
0.88	0.0490	0.7086	0.4744	1.1269
1.10	0.0717	0.9104	0.4245	1.3314
1.32	0.0910	1.1195	0.3895	1.5453
1.54	0.1068	1.3328	0.3644	1.7629
2.20	0.1385	1.9838	0.3199	2.422

$\lambda = 0.4$

β	a	b	c	d
0.11	0.0004	0.0562	0.7010	0.7132
0.22	0.0030	0.1332	0.6813	0.7344
0.44	0.0224	0.2813	0.6134	0.8253
0.66	0.0590	0.4484	0.5376	0.9722
0.88	0.0965	0.6399	0.4791	1.1725
1.10	0.1261	0.8459	0.4389	1.3820
1.32	0.1482	1.0586	0.4112	1.5976
1.54	0.1649	1.2746	0.3912	1.8156
2.20	0.1964	1.9297	0.3553	2.4739

$\lambda = 0.2$

β	a	b	c	d
0.11	0.0008	0.0412	0.7003	0.7139
0.22	0.0074	0.0955	0.6784	0.7380
0.44	0.0493	0.2162	0.6069	0.8414
0.66	0.1073	0.3835	0.5358	1.0098
0.88	0.1521	0.5825	0.4861	1.2108
1.10	0.1828	0.7938	0.4532	1.4234
1.32	0.2043	1.0097	0.4306	1.6402
1.54	0.2199	1.2276	0.4143	1.8586
2.20	0.2485	1.8853	0.3847	2.5170

$\lambda = 0.0$

β	a	b	c	d
0.11	0.0073	0.0075	0.6996	0.7146
0.22	0.0314	0.0345	0.6756	0.7416

Table 15 (*cont.*)

$\lambda = 0.0$				
β	a	b	c	d
0.44	0.1054	0.1501	0.6016	0.8572
0.66	0.1707	0.3303	0.5363	1.0374
0.88	0.2131	0.5368	0.4939	1.2440
1.10	0.2404	0.7517	0.4666	1.4588
1.32	0.2590	0.9695	0.4480	1.6766
1.54	0.2724	1.1883	0.4346	1.8954
2.20	0.2967	1.8472	0.4103	2.5543

$\lambda = -0.2$				
β	a	b	c	d
0.11	0.0440	0.0017	0.6989	0.7154
0.22	0.0955	0.0162	0.6728	0.7453
0.44	0.1836	0.1174	0.5976	0.8726
0.66	0.2418	0.2964	0.5380	1.0624
0.88	0.2762	0.5033	0.5018	1.2732
1.10	0.2978	0.7183	0.4791	1.4900
1.32	0.3124	0.9360	0.4638	1.7087
1.54	0.3230	1.1549	0.4528	1.9281
2.20	0.3421	1.8137	0.4330	2.5876

$\lambda = -0.4$				
β	a	b	c	d
0.11	0.0641	0.0016	0.6983	0.7161
0.22	0.1354	0.0155	0.6702	0.7490
0.44	0.2495	0.1112	0.5946	0.8875
0.66	0.3093	0.2804	0.5406	1.0852
0.88	0.3378	0.4808	0.5097	1.2996
1.10	0.3538	0.6924	0.4908	1.5180
1.32	0.3643	0.908	0.4782	1.7376
1.54	0.3718	1.1262	0.4693	1.9575
2.20	0.3854	1.7837	0.4536	2.6178

Table 16. Integrals of trigonometric functions.

(1) $$H_{2n} = \int_0^{\pi/2} \cos(\gamma x \cos\varphi)\cos(\gamma y \sin\varphi)\sin^{2n}\varphi \, d\varphi$$

$$= \frac{\pi}{2}\left[\varrho^{2n,n}\frac{J_n(\gamma r)}{(\gamma r)^n} + \varrho^{2n,n-1}\frac{J_{n-1}(\gamma r)}{(\gamma r)^{n-1}} + \dots + \varrho^{2n,0}J_0(\gamma r)\right],$$

where

$$\varrho^{2n,n} = 1 \cdot 3 \dots (2n-1)\cos 2n\vartheta,$$

$$\varrho^{2n,n-m} = 1 \cdot 3 \dots (2n-2m-1)\left[\frac{2n(2n-1)\dots(2n-2m+1)}{(2m)!}\times\right.$$

$$\times \cos^{2(n-m)}\vartheta \sin^{2m}\vartheta - \frac{2n(2n-1)\dots(2n-2m-1)}{(2m+2)!}(m+1)\cos^{2(n-m-1)}\vartheta \times$$

$$\times \sin^{2(m+1)}\vartheta + \dots + (-1)^i \frac{2n(2n-1)\dots(2n-2m-2i+1)}{(2m+2i)!}\times$$

$$\times \frac{(m+i)(m+i-1)\dots(i+1)}{m!}\cos^{2(n-m-i)}\vartheta \sin^{2(m+i)}\vartheta + \dots$$

$$\left.\dots + (-1)^{n-m}\frac{n(n-1)\dots(n-m+1)}{m!}\sin^{2n}\vartheta\right],$$

$$i = 0, 1, 2, \dots n-m+1, \quad m = 1, 2, \dots, \quad m < n,$$

$$\varrho^{2n,0} = \sin^{2n}\vartheta,$$

$$\varrho^{0,0} = 1, \quad \vartheta = \arcsin\frac{y}{r}, \quad r = \sqrt{x^2+y^2}.$$

(2) $$D_{2n} = \int_0^{\pi/2} \sin(\gamma x \cos\varphi)\sin(\gamma y \sin\varphi)\cos\varphi \sin^{2n+1}\varphi \, d\varphi$$

$$= \frac{\pi}{2}\left[\mu^{2n,n+1}\frac{J_{n+1}(\gamma r)}{(\gamma r)^{n+1}} + \mu^{2n,n}\frac{J_n(\gamma r)}{(\gamma r)^n} + \dots + \mu^{2n,0}J_0(\gamma r)\right],$$

$$\mu^{2n,n+1} = 1 \cdot 3 \dots (2n+1)\sin(2n+2)\vartheta,$$

$$\mu^{2n,n} = 1 \cdot 3 \dots (2n-1)\frac{1}{2}[n\sin 2n\vartheta + (-1)^n(n+1)\sin(2n+2)\vartheta],$$

$$\mu^{2n,n-m} = 1 \cdot 3 \dots (2n-m+1)\times$$

$$\times\left\{-\frac{(2n+1)2n\dots(2n-2m+1)}{(2m+1)!}\cos^{2(n-m)+1}\vartheta \sin^{2m+1}\vartheta +\right.$$

$$+\left[\frac{(2n+1)2n\dots(2n-2m)}{(2m+2)!}1 + \frac{(2n+1)2n\dots(2n-2m-1)}{(2m+3)!}(m+2)\right]\times$$

$$\times \cos^{2(n-m-1)+1}\vartheta \sin^{2(m+1)+1}\vartheta + \dots$$

$$\dots + (-1)^{i+1}\left[\frac{(2n+1)2n\dots(2n-2m-2i+2)}{(2m+2i)!}\cdot\frac{(m+i)(m+i-1)\dots i}{(m+1)!}+\right.$$

Table 16. (*cont.*)

$$+\frac{(2n+1)2n\,\dots\,(2n-2m-2i+1)}{(2m+2i+1)!}\cdot\frac{(m+i+1)(m+i)\,\dots\,(i+1)}{(m+1)!}\Bigg]\times$$

$$\times\cos^{2(n-m-i)+1}\vartheta\,\sin^{2(m+i)+1}\vartheta+\;\dots+(-1)^{n-m-1}\Bigg[(2n+1)\,\frac{n(n-1)\,\dots\,(n-m)}{(m+1)!}+$$

$$+1\,\frac{(n+1)n\,\dots\,(n-m+1)}{(m+1)}\Bigg]\cos\vartheta\sin^{2n+1}\vartheta\Bigg\},$$

$m=1,2,\dots;\quad n=0,1,2,\dots;\quad i=0,1,2,\dots,n-m;\quad m<n,$

$\mu^{2n,0}=-\cos\vartheta\sin^{2n+1}\vartheta.$

(3) $$G_{2n}=\int_0^{\pi/2}\sin(\gamma x\cos\varphi)\cos(\gamma y\sin\varphi)\cos\varphi\,\sin^{2n}\varphi\,\mathrm{d}\varphi$$

$$=\frac{\pi}{2}\,[\nu^{2n,\,2n+1}J_{2n+1}(\gamma r)+\nu^{2n,\,2n-1}J_{2n-1}(\gamma r)+\;\dots\;+\nu^{2n,\,1}J_1(\gamma r)],$$

$$\nu^{2n,\,2n+1}=\frac{1}{2^{2n}}\cos(2n+1)\vartheta,$$

$$\nu^{2n,\,2n-1}=\frac{1}{2^{2n}}(2n-1)\cos(2n-1)\vartheta,$$

$$\nu^{2n,\,2n-m}=\frac{1}{2^{2n}}\,\frac{2n(2n-1)\,\dots\left(2n-\dfrac{m-3}{2}\right)}{\left(\dfrac{m+1}{2}\right)!}\,(2n-m)\cos(2n-m)\vartheta,$$

$m=3,5,7,\dots;\quad m<2n.$

(4) $$F_{2n}=\int_0^{\pi/2}\cos(\gamma x\cos\varphi)\sin(\gamma y\sin\varphi)\sin^{2n+1}\varphi\,\mathrm{d}\varphi$$

$$=\frac{\pi}{2}\,[\eta^{2n,\,2n+1}J_{2n+1}(\gamma r)+\eta^{2n,\,2n-1}J_{2n-1}(\gamma r)+\;\dots\;+\eta^{2n,\,1}J_1(\gamma r)],$$

$$\eta^{2n,\,2n+1}=\frac{1}{2^{2n}}\sin(2n+1)\vartheta,$$

$$\eta^{2n,\,2n-}=\frac{1}{2^{2n}}(2n+1)\sin(2n-1)\vartheta,$$

$$\eta^{2n,\,2n-m1}=\frac{1}{2^{2n}}\,\frac{(2n+1)2n(2n-1)\,\dots\left(2n-\dfrac{m-3}{2}\right)}{\left(\dfrac{m+1}{2}\right)!}\,\sin(2n-m)\vartheta,$$

$m=3,5,7\,\dots<2n.$

Table 17. Integrals of Bessel's functions.

$$(1)\quad I^k_{m,n} = \int_0^\infty \frac{\gamma^{k+1}}{(\gamma^4+1)^{1+n}} \frac{J_m(\gamma r)}{\gamma^m}\,d\gamma, \qquad \frac{d}{dr}(I^k_{m,n}) = I^k_{m-1,n} - \frac{m}{r} I^k_{m,n},$$

$$I^0_{0,0} = -\operatorname{kei} r, \quad I^0_{1,0} = \frac{1}{r} + \operatorname{ker}' r,$$

$$I^0_{2,0} = \frac{1}{2} - \frac{2}{r}\operatorname{kei}' r + \operatorname{ker} r,$$

$$I^0_{3,0} = \frac{r}{2^2 2!} - \frac{4}{r}\left(\frac{2}{r} I^0_{1,0} - I^0_{0,0}\right) + \operatorname{kei}' r$$

$$= \frac{r}{2^2 2!} - \frac{4}{r}\left[\frac{2}{r}\left(\frac{1}{r} + \operatorname{ker}' r\right) + \operatorname{kei} r\right] + \operatorname{kei}' r,$$

$$I^0_{m,0} = \frac{r^{m-2}}{2^{m-1}(m-1)!} - \frac{2(m-1)}{r}\left[\frac{2(m-2)}{r} I^0_{m-2,0} - I^0_{m-3,0}\right] +$$

$$+ \frac{2(m-3)}{r} I_{n-3,0} - I_{m-4,0}, \quad m \geqslant 4,$$

$$I^0_{0,1} = \left(1 - \frac{2}{4}\right) I^0_{0,0} - \frac{r}{4} I^{0\prime}_{0\,0} = -\frac{1}{2}\operatorname{kei} r + \frac{r}{4}\operatorname{kei}' r,$$

$$I^0_{m,1} = \left(1 - \frac{-(m-2)}{4}\right) I^0_{m,0} - \frac{r}{4} I^{0\prime}_{m,0},$$

$$I^0_{m,n} = \left(1 - \frac{-(m-2)}{4n}\right) I^0_{m,n-1} - \frac{r}{4n} I^{0\prime}_{m,n-1}, \; n > 1\,.$$

$$(2)\quad I^2_{m,n} = \int_0^\infty \frac{\gamma^3}{(\gamma^4+1)^{1+n}} \frac{J_m(\gamma, r)}{\gamma^m}\,d\gamma,$$

$$I^2_{0,0} = \operatorname{ker} r, \quad I^2_{1,0} = \operatorname{kei}' r,$$

$$I^2_{2,0} = \frac{2}{r} I^0_{1,0} - I^0_{0,0} = \frac{2}{r^2} + \frac{2}{r}\operatorname{ker}' r + \operatorname{kei} r,$$

$$I^2_{m,0} = \left[\frac{2(m-1)}{r} I^0_{m-1,0} - I^0_{m-2,0}\right], \quad m \geqslant 2,$$

$$I^2_{0,1} = \left(1 - \frac{4}{4}\right) I^2_{0,0} - \frac{r}{4} I^{2\prime}_{0,0} = -\frac{r}{4}\operatorname{ker}' r,$$

$$I^2_{1,1} = \left(1 - \frac{3}{4}\right) I^2_{1,0} - \frac{r}{4} I^{2\prime}_{1,0} = \frac{1}{2}\operatorname{kei}' r - \frac{1}{4} r \operatorname{ker} r,$$

$$I^2_{m,1} = \left[1 - \frac{-(m-4)}{4}\right] I^2_{m,0} - \frac{r}{4} I^{2\prime}_{m,0},$$

Table 17. (*cont.*)

$$I^2_{m,n} = \left[1 - \frac{-(m-4)}{4n}\right] I^2_{m,n-1} - \frac{r}{4n} I^{2\prime}_{m,n-1}.$$

$$\text{(3)} \quad U^k_{m,n} = \int_0^\infty \frac{\gamma^k J_m(\gamma_r) d\gamma}{(\gamma^4+1)^{1+n}},$$

$$U^0_{1,n} = I^0_{1,n},$$

$$U^0_{3,n} = I^2_{3,n},$$

$$U^0_{5,n} = \frac{8}{r} I^2_{4,n} - I^2_{3,n},$$

$$U^0_{7,n} = \frac{12}{r}\left(\frac{10}{r} I^2_{5,n} - I^2_{4,n}\right) - U^5_{0,n}, \qquad n = 1, 3, 5, \ldots$$

. . . .

$$\text{(4)} \quad U^2_{m,n} = \int_0^\infty \frac{\gamma^2 J_m(\gamma r) \mathrm{d}\gamma}{(\gamma^4+1)^{n+1}},$$

$$U^2_{1,n} = I^2_{1,n},$$

$$U^2_{3,n} = \frac{4}{r}(I^2_{2,n} - I^2_{1,n}),$$

$$U^2_{5,n} = \frac{8}{r}\left(\frac{6}{r} I^2_{3,n} - I^2_{2,n}) - U^2_{3,n},$$

$$U^2_{7,n} = \frac{12}{r}\left[\frac{10}{r}\left(\frac{8}{r} I^2_{4,n} - I^2_{3,n}\right) - \left(\frac{6}{r} I^2_{3,n} - I^2_{2,n}\right)\right] - U^2_{5,n},$$

. . . .

$$\text{(5)} \quad U^4_{m,n} = \int_0^\infty \frac{\gamma^4 J_m(\gamma r) \mathrm{d}\gamma}{(\gamma^4+1)^{n+1}},$$

$$U^4_{1,0} = -\mathrm{ker}' r, \qquad U^4_{1,1} = -\frac{1}{4} r \,\mathrm{kei}\, r,$$

$$U^4_{1,n} = I^0_{1,n-1} - I^0_{1,n},$$

$$U^4_{3,n} = I^2_{3,n-1} - I^2_{3,n},$$

$$U^4_{5,n} = U^0_{5,n-1} - U^0_{5,n},$$

$$U^4_{m,n} = U^0_{m,n-1} - U^0_{m,n},$$

$$U^k_{m,n} = I^{k+m-1}_{m,n}; \qquad \frac{\mathrm{d}}{\mathrm{d}r}(U^k_{m,n}) = U^k_{m-1,n} - \frac{m}{r} U^k_{m,n}.$$

Table 17. (*cont.*)

(6) $$\int_0^\infty \frac{J_0(\gamma r)}{1+\gamma^4}\,\gamma\,\mathrm{d}\gamma = -\operatorname{kei} r,$$

$$\int_0^\infty \frac{J_0(\gamma r)}{1+\gamma^4}\,\gamma^3\,\mathrm{d}\gamma = \operatorname{ker} r,$$

$$\int_0^\infty \frac{J_0(\gamma r)}{1+\gamma^4}\,\gamma^5\,\mathrm{d}\gamma = \operatorname{kei} r,$$

$$\int_0^\infty \frac{\gamma J_0(\gamma r)\,\mathrm{d}\gamma}{\gamma^2-2k} = -\frac{\pi}{2}\,Y_0(r\sqrt{2k}).$$

(7) $$\int_0^\infty \frac{J_1(\gamma r)}{1+\gamma^4}\,\mathrm{d}\gamma = \frac{1}{r}+\operatorname{ker}' r,$$

$$\int_0^\infty \frac{J_1(\gamma r)\gamma^2\,\mathrm{d}\gamma}{1+\gamma^4} = \operatorname{kei}' r,$$

$$\int_0^\infty \frac{J_1(\gamma r)\gamma^4\,\mathrm{d}\gamma}{1+\gamma^4} = -\operatorname{ker}' r,$$

$$\int_0^\infty \frac{J_1(\gamma r)\gamma^6\,\mathrm{d}\gamma}{1+\gamma^4} = -\operatorname{kei}' r.$$

(8) $$\int_0^\infty\int_0^\infty \frac{\cos\alpha x\cos\beta y\,\mathrm{d}\alpha\,\mathrm{d}\beta}{(\alpha^2+\beta^2)^2+1} = -\frac{\pi}{2}\operatorname{kei} r,$$

$$\int_0^\infty\int_0^\infty \frac{(\alpha^2+\beta^2)\cos\alpha x\cos\beta y\,\mathrm{d}\alpha\,\mathrm{d}\beta}{(\alpha^2+\beta^2)^2+1} = \frac{\pi}{2}\operatorname{ker} r,$$

$$\int_0^\infty\int_0^\infty \frac{\cos\alpha x\cos\beta y\,\mathrm{d}\alpha\,\mathrm{d}\beta}{\alpha^2+\beta^2} = -\frac{\pi}{2}\ln r,$$

$$\int_0^\infty\int_0^\infty \frac{(\alpha^2-\beta^2)\cos\alpha x\cos\beta y\,\mathrm{d}\alpha\,\mathrm{d}\beta}{(\alpha^2+\beta^2)^2+1} = -\frac{\pi}{2}\cos 2\vartheta\operatorname{ker}_2 r,$$

$$\int_0^\infty\int_0^\infty \frac{(\alpha^2-\beta^2)(\alpha^2+\beta^2)\cos\alpha x\cos\beta y\,\mathrm{d}\alpha\,\mathrm{d}\beta}{(\alpha^2+\beta^2)^2+1} = -\frac{\pi}{2}\cos 2\vartheta\operatorname{kei}_2 r,$$

Table 17. *(cont.)*

$$\int_0^\infty\int_0^\infty \frac{(\alpha^2-\beta^2)\cos\alpha x\cos\beta y\,\mathrm{d}\alpha\,\mathrm{d}\beta}{(\alpha^2+\beta^2)^2} = -\frac{\pi}{r^2}\cos 2\vartheta,$$

$$\int_0^\infty\int_0^\infty \frac{\alpha^4\cos\alpha x\cos\beta y\,\mathrm{d}\alpha\,\mathrm{d}\beta}{(\alpha^2+\beta^2)^4+\alpha^4} = \frac{\pi}{4}\Big\{-\ker\frac{r}{2}\sin\frac{x}{2\sqrt{2}}\sinh\frac{x}{2\sqrt{2}}-$$

$$-\mathrm{kei}\frac{r}{2}\cos\frac{x}{2\sqrt{2}}\cosh\frac{x}{2\sqrt{2}}+$$

$$+\ker_1\frac{r}{2}\cos\frac{x}{2\sqrt{2}}\sinh\frac{x}{2\sqrt{2}}\cos\vartheta-$$

$$-\mathrm{kei}_1\frac{r}{2}\sin\frac{x}{2\sqrt{2}}\cosh\frac{x}{2\sqrt{2}}\cos\vartheta\Big\}.$$

Table 18. Exponential and trigonometric functions.

(1) $$\int_B^\infty \mathrm{e}^{-\beta x}\cos\beta y\,\mathrm{d}\beta = \frac{\mathrm{e}^{-Bx}}{x^2+y^2}(x\cos By-y\sin By) = \frac{\mathrm{e}^{-Bx}}{r}\cos(By+\vartheta).$$

(2) $$\int_0^\infty \mathrm{e}^{-\beta x}\cos\beta y\,\mathrm{d}\beta = \frac{x}{r^2} = \frac{\cos\vartheta}{r},\qquad r=\sqrt{x^2+y^2};\qquad \vartheta=\arcsin\frac{y}{r}.$$

(3) $$\int_B^\infty \mathrm{e}^{-\beta x}\sin\beta y\,\mathrm{d}\beta = \frac{\mathrm{e}^{-\beta x}}{x^2+y^2}(x\sin By+y\cos By) = \frac{\mathrm{e}^{-Bx}}{r}\sin(By+\vartheta).$$

(4) $$\int_0^\infty \mathrm{e}^{-\beta x}\sin\beta y\,\mathrm{d}\beta = \frac{y}{r^2} = \frac{\sin\vartheta}{r}.$$

(5) $$\int_B^\infty \beta\mathrm{e}^{-\beta x}\cos\beta y\,\mathrm{d}\beta = \frac{\mathrm{e}^{-Bx}}{x^2+y^2}B(x\cos By-y\sin By)+$$

$$+\frac{\mathrm{e}^{-Bx}}{(x^2+y^2)^2}[(x^2-y^2)\cos By-2xy\sin By]$$

$$=\frac{B\mathrm{e}^{-Bx}}{r}\cos(By+\vartheta)+\frac{\mathrm{e}^{-Bx}}{r^2}\cos(By+2\vartheta).$$

(6) $$\int_0^\infty \beta\mathrm{e}^{-\beta x}\cos\beta y\,\mathrm{d}\beta = \frac{x^2-y^2}{(x^2+y^2)^2} = \frac{1}{r^2}\cos 2\vartheta.$$

Table 18. (*cont.*)

(7) $$\int_B^\infty \beta e^{-\beta x} \sin\beta y \, d\beta = \frac{Be^{-Bx}}{r} \sin(By+\vartheta) + \frac{e^{-Bx}}{r^2} \sin(By+2\vartheta).$$

(8) $$\int_0^\infty \beta e^{-\beta x} \sin\beta y \, d\beta = \frac{\sin 2\vartheta}{r^2}.$$

(9) $$\int_B^\infty \beta^2 e^{-Bx} \cos\beta y \, dy = e^{-Bx}\left[\frac{B^2}{r}\cos(By+\vartheta) + \frac{2B}{r^2}\cos(By+2\vartheta) + \frac{2}{r^3}\cos(By+3\vartheta)\right].$$

(10) $$\int_0^\infty \beta^2 e^{-\beta x} \cos\beta y \, d\beta = \frac{2}{r^3}\cos 3\vartheta.$$

(11) $$\int_B^\infty \beta^2 e^{-\beta x} \sin\beta y \, d\beta = \frac{e^{-Bx}}{r}\left[B^2 \sin(By+\vartheta) + \frac{2B}{r}\sin(By+2\vartheta) + \frac{2}{r^2}\sin(By+3\vartheta)\right].$$

(12) $$\int_0^\infty \beta^2 e^{-\beta x} \sin\beta y \, d\beta = \frac{2\sin 3\vartheta}{r^3}.$$

(13) $$\int_B^\infty \frac{e^{-Bx}}{\beta} \cos\beta y \, d\beta = -\mathrm{re}\, Ei[-(x-iy)B] = -\left[\gamma_0 + \ln r + \ln B + \sum_{k=1}^{\infty} (-B)^k \frac{r^k \cos k\vartheta}{kk!}\right],$$

$\gamma_0 = 0.5772.$

(14) $$\int_0^\infty \frac{e^{-\beta x}}{\beta} \cos\beta y \, d\beta = \infty.$$

(15) $$\int_B^\infty \frac{e^{-\beta x}}{\beta} \sin\beta y \, d\beta = -\mathrm{im}\, Ei[-(x-iy)B] = \left[\vartheta + \sum_{k=1}^{\infty} (-B)^k \frac{r^k \sin\vartheta}{kk!}\right].$$

(16) $$\int_0^\infty \frac{e^{-\beta x}}{\beta} \sin\beta y \, d\beta = \arctan\frac{y}{x} = \vartheta.$$

(17) $$\int_B^\infty \frac{\cos\beta y}{\beta} \, d\beta = -\mathrm{Ci}(By).$$

(18) $$\int_0^\infty \frac{\cos\beta y}{\beta} \, d\beta = \infty.$$

Table 18. (*cont.*)

(19) $$\int_B^\infty \frac{\sin \beta y}{\beta} d\beta = -\mathrm{Si}(By).$$

(20) $$\int_0^\infty \frac{\sin \beta y}{\beta} d\beta = \frac{\pi}{2}.$$

(21) $$\int_B^\infty \frac{e^{-\beta x}}{\beta^2} \cos \beta y d\beta = \mathrm{re}\left\{\frac{e^{-(x-iy)B}}{B} + (x-iy)Ei[-(x-iy)B]\right\}$$

$$= \frac{e^{-Bx}}{B} \cos By + x[\ln r + \ln B + \gamma] - y\vartheta + \sum_{k=1}^\infty (-B)^k \frac{r^{k+1}\cos(k+1)\vartheta}{kk!}.$$

(22) $$\int_0^\infty \frac{e^{-\beta x}}{\beta^2} \cos \beta y d\beta = \infty.$$

(23) $$\int_B^\infty \frac{e^{-\beta x}}{\beta^2} \sin \beta y d\beta = \mathrm{im}\left\{\frac{e^{-(x-iy)}}{B} + (x-iy)Ei[-(x-iy)B]\right\}$$

$$= \frac{e^{-Bx}}{B} \sin By - x\vartheta - y(\ln r + \ln B + \gamma) - \sum_{k=1}^\infty (-B)^k \frac{r^{k+1}\sin(k+1)\vartheta}{kk!}.$$

(24) $$\int_0^\infty \frac{e^{-\beta x}}{\beta^2} \sin \beta y d\beta = \infty.$$

(25) $$\int_B^\infty \frac{\cos \beta y}{\beta^2} d\beta = \frac{\cos By}{B} + \mathrm{Si}(By).$$

(26) $$\int_B^\infty \frac{\sin \beta y}{\beta^2} d\beta = \frac{\sin By}{B} - \mathrm{Ci}(By).$$

(27) $$\int_B^\infty \frac{e^{-\beta x}}{\beta^3} \cos \beta y d\beta = \frac{e^{-Bx}}{2B^2} \cos By - x\frac{e^{-Bx}}{2B} \cos By - y\frac{e^{-Bx}}{2B} \sin By -$$

$$-\frac{1}{2}(x^2-y^2)\left[\gamma + \ln r + \ln B + \sum_{k=1}^\infty (-B)^k \frac{r^k \cos k\vartheta}{kk!}\right] +$$

$$+xy\left[\vartheta + \sum_{k=1}^\infty (-B)^k \frac{r^k \sin k\vartheta}{kk!}\right].$$

Table 18. (*cont.*)

$$(28)\quad \int_B^\infty \frac{\sin\beta c}{\beta c} e^{-\beta x}\cos\beta y \, d\beta = \frac{1}{2c}\left[\vartheta_2-\vartheta_1+\sum_{k=1}^{\infty}(-B)^k \frac{r_2^k \sin k\vartheta_2 - r_1^k \sin k\vartheta_1}{kk!}\right]$$

where for ϑ_1 i ϑ_2 see Fig. 183, $r_1 = \sqrt{x^2+(y-c)^2}$, $r_3 = \sqrt{x^2+(y+c)^2}$.

$$(29)\quad \int_0^\infty \frac{\sin\beta c}{\beta c} e^{-\beta x}\cos\beta y \, d\beta = \frac{1}{2c}(\vartheta_2-\vartheta_1).$$

$$(30)\quad \int_0^\infty \frac{\sin\beta c}{\beta c} \cos\beta y \, d\beta = \begin{cases} \dfrac{\pi}{2c} & \text{for} \quad |y| < |c|, \\ \dfrac{\pi}{4c} & \text{for} \quad |y| = |c|, \\ 0 & \text{for} \quad |y| > |c|. \end{cases}$$

$$(31)\quad \int_B^\infty \frac{\sin\beta c}{\beta c} e^{-\beta x}\sin\beta y \, d\beta = \frac{1}{2c}\left[\ln\frac{r_2}{r_1}+\sum_{k=1}^{\infty}(-B)^k \frac{r_2^k \cos k\vartheta_2 - r_1^k \cos k\vartheta_1}{kk!}\right].$$

$$(32)\quad \int_0^\infty \frac{\sin\beta c}{\beta c} e^{-\beta x}\sin\beta y \, d\beta = \frac{1}{2c}\ln\frac{r_2}{r_1}.$$

$$(33)\quad \int_B^\infty \frac{\sin\beta c}{\beta c} \beta e^{-\beta x}\cos\beta y \, d\beta = \frac{1}{2c}\left[\frac{e^{-Bx}}{r_2}\sin[B(y+c)+\vartheta_2]-\frac{e^{-Bx}}{r_1}\sin[B(y-c)+\vartheta_1]\right].$$

$$(34)\quad x\int_0^\infty \frac{\sin\beta x}{\beta c} \beta e^{-\beta x}\cos\beta y \, d\beta = \frac{1}{4c}[\sin 2\vartheta_2 - \sin 2\vartheta_1],$$

$$\int_0^\infty \frac{\sin\beta x}{\beta c} \beta e^{-\beta x}\cos\beta \, d\beta = \frac{1}{2c}\left[\frac{\sin\vartheta_2}{r_2}-\frac{\sin\vartheta_1}{r_1}\right].$$

$$(35)\quad \int_B^\infty \frac{\sin\beta c}{\beta c} \beta e^{-\beta x}\sin\beta y \, d\beta$$

$$= \frac{1}{2c}\left\{\frac{e^{-Bx}}{r_1}\cos[B(y-c)+\vartheta_1]-\frac{e^{-Bx}}{r_2}\cos[B(y+c)+\vartheta_2]\right\}.$$

$$(36)\quad x\int_0^\infty \frac{\sin\beta c}{\beta c} \beta e^{-\beta x}\sin\beta y \, d\beta = -\frac{1}{4c}[\cos 2\vartheta_2 - \cos 2\vartheta_1],$$

Table 18. (*cont.*)

$$\int_0^\infty \frac{\sin\beta c}{\beta c}\,\beta e^{-\beta x}\sin\beta y\,d\beta = \frac{1}{2c}\left(\frac{\cos\vartheta_1}{r_1}-\frac{\cos\vartheta_2}{r_2}\right).$$

(37) $$\int_B^\infty \frac{\sin\beta c}{\beta c}\,\beta^2 e^{-\beta x}\cos\beta y\,d\beta = \frac{Be^{-Bx}}{2c}\left[\frac{\sin(By+\vartheta_2)}{r_2}-\frac{\sin(By+\vartheta_1)}{r_1}\right]+ \\ +\frac{e^{-Bx}}{2c}\left[\frac{\sin(By+2\vartheta_2)}{r_2^2}-\frac{\sin(By+2\vartheta_1)}{r_1^2}\right].$$

(38) $$\int_0^\infty \frac{\sin\beta c}{\beta c}\,\beta^2 e^{-\beta x}\cos\beta y\,d\beta = \frac{1}{2c}\left[\frac{\sin 2\vartheta_2}{r_2^2}-\frac{\sin 2\vartheta_1}{r_1^2}\right].$$

(39) $$\int_B^\infty \frac{\sin\beta c}{\beta c}\,\beta^2 e^{-\beta x}\sin\beta y\,d\beta = \frac{Be^{-Bx}}{2c}\left[\frac{\cos(By+\vartheta_1)}{r_1}-\frac{\cos(By+\vartheta_2)}{r_2}\right]+ \\ +\frac{e^{-Bx}}{c}\left[\frac{\cos(By+2\vartheta_1)}{r_1^2}-\frac{\cos(By+2\vartheta_2)}{r_2^2}\right].$$

(40) $$\int_0^\infty \frac{\sin\beta c}{\beta c}\,\beta^2 e^{-\beta x}\sin\beta y\,d\beta = \frac{1}{2c}\left[\frac{\cos 2\vartheta_1}{r_1^2}-\frac{\cos 2\vartheta_2}{r_2^2}\right].$$

(41) $$\int_B^\infty \frac{\sin\beta c}{\beta c}\,\frac{e^{-\beta x}}{\beta}\cos\beta y\,d\beta$$

$$= \frac{1}{2c}\Bigg\{\frac{e^{-Bx}}{B}\sin B(y+c)-\sin B(y-c)]- \\ -x(\vartheta_2-\vartheta_1)-(y+c)\ln r_2+(y-c)\ln r_1-2c(\ln B+\gamma)- \\ -\sum_{k=1}^\infty (-B)^k\frac{r_2^{k+1}\sin(k+1)\vartheta_2-r_1^{k+1}\sin(k+1)\vartheta_1}{kk!}\Bigg\}.$$

(42) $$\int_0^\infty \frac{\sin\beta c}{\beta c}\,\frac{e^{-\beta x}}{\beta}\cos\beta y\,d\beta = \infty.$$

(43) $$\int_B^\infty \frac{\sin\beta c}{\beta c}\,\frac{e^{-\beta x}}{\beta}\sin\beta y\,d\beta = \frac{1}{2c}\Bigg\{\frac{e^{-Bx}}{B}[\cos B(y-c)-\cos B(y+c)]- \\ -x\ln\frac{r_2}{r_1}-[(y-c)\vartheta_1-(y+c)\vartheta_2]+\sum_{k=1}^\infty (-B)^k\frac{r_1^{k+1}\cos(k+1)\vartheta_1-r_2^{k+1}\cos(k+1)\vartheta_2}{kk!}\Bigg\}.$$

Table 18. (*cont.*)

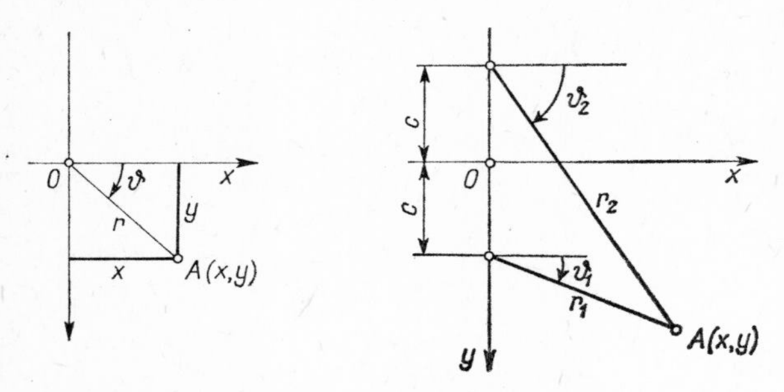

Fig. 183. Notation of axes of coordinates.

(44) $$\int_0^\infty \frac{\sin\beta c}{\beta c}\,\frac{e^{-\beta x}}{\beta}\sin\beta y\,d\beta = \frac{1}{2c}\left[-x\ln\frac{r_2}{r_1}+(y+c)\vartheta_2-(y-c)\vartheta_1\right].$$

If $x = 0$ $y \neq 0$, then:

(45) $$\int_B^\infty \frac{\sin\beta c}{\beta c}\,\frac{\cos\beta y}{\beta^2}\,d\beta = \frac{1}{2cB}\,[\sin B(y+c)-\sin B(y-c)-$$

$$-\frac{1}{2c}\,[(y+c)\operatorname{Ci}(y+c)B-(y-c)\operatorname{Ci}(y-c)B].$$

(46) $$\int_B^\infty \frac{\sin\beta c}{\beta c}\,\frac{\sin\beta y}{\beta^2}\,d\beta = \frac{1}{2cB}\,[\cos B(y-c)-\cos B(y+c)]+$$

$$+\frac{1}{2c}\,[(y+c)\operatorname{Si}(y+c)B-(y-c)\operatorname{Si}(y-c)B],$$

$$\int_0^\infty \frac{\sin\alpha x\,d\alpha}{\alpha(\alpha^4+4\varkappa^4)} = \frac{\pi}{8\varkappa^4}\,(1-e^{-\varkappa x}\cos\varkappa x),$$

$$\int_0^\infty \frac{\alpha\sin\alpha x\,d\alpha}{\alpha^4+4\varkappa^4} = \frac{\pi}{4\varkappa^2}\,e^{-\varkappa x}\sin\varkappa x.$$

Author index

Subject index